APPLYING CONTEMPORARY STATISTICAL TECHNIQUES

APPLYING CONTEMPORARY STATISTICAL TECHNIQUES

Rand R. Wilcox

University of Southern California
Los Angeles, CA

ACADEMIC PRESS
An imprint of Elsevier Science

Amsterdam Boston London New York Oxford Paris
San Diego San Francisco Singapore Sydney Tokyo

Senior Editor, Mathematics	Barbara A. Holland
Senior Project Manager	Angela Dooley
Editorial Coordinator	Tom Singer
Product Manager	Anne O'Mara
Cover Design	Shawn Girsberger
Copyeditor	Elliot Simon
Composition	CEPHA
Printer	Maple-Vail

This book is printed on acid-free paper. ∞

Cover image: Frenzied People © Diana Ong/SuperStock

Academic Press
An imprint of Elsevier Science
525 B Street, Suite 1900, San Diego, California 92101-4495, USA
http://www.academicpress.com

Academic Press
An imprint of Elsevier Science
84 Theobald's Road, London WC1X 8RR, UK
http://www.academicpress.com

Academic Press
An imprint of Elsevier Science
200 Wheeler Road, Burlington, Massachusetts 01803, USA
http://www.academicpressbooks.com

Library of Congress Catalog Card Number: 2002111027

International Standard Book Number: 0-12-751541-0

PRINTED IN THE UNITED STATES OF AMERICA
02 03 04 05 06   9 8 7 6 5 4 3 2 1

CONTENTS

PREFACE

Overview

The goals in this book are: (1) to describe fundamental principles in a manner that takes into account many new insights and advances that are often ignored in an introductory course, (2) to summarize basic methods covered in a graduate level, applied statistics course dealing with ANOVA, regression, and rank-based methods, (3) to describe how and why conventional methods can be unsatisfactory, and (4) to describe recently developed methods for dealing with the practical problems associated with standard techniques. Another goal is to help make contemporary techniques more accessible by supplying and describing easy-to-use S-PLUS functions. Many of the S-PLUS functions included here have not appeared in any other book. (Chapter 1 provides a brief introduction to S-PLUS so that readers unfamiliar with S-PLUS can employ the methods covered in the book.) Problems with standard statistical methods are well known among quantitative experts but are rarely explained to students and applied researchers. The many details are simplified and elaborated upon in a manner that is not available in any other book. No prior training in statistics is assumed.

Features

The book contains many methods beyond those in any other book and provides a much more up-to-date look at the strategies used to address nonnormality and heteroscedasticity. The material on regression includes several estimators that have recently been found to have practical value. Included is the deepest regression line estimator recently proposed by Rousseeuw and his colleagues. The last chapter covers rank-based methods, but unlike any other book, the latest information on handling tied values is described. (Brunner and Cliff describe different strategies for dealing with ties and both are considered.) Recent results on two-way designs are covered, including repeated measures designs.

Chapter 7 provides a simple introduction to bootstrap methods, and chapters 8–14 include the latest information on the relative merits of different bootstrap techniques when dealing with ANOVA and regression. The best non-bootstrap methods are covered as well. Again, methods and advances not available in any other book are described.

Chapters 13–14 include many new insights about robust regression that are not available in any other book. For example, many estimators often provide substantial improvements over ordinary least squares, but recently it has been found that some of these estimators do not always correct commonly occurring problems. Improved methods are covered in this book. Smoothers are described and recent results on checking for linearity are included.

Acknowledgments

The author is grateful to Sam Green, Philip Ramsey, Jay Devore, E. D. McCune, Xuming He, and Christine Anderson-Cook for their helpful comments on how to improve this book. I am also grateful to Pat Goeters and Matt Carlton for their checks on accuracy, but of course I am responsible for any remaining errors. I'm especially grateful to Harvey Keselman for many stimulating conversations regarding this book as well as inferential methods in general.

Rand R. Wilcox
Los Angeles, California

INTRODUCTION

The goals of this book are to describe the basics of applied statistical methods in a manner that takes into account the many insights from the last half century, to describe contemporary approaches to commonly encountered statistical problems, to provide an intuitive sense of why modern methods—developed after the year 1960—have substantial advantages over conventional techniques, and to make these new methods practical and accessible. Once basic concepts are covered, the main goal will be to address two general types of statistical problems that play a prominent role in applied research. The first is finding methods for comparing groups of individuals or things, and the second has to do with studying how two or more variables are related.

To elaborate on the first general problem to be considered, imagine that you give one group of 20 individuals a drug for lowering their cholesterol level and that a second group of 20 gets a placebo. Suppose the average decrease for the first group is 9.5 and for the second group is 7.2. What can we say about the population of all individuals who might take this drug? A natural guess is that if all individuals of interest took the drug, the average drop in cholesterol will be lower versus using the placebo. But obviously this conclusion might be wrong, because for each group we are attempting to generalize from a sample of 20 individuals to the millions of people who might take the drug. A general goal in statistics is to describe conditions and methods where the precision of generalizations can be assessed.

The most common approach to the problem just posed is based on a general strategy developed by Pierre-Simon Laplace about two centuries ago. In the drug example, 9.5 is the average based on the 20 available participants. But as an estimate of the average we would get if millions of people took the drug, chances are that 9.5 is inaccurate. That is, it differs from the *population* average we would get if all potential individuals took the drug. So a natural question is whether we can find some way of measuring the precision of the estimate, 9.5. That is, can we rule out certain values for the population average based on the data at hand, and can we specify a range of values that is likely to contain it?

Laplace actually developed two general approaches to the problem of assessing precision. His first approach was based on what we now call a Bayesian method.[1] His second approach is now called the *frequentist* approach to statistical problems. It is described in Chapter 4, it is covered in almost all introductory statistics books, and currently it forms the backbone of statistical methods routinely used in applied research. Laplace's method is based in part on assuming that if all individuals of interest could be measured and the results were plotted, we would get a particular bell-shaped curve called a *normal distribution*. Laplace realized that there was no particular reason to assume normality, and he dealt with this issue by using his central limit theorem, which he publicly announced in 1810. Simply put, the *central limit theorem* says that if a sufficiently large number of observations is randomly sampled, then normality can be assumed when using Laplace's method for making inferences about a population of people (or things) based on the data available to us. (Details about the central limit theorem are covered in Chapter 4.)

One obvious concern about the central limit theorem is the phrase *sufficiently large*. Just how many observations do we require so that normality can be assumed? Some books claim that the answer is 40, and others state that even 25 observations suffice. These statements are not wild speculations; they stem from results discussed in Chapter 4. But we now know that this view can be highly misleading and inaccurate. For some of the simplest problems to be covered, practical situations arise where hundreds of observations are needed. For other routinely used techniques, inaccurate inferences are possible no matter how many observations happen to be available. Yet it seems fair to say that despite the insights made during the last 40 years, conventional wisdom still holds that the most frequently used techniques perform in a satisfactory manner for the majority of situations that arise in practice. Consequently, it is important to understand why the methods typically taught in an introductory statistics course can be highly unsatisfactory and how modern technology can be used to address this problem.

In our earlier illustration, two groups of individuals are being compared; but in many situations multiple groups are compared instead. For example, there might be interest in three experimental drugs for lowering cholesterol and how they compare to a placebo. So now a total of four experimental groups might take part in an experiment. Another common view is that the more groups of individuals we compare, the more certain we can be that conventional methods (methods developed prior to 1960 and routinely used today) perform in a satisfactory manner. Unfortunately, this speculation is incorrect as well, and again it is important to understand why in order to appreciate the modern techniques described in this book.

The other general problem covered in this book has to do with discovering and describing associations among variables of interest. Two examples will help clarify what this means. The first has to do with a classic problem in astronomy: Is the universe expanding? At one point Albert Einstein assumed that the answer is no — all stars are fixed in space. This view was based on a collective intuition regarding the nature of space and time built up through everyday experiences over thousands of years. But one implication of Einstein's general theory of relativity is that the

---

1 The Reverend Thomas Bayes was the first to propose what we now call the Bayesian approach to statistics. But it appears that Laplace invented this approach independent of Bayes, and it was certainly Laplace who developed and extended the method so that it could be used for a wide range of problems.

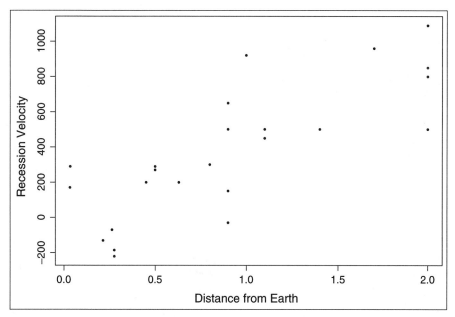

FIGURE 1.1   A scatterplot of Hubble's data.

universe cannot be static. In fact, during the early 1920s, the Russian meteorologist Alexander Friedmann provided the details showing that Einstein's theory implied an expanding universe. But during the early years of the twentieth century, the notion of a never-changing universe was so ingrained that even Einstein could not accept this implication of his theory. For this reason, he revisited his equations and introduced what is known as the cosmological constant, a term that avoids the prediction of a changing universe.

But 12 years later, Edwin Hubble made some astronomical measurements indicating that galaxies are either approaching or receding from our own Milky Way Galaxy. Moreover, Hubble concluded that typically, the further away a galaxy happens to be from our own, the faster it is moving away. A scatterplot of his observations is shown in Figure 1.1 and shows the rate (in kilometers per second) at which some galaxies are receding from our own galaxy versus its distance (in megaparsecs) from us. (The data are given in Table 6.1.) Hubble's empirical evidence convinced Einstein that the universe is generally expanding, and there has been considerable confirmation during the ensuing years (but alternative views cannot be completely ruled out, for reasons reviewed by Clark, 1999). Based on Hubble's data, is the conclusion of an expanding universe reasonable? After all, there are billions of galaxies, and his observations reflect only a very small proportion of the potential measurements he might make. In what sense can we use the data available to us to generalize to all the galaxies in our universe?

Here is another example where we would like to understand how two variables are related: Is there an association between breast cancer rates (per 100,000 women) and solar radiation (in calories per square centimeter)? Figure 1.2 shows a scatterplot, based on 24 cities in the United States, of the breast cancer rate among

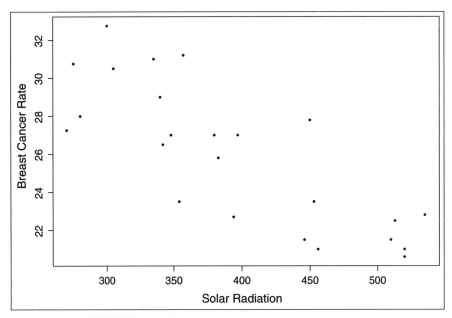

FIGURE 1.2   Breast cancer rates versus solar radiation.

100,000 women versus the average daily amount of solar radiation in calories per square centimeter. Can we make reasonable inferences about the association between these two variables regarding all geographical regions we might measure? What must be assumed to make such inferences? To what extent can we violate these assumptions and still arrive at reasonably accurate conclusions? Again it was Laplace who laid down the basic tools and assumptions that are used today. The great mathematician Carl Gauss extended and refined Laplace's techniques in ways that will be described in subsequent chapters. More refinements would come about a century later that are routinely used today — methods that are in some sense dictated by a lack of access to high-speed computers. But two fundamental assumptions routinely made in the applied work of both Laplace and Gauss are at the heart of the conventional methods that play a dominant role in modern research. Now that we are about two centuries beyond Laplace's great insight, what can be said about the accuracy of his approach and the conventional modifications routinely used today? They are, after all, mere approximations of reality. How does access to high-speed computers help us analyze data? Do modern methods and computers open the door to new ways of analyzing data that have practical value?

   The answer to the last question is an unequivocal yes. Nearly a half century ago it became obvious from a theoretical point of view that conventional methods have an inherent problem with potentially devastating implications for applied researchers. And more recently, new insights have raised additional concerns of great practical importance. In simple terms, if groups differ or variables are related in some manner, conventional methods might be poorly designed to discover this. Moreover, the precision and accuracy of conventional methods can be relatively poor unless sample sizes are fairly large. One strategy for dealing with these problems is simply to hope they never arise in practice. But all indications are that such situations are rather common. Interestingly, even Laplace had derived theoretical results hinting of serious problems associated with techniques routinely used today. But because of both technical and computational difficulties, finding practical alternatives proved to be extremely difficult until very recently.

In addition to theoretical concerns regarding standard statistical methods are empirical studies indicating practical difficulties. The first such study was conducted by Bessell in 1818 with the goal to determine whether the normal curve provides a good approximation of what we find in nature. Bessell's data reflected a property that is frequently encountered and poorly handled by conventional techniques. But unfortunately, Bessell did not have the mathematical tools needed to understand and appreciate the possible importance of what he saw in his data. Indeed, it would be nearly 150 years before the importance of Bessell's observation would be appreciated. Today, a variety of empirical studies support concerns about traditional techniques used to analyze data, as will be illustrated in subsequent chapters.

1.1 Software

One goal in this book is to provide easy access to many of the modern statistical methods that have not yet appeared in popular commercial software. This is done by supplying S-PLUS[2] functions that are very easy to use and can be downloaded, as described in Section 1.2. For most situations, you simply input your data, and a single call to some function will perform the computations described in subsequent chapters. S-PLUS is a powerful and vast software package that is described in various books (e.g., Krause and Olson, 2000) and manuals.[3] Included are a wide range of built-in functions not described in this book.[4]

An alternative to S-PLUS is R, which is nearly identical to S-PLUS and can be downloaded for free from www.R-project.org. Both zipped and unzipped files containing R are available. (Files ending in .tgz are zipped.) The zipped file can be downloaded more quickly, but it requires special software to unzip it so that it can be used. Also available from this Web site is a free manual explaining how to use R that can serve as a guide to using S-PLUS as well. Unfortunately, S-PLUS has a few built-in functions that are not standard in R but that are used in subsequent chapters.

The goal in the remainder of this section is to describe the basic features of S-PLUS that are needed to apply the statistical methods covered in subsequent chapters. An exhaustive description of the many features and nuances of S-PLUS go well beyond the scope of this book.

Once you start S-PLUS you will see this prompt:

```
>
```

---

2 S-PLUS is a registered trademark of Insightful Corporation, which can be contacted at www.insightful.com

3 See, in particular, the *S-PLUS User's Guide* as well as *S-PLUS 4 Guide to Statistics*, Data Analysis Products Division, Mathsoft, Seattle, WA.

4 For software that links the S-PLUS functions in this book to SPSS, see the Web site zumastat.com

It means that S-PLUS is waiting for a command. To quit S-PLUS, use the command

```
> q()
```

1.1.1 Entering Data

To begin with the simplest case, imagine you want to store the value 5 in an S-PLUS variable called *dat*. This can be done with the command

```
> dat < -5,
```

where < − is a "less than" sign followed by a minus sign. Typing *dat* and hitting Return will produce the value 5 on the computer screen.

To store the values 2, 4, 6, 8, 12 in the S-PLUS variable dat, use the c command, which stands for "combine." That is, the command

```
> dat < -c(2,4,6,8,12)
```

will store these values in the S-PLUS variable dat.

To read data stored in a file into an S-PLUS variable, use the scan command. The simplest method assumes that values are separated by one or more spaces. *Missing values* are recorded as NA, for "not available." For example, imagine that a file called *ice.dat* contains

```
6 3 12 8 9
```

Then the command

```
> dat < -scan(file="ice.dat")
```

will read these values from the file and store them in the S-PLUS variable dat. When using the scan command, the file name must be in quotes. If instead you have a file called *dis.data* that contains

```
12   6   4
 7  NA   8
 1  18   2
```

then the command

```
> dat2 < -scan(file="dis.data")
```

will store the data in the S-PLUS variable dat2. Typing dat2 and hitting Enter returns

```
12 6 4 7 NA 8 1 18 2
```

Values stored in S-PLUS variables stay there until they are removed. (On some systems, enabling this feature might require the command !SPLUS CHAPTER.) So in this last example, if you turn off your computer and then turn it back on, typing dat2 will again return the values just displayed. To remove data, use the rm command. For example,

```
> rm(dat)
```

would remove the data stored in dat.

S-PLUS variables are case sensitive. So, for example, the command

```
> Dat2 < -5
```

would store the value 5 in Dat2, but the S-PLUS variable dat2 would still contain the nine values listed previously, unless of course they had been removed.

S-PLUS has many built-in functions, and generally it is advisable not to store data in an S-PLUS variable having the same name as a built-in function. For instance, S-PLUS has a built-in function called *mean* that computes the average of the values stored in some S-PLUS variable. For example, the command

```
> mean(x)
```

will compute the average of the values stored in x and print it on the screen. In some situations S-PLUS will tell you that a certain variable name is reserved for special purposes and will not allow you to use it for your own data. In other situations it is allowed even when the variable name also corresponds to a built-in function. For example, the command

```
> mean < -2
```

will store the value 2 in an S-PLUS variable called *mean*, but mean(x) will still compute the average values stored in x. However, to avoid problems, particularly when using the functions written for this book, it is suggested that you do not use a built-in function name as an S-PLUS variable for storing data. A simple way to find out whether something is a built-in function is to type the name and hit Return. For instance, typing

```
> mean
```

will return

```
function(x, trim = 0, na.rm = F)
```

That is, mean is a built-in function with three arguments. The latter two arguments are optional, with default values if not specified. For example, na.rm indicates whether missing values are to be removed. By default, na.rm=F (for false), meaning that missing values are not removed. So if there are any missing values stored in x, mean(x) will result in the value NA. If, for example, you use the command mean(z,na.rm=T), any missing values will be removed and the average of the remaining values is computed. (Some details about built-in functions are provided by the help command. For example, help(mean) provides details about the function mean.)

If you type

```
> blob
```

and hit Enter, S-PLUS returns

```
Object "blob" not found
```

because data were never stored in the S-PLUS variable blob and there is no built-in function with this name.

One of the optional arguments associated with the scan command is called *skip*. It allows you to skip one or more lines in a file before beginning to read your data. For example, if a file called *dis1.dat* contains

```
This is Phase One
for my Dissertation
 12    6              4
  7    3              8
  1   18              2
```

the command

```
> dat1 < -scan(file="dis1.dat",skip=2)
```

will skip the first two lines in the file dis1.dat before beginning to read the data.

1.1.2 Storing Data in a Matrix

For many purposes it is convenient to store data in a matrix. Imagine, for example, that for each of five individuals you have measures taken at three different times. For instance, you might be interested in how blood pressure changes during the day, so you measure diastolic blood pressure in the morning, in the afternoon, and in the evening. One convenient way of storing these data is in a matrix having five rows and three columns. If the data are stored in the file bp.dat in the form

```
140   120   115
 95   100   100
110   120   115
 90    85    80
 85    90    85
```

then the command

```
> m < -matrix(scan(file="bp.dat"),ncol=3,byrow=T)
```

will read the data from the file into a matrix called *m* having three columns. Here the argument ncol indicates how many columns the matrix is to have. (The number of rows can be specified as well with the argument nrow.) Typing m and hitting Return outputs

```
        [,1]   [,2]   [,3]
[1,]    140    120    115
[2,]     95    100    100
[3,]    110    120    115
[4,]     90     85     80
[5,]     85     90     85
```

on the computer screen.

The argument byrow=T (where T is for "true") means that data will be read by rows. That is, the first row of the matrix will contain 140, 120, and 115, the second row will contain 95, 100, and 100, and so forth. If not specified, byrow defaults to F (for "false"), meaning that the data will be read by columns instead. In the example, the first row of data would now contain 140, 95, 110 (the first three values stored in column 1), the second row would contain 90, 85, and 120, and so on.

Once stored in a matrix, it is a simple matter to access a subset of the data. For example, m[1,1] contains the value in the first row and first column, m[1,3] contains the value in the first row and third column, and m[2,4] contains the value in row 2 and column 4. The symbol [,1] refers to the first column and [2,] is the second row. So typing m[,2] and hitting Enter returns

```
[1] 120 100 120 85 90
```

which is the data in the second column.

As before, when reading data from a file, you can skip lines using the skip command. For example, if the data in your file were

```
My   data   on snakes
21   45
67   81
32   72
```

then the command

```
> fdat <-matrix(scan("data.dat",skip=1),ncol=2,byrow=T)
```

would skip the first line and begin reading data.

1.1.3 Storing Data in List Mode

For certain purposes it is convenient to store data in what is called *list mode*. As a simple example, imagine you have three groups of individuals who are treated for anorexia via different methods. For illustrative purposes, suppose a rating method has been devised and that the observations are

```
G1:   36   24   82   12   90   33   14   19
G2:    9   17    8   22   15
G3:   43   56   23   10
```

In some situations it is convenient to have the data stored under one variable name, and this can be done using list mode. One way of storing data in list mode is as follows. First create a variable having list mode. If you want the variable to be called *gdat*, use the command

```
> gdat <-list()
```

Then the data for group 1 can be stored via the command

```
> gdat[[1]] <-c(36, 24, 82, 12, 90, 33, 14, 19),
```

the group 2 data would be stored via the command

```
> gdat[[2]] <-c(9, 17, 8, 22, 15),
```

and group 3 data would be stored by using the command

```
> gdat[[3]] <-c(43, 56, 23, 10)
```

Typing the command gdat and hitting Enter returns

```
[[1]]:
[1] 36 24 82 12 90 33 14 19

[[2]]:
[1] 9 17 8 22 15

[[3]]:
[1] 43 56 23 10
```

That is, gdat contains three vectors of numbers corresponding to the three groups under study.

Another way to store data in list mode is with a variation of the scan command. Suppose the data are stored in a file called *mydata.dat* and are arranged as follows:

```
36    9   43
24   17   56
82    8   23
12   22   10
90   15   NA
33   NA   NA
14   NA   NA
19   NA   NA
```

Then the command

```
> gdat <-scan("mydata.dat",list(g1=0,g2=0,g3=0))
```

will store the data in gdat in list mode. Typing gdat and hitting Enter returns

```
$g1:
 [1]    36   24   82   12   90   33   14   19

$g2:
 [1]     9   17    8   22   15   NA   NA   NA

$g3:
 [1]    43   56   23   10   NA   NA   NA   NA
```

So the data for group 1 are stored in gdat$g1, for group 2 they are in gdat$g2, and for group 3 they are in gdat$g3. An alternative way of accessing the data in group 1 is with gdat[[1]]. Note that as used, scan assumes that the data for group 1 are stored in column 1, group 2 data are stored in column 2, and group 3 data are in column 3.

1.1.4 Arithmetic Operations

In the simplest case, arithmetic operations can be performed on numbers using the operators $+$, $-$, $*$ (multiplication), $/$ (division), and $\wedge$ (exponentiation). For example, to compute 1 plus 5 squared, use the command

```
> 1+5^2,
```

TABLE 1.1 Some Basic S-PLUS Functions.

Function	Description
exp	exponential
log	natural logarithm
sqrt	square root
cor	correlation
mean	arithmetic mean (with a trimming option)
median	median
min	smallest value
max	largest value
quantile	quantiles
range	max value minus the min value
sum	arithmetic sum
var	variance and covariance

which returns

`[1] 26.`

To store the answer in an S-PLUS variable — say, ans — use the command

`> ans < -1+ 5^2.`

If a vector of observations is stored in an S-PLUS variable, arithmetic operations applied to the variable name will be performed on all the values. For example, if the values 2, 5, 8, 12, and 25 are stored in the S-PLUS variable vdat, then the command

`> vinv < -1/vdat`

will compute 1/2, 1/5, 1/8, 1/12, and 1/25 and store the results in the S-PLUS variable vinv.

Most S-PLUS commands consist of a name of some function followed by one or more arguments enclosed in parentheses. There are hundreds of functions that come with S-PLUS, and Section 1.2 describes how to obtain the library of functions written for this book and described in subsequent chapters. For convenience, some of the more basic functions are listed in Table 1.1.

EXAMPLE. If the values 2, 7, 9, and 14 are stored in the S-PLUS variable x, the command

`> min(x)`

returns 2, the smallest of the four values stored in x. The average of the numbers is computed with the command mean(x) and is 8. The command range(x) returns the difference between the largest and smallest values stored in x and is $14 - 2 = 12$, and sum(x) returns the value $2 + 7 + 9 + 14 = 32$.

Continued

> **EXAMPLE.** (*Continued*) Suppose you want to subtract the average from each value stored in the S-PLUS variable blob. The command
>
> ```
> > blob-mean(blob)
> ```
>
> accomplishes this goal. If in addition you want to square each of these differences and then sum the results, use the command
>
> ```
> > sum((blob-mean(blob))^2).
> ```
>
> You can apply arithmetic operations to specific rows or columns of a matrix. For example, to compute the average of all values in column 1 of the matrix m, use the command
>
> ```
> > mean(m[,1]).
> ```
>
> The command
>
> ```
> > mean(m[2,])
> ```
>
> will compute the average of all values in row 2. In contrast, the command mean(m) will average all of the values in m. In a similar manner, if x has list mode, then
>
> ```
> > mean(x[[2]])
> ```
>
> will average the values in x[[2]].   ■

1.1.5 Data Management

There are many ways to manipulate data in S-PLUS. Here attention is focused on those methods that are particularly useful in subsequent chapters.

For certain purposes it is common to want to split data into two groups. For example, situations might arise where you want to focus on those values stored in x that are less than or equal to 6. One way to do this is with the command

```
> z <-x[x<=6],
```

which will take all values stored in x that are less than or equal to 6 and store them in z. More generally, S-PLUS will evaluate any logical expression inside the brackets and operate only on those for which the condition is true. The basic conditions are: == (equality), != (not equal to), < (less than), <= (less than or equal to) > (greater than), >= (greater than or equal to), && (and), || (or). So the command

```
> z <-x[x<=6 || x > 32]
```

will take all values in x that are less than or equal to 6 or greater than 32 and store them in z. The command

```
> z <-x[x>=4 && x <=40]
```

will store all values between 4 and 40, inclusive, in z.

Now suppose you have two measures for each of 10 individuals that are stored in the variables x and y. To be concrete, it is assumed the values are:

x	y
9	23
14	19
23	36
29	24
36	32
42	45
49	39
50	60
63	71
88	92

Situations arise where there is interest in those y values for which the x values satisfy some condition. If you want to operate on only those y values for which x is less than 42, say, use y[x<42]. So the command

```
mean(y[x<42])
```

would average all of the y values for which the corresponding x value is less than 42. In the example, this command would compute

$$(23 + 19 + 36 + 24 + 32)/5.$$

To compute the average of the y values for which the corresponding x value is less than or equal to the average of the x values, use the command

```
mean(y[x<=mean(x)]).
```

To compute the average of the y values for which the corresponding x value is less than or equal to 14 or greater than or equal to 50, use

```
mean(y[x<=14 || x>=50]).
```

Situations also arise where you might need to change the storage mode used. For example, Chapter 13 describes methods for detecting outliers (points that are unusually far from the majority of points) in multivariate data. Some of the functions for accomplishing this important goal assume data are stored in a matrix. For the data in the example, the values in x and y can be stored in a 10×2 matrix called *m* via the command

```
> m <-cbind(x,y)
```

That is, cbind combines columns of data. (The command rbind combines rows of data instead.)

1.1.6 S-PLUS Function selby

A common situation is where one column of data indicates group membership. For example, imagine a file called *dis.dat* with the following values:

G	OUTCOME
1	34
1	23
1	56
2	19
2	32
1	41
3	29
3	62

There are three groups of individuals corresponding to the values stored under G. The first group, for example, has four individuals with the values 34, 23, 56, and 41. The problem is storing the data in a manner that can be used by the functions described in subsequent chapters. To facilitate matters, the function

```
selby(m,grpc,coln)
```

has been supplied for separating the data by groups and storing it in list mode. (This function is part of the library of functions written for this book.) The first argument, m, can be any S-PLUS variable containing data stored in a matrix. The second argument (grpc) indicates which column indicates group membership, and coln indicates which column contains the measures to be analyzed. In the example, if the data are stored in the S-PLUS matrix dis, the command

```
> selby(dis,1,2)
```

will return

```
$x:
$x[[1]]:
[1] 34 23 56 41

$x[[2]]:
[1] 19 32

$x[[3]]:
[1] 29 62

$grpn:
[1] 1 2 3
```

If the command

```
> ddat<-selby(dis,1,2)
```

is used, ddat$x[[1]] contains the data for group 1, ddat$x[[2]] contains the data for group 2, and so forth. More generally, the data are now stored in list mode in a variable called *ddat$x* — not *ddat*, as might be thought. The command

```
> tryit <- selby(dis,1,2)
```

would store the data in a variable called *tryit$x* instead.

1.2 R and S-PLUS Functions Written for This Book

A rather large library of S-PLUS functions has been written for this book. They can be obtained via anonymous ftp at ftp.usc.edu. That is, use the login name anonymous and use your e-mail address as the password. Once connected, change directories to pub/wilcox; on a UNIX system you can use the command

```
cd pub/wilcox
```

The functions are stored in two files called *allfunv1* and *allfunv2*. Alternatively, these files can be downloaded from www-rcf.usc.edu/~rwilcox/ using the Save As command. When using this Web site, on some systems the file allfunv1 will be downloaded into a file called allfunv1.txt rather than just allfunv1, and of course the same will be true with allfunv2. On other systems, allfunv1 will be downloaded into the file allfunv1.html. When using R, download the files Rallfunv1 and Rallfunv2 instead. (They are nearly identical to the S-PLUS functions, but a few changes were needed to make them run under R.) When using ftp on a Unix machine, use the get command to download them to your computer. For example, the command

```
get allfunv1
```

will download the first file.

The files allfunv1 and allfunv2 should be stored in the same directory where you are using S-PLUS. To make these functions part of your version of S-PLUS, use the command

```
> source("allfunv1").
```

When running under a UNIX system, this command assumes that the file allfunv1 is stored in the directory from which S-PLUS was invoked. When using a PC, the easiest method is to store allfunv1 in the directory being used by S-PLUS. For example, when running the Windows 2000 version, the top of the window indicates that S-PLUS is using the directory

```
C: Program Files\sp2000\users\default
```

Storing allfunv1 in the subdirectory default, the source command given earlier will cause the library of functions stored in allfunv1 to become a part of your version of S-PLUS, until you remove them. Of course, for the remaining functions in allfunv2, use the command

```
source("allfunv2")
```

The arguments used by any of these functions can be checked with the args command. For example, there is a function called *yuen*, and the command

```
> args(yuen)
```

returns

```
function(x, y, tr = 0.2, alpha = 0.05).
```

The first two arguments are mandatory and are assumed to contain data. Arguments with an = are optional and default to the value shown. Here, the argument tr defaults to .2 and alpha defaults to .05. The command

```
yuen(x,y,tr=0,alpha=.1)
```

would use tr=0 and alpha=.1 (the meaning of which is described in Chapter 8).

Each function also contains a brief description of itself that can be read by typing the function name only (with no parentheses) and hitting Enter. For example, the first few lines returned by the command

```
> yuen
```

are

```
#
# Perform Yuen's test for trimmed means on the data in
  x and y.
# The default amount of trimming is 20%.
# Missing values (values stored as NA) are
  automatically removed.
#
# A confidence interval for the trimmed mean of x
  minus the
# the trimmed mean of y is computed and returned in
  yuen$ci.
# The significance level is returned in yuen$siglevel.
#
# For an omnibus test with more than two independent
  groups,
# use t1way.
#
```

The remaining lines are the S-PLUS commands used to perform the analysis, which presumably are not of interest to most readers.

Many of the data sets used in this book can be downloaded as well. You would proceed as was described when downloading allfunv1 and allfunv2, only download the files ending in .dat. For example, read.dat contains data from a reading study that is used in various chapters.

PROBABILITY AND RELATED CONCEPTS

This chapter covers the fundamentals of probability and some related concepts that will be needed in this book. Some ideas are basic and in all likelihood familiar to most readers. But some concepts are not always covered or stressed in an introductory statistics course, whereas other features are rarely if ever discussed, so it is suggested that even if the reader has had some training in basic statistics and probability, the information in this chapter should be scrutinized carefully, particularly Section 2.7.

2.1 Basic Probability

The term *probability* is of course routinely used; all of us have some vague notion of what it means. Yet there is disagreement about the philosophy and interpretation of probability. Devising a satisfactory definition of the term is, from a technical point of view, a nontrivial issue that has received a great deal of scrutiny from stellar mathematicians. Here, however, consideration of these issues is not directly relevant to the topics covered. For present purposes it suffices to think about probabilities in terms of proportions associated with some population of people or things that are of interest. For example, imagine you are a psychologist interested in mental health and one of your goals is to assess feelings of loneliness among college students. Further assume that a measure of loneliness has been developed where an individual can get one of five scores consisting of the integers 1 through 5. A score of 1 indicates relatively no feelings of loneliness and a score of 5 indicates extreme feelings of loneliness. Among the entire population of college students, imagine that 15% would get a loneliness score of 1. Then we say that the probability of the score 1 is .15. Again, when dealing with the mathematical foundations of probability, this view is not completely satisfactory, but attaching a probabilistic interpretation to proportions is all that is required in this book.

In statistics, an uppercase roman letter is typically used to represent whatever measure happens to be of interest, the most common letter being X. For the loneliness study, X represents a measure of loneliness, and the possible values of X are the integers 1 through 5. But X could just as well represent how tall someone is, how much she weighs, her IQ, and so on. That is, X represents whatever happens to be of interest in a given situation. In the illustration, we write $X = 1$ to indicate the event that a college student receives a score of 1 for loneliness, $X = 2$ means a student got a score of 2, and so on.

In the illustration there are five possible events: $X = 1, X = 2, \ldots, X = 5$, and the notation

$$p(x) \tag{2.1}$$

is used to indicate the probability assigned to the value x. So $p(1)$ is the probability that a college student will have a loneliness score of 1, $p(2)$ is the probability of a score of 2, and so forth. Generally, $p(x)$ is called the *probability function* associated with the variable X.

Unless stated otherwise, it is assumed that the possible responses we might observe are mutually exclusive and exhaustive. In the illustration, describing the five possible ratings of loneliness as being *mutually exclusive* means that a student can get one and only one rating. By assumption, it is impossible, for example, to have ratings of both 2 and 3. *Exhaustive* means that a complete list of the possible values we might observe has been specified. If we consider only those students who get a rating between 1 and 5, meaning, for example, that we exclude the possibility of no response, then the ratings 1–5 are exhaustive. If instead we let 0 represent no response, then an exhaustive list of the possible responses would be 0, 1, 2, 3, 4, and 5.

The set of all possible responses is called a *sample space*. If in our ratings illustration the only possible responses are the integers 1–5, then the sample space consists of the numbers 1, 2, 3, 4, and 5. If instead we let 0 represent no response, then the sample space is 0, 1, 2, 3, 4, and 5. If our goal is to study birth weight among humans, the sample space can be viewed as all numbers greater than or equal to zero. Obviously some birth weights are impossible — there seems to be no record of someone weighing 100 pounds at birth — but for convenience the sample space might contain outcomes that have zero probability of occurring.

It is assumed that the reader is familiar with the most basic principles of probability. But as a brief reminder, and to help establish notation, these basic principles are illustrated with the ratings example assuming that the outcomes 1, 2, 3, 4, and 5 are mutually exclusive and exhaustive. The basic principle is that in order for $p(x)$ to qualify as a *probability function*, it must be the case that

- $p(x) \geq 0$ for any x.
- For any two mutually exclusive outcomes — say, x and y — $p(x \text{ or } y) = p(x) + p(y)$.
- $\sum p(x) = 1$, where the notation $\sum p(x)$ means that $p(x)$ is evaluated for all possible values of x and the results are summed. In the loneliness example where the sample space is x: 1, 2, 3, 4, 5, $\sum p(x) = p(1) + p(2) + p(3) + p(4) + p(5) = 1$.

In words, the first criterion is that any probability must be greater than or equal to zero. The second criterion says, for example, that if the responses 1 and 2 are mutually exclusive, then the probability that a student gets a rating of 1 or 2 is equal to the probability of a 1 plus the probability of a 2. Notice that this criterion makes perfect sense when probabilities are viewed as relative proportions. If, for example, 15% of students have a rating of 1, and 20% have a rating of 2, then the probability of a rating of 1 or 2 is just the sum of the proportions: $.15 + .20 = .35$. The third criterion is that if we sum the probabilities of all possible events that are mutually exclusive, we get 1. (In more formal terms, the probability that an observation belongs to the sample space is 1.)

2.2 Expected Values

A fundamental tool in statistics is the notion of *expected values*. Most of the concepts and issues in this book can be understood without employing expected values, but *expected values* is a fairly simple idea that might provide a deeper understanding of some important results to be covered. Also, having an intuitive understanding of expected values facilitates communications with statistical experts, so this topic is covered here.

To convey the basic principle, it helps to start with a simple but unrealistic situation. Still using our loneliness illustration, imagine that the entire population of college students consists of 10 people; that is, we are interested in these 10 individuals only. So in particular we have no desire to generalize to a larger group of college students. Further assume that two students have a loneliness rating of 1, three a rating of 2, two a rating of 3, one a rating of 4, and two a rating of 5. So for this particular population of individuals, the probability of the rating 1 is 2/10, the proportion of individuals who have a rating of 1. Written in a more formal manner, $p(1) = 2/10$. Similarly, the probability of the rating 2 is $p(2) = 3/10$. As is evident, the average of these 10 ratings is

$$\frac{1 + 1 + 2 + 2 + 2 + 3 + 3 + 4 + 5 + 5}{10} = 2.8.$$

Notice that the left side of this last equation can be written as

$$\frac{1(2) + 2(3) + 3(2) + 4(1) + 5(2)}{10} = 1\frac{2}{10} + 2\frac{3}{10} + 3\frac{2}{10} + 4\frac{1}{10} + 5\frac{2}{10}.$$

But the fractions in this last equation are just the probabilities associated with the possible outcomes. That is, the average rating for all college students, which is given by the right side of this last equation, can be written as

$$1p(1) + 2p(2) + 3p(3) + 4p(4) + 5p(5).$$

EXAMPLE. If there are a million college students, and the proportion of students associated with the five possible ratings 1, 2, 3, 4, and 5 are .1, .15, .25, .3, and .2, respectively, then the average rating for all 1 million students is

$$1(.1) + 2(.15) + 3(.25) + 4(.3) + 5(.2) = 3.35$$

■

EXAMPLE. If there are a billion college students, and the probabilities associated with the five possible ratings are .15, .2, .25, .3, and .1, respectively, then the average rating of all 1 billion students is

$$1(.15) + 2(.2) + 3(.25) + 4(.3) + 5(.1) = 3.$$

■

Next we introduce some general notation for computing an average based on the view just illustrated. Again let a lowercase x represent a particular value you might observe associated with the variable X. The *expected value* of X, written $E(X)$, is

$$E(X) = \sum xp(x),\tag{2.2}$$

where the notation $\sum xp(x)$ means that you compute $xp(x)$ for every possible value of x and sum the results. So if, for example, the possible values for X are the integers 0, 1, 2, 3, 4, and 5, then

$$\sum xp(x) = 0p(0) + 1p(1) + 2p(2) + 3p(3) + 4p(4) + 5p(5).$$

The expected value of X is so fundamental it has been given a special name: the *population mean*. Typically the population mean is represented by μ. So

$$\mu = E(X)$$

is the average value for all individuals in the population of interest.

EXAMPLE. Imagine that an auto manufacturer wants to evaluate how potential customers will rate handling for a new car being considered for production. So here, X represents ratings of how well the car handles, and the population of individuals who are of interest consists of all individuals who might purchase it. If all potential customers were to rate handling on a four-point scale, 1 being poor and 4 being excellent, and if the corresponding probabilities associated with these ratings are $p(1) = .2$, $p(2) = .4$, $p(3) = .3$, and $p(4) = .1$, then the population mean is

$$\mu = E(X) = 1(.2) + 2(.4) + 3(.3) + 4(.1) = 2.3.$$

That is, the average rating is 2.3.   ■

2.3 Conditional Probability and Independence

Conditional probability refers to the probability of some event given that some other event has occurred; it plays a fundamental role in statistics. The notion of conditional probability is illustrated in two ways. The first is based on what is called a *contingency table*,

TABLE 2.1 Hypothetical Probabilities for Sex and Political Affiliation

Sex	Democrat (D)	Republican (R)	
M	0.25	0.20	0.45
F	0.28	0.27	0.55
	0.53	0.47	1.00

an example of which is shown in Table 2.1. In the contingency table are the probabilities associated with four mutually exclusive groups: individuals who are (1) both male and belong to the Republican party, (2) male and belong to the Democratic party, (3) female and belong to the Republican party, and (4) female and belong to the Democratic party. So according to Table 2.1, the proportion of people who are both female and Republican is 0.27. The last column shows what are called the *marginal probabilities*. For example, the probability of being male is $0.20 + 0.25 = 0.45$, which is just the proportion of males who are a Democrat plus the proportion who are Republican. The last line of Table 2.1 shows the marginal probabilities associated with party affiliation. For example, the probability of being a Democrat is $0.25 + 0.28 = 0.53$.

Now consider the probability of being a Democrat given that the individual is male. According to Table 2.1, the proportion of people who are male is 0.45. So among the people who are male, the proportion who belong to the Democratic party is $0.25/0.45 = 0.56$. Put another way, the probability of being a Democrat, given that the individual is male, is 0.56.

Notice that a conditional probability is determined by altering the sample space. In the illustration, the proportion of all people who belong to the Democratic party is 0.53. But restricting attention to males, meaning that the sample space has been altered to include males only, the proportion is $0.25/0.45 = 0.56$. In a more general notation, if A and B are any two events, and if we let $P(A)$ represent the probability of event A and $P(A \text{ and } B)$ represent the probability that events A and B occur simultaneously, then the conditional probability of A, given that B has occurred, is

$$P(A|B) = \frac{P(A \text{ and } B)}{P(B)}. \tag{2.3}$$

In the illustration, A is the event of being a Democrat, B is the event that a person is male. According to Table 2.1, $P(A \text{ and } B) = 0.25$, $P(B) = 0.45$, so $P(A|B) = 0.25/0.45$, as previously indicated.

EXAMPLE. From Table 2.1, the probability that someone is a female, given that she is Republican, is

$$0.27/0.47 = 0.5745.$$

■

Roughly, two events are *independent* if the probability associated with the first event is not altered when the second event is known. If the probability is altered, the events are *dependent*.

EXAMPLE. According to Table 2.1, the probability that someone is a Democrat is 0.53. The event that someone is a Democrat is independent of the event someone is male if when we are told that someone is male, the probability of being a Democrat remains 0.53. We have seen, however, that the probability of being a Democrat, given that the person is male, is 0.56, so these two events are dependent.   ■

Consider any two variables — say, X and Y — and let x and y be any two possible values corresponding to these variables. We say that the variables X and Y are independent if for any x and y we might pick

$$P(Y = y | X = x) = P(Y = y). \qquad (2.4)$$

Otherwise they are said to be dependent.

EXAMPLE. Imagine that married couples are asked to rate the effectiveness of the President of the United States. To keep things simple, assume that both husbands and wives rate effectiveness with the values 1, 2, and 3, where the values stand for fair, good, and excellent, respectively. Further assume that the probabilities associated with the possible outcomes are as shown in Table 2.2. We see that the probability a wife (Y) gives a rating of 1 is 0.2. In symbols, $P(Y = 1) = 0.2$. Furthermore, $P(Y = 1 | X = 1) = .02/.1 = .2$, where $X = 1$ indicates that the wife's husband gave a rating of 1. So the event $Y = 1$ is independent of the event $X = 1$. If the probability had changed, we could stop and say that X and Y are dependent. But to say that they are independent requires that we check all possible outcomes. For example, another possible outcome is $Y = 1$ and $X = 2$. We see that $P(Y = 1 | X = 2) = .1/.5 = .2$, which again is equal to $P(Y = 1)$. Continuing in this manner, it can be seen that for any possible values for Y and X, the corresponding events are independent, so we say that X and Y are independent. That is, they are independent regardless of what their respective values might be.   ■

Now, the notion of *dependence* is described and illustrated in another manner. A common and fundamental question in applied research is whether information about one variable influences the probabilities associated with another variable. For example, in a study dealing with diabetes in children, one issue of interest was the association between a child's age and the level of serum C-peptide at diagnosis. For convenience, let X represent age and Y represent C-peptide concentration. For any child we

TABLE 2.2   Hypothetical Probabilities for Presidential Effectiveness

Wife (Y)	Husband (X)			
	1	2	3	
1	.02	.10	.08	0.2
2	.07	.35	.28	0.7
3	.01	.05	.04	0.1
	0.1	0.5	0.4	

might observe, there is some probability that her C-peptide concentration is less than 3, or less than 4, or less than c, where c is any constant we might pick. The issue at hand is whether information about X (a child's age) alters the probabilities associated with Y (a child's C-peptide level). That is, does the conditional probability of Y, given X, differ from the probabilities associated with Y when X is not known or ignored. If knowing X does not alter the probabilities associated with Y, we say that X and Y are independent. Equation (2.4) is one way of providing a formal definition. An alternative way is to say that X and Y are independent if

$$P(Y \le y | X = x) = P(Y \le y) \qquad (2.5)$$

for any x and y values we might pick. Equation (2.5) implies Equation (2.4). Yet another way of describing independence is that for any x and y values we might pick,

$$\frac{P(Y = y \text{ and } X = x)}{P(X = x)} = P(Y = y), \qquad (2.6)$$

which follows from Equation (2.4). From this last equation it can be seen that if X and Y are independent, then

$$P(X = x \text{ and } Y = y) = P(X = x)P(Y = y). \qquad (2.7)$$

Equation (2.7) is called the *product rule* and says that if two events are independent, the probability that they occur simultaneously is equal to the product of their individual probabilities.

EXAMPLE. If two wives rate presidential effectiveness according to the probabilities in Table 2.2, and if their responses are independent, then the probability that both give a response of 2 is $.7 \times .7 = .49$.   ■

EXAMPLE. Suppose that for all children we might measure, the probability of having a C-peptide concentration less than or equal to 3 is $P(Y \le 3) = .4$.

Continued

EXAMPLE. (*Continued*) Now consider only children who are 7 years old and imagine that for this subpopulation of children, the probability of having a C-peptide concentration less than 3 is 0.2. In symbols, $P(Y \leq 3|X = 7) = 0.2$. Then C-peptide concentrations and age are said to be dependent, because knowing that the child's age is 7 alters the probability that the child's C-peptide concentration is less than 3. If instead $P(Y \leq 3|X = 7) = 0.4$, the events $Y \leq 3$ and $X = 7$ are independent. More generally, if, for any x and y we pick, $P(Y \leq y|X = x) = P(Y = y)$, then C-peptide concentration and age are independent. ■

Attaining a graphical intuition of independence will be helpful in subsequent chapters. To be concrete, imagine a study where the goal is to study the association between a person's general feeling of well-being (Y) and the amount of chocolate they consume (X). Assume that an appropriate measure for these two variables has been devised and that the two variables are independent. If we were to measure these two variables for a very large sample of individuals, what would a plot of the results look like? Figure 2.1 shows a scatterplot of observations where values were generated on a computer with X and Y independent. As is evident, there is no visible pattern.

If X and Y are dependent, generally — but not always — there is some discernible pattern. But it is important to keep in mind that there are many types of patterns that can and do arise. (Section 6.5 describes situations where patterns are not evident based on a scatterplot, yet X and Y are dependent.) Figure 2.2 shows four types of patterns where feelings of well-being and chocolate consumption are dependent.

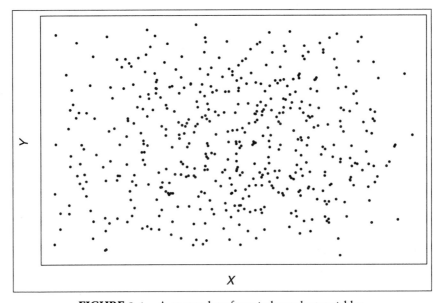

FIGURE 2.1   A scatterplot of two independent variables.

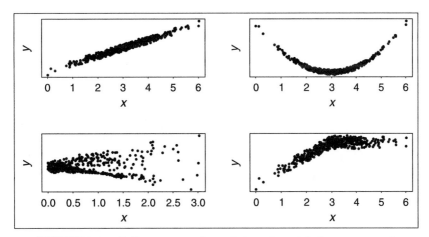

FIGURE 2.2 Different types of associations that might be encountered.

The two upper scatterplots show some rather obvious types of dependence that might arise. The upper left scatterplot, for example, shows a linear association where feelings of well-being increase with chocolate consumption. The upper right scatterplot shows a curved, nonlinear association. The type of dependence shown in the lower two scatterplots are, perhaps, less commonly considered when describing dependence, but in recent years both have been found to be relevant and very important in applied work, as we shall see. In the lower left scatterplot we see that the variation in feelings of well-being differs depending on how much chocolate is consumed. The points in the left portion of this scatterplot are more tightly clustered together. For the left portion of this scatterplot there is, for example, virtually no possibility that someone's feeling of well-being exceeds 1. But for the right portion of this scatterplot, the data were generated so that among individuals with a chocolate consumption of 3, there is a .2 probability that the corresponding value of well-being exceeds 1. That is, $P(Y \leq 1|X)$ decreases as X gets large, so X and Y are dependent. Generally, any situation where the variation among the Y values changes with X implies that X and Y are dependent. Finally, the lower right scatterplot shows a situation where feelings of well-being tend to increase for consumption less than 3, but for $X > 3$ this is no longer the case. Considered as whole, X and Y are dependent, but in this case, if attention is restricted to $X > 3$, X and Y are independent.

The lower left scatterplot of Figure 2.2 illustrates a general principle that is worth stressing: If knowing the value of X alters the range of possible values for Y, then X and Y are dependent. In the illustration, the range of possible values for well-being increases as chocolate consumption increases, so they must be dependent.

2.4 Population Variance

Associated with every probability function is a quantity called the *population variance*. The *population variance* reflects the average squared difference between the population mean and an observation you might make.

Consider, for example, the following probability function:

x:	0	1	2	3
$p(x)$:	.1	.3	.4	.2

The population mean is $\mu = 1.7$. If, for instance, we observe the value 0, its squared distance from the population mean is $(0 - 1.7)^2 = 2.89$ and reflects how far away the value 0 is from the population mean. Moreover, the probability associated with this squared difference is .1, the probability of observing the value 0. In a similar manner, the squared difference between 1 and the population mean is .49, and the probability associated with this squared difference is .3, the same probability associated with the value 1. More generally, for any value x, it has some squared difference between it and the population mean, namely, $(x - \mu)^2$, and the probability associated with this squared difference is $p(x)$. So if we know the probability function, we know the probabilities associated with all squared differences from the population mean. For the probability function considered here, we see that the probability function associated with all possible values of $(x - \mu)^2$ is

$(x - \mu)^2$:	2.89	0.49	0.09	1.69
$p(x)$:	.1	.3	.4	.2

Because we know the probability function associated with all possible squared differences from the population mean, we can determine the average squared difference as well. This average squared difference, called the *population variance*, is typically labeled σ^2. More succinctly, the population variance is

$$\sigma^2 = E[(X - \mu)^2], \tag{2.8}$$

the expected value of $(X - \mu)^2$. Said another way,

$$\sigma^2 = \sum (x - \mu)^2 p(x).$$

The *population standard deviation* is σ, the (positive) square root of the population variance. (Often it is σ, rather than σ^2, that is of interest in applied work.)

EXAMPLE. Suppose that for a five-point scale of anxiety, the probability function for all adults living in New York City is

x:	1	2	3	4	5
$p(x)$:	.05	.1	.7	.1	.05

Continued

EXAMPLE. (*Continued*) The population mean is

$$\mu = 1(.05) + 2(.1) + 3(.7) + 4(.1) + 5(.05) = 3,$$

so the population variance is

$$\sigma^2 = (1-3)^2(.05) + (2-3)^2(.1) + (3-3)^2(.7) + (4-3)^2(.1) + (5-3)^2(.05) = .6,$$

and the population standard deviation is $\sigma = \sqrt{.6} = .775$. ■

Understanding the practical implications associated with the magnitude of the population variance is a complex task that is addressed at various points in this book. There are circumstances where knowing σ is very useful, but there are common situations where it can mislead and give a highly distorted view of what a variable is like. For the moment, complete details must be postponed. But to begin to provide some sense of what σ tells us, consider the following probability function:

x:	1	2	3	4	5
$p(x)$:	.2	.2	.2	.2	.2

It can be seen that $\mu = 3$, the same population mean associated with the probability function in the last example, but the population variance is

$$\sigma^2 = (1 - 3)^2(.2) + (2 - 3)^2(.2) + (3 - 3)^2(.2) + (4 - 3)^2(.2) + (5 - 3)^2(.2) = 2.$$

Notice that this variance is larger than the variance in the previous example, where $\sigma^2 = .6$. The reason is that in the former example, it is much less likely for a value to be far from the mean than is the case for the probability function considered here. Here, for example, there is a .4 probability of getting the value 1 or 5. In the previous example, this probability is only .1. Here the probability that an observation differs from the population mean is .8, but in the previous example it was only .3. This illustrates the crude rule of thumb that larger values for the population variance reflect situations where observed values are likely to be far from the mean, and small population variances indicate the opposite.

For discrete data, it is common to represent probabilities graphically with the height of spikes. Figure 2.3 illustrates this approach with the last two probability functions used to illustrate the variance. The left panel shows the probability function

x:	1	2	3	4	5
$p(x)$:	.05	.1	.7	.1	.05

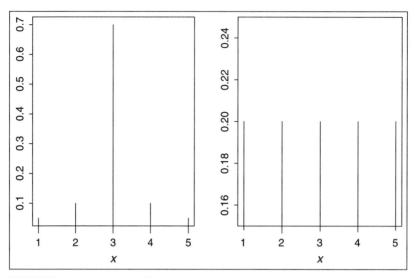

FIGURE 2.3   Examples of how probabilities associated with discrete variables are graphed.

The right panel graphically shows the probability function

x:	1	2	3	4	5
$p(x)$:	.2	.2	.2	.2	.2

Look at the graphed probabilities in Figure 2.3 and notice that the graphed probabilities in the left panel indicate that an observed value is more likely to be close to the mean; in the right panel they are more likely to be further from the mean. That is, the graphs suggest that the variance is smaller in the left panel because, probabilistically, observations are more tightly clustered around the mean.

2.5 The Binomial Probability Function

The most important discrete distribution is the binomial. It arises in situations where only two possible outcomes are possible when making a single observation. The outcomes might be yes and no, success and failure, agree and disagree. Such random variables are called *binary*. Typically the number 1 is used to represent a success, and a failure is represented by 0. A common convention is to let p represent the probability of success and to let $q = 1 - p$ be the probability of a failure.

Before continuing, a comment about notation might help. Consistent with Section 2.1, we follow the common convention of letting X be a variable that represents the number of successes among n observations. The notation $X = 2$, for example, means we observed two successes; more generally, $X = x$ means we observed x successes, where the possible values for x are 0, 1, ..., n.

In applied work, often the goal is to estimate p, the probability of success, given some data. But before taking up this problem, we must first consider how to compute probabilities given p. For example, suppose you ask five people whether they approve of a certain political leader. If these five people are allowed to respond only yes or no, and if the probability of a yes response is $p = .6$, what is the probability that exactly three of five randomly sampled people will say yes? There is a convenient formula for solving this problem based on the *binomial probability function*. It says that among n observations, the probability of exactly x successes, $P(X = x)$, is given by

$$p(x) = \binom{n}{x} p^x q^{n-x}. \tag{2.9}$$

The first term on the right side of this equation, called the *binomial coefficient*, is defined to be

$$\binom{n}{x} = \frac{n!}{x!(n-x)!},$$

where $n!$ represents n factorial. That is,

$$n! = 1 \times 2 \times 3 \times \cdots \times (n-1) \times n.$$

For example, $1! = 1$, $2! = 2$, and $3! = 6$. By convention, $0! = 1$.

In the illustration, you have $n = 5$ randomly sampled people and you want to know the probability that exactly $x = 3$ people will respond yes when the probability of a yes is $p = .6$. To solve this problem, compute

$$n! = 1 \times 2 \times 3 \times 4 \times 5 = 120,$$
$$x! = 1 \times 2 \times 3 = 6,$$
$$(n-x)! = 2! = 2,$$

in which case

$$p(3) = \frac{120}{6 \times 2} \left(.6^3\right)\left(.4^2\right) = .3456.$$

As another illustration, suppose you randomly sample 10 couples who recently got married, and your experiment consists of assessing whether they are happily married at the end of 1 year. If the probability of success is $p = .3$, the probability that exactly $x = 4$ couples will report that they are happily married is

$$p(4) = \frac{10!}{4! \times 6!} \left(.3^4\right)\left(.7^6\right) = .2001.$$

Often attention is focused on the probability of *at least* x successes in n trials or *at most* x successes, rather than the probability of getting *exactly* x successes. In the last illustration, you might want to know the probability that four couples or fewer are happily married as opposed to exactly four. The former probability consists of five mutually exclusive events, namely, $x = 0$, $x = 1$, $x = 2$, $x = 3$, and $x = 4$. Thus, the probability that four couples or fewer are happily married is

$$P(X \leq 4) = p(0) + p(1) + p(2) + p(3) + p(4).$$

In summation notation,

$$P(X \leq 4) = \sum_{x=0}^{4} p(x).$$

More generally, the probability of k successes or less in n trials is

$$P(X \leq k) = \sum_{x=0}^{k} p(x)$$

$$= \sum_{x=0}^{k} \binom{n}{x} p^x q^{n-x}.$$

Table 2 in Appendix B gives the values of $P(X \leq k)$ for various values of n and p. Returning to the illustration where $p = .3$ and $n = 10$, Table 2 reports that the probability of four successes or less is .85. Notice that the probability of five successes or more is just the complement of getting four successes or less, so

$$P(X \geq 5) = 1 - P(X \leq 4) = 1 - .85$$

$$= .15.$$

In general,

$$P(X \geq k) = 1 - P(X \leq k - 1),$$

so $P(X \geq k)$ is easily evaluated with Table 2.

Expressions like

$$P(2 \leq x \leq 8),$$

meaning you want to know the probability that the number of successes is between 2 and 8, inclusive, can also be evaluated with Table 2 by noting that

$$P(2 \leq x \leq 8) = P(x \leq 8) - P(x \leq 1).$$

In words, the event of eight successes or less can be broken down into the sum of two mutually exclusive events: the event that the number of successes is less than or equal to 1 and the event that the number of successes is between 2 and 8, inclusive. Rearranging terms yields the last equation. The point is that $P(2 \leq x \leq 8)$ can be written in terms of two expressions that are easily evaluated with Table 2 in Appendix B.

EXAMPLE. Assume $n = 10$ and $p = .5$. From Table 2 in Appendix B, $P(X \leq 1) = .011$ and $P(X \leq 8) = .989$, so

$$P(2 \leq X \leq 8) = .989 - .011 = .978.$$

■

A related problem is determining the probability of one success or less or nine successes or more. The first part is simply read from Table 2 and can be seen to be .011. The probability of nine successes or more is the complement of eight successes or less, so

$$P(X \geq 9) = 1 - P(X \leq 8) = 1 - .989 = .011,$$

again assuming that $n = 10$ and $p = .5$. Thus, the probability of one success or less or nine successes or more is $.011 + .011 = .022$. In symbols,

$$P(X \leq 1 \text{ or } X \geq 9) = .022.$$

There are times when you will need to compute the mean and variance of a binomial probability function once you are given n and p. It can be shown that the mean and variance are given by

$$\mu = E(X)$$
$$= np,$$

and

$$\sigma^2 = npq.$$

For example, if $n = 16$ and $p = .5$, the mean of the binomial probability function is $\mu = np = 16(.5) = 8$. That is, on average, 8 of the 16 observations in a random sample will be a success, while the other 8 will not. The variance is $\sigma^2 = npq = 16(.5)(.5) = 4$, so the standard deviation is $\sigma = \sqrt{4} = 2$. If, instead, $p = .3$, then $\mu = 16(.3) = 4.8$. That is, the average number of successes is 4.8.

In most situations, p, the probability of a success, is not known and must be estimated based on x, the observed number of successes. The result, $E(X) = np$, suggests that x/n be used as an estimator of p; and indeed this is the estimator that is typically used. Often this estimator is written as

$$\hat{p} = \frac{x}{n}.$$

Note that $\hat{p}$ is just the proportion of successes in n trials. It can be shown (using the rules of expected values covered in Section 2.9) that

$$E(\hat{p}) = p.$$

That is, if you were to repeat an experiment infinitely many times, each time randomly sampling n observations, the average of these infinitely many $\hat{p}$ values is p. It can also be shown that the variance of $\hat{p}$ is

$$\sigma_{\hat{p}}^2 = \frac{pq}{n}.$$

EXAMPLE. If you sample 25 people and the probability of success is .4, then the variance of $\hat{p}$ is

$$\sigma_{\hat{p}}^2 = \frac{.4 \times .6}{25} = .098.$$

■

The characteristics and properties of the binomial probability function can be summarized as follows:

- The experiment consists of exactly n independent trials.
- Only two possible outcomes are possible on each trial, usually called *success* and *failure*.
- Each trial has the same probability of success, p.
- $q = 1 - p$ is the probability of a failure.
- There are x successes among the n trials.
- $p(x) = \binom{n}{x} p^x q^{n-x}$ is the probability of x successes in n trials, $x = 0, 1, \ldots, n$.
- $\binom{n}{x} = (n!/x!(n-x)!)$.
- You estimate p with $\hat{p} = x/n$, where x is the total number of successes.
- $E(\hat{p}) = p$.
- The variance of $\hat{p}$ is $\sigma^2 = pq/n$.
- The average or expected number of successes in n trials is $\mu = E(X) = np$.
- The variance of X is $\sigma^2 = npq$.

2.6 Continuous Variables and the Normal Curve

For various reasons (described in subsequent chapters), *continuous* variables, meaning that the variables can have any value over some range of values, play a fundamental and useful role in statistics. In contrast to discrete variables, probabilities associated with continuous variables are given by the area under a curve. The equation for this curve is called a *probability density function*. If, for instance, we wanted to know the probability that a variable has a value between 2 and 5, say, this is represented by the area under the curve and between 2 and 5.

EXAMPLE. Suppose X represents the proportion of time someone spends on pleasant tasks at their job. So, of course, for any individual we observe, X has some value between 0 and 1. Assume that for the population of all working adults, the probability density function is as shown in Figure 2.4. Further assume that we want to know the probability that the proportion of time spent working on pleasant tasks is less than or equal to .4. In symbols,

Continued

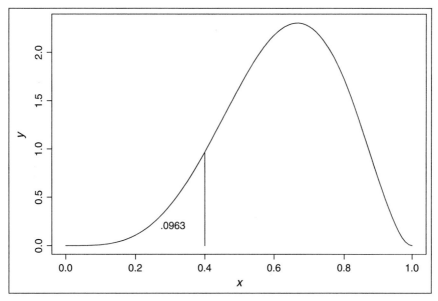

FIGURE 2.4   Probabilities associated with continuous variables are represented by the area under a curve. Here the area under the curve and to the left of 0.4 is .0963. That is, according to this probability curve, $P(X \leq .4) = .0963$.

EXAMPLE. (*Continued*) we want to know $P(X \leq .4)$. In Figure 2.4, the area under the curve and to the left of .4 is .096. That is, the probability we seek is .096. In symbols, $P(X \leq .4) = .096$.   ■

If $P(X \leq 5) = .8$ and X is a continuous variable, then the value 5 is called the .8 quantile. If $P(X \leq 3) = .4$, then 3 is the .4 quantile. In general, if $P(X \leq c) = q$, then c is called the qth *quantile*. In Figure 2.4, for example, .4 is the .096 quantile. *Percentiles* are just quantiles multiplied by 100. So in Figure 2.4, .4 is the 9.6 percentile. There are some mathematical difficulties when defining quantiles for discrete data; there is a standard method for dealing with this issue (e.g., Serfling, 1980, p. 3), but the details are not important here.

The .5 quantile is called the *population median*. If $P(X \leq 6) = .5$, then 6 is the population median. The median is centrally located in a probabilistic sense, because there is a .5 probability that a value is less than the median and there is a .5 probability that a value is greater than the median instead.

2.6.1 The Normal Curve

The best-known and most important probability density function is the normal curve, an example of which is shown in Figure 2.5. Normal curves have the following important properties:

1. The total area under the curve is 1. (This is a requirement of any probability density function.)

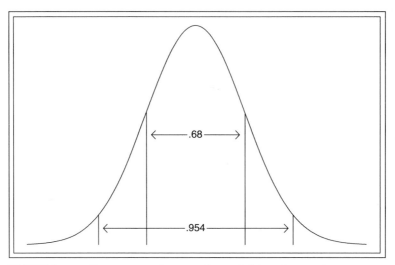

FIGURE 2.5   For any normal curve, the probability that an observation is within one standard deviation of the mean is always .68. The probability of being within two standard deviations is always .954.

2. All normal distributions are bell shaped and symmetric about their mean, μ. It follows that the population mean and median are identical.

3. Although not indicated in Figure 2.5, all normal curves extend from $-\infty$ to ∞ along the x-axis.

4. If the variable X has a normal distribution, the probability that X has a value within one standard deviation of the mean is .68, as indicated in Figure 2.5. In symbols, if X has a normal distribution, then

$$P(\mu - \sigma < X < \mu + \sigma) = .68$$

regardless of what the population mean and variance happen to be. The probability of being within two standard deviations is approximately .954. In symbols,

$$P(\mu - 2\sigma < X < \mu + 2\sigma) = .954.$$

The probability of being within three standard deviations is

$$P(\mu - 3\sigma < X < \mu + 3\sigma) = .9975.$$

5. The probability density function of a normal distribution is

$$f(x) = \frac{1}{\sigma\sqrt{2\pi}} \exp\left[-\frac{(x - \mu)^2}{2\sigma^2}\right], \tag{2.10}$$

where, as usual, μ and σ^2 are the mean and variance. This rather complicated-looking equation does not play a direct role in applied work, so no illustrations are given on how it is evaluated. Be sure to notice, however, that the probability density function is determined by the mean and variance. If, for example, we want to determine the probability that a variable is less than 25, this probability is completely determined by the mean and variance *if* we assume normality.

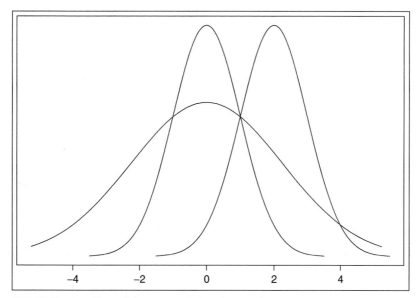

FIGURE 2.6 Two of these normal distributions have equal means and two have equal variances. Note that for normal distributions, increasing the standard deviation from 1 to 1.5 results in a substantial change in the probability curve. (Compare this with Figure 2.8.)

Figure 2.6 shows three normal distributions, two of which have equal means of zero but standard deviations $\sigma = 1$ and $\sigma = 1.5$, and the third again has standard deviation $\sigma = 1$ but with a mean of $\mu = 2$. There are two things to notice. First, if two normal distributions have equal variances but unequal means, the two probability curves are centered around different values but otherwise are identical. Second, for *normal* distributions, there is a distinct and rather noticeable difference between the two curves when the standard deviation increases from 1 to 1.5.

2.6.2 Computing Probabilities Associated with Normal Curves

Assume that human infants have birth weights that are normally distributed with a mean of 3700 grams and a standard deviation of 200 grams. What is the probability that a baby's birth weight will be less than or equal to 3000 grams? As previously explained, this probability is given by the area under the normal curve, but simple methods for computing this area are required. Today the answer is easily obtained on a computer. (For example, the S-PLUS function pnorm can be used.) But for pedagogical reasons a more traditional method is covered here. We begin by considering the special case where the mean is zero and the standard deviation is 1 ($\mu = 0$, $\sigma = 1$), after which we illustrate how to compute probabilities for any mean and standard deviation.

Standard Normal

The *standard normal distribution* is a normal distribution with mean $\mu = 0$ and standard deviation $\sigma = 1$; it plays a central role in many areas of statistics. As is typically done,

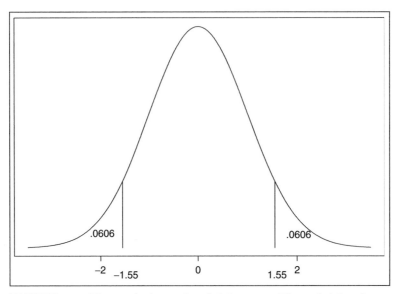

FIGURE 2.7   The standard normal probability curve. The probability that an observation is less than or equal to -1.55 is .0606, and the probability that an observation is greater than or equal to 1.55 is .0606.

Z is used to represent a variable that has a standard normal distribution. Our immediate goal is to describe how to determine the probability that an observation randomly sampled from a standard normal distribution is less than any constant c we might choose.

These probabilities are easily determined using Table 1 in Appendix B, which reports the probability that a standard normal random variable has probability less than or equal to c for $c = -3.00, -2.99, -2.98, \ldots, -0.01, 0, .01, \ldots, 3.00$. The first entry in the first column shows -3. The column next to it gives the corresponding probability, .0013. That is, the probability that a standard normal random variable is less than or equal to -3 is $P(Z \leq -3) = .0013$. Put another way, -3 is the .0013 quantile of the standard normal distribution. Going down the first column we see the entry -2.08; the column next to it indicates that the probability that a standard normal variable is less than or equal to -2.08 is .0188. This says that -2.08 is the .0188 quantile. Looking at the last entry in the third column, we see -1.55; the entry just to the right, in the fourth column, is .0606, so $P(Z \leq -1.55) = .0606$. This probability corresponds to the area in the left portion of Figure 2.7. Because the standard normal curve is symmetric about zero, the probability that X is greater than 1.55 is also .0606, which is shown in the right portion of Figure 2.7. Again looking at the first column of Table 1 in Appendix B, we see the value $z = 1.53$; next to it is the value .9370, meaning that $P(Z \leq 1.53) = .9370$.

In applied work, there are three types of probabilities that need to be determined:

1. $P(Z \leq c)$, the probability that a standard normal random variable is less than or equal to c

2. $P(Z \geq c)$, the probability that a standard normal random variable is greater than or equal to c

3. $P(a \leq Z \leq b)$, the probability that a standard normal random variable is between the values a and b

The first of these is determined from Table 1 in Appendix B, as already indicated. Because the area under the curve is 1, the second is given by

$$P(Z \geq c) = 1 - P(Z \leq c).$$

The third is given by

$$P(a \leq Z \leq b) = P(Z \leq b) - P(Z \leq a).$$

EXAMPLE. Determine $P(Z \geq 1.5)$, the probability that a standard normal random variable is greater than 1.5. From Table 1 in Appendix B, $P(Z \leq 1.5) = .9332$. Therefore, $P(Z \geq 1.5) = 1 - .9332 = .0668$. ■

EXAMPLE. Next we determine $P(-1.96 \leq Z \leq 1.96)$, the probability that a standard normal random variable is between -1.96 and 1.96. From Table 1 in Appendix B, $P(Z \leq 1.96) = .975$. Also, $P(Z \leq -1.96) = .025$, so

$$P(-1.96 \leq Z \leq 1.96) = .975 - .025 = .95.$$

■

In some situations it is necessary to use Table 1 (in Appendix B) backwards. That is, we are given a probability and the goal is to determine c. For example, if we are told that $P(Z \leq c) = .99$, what is c? We simply find where .99 happens to be in Table 1 under the columns headed by $P(Z \leq z)$ and then read the number to the left, under the column headed by z. The answer is 2.33.

Two related problems also arise. The first is determining c given the value of

$$P(Z \geq c).$$

A solution is obtained by noting that the area under the curve is 1, so $P(Z \geq c) = 1 - P(Z \leq c)$, which involves a quantity we can determine from Table 1. That is, you compute $d = 1 - P(Z \geq c)$ and then determine c such that

$$P(Z \leq c) = d.$$

EXAMPLE. To determine c if $P(Z \geq c) = .9$, first compute $d = 1 - P(Z \leq c) = 1 - .9 = .1$. Then c is given by $P(Z \leq c) = .1$. Referring to Table 1 in Appendix B, $c = -1.28$. ■

The other type of problem is determining c given

$$P(-c \leq Z \leq c).$$

Letting $d = P(-c \leq Z \leq c)$, the answer is given by

$$P(Z \leq c) = \frac{1 + d}{2}.$$

> **EXAMPLE.** To determine c if $P(-c \leq Z \leq c) = .9$, let $d = P(-c \leq Z \leq c) = .9$ and then compute $(1 + d)/2 = (1 + .9)/2 = .95$. Then c is given by $P(Z \leq c) = .95$. Referring to Table 1 in Appendix B, $c = 1.645$. ■

Solution for Any Normal Distribution

Now consider any normal random variable having mean μ and standard deviation σ. The next goal is to describe how to determine the probability that an observation is less than c, where, as usual, c is any constant that might be of interest. The solution is based on *standardizing* a normal random variable, which means that we subtract the population mean μ and divide by the standard deviation, σ. In symbols, we standardize a normal random variable X by transforming it to

$$Z = \frac{X - \mu}{\sigma}. \tag{2.11}$$

It can be shown that if X has a normal distribution, then the distribution of Z is standard normal. In particular, the probability that a normal random variable X is less than or equal to c is

$$P(X \leq c) = P\left(Z \leq \frac{c - \mu}{\sigma}\right). \tag{2.12}$$

> **EXAMPLE.** Suppose it is claimed that the cholesterol levels in adults have a normal distribution with mean $\mu = 230$ and standard deviation $\sigma = 20$. If this is true, what is the probability that an adult will have a cholesterol level less than or equal to $c = 200$? Referring to Equation (2.12), the answer is
>
> $$P(X \leq 200) = P\left(Z \leq \frac{200 - 230}{20}\right) = P(Z < -1.5) = .0668,$$
>
> where .0668 is read from Table 1 in Appendix B. This means that the probability that an adult has a cholesterol level less than 200 is .0668. ■

In a similar manner, we can determine the probability that an observation is greater than or equal to 240 or between 210 and 250. More generally, for any constant c that is of interest, we can determine the probability that an observation is greater than c with the equation

$$P(X \geq c) = 1 - P(X \leq c),$$

the point being that the right side of this equation can be determined with Equation (2.12). In a similar manner, for any two constants a and b,

$$P(a \leq X \leq b) = P(X \leq b) - P(X \leq a).$$

EXAMPLE. Continuing the last example, determine the probability of observing an adult with a cholesterol level greater than or equal to 240. We have that

$$P(X \geq 240) = 1 - P(X \leq 240).$$

Referring to Equation (2.12),

$$P(X \leq 240) = P\left(Z \leq \frac{240 - 230}{20}\right) = P(Z < .5) = .6915,$$

so

$$P(X \geq 240) = 1 - .6915 = .3085.$$

In words, the probability that an adult has a cholesterol level greater than or equal to 240 is .3085.   ■

EXAMPLE. Continuing the cholesterol example, we determine

$$P(210 \leq X \leq 250).$$

We have that

$$P(210 \leq X \leq 250) = P(X \leq 250) - P(X \leq 210).$$

Now,

$$P(X \leq 250) = P\left(Z < \frac{250 - 230}{20}\right) = P(Z \leq 1) = .8413$$

and

$$P(X \leq 210) = P\left(Z < \frac{210 - 230}{20}\right) = P(Z \leq -1) = .1587,$$

so

$$P(210 \leq X \leq 250) = .8413 - .1587 = .6826,$$

meaning that the probability of observing a cholesterol level between 210 and 250 is .6826.   ■

2.7 Understanding the Effects of Nonnormality

Conventional statistical methods are based on the assumption that observations follow a normal curve. It was once thought that violating the normality assumption rarely had a detrimental impact on these methods, but theoretical and empirical advances have made it clear that two general types of nonnormality cause serious practical problems in a wide range of commonly occurring situations. Indeed, even very slight departures from normality can be a source of concern. To appreciate the practical utility of modern statistical techniques, it helps to build an intuitive sense of how nonnormality influences the population mean and variance and how this effect is related to determining probabilities.

The so-called contaminated, or mixed normal, distribution is a classic way of illustrating some of the more important effects of nonnormality. Consider a situation where we have two subpopulations of individuals or things. Assume each subpopulation has a normal distribution but that they differ in terms of their means or variances or both. When we mix the two populations together we get what is called a *mixed*, or *contaminated*, normal. Generally, mixed normals fall outside the class of normal distributions. That is, for a distribution to qualify as normal, the equation for its curve must have the form given by Equation (2.10), and the mixed normal does not satisfy this requirement. When the two normals mixed together have a common mean but unequal variances, the resulting probability curve is again symmetric about the mean, but even then the mixed normal is not a normal curve.

To provide a more concrete description of the mixed normal, consider the entire population of adults living around the world and let X represent the amount of weight they have gained or lost during the last year. Let's divide the population of adults into two groups: those who have tried some form of dieting to lose weight and those that have not. For illustrative purposes, assume that for adults who have not tried to lose weight, the distribution of their weight loss is standard normal (so $\mu = 0$ and $\sigma = 1$). As for adults who have dieted to lose weight, assume that their weight loss is normally distributed, again with mean $\mu = 0$ but with standard deviation $\sigma = 10$. Finally, suppose that 10% of all adults went on a diet last year to lose weight. So if we were to randomly pick an adult, there is a 10% chance of selecting someone who has dieted. That is, there is a 10% chance of selecting an observation from a normal distribution having standard deviation 10, so there is a 90% chance of selecting an observation from a normal curve having a standard deviation of 1.

Now, if we mix these two populations of adults together, the exact distribution of X (the weight loss for a randomly sampled adult) can be derived and is shown in Figure 2.8. Also shown is the standard normal distribution, and as is evident there is little separating the two curves. Let $P(X \leq c)$ be the probability that an observation is less than c when sampling from the mixed normal, and let $P(Z \leq c)$ be the probability when sampling from the standard normal instead. For any constant c we might pick, it can be shown that $P(X \leq c)$ does not differ from $P(Z \leq c)$ by more than .04. For example, for a standard normal curve, we see from Table 1 in Appendix B that $P(Z \leq 1) = .8413$. If X has the mixed normal distribution considered here, then the probability that X has a value less than or equal to 1 will not differ from .8413 by more than .04; it will be between .8013 and .8813. The exact value happens to be .81.

Here is the point: Very small departures from normality can greatly influence the value of the population variance. For the standard normal in Figure 2.8 the variance is 1, but for the mixed normal it is 10.9. The full implications of this result are impossible to appreciate at this point, but they will become clear in subsequent chapters.

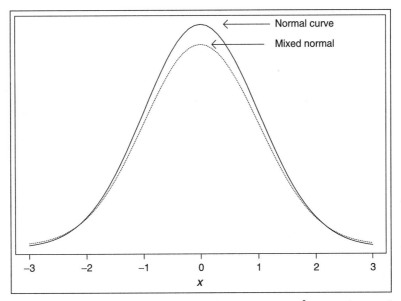

FIGURE 2.8 A standard normal curve having variance $\sigma^2 = 1$ and a mixed normal curve having variance $\sigma^2 = 10.9$. Figure 2.6 illustrated that slight increases in σ result in a substantial change in a normal curve. But the contaminated normal illustrates that two distributions can have substantially different variances even though their probability curves are very similar.

The main goal now is to lay the foundation for understanding some of the problems associated with conventional methods to be described.

To illustrate one of the many implications associated with the mixed normal, consider the following problem: Given the population mean and variance, how can we determine the probability that an observation is less than c. If, for example, $\mu = 0$ and $\sigma^2 = 10.9$, and if we want to know the probability that an observation is less than 1, we get an answer if we assume normality and use the method described in Section 2.6.2. The answer is .619. But for the mixed normal having the same mean and variance, the answer is .81, as previously indicated. So determining probabilities assuming normality when in fact a distribution is slightly nonnormal can lead to a fairly inaccurate result. Figure 2.9 graphically illustrates the problem. Both curves have equal means and variances, yet there is a very distinct difference.

Figure 2.9 illustrates another closely related point. As previously pointed out, normal curves are completely determined by their mean and variance, and Figure 2.6 illustrated that under normality, increasing the variance from 1 to 1.5 results in a very noticeable difference in the graphs of the probability curves. If we assume that curves are normal, or at least approximately normal, this might suggest that in general, if two distributions have equal variances, surely they will appear very similar in shape. But this is not necessarily true even when the two curves are symmetric about the population mean and are bell-shaped. Again, knowing σ is useful in some situations to be covered, but there are many situations where it can mislead.

Figure 2.10 provides another illustration that two curves can have equal means and variances yet differ substantially.

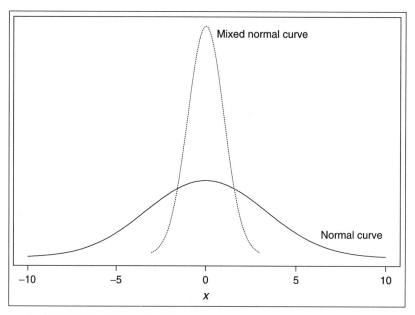

FIGURE 2.9   Two probability curves having equal means and variances.

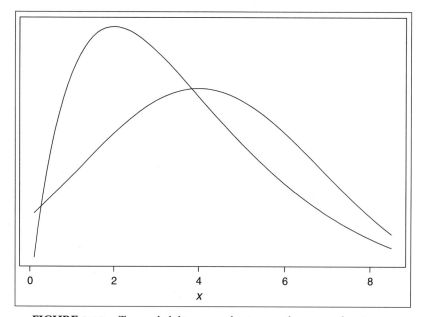

FIGURE 2.10   Two probability curves having equal means and variances.

Here is another way σ might mislead. We saw that for a normal distribution, there is a .68 probability that an observation is within one standard deviation of the mean. It is incorrect to conclude, however, that for nonnormal distributions, this rule always applies. The mixed normal is approximately normal in a sense already described, yet the probability of being within one standard deviation of the mean now exceeds .999.

One reason this last point is important in applied work is related to the notion of outliers. *Outliers* are values that are unusually large or small. For a variety of reasons to be described in subsequent chapters, detecting outliers is important. Assuming normality, a common rule is to declare a value an outlier if it is more than two standard deviations from the mean. In symbols, declare X an outlier if

$$|X - \mu| > 2\sigma. \tag{2.13}$$

So, for example, if $\mu = 4$ and $\sigma = 3$, the value $X = 5$ would not be declared an outlier because $|5 - 4|$ is less than $2 \times 3 = 6$. In contrast the value 12 would be labeled an outlier. The idea is that if a value lies more than two standard deviations from the mean, then probabilistically it is unusual. For normal curves, the probability that an observation is more than two standard deviations from the mean is .046.

To illustrate a concern about this rule, consider what happens when the probability density function is the mixed normal in Figure 2.8. Because the variance is 10.9, we would declare X an outlier if

$$|X - \mu| > 2\sqrt{10.9} = 6.6.$$

But $\mu = 0$, so we declare X to be an outlier if $|X| > 6.6$. It can be seen that now the probability of declaring a value an outlier is 4×10^{-11} — it is virtually impossible. (The method used to derive this probability is not important here.) The value 6, for example, would not be declared an outlier, even though the probability of getting a value greater than or equal to 6 is 9.87×10^{-10}. That is, from a probabilistic point of view, 6 is unusually large, because the probability of getting this value or larger is less than 1 in a billion, yet Equation (2.13) does not flag it as being unusual.

Note that in Figure 2.8, the tails of the mixed normal lie above the tails of the normal. For this reason, the mixed normal is often described as being *heavy-tailed*. Because the area under the extreme portions of a heavy-tailed distribution is larger than the area under a normal curve, extreme values or outliers are more likely when sampling from the mixed normal. Generally, very slight changes in the tail of any probability density function can inflate the variance tremendously, which in turn can make it difficult and even virtually impossible to detect outliers using the rule given by Equation (2.13), even though outliers are relatively common. There are very effective methods for dealing with this problem, but the details are postponed until Chapter 3.

2.7.1 Skewness

Heavy-tailed distributions are one source of concern when employing conventional statistical techniques. Another is *skewness*, which generally refers to distributions that are not exactly symmetric. It is too soon to discuss all the practical problems associated with skewed distributions, but one of the more fundamental issues can be described here.

Consider how we might choose a single number to represent the typical individual or thing under study. A seemingly natural approach is to use the population mean. If a distribution is symmetric about its mean, as is the case when a distribution is normal, there is general agreement that the population mean is indeed a reasonable reflection of what is typical. But when distributions are skewed, at some point doubt begins to

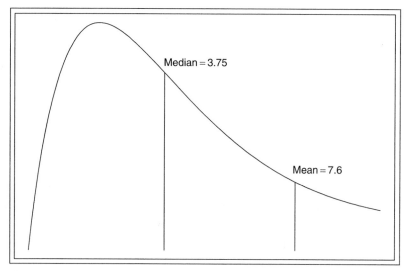

FIGURE 2.11   The population mean can be located in the extreme portion of the tail of a probability curve.

arise as to whether the mean is a good choice. Consider, for example, the distribution shown in Figure 2.11, which is skewed to the right. In this particular case the population mean is located in the extreme right portion of the curve. In fact, the probability that an observation is less than the population mean is 0.74. So from a probabilistic point of view, the population mean is rather atypical. In contrast, the median is located near the more likely outcomes and would seem to better reflect what is typical.

One strategy is to routinely use means and hope that you never encounter a situation where extreme skewness occurs. We will see empirical evidence, however, that such situations arise in practice. Another strategy is simply to switch to the median. If a distribution is symmetric, the population mean and median are identical; but if a distribution is skewed, the median can be argued to be a better indication of what is typical. In some applied settings, the median is a good choice, but unfortunately the routine use of the median can be rather unsatisfactory as well. The reasons are rather involved, but they will be made clear in subsequent chapters. For the moment it is merely remarked that dealing with skewness is a complex issue that has received a great deal of attention. In addition to the concern illustrated by Figure 2.11, there are a variety of other problems, which become evident in Chapter 5. Yet another strategy is to use some simple transformation of the data in an attempt to deal with skewness. A common method is to take logarithms, but this can fail as well, for reasons described in Chapters 3 and 4.

2.8 Pearson's Correlation

This section introduces Pearson's correlation and some of its properties. Pearson's correlation is covered in detail in Chapter 6, but to address certain technical issues it is convenient to introduce it here. The goal is to provide a slightly deeper understanding of why some seemingly natural strategies for analyzing data are theoretically unsound and cannot be recommended. Verbal and graphical explanations are provided as well, and so this section can be skipped by readers who are willing to accept certain results.

Imagine any situation where we have two measures. For example, for each individual among a population of individuals, the two measures might be height and weight or measures of gregariousness and severity of heart disease. Or another situation might be where we sample married couples and the two measures are the cholesterol levels of the wife and husband. For convenience we label the two measures X and Y. Associated with these two variables are two quantities that play a major role in statistics: their covariance and Pearson's correlation.

The *covariance* between X and Y is

$$\sigma_{xy} = E[(X - \mu_x)(Y - \mu_y)].$$

In words, if for some population of individuals we subtract the mean of X from every possible value for X and do the same for Y, then the covariance between X and Y is defined to be the average of the products of these differences. It might help to note that the covariance of X with itself is just its variance, and the same is true for Y. That is, the idea of covariance generalizes the notion of variance to two variables. Pearson's correlation is the covariance divided by the product of the standard deviations and is typically labeled ρ. That is,

$$\rho = \frac{\sigma_{xy}}{\sigma_x \sigma_y}. \tag{2.14}$$

Here are the properties that will be important in some of the chapters to follow:

- $-1 \le \rho \le 1$ (Pearson's correlation always has a value between -1 and 1.)
- If X and Y are independent, then $\rho = \sigma_{xy} = 0$.
- If $\rho \ne 0$, then X and Y are dependent.
- For any two variables, the variance of their sum is

$$\text{VAR}(X + Y) = \sigma_x^2 + \sigma_y^2 + 2\rho\sigma_x\sigma_y. \tag{2.15}$$

This last property is important when explaining why some seemingly reasonable strategies for analyzing data (covered in Chapter 4) are technically incorrect. Note that when we add any two measures together, their sum will have some average value. Using the rules of expected values (covered in Section 2.9), the average of this sum is simply $\mu_x + \mu_y$, the sum of the means. Equation (2.15) says that if we add two measures together, the average squared difference between any sum we might observe, and the mean of the sum, is completely determined by the individual variances plus the correlation.

EXAMPLE. There are two variables in Table 2.2: a wife's rating and a husband's rating. The variance associated with the wives can be seen to be $\sigma_y^2 = 0.29$. As for the husband's, $\sigma_x^2 = .41$. It was already pointed out that the ratings for

Continued

> **EXAMPLE.** (*Continued*) husbands and wives are independent, so $\rho = 0$. Consequently, the variance of the sum of the ratings is $.29 + .41 = .7$. That is, without even determining the probability function associated with this sum, its variance can be determined. ■

A cautionary note should be added. Although independence implies $\rho = 0$, it is *not* necessarily the case that $\rho = 0$ implies independence. In the lower left scatterplot of Figure 2.2, for example, points were generated on a computer with $\rho = 0$, yet there is dependence because, as already indicated, the variation in the Y values increases with X. (This issue is discussed in more detail in Chapter 6.)

2.8.1 Computing the Population Covariance

This subsection is added for readers interested in understanding how the population covariance is computed when probabilities are known. These details are not crucial in what follows, but they might provide a better sense of how the covariance is defined, which in turn provides some details about how the population correlation is defined as well. But readers not interested in technical issues can skip this section.

Consider the probabilities shown in Table 2.3. It can be seen that the expected value of X is $\mu_x = 2.08$ and that the expected value of Y is $\mu_y = 2.02$. Let $p(x, y)$ be the probability of observing the values $X = x$ and $Y = y$ simultaneously. So, according to Table 2.3, the probability that $Y = 1$ and $X = 1$ is $p(1, 1) = .13$; and $p(2, 3) = .18$. To compute the population covariance, you simply perform the calculations shown in Table 2.4. That is, for every combination of values for X and Y, you subtract the corresponding means, yielding the values in columns three and four of Table 2.4. The probabilities associated with all possible pairs of values are shown in column five. Column six shows the product of the values in columns three, four, and five. The population covariance is the sum of the values in column six, which is .0748. Under independence, X and Y must have a covariance of zero, so we have established that the variables considered here are dependent, because the covariance differs from zero.

TABLE 2.3 Hypothetical Probabilities for Pearson's Correlation

		X		
Y	1	2	3	
1	.13	.15	.06	0.34
2	.04	.08	.18	0.30
3	.10	.15	.11	0.36
	.27	.38	.35	

TABLE 2.4   How to Compute the Covariance for the Probabilities in Table 2.3

x	y	$x - \mu_x$	$y - \mu_y$	$p(x,y)$	$(x - \mu_x)(y - \mu_y)p(x,y)$
1	1	−1.08	−1.02	.13	0.143208
1	2	−1.08	−0.02	.04	0.000864
1	3	−1.08	0.98	.10	−0.063504
2	1	−0.08	−1.02	.15	0.003264
2	2	−0.08	−0.02	.08	0.000128
2	3	−0.08	0.98	.15	−0.014112
3	1	0.92	−1.02	.06	−0.093840
3	2	0.92	−0.02	.18	−0.002760
3	3	0.92	0.98	.11	0.099176
					0.0748

EXAMPLE. To compute the population correlation for the values and probabilities shown in Table 2.3, first compute the covariance, which we just saw is $\sigma_{xy} = 0.0748$. It is left as an exercise to show that the variances are $\sigma_x^2 = 0.6136$ and $\sigma_y^2 = .6996$. Consequently,

$$\rho = \frac{.0748}{\sqrt{0.6136} \times \sqrt{.6996}} = .11.$$

So according to Equation (2.15), the variance of the sum, $X + Y$, is

$$0.6136 + .6996 + 2 \times 0.0748 = 1.46.$$

■

2.9 Some Rules About Expected Values

This section summarizes some basic rules about expected values that will be useful in subsequent chapters. The first rule is that if we multiply a variable by some constant c, its expected value is multiplied by c as well. In symbols,

$$E(cX) = cE(X). \tag{2.16}$$

This is just a fancy way of saying, for example, that if the average height of some population of children is five feet ($\mu = 5$), then the average in inches is 60 (the average in feet multiplied by 12). More formally, if $E(X) = \mu = 5$, then $E(12X) = 12 \times 5 = 60$.

The second rule is that if we add c to every possible value, the expected value increases by c as well. That is,

$$E(X + c) = E(X) + c. \tag{2.17}$$

So if $\mu = 6$ and 4 is added to every possible value for X, the average becomes 10. Or, in terms of Equation (2.17), $E(X + 4) = E(X) + 4 = 6 + 4 = 10$.

EXAMPLE. Because μ is a constant, $E(X - \mu) = E(X) - \mu = \mu - \mu = 0$. That is, if we subtract the population mean from every possible value we might observe, the average value of this difference is always zero. If, for instance, the average height of all adult men is 5.9 feet and we subtract 5.9 from everyone's height, the average will be zero.   ■

To provide some intuition about the next rule, imagine that for the population of all married women, if they were to rate their marital satisfaction, the probability function would be

x:	1	2	3	4	5
$p(x)$:	.2	.1	.4	.2	.1

Now consider two individuals, Mary and Jane, and suppose they are asked about how they would rate their level of satisfaction regarding their married life. For convenience, label Mary's response X_1 and Jane's response X_2. So before Mary rates her marriage, there is a .2 probability that she will rate her marriage satisfaction as 1, a .1 probability she will rate it as 2, and so on. The same is assumed to be true for Jane. Now consider the sum of their two ratings, which we label $X = X_1 + X_2$. What is the expected value of this sum? That is, on average, what is the value of X?

One way of solving this problem is to attempt to derive the probability function of the sum, X. The possible values for X are $1 + 1 = 2$, $1 + 2 = 3, \ldots, 6 + 6 = 12$, and if we could derive the probabilities associated with these values, we could determine the expected value of X. But there is a much simpler method, because it can be shown that the expected value of a sum is just the sum of the expected values. That is,

$$E(X_1 + X_2) = E(X_1) + E(X_2), \tag{2.18}$$

so the expected value of the sum can be determined if we know the probability function associated with each of the observations we make. But given the probability function, we can do just that. We see that $E(X_1) = 2.9$; in a similar manner $E(X_2) = 2.9$. So the expected value of their sum is 5.8. That is, if two women are asked to rate their marital satisfaction, the average sum of their ratings, over all pairs of women we might interview, is 5.8.

This last illustration demonstrates a more general principle that will be helpful. If X_1 and X_2 have identical probability functions, so in particular the variables have a common mean, μ, then the expected value of their sum is 2μ. So using our rule for constants, we see that the average of these two ratings is μ. That is,

$$E\left[\frac{1}{2}(X_1 + X_2)\right] = \frac{1}{2}(\mu + \mu) = \mu.$$

Here is a summary of the rules for expected values, where c is any constant:

- $E(cX) = cE(X) = c\mu$.

- $E(X + c) = E(X) + c = \mu + c$.
- $E(X_1 + X_2) = E(X_1) + E(X_2)$. For the special case where X_1 and X_2 have a common mean μ, which occurs when they have identical probability functions, $E(X_1 + X_2) = 2\mu$.

2.10 Chi-Squared Distributions

There is an important family of distributions related to the standard normal distribution that will play a role in subsequent chapters. It is called the family of chi-squared distributions and arises as follows. Suppose Z has a standard normal distribution and let $Y = Z^2$. The distribution of Y is so important, it has been given a special name: a chi-squared distribution with one degree of freedom. Next, suppose two independent observations are made, Z_1 and Z_2, both of which have standard normal distributions. Then the distribution of

$$Y = Z_1^2 + Z_2^2$$

is called chi-square distribution with two degrees of freedom. More generally, for n independent standard normal variables, $Z_1, \ldots, Z_n$,

$$Y = Z_1^2 + \cdots + Z_n^2$$

is said to have a chi-square distribution with n degrees of freedom. There are many statistical methods that utilize the family of chi-squared distributions. The only goal now is to introduce the distribution and to emphasize that any use of a chi-squared distribution is intimately connected to normality.

2.11 Exercises

1. For the probability function

x:	0,	1
$p(x)$:	.7,	.3

   verify that the mean and variance are .3 and .21. What is the probability of getting a value less than the mean?
2. Standardizing the possible values in Exercise 1 means that we transform the possible values (0 and 1) by subtracting the population mean and dividing by the population standard deviation. Here this yields $(1 - .3)/\sqrt{.21} = .7/\sqrt{.21}$ and $(0 - .3)/\sqrt{.21} = -.3/\sqrt{.21}$, respectively. The probabilities associated with these two values are .3 and .7. Verify that the expected value of the standardized values is zero and the variance is 1.
3. For the probability function

x:	1,	2,	3,	4,	5
$p(x)$:	.15,	.2,	.3,	.2,	.15

   determine the mean, the variance, and $P(X \leq \mu)$.

4. For the probability function

x:	1,	2,	3,	4,	5
$p(x)$:	.1,	.25,	.3,	.25,	.1

would you expect the variance to be larger or smaller than the variance associated with the probability function used in the previous exercise? Verify your answer by computing the variance for the probability function given here.

5. For the probability function

x:	1,	2,	3,	4,	5
$p(x)$:	.2,	.2,	.2,	.2,	.2

would you expect the variance to be larger or smaller than the variance associated with the probability function used in the previous exercise? Verify your answer by computing the variance.

6. Verify that if we standardize the possible values in Exercise 5, the resulting mean is zero and the variance is 1.

7. For the following probabilities, determine (a) the probability that someone is under 30, (b) the probability that someone has a high income given that he or she is under 30, (c) the probability that someone has a low income given that he or she is under 30, and (d) the probability that someone has a medium income given that he or she is over 50.

	Income		
Age	High	Medium	Low
<30	.030	.180	.090
30–50	.052	.312	.156
>50	.018	.108	.054

8. For Exercise 7, are income and age independent?

9. Coleman (1964) interviewed 3,398 schoolboys and asked them about their self-perceived membership in the "leading crowd." Their response was either yes, they were a member, or no, they were not. The same boys were also asked about their attitude concerning the leading crowd. In particular, they were asked whether membership meant that it does not require going against one's principles sometimes or whether they think it does. Here, the first response will be indicated by a 1 and the second will be indicated by a 0. The results were as follows:

	Attitude	
Member?	1	0
Yes	757	496
No	1071	1074

These values, divided by the sample size, 3,398, are called *relative frequencies*. For example, the relative frequency of the event (Yes, 1) is 757/3398. Treat the relative frequencies as probabilities and determine (a) the probability that an arbitrarily chosen boy responds Yes, (b) $P(\text{Yes}|1)$, (c) $P(1|\text{Yes})$, (d) whether the response Yes is independent of the attitude 0, (e) the probability of a (Yes and 1) or a (No and 0) response, (f) the probability of not responding (Yes and 1), (g) the probability of responding Yes or 1.

10. The probability density function associated with a so-called *uniform distribution* is given by $f(x) = 1/(b-a)$, where a and b are given constants and $a \leq x \leq b$. That is, the possible values you might observe range between the constants a and b, with every value between a and b equally likely. If, for example, $a = 0$ and $b = 1$, the possible values you might observe lie between 0 and 1. For a uniform distribution over the interval $a = 1$ and $b = 4$, draw the probability density function and determine the median and the .1 and .9 quantiles.

11. For the uniform distribution over the interval -3 to 2, determine (a) $P(X \leq 1)$, (b) $P(X < -1.5)$, (c) $P(X > 0)$, (d) $P(-1.2 \leq X \leq 1)$, (e) $P(X = 1)$.

12. For the uniform distribution in Exercise 11, determine the median and the .25 and .9 quantiles.

13. For the uniform distribution with $a = -1$ and $b = 1$, determine c such that (a) $P(X \leq c) = .9$, (b) $P(X \leq c) = .95$, (c) $P(X > c) = .99$.

14. For the uniform distribution with $a = -1$ and $b = 1$, determine c such that (a) $P(-c \leq X \leq c) = .9$, (b) $P(-c \leq X \leq c) = .95$, (c) $P(-c \leq X \leq c) = .99$.

15. Suppose the waiting time at a traffic light has a uniform distribution from 0 to 20 seconds. Determine the probability of waiting (a) exactly 12 seconds, (b) less than 5 seconds, (c) more than 10 seconds.

16. When you look at a clock, the number of minutes past the hour — say, X — is some number between 0 and 60. Assume the number of minutes past the hour has a uniform distribution. Determine (a) $P(X = 30)$, (b) $P(X \leq 10)$, (c) $P(X \geq 20)$, (d) $P(10 \leq X < 20)$.

17. For Exercise 16, determine the .8 quantile.

18. Given that Z has a standard normal distribution, use Table 1 in Appendix B to determine (a) $P(Z \geq 1.5)$, (b) $P(Z \leq -2.5)$, (c) $P(Z < -2.5)$, (d) $P(-1 \leq Z \leq 1)$.

19. If Z has a standard normal distribution, determine (a) $P(Z \leq .5)$, (b) $P(Z > -1.25)$, (c) $P(-1.2 < Z < 1.2)$, (d) $P(-1.8 \leq Z < 1.8)$.

20. If Z has a standard normal distribution, determine (a) $P(Z < -.5)$, (b) $P(Z < 1.2)$, (c) $P(Z > 2.1)$, (d) $P(-.28 < Z < .28)$.

21. If Z has a standard normal distribution, find c such that (a) $P(Z \leq c) = .0099$, (b) $P(Z < c) = .9732$, (c) $P(Z > c) = .5691$, (d) $P(-c \leq Z \leq c) = .2358$.

22. If Z has a standard normal distribution, find c such that (a) $P(Z > c) = .0764$, (b) $P(Z > c) = .5040$, (c) $P(-c \leq Z < c) = .9108$, (d) $P(-c \leq Z \leq c) = .8$.

23. If X has a normal distribution with mean $\mu = 50$ and standard deviation $\sigma = 9$, determine (a) $P(X \leq 40)$, (b) $P(X < 55)$, (c) $P(X > 60)$, (d) $P(40 \leq X \leq 60)$.

24. If X has a normal distribution with mean $\mu = 20$ and standard deviation $\sigma = 9$, determine (a) $P(X < 22)$, (b) $P(X > 17)$, (c) $P(X > 15)$, (d) $P(2 < X < 38)$.

25. If X has a normal distribution with mean $\mu = .75$ and standard deviation $\sigma = .5$, determine (a) $P(X < .25)$, (b) $P(X > .9)$, (c) $P(.5 < X < 1)$, (d) $P(.25 < X < 1.25)$.

26. If X has a normal distribution, determine c such that

$$P(\mu - c\sigma < X < \mu + c\sigma) = .95.$$

27. If X has a normal distribution, determine c such that

$$P(\mu - c\sigma < X < \mu + c\sigma) = .8.$$

28. Assuming that the scores on a math achievement test are normally distributed with mean $\mu = 68$ and standard deviation $\sigma = 10$, what is the probability of getting a score greater than 78?

29. In Exercise 28, how high must someone score to be in the top 5%? That is, determine c such that $P(X > c) = .05$.

30. A manufacturer of car batteries claims that the life of their batteries is normally distributed with mean $\mu = 58$ months and standard deviation $\sigma = 3$. Determine the probability that a randomly selected battery will last at least 62 months.

31. Assume that the income of pediatricians is normally distributed with mean $\mu = \$100,000$ and standard deviation $\sigma = 10,000$. Determine the probability of observing an income between $85,000 and $115,000.

32. Suppose the winnings of gamblers at Las Vegas are normally distributed with mean $\mu = -300$ (the typical person loses $300) and standard deviation $\sigma = 100$. Determine the probability that a gambler does not lose any money.

33. A large computer company claims that their salaries are normally distributed with mean $50,000 and standard deviation 10,000. What is the probability of observing an income between $40,000 and $60,000?

34. Suppose the daily amount of solar radiation in Los Angeles is normally distributed with mean 450 calories and standard deviation 50. Determine the probability that for a randomly chosen day, the amount of solar radiation is between 350 and 550.

35. If the cholesterol levels of adults are normally distributed with mean 230 and standard deviation 25, what is the probability that a randomly sampled adult has a cholesterol level greater than 260?

36. If after one year, the annual mileage of privately owned cars is normally distributed with mean 14,000 miles and standard deviation 3,500, what is the probability that a car has mileage greater than 20,000 miles?

37. Can small changes in the tails of a distribution result in large changes in the population mean, μ, relative to changes in the median?

38. Explain in what sense the population variance is sensitive to small changes in a distribution.

39. For normal random variables, the probability of being within one standard deviation of the mean is .68. That is, $P(\mu - \sigma \leq X \leq \mu + \sigma) = .68$ if X has a normal distribution. For nonnormal distributions, is it safe to assume that this probability is again .68? Explain your answer.

40. If a distribution appears to be bell-shaped and symmetric about its mean, can we assume that the probability of being within one standard deviation of the mean is .68?

41. Can two distributions differ by a large amount yet have equal means and variances?

42. If a distribution is skewed, is it possible that the mean exceeds the .85 quantile?

43. Determine $P(\mu - \sigma \leq X \leq \mu + \sigma)$ for the probability function

x:	1,	2,	3,	4
$p(x)$:	.2,	.4,	.3,	.1

44. The U.S. Department of Agriculture reports that 75% of people who invest in the futures market lose money. Based on the binomial probability function with $n = 5$, determine:

   (a) The probability that all five lose money

   (b) The probability that all five make money

   (c) The probability that at least two lose money

45. If for a binomial, $p = .4$ and $n = 25$, determine (a) $P(X < 11)$, (b) $P(X \leq 11)$, (c) $P(X > 9)$, (d) $P(X \geq 9)$.

46. In Exercise 45, determine $E(X)$, the variance of X, $E(\hat{p})$, and the variance of $\hat{p}$.

3

SUMMARIZING DATA

Chapter 2 covered some ways of describing and summarizing a population of individuals (or things) when we know the probabilities associated with some variable of interest. For example, the population mean and median can be used to reflect the typical individual, and σ provides some indication of the variation among the individuals under study. But of course in most situations we do not know the probabilities, and often we have little or no information about the probability density function, so the population mean and median are not known. If we could measure every individual of interest, then the probabilities would be known, but obviously measuring every individual can be difficult or impossible to do. However, suppose we are able to obtain a sample of individuals, meaning a subset of the population of individuals under study. One of our main goals is to find ways of making inferences about the entire population of individuals based on this sample. Simultaneously, we need to describe conditions under which accurate inferences can be made. But before addressing these important problems, we first describe some methods for summarizing a sample of observations. We begin with standard methods typically covered in an introductory course, and then we introduce some nonstandard techniques that play an important role in this book.

3.1 Basic Summation Notation

To make this book as self-contained as possible, basic summation is briefly described for the benefit of any readers not familiar with it. Imagine that 15 college students are asked to rate their feelings of optimism about their future on a six-point scale. If the first student gives a rating of 6, this result is typically written $X_1 = 6$, where the subscript 1 indicates that this is the first student interviewed. If you sample a second student, who gets a score of 4, you write this as $X_2 = 4$, where now the subscript 2 indicates that this is the second student you measure. Here we assume 15 students are interviewed, and their ratings are represented by $X_1, \ldots, X_{15}$. The notation X_i is used to represent the ith subject. In the example with a total of 15 subjects, the possible values for i are the integers $1, 2, \ldots, 15$. Typically, the sample size is represented by n. In the illustration, there are 15 subjects, and this is written as $n = 15$. Table 3.1 illustrates the notation, along with the ratings you might get. The first subject ($i = 1$) got a score of 3, so $X_1 = 3$. The next subject ($i = 2$) got a score of 7, so $X_2 = 7$.

TABLE 3.1 Hypothetical Data Illustrating Commonly Used Notation.

Subject's name	i	X_i
Tom	1	3
Alice	2	7
Dick	3	6
Harry	4	4
Quinn	5	8
Bryce	6	9
Bruce	7	10
Nancy	8	4
Linda	9	5
Karen	10	4
George	11	5
Peter	12	6
Adrian	13	5
Marsha	14	7
Jean	15	6

The notation $\sum X_i$ is a shorthand way of indicating that the observations are to be summed. That is,

$$\sum X_i = X_1 + X_2 + \cdots + X_n. \tag{3.1}$$

For the data in Table 3.1,

$$\sum X_i = 3 + 7 + 6 + \cdots + 7 + 6 = 89.$$

3.2 Measures of Location

One of the most common approaches to summarizing a sample of subjects or a batch of numbers, is to use a so-called measure of location. Roughly, a *measure of location* is a number intended to reflect the typical individual or thing under study. Measures of location are also called *measures of central tendency*, the idea being that they are intended to reflect the middle portion of a set of observations. Examples of population measures of location are the population mean (μ) and the population median. Here attention is focused on sample analogs of these measures plus some additional measures of location that play a prominent role in this book.

3.2.1 The Sample Mean

A natural and very common way of summarizing a batch of numbers is to compute their average. In symbols, the average of n numbers, $X_1, \ldots, X_n$, is

$$\bar{X} = \frac{1}{n} \sum X_i, \tag{3.2}$$

where the notation $\bar{X}$ is read "X bar." In statistics, $\bar{X}$ is called the *sample mean*. As is probably evident, the sample mean is intended as an estimate of the population mean, μ. Of course, a fundamental problem is determining how well the sample mean estimates the population mean, and we begin to discuss this issue in Chapter 4. For now, attention is restricted to other important properties of the sample mean.

> **EXAMPLE.** You sample 10 married couples and determine the number of children they have. The results are 0, 4, 3, 2, 2, 3, 2, 1, 0, 8, and the sample mean is $\bar{X} = 2.5$. Based on this result, it is estimated that if we could measure all married couples, the average number of children would be 2.5. In more formal terms, $\bar{X} = 2.5$ is an estimate of μ. In all likelihood the population mean is not 2.5, so there is the issue of how close the sample mean is likely to be to the population mean. Again, we get to this topic in due course. ■

To elaborate on how the population mean and the sample mean are related, it helps to describe how the sample mean can be computed based on the frequencies of the observations, particularly when the number of observations is large. Here we let f_x represent the number of times the value x was observed among a sample of n observations. That is, f_x represents the *frequency* associated with x. In the last example, the frequencies associated with the number of children are $f_0 = 2, f_1 = 1, f_2 = 3, f_3 = 2, f_4 = 1$, and $f_8 = 1$. So there were two couples with 0 children, one couple had 1 child, three had 2 children, and so forth.

The summation notation introduced in Section 3.1 is used almost exclusively in this book. But in this subsection it helps to introduce a variation of this notation:

$$\sum_x$$

This indicates that a sum is to be computed over all possible values of x. For example,

$$\sum_x f_x.$$

means that we sum the frequencies for all the x values available. Continuing the illustration in the last paragraph, the observed values for x are 0, 1, 2, 3, 4, and 8, so

$$\sum_x f_x = f_0 + f_1 + f_2 + f_3 + f_4 + f_8 = 10.$$

The sum of the observations is just the sum of every possible value multiplied by its frequency. In the present notation, the sum of the observations is

$$\sum x f_x = 0 f_0 + 1 f_1 + 2 f_2 + 3 f_3 + 4 f_4 + 8 f_8$$

$$= 0(2) + 1(1) + 2(3) + 3(2) + 4(1) + 8(1)$$

$$= 25.$$

Dividing this sum by the sample size, n, gives the sample mean. In symbols, another way of writing the sample mean is

$$\frac{1}{n}\sum_x x f_x = \sum_x x \frac{f_x}{n}. \tag{3.3}$$

In words, the sample mean can be computed by multiplying every observed value x by its frequency, f_x, dividing by the sample size, n, and then summing the results.

Note that a natural way of estimating the probability associated with the value x is with the proportion of times it is observed among a sample of observations. In more formal terms, f_x/n, the relative frequency of the value x, is used to estimate $p(x)$. This reveals a close connection between the description of the sample mean just given and the description of the population mean given in Chapter 2. The main difference is that the population mean, $\mu = \sum x p(x)$, is defined in terms of $p(x)$, the proportion of all individuals among the entire population of individuals having the response x, whereas the sample mean uses f_x/n in place of $p(x)$.

EXAMPLE. Consider a sample of $n = 1000$ couples where the proportions of couples having 0, 1, 2, 3, 4, or 5 children are .12, .18, .29, .24, .14, and .02, respectively. In symbols, the relative frequencies for the number of children are $f_0/n = .12, f_1/n = .18, f_2/n = .29, f_3/n = .24, f_4/n = .14,$ and $f_5/n = .02$. Then the sample mean is easily determined by substituting the appropriate values into Equation (3.3). This yields

$$\bar{X} = 0(.12) + 1(.18) + 2(.29) + 3(.24) + 4(.14) + 5(.02) = 2.14.$$

That is, based on these 1000 couples, the estimate of the population mean, the average number of children among all couples, is 2.14.   ■

Chapter 2 demonstrated that the population mean can lie in the extreme tails of a distribution and can be argued to provide a misleading reflection of what is typical in some situations. That is, the population mean can in fact be an extreme value that is relatively atypical. A similar argument can be made about the sample mean, as demonstrated by the following example, based on data from an actual study.

EXAMPLE. Why is it that so many marriages in the United States end in divorce? One proposed explanation is that humans, especially men, seek multiple sexual partners and that this propensity is rooted in our evolutionary past. In support of this view, some researchers have pointed out that when young

Continued

TABLE 3.2 Desired Number of Sexual Partners for 105 Males

x:	0	1	2	3	4	5	6	7	8	9
f_x:	5	49	4	5	9	4	4	1	1	2
x:	10	11	12	13	15	18	19	30	40	45
f_x:	3	2	3	1	2	1	2	2	1	1
x:	150	6000								
f_x:	2	1								

EXAMPLE. (*Continued*) males are asked how many sexual partners they desire over their lifetime, the average number has been found to be substantially higher than the corresponding responses given by females. Pedersen, Miller, Putcha-Bhagavatula, and Yang (2002) point out, however, that the data are typically skewed. In one portion of the study by Pedersen et al., the responses given by 105 males, regarding the desired number of sexual partners over the next 30 years, were as shown in Table 3.2. The sample mean is 64.9. This is, however, a dubious indication of the desired number of sexual partners, because 97% of the observations fall below the sample mean. Notice, for example, that 49 of the males said they wanted one sexual partner, and more than half gave a response of zero or one. In fact 5 gave a response of zero. ■

A criticism of the sample mean is that a single outlier can greatly influence its value. In the last example, one individual responded that he wanted 6000 sexual partners over the next 30 years. This response is unusually large and has an inordinate influence on the sample mean. If, for example, it is removed, the mean of the remaining observations is 7.9. But even 7.9 is rather misleading, because over 77% of the remaining observations fall below 7.9.

One way of quantifying the sensitivity of the sample mean to outliers is with the so-called finite-sample breakdown point. The *finite-sample breakdown point* of the sample mean is the smallest proportion of observations that can make it arbitrarily large or small. Said another way, the finite-sample breakdown point of the sample mean is the smallest proportion of n observations that can render it meaningless. A single observation can make the sample mean arbitrarily large or small, regardless of what the other values might be, so its finite-sample breakdown point is $1/n$.

3.2.2 The Sample Median

Another important measure of location is the *sample median*, which is intended as an estimate of the population median. Simply put, if the sample size is odd, the sample median is the middle value after putting the observations in ascending order. If the sample size is even, the sample median is the average of the two middle values.

Chapter 2 noted that for symmetric distributions, the population mean and median are identical, so for this special case the sample median provides another way of estimating the population mean. But for skewed distributions the population mean and median differ, so generally the sample mean and median are attempting to estimate different quantities.

It helps to describe the sample median in a more formal manner in order to illustrate a commonly used notation. For the observations $X_1, \ldots, X_n$, let $X_{(1)}$ represent the smallest number, $X_{(2)}$ the next smallest, and $X_{(n)}$ the largest. More generally,

$$X_{(1)} \leq X_{(2)} \leq X_{(3)} \leq \cdots \leq X_{(n)}$$

is the notation used to indicate that n values are to be put in ascending order. The sample median is computed as follows:

1. If the number of observations, n, is odd, compute $m = (n + 1)/2$. Then the sample median is

$$M = X_{(m)},$$

the mth value after the observations are put in order.

2. If the number of observations, n, is even, compute $m = n/2$. Then the sample median is

$$M = (X_{(m)} + X_{(m+1)})/2,$$

the average of the mth and $(m + 1)$th observations after putting the observed values in ascending order.

EXAMPLE. Consider the values 1.1, 2.3, 1.7, 0.9, and 3.1. The smallest of the five observations is 0.9, so $X_{(1)} = 0.9$. The smallest of the remaining four observations is 1.1, and this is written as $X_{(2)} = 1.1$. The smallest of the remaining three observations is 1.7, so $X_{(3)} = 1.7$; the largest of the five values is 3.1, and this is written as $X_{(5)} = 3.1$. ■

EXAMPLE. Seven subjects are given a test that measures depression. The observed scores are

34, 29, 55, 45, 21, 32, 39.

Because the number of observations is $n = 7$, which is odd, $m = (7 + 1)/2 = 4$. Putting the observations in order yields

21, 29, 32, 34, 39, 45, 55.

The fourth observation is $X_{(4)} = 34$, so the sample median is $M = 34$. ■

> **EXAMPLE.** We repeat the last example, only with six subjects having test scores
>
> $$29, 55, 45, 21, 32, 39.$$
>
> Because the number of observations is $n = 6$, which is even, $m = 6/2 = 3$. Putting the observations in order yields
>
> $$21, 29, \mathbf{32}, \mathbf{39}, 45, 55.$$
>
> The third and fourth observations are $X_{(3)} = 32$ and $X_{(4)} = 39$, so the sample median is $M = (32 + 39)/2 = 35.5$. ■

Notice that nearly half of any n values can be made arbitrarily large without making the value of the sample median arbitrarily large as well. Consequently, the finite-sample breakdown point is approximately .5, the highest possible value. So the mean and median lie at two extremes in terms of their sensitivity to outliers. The sample mean can be affected by a single outlier, but nearly half of the observations can be outliers without affecting the median. For the data in Table 3.2, the sample median is $M = 1$, which gives a decidedly different picture of what is typical as compared to the mean, which is 64.9.

Based on the single criterion of having a high breakdown point, the median beats the mean. But it is stressed that this is not a compelling reason to routinely use the median over the mean. Chapter 4 describes other criteria for judging measures of location, and situations will be described where both the median and mean are unsatisfactory.

Although we will see several practical problems with the mean, it is not being argued that the mean is always inappropriate. Imagine that someone invests $200,000 and reports that the median amount earned per year, over a 10-year period, is $100,000. This sounds good, but now imagine that the earnings for each year are:

$100,000, $200,000, $200,000, $200,000, $200,000, $200,000, $200,000, $300,000, $300,000, $-1,800,000.

So at the end of 10 years this individual has earned nothing and in fact has lost the initial $200,000 investment. Certainly the long-term total amount earned is relevant, in which case the sample mean provides a useful summary of the investment strategy that was followed.

3.2.3 A Weighted Mean

A general goal of this book is to build an understanding of and appreciation for the practical benefits of contemporary statistical methods. To do this requires some understanding of the circumstances under which more conventional methods are optimal. Many introductory books make it clear that the sample mean is optimal when sampling from a normal distribution, but there are some additional details that will be important in this book, which we begin to discuss in Chapter 4. With this goal in mind, this subsection introduces the weighted mean.

Let $w_1, \ldots, w_n$ be any n constants. A *weighted mean* is just

$$\sum w_i X_i = w_1 X_1 + \cdots + w_n X_n. \tag{3.4}$$

An important special case of a weighted mean is where the constants $w_1, \ldots, w_n$ sum to 1. That is,

$$\sum w_i = 1.$$

In this case, any weighted mean provides a reasonable estimate of the population mean. So an important issue is determining which weights should be used in applied work; we begin to examine this problem in Chapter 4. Weighted means where the weights do not sum to 1 are in common use (as we shall see in connection with least squares regression). But for now attention is focused on situations where the weights sum to 1. The sample mean is a special case where

$$w_1 = w_2 = \cdots = w_n = \frac{1}{n}.$$

EXAMPLE. For the weights $w_1 = .2$, $w_2 = .1$, $w_3 = .4$, $w_4 = .3$, the weighted mean corresponding to $X_1 = 6$, $X_2 = 12$, $X_3 = 14$, and $X_4 = 10$ is

$$.2(6) + .1(12) + .4(14) + .3(10) = 11.$$

∎

3.2.4 A Trimmed Mean

A *trimmed mean* refers to a situation where a certain proportion of the largest and smallest observations are removed and the remaining observations are averaged. Trimmed means contain as special cases the sample mean, where no observations are trimmed, and the median, where the maximum possible amount of trimming is used.

EXAMPLE. Consider the values

$$37, 14, 26, 17, 21, 43, 25, 6, 9, 11.$$

When computing a trimmed mean it is convenient first to put the observations in order. Here this yields

$$6, 9, 11, 14, 17, 21, 25, 26, 37, 43.$$

A 10% trimmed mean indicates that 10% of the smallest observations are removed, as are 10% of the largest, and the remaining values are averaged.

Continued

EXAMPLE. (*Continued*) Here there are 10 observations, so the 10% trimmed mean removes the smallest and largest observations and is given by

$$\bar{X}_t = \frac{9 + 11 + 14 + 17 + 21 + 25 + 26 + 37}{8} = 20.$$

The 20% trimmed mean removes 20% of the largest and smallest values and is

$$\bar{X}_t = \frac{11 + 14 + 17 + 21 + 25 + 26}{6} = 19.$$

■

A more general notation for describing a trimmed mean is useful, to avoid any ambiguity about how it is computed. Let γ be some constant, $0 \leq \gamma < .5$, and set $g = [\gamma n]$, where the notation $[\gamma n]$ means that γn is rounded down to the nearest integer. For example, if $\gamma = .1$ and $n = 99$, then $g = [\gamma n] = [9.9] = 9$. The γ-trimmed mean is just the average of the values after the g smallest and g largest observations are removed. In symbols, the γ-trimmed mean is

$$\bar{X}_t = \frac{1}{n - 2g}(X_{(g+1)} + X_{(g+2)} + \cdots + X_{(n-g)}). \tag{3.5}$$

Setting $\gamma = .1$ yields the 10% trimmed mean, and $\gamma = .2$ is the 20% trimmed mean.

The finite-sample breakdown point of the γ-trimmed mean is γ. So, in particular, the 10% trimmed mean has a breakdown point of .1, and the 20% trimmed mean has a breakdown point of .2. This says that when using the 20% trimmed mean, for example, more than 20% of the values must be altered to make the 20% trimmed mean arbitrarily large or small.

A fundamental issue is deciding how much to trim. At some level it might seem that no trimming should be done in most cases; otherwise, information will be lost somehow. We will see, however, that when addressing a variety of practical goals, 20% trimming often offers a considerable advantage over no trimming and the median. Moreover, Huber (1993) has argued that any estimator with a breakdown point less than or equal to .1 is dangerous and should be avoided. Eventually some of the reasons for this remark will become clear, but for now the details must be postponed.

3.2.5 S-PLUS Function for the Trimmed Mean

S-PLUS has a built-in function for computing a trimmed mean that has the general form

```
mean(x,tr=0),
```

where tr indicates the amount of trimming and x is now an S-PLUS variable containing a batch of numbers. Following the standard conventions used by S-PLUS functions, the notation tr=0 indicates that the amount of trimming defaults to 0, meaning that the S-PLUS function mean will compute the sample mean if a value for tr is not specified. For example, if the values 2, 6, 8, 12, 23, 45, 56, 65, 72 are stored in the S-PLUS variable x, then mean(x) will return the value 32.111, which is the sample mean. The command mean(x,.2) returns the 20% trimmed mean, which is 30.71. [For convenience the S-PLUS function tmean(x,tr = .2) has been supplied, which by default computes a 20% trimmed mean.]

3.2.6 A Winsorized Mean

In order to deal with some technical issues described in Chapter 4, we will need the so-called *Winsorized mean*. The Winsorized mean is similar to the trimmed mean, only the smallest and largest observations are not removed, but rather transformed by "pulling them in." To explain, we first describe what it means to Winsorize a batch of numbers.

Recall that when computing the 10% trimmed mean, you remove the smallest 10% of the observations. Winsorizing the observations by 10% simply means that rather than remove the smallest 10%, their values are set equal to the smallest value not trimmed when computing the 10% trimmed mean. Simultaneously, the largest 10% are reset to the largest value not trimmed. In a similar manner, 20% Winsorizing means that the smallest 20% of the observations are pulled up to the smallest value not trimmed when computing the 20% trimmed mean, and the largest 20% are pulled down to the largest value not trimmed. The Winsorized mean is just the average of the Winsorized values, which is labeled $\bar{X}_w$. The finite-sample breakdown point of the 20% Winsorized mean is .2, the same as the 20% trimmed mean. More generally, the finite-sample breakdown point of the γ-Winsorized mean is γ.

EXAMPLE. Consider again the 10 values

$$37, 14, 26, 17, 21, 43, 25, 6, 9, 11.$$

Because $n = 10$, with 10% trimming $g = [.1(10)] = 1$, meaning that the smallest and largest observations are removed when computing the 10% trimmed mean. The smallest value is 6, and the smallest value not removed when computing the 10% trimmed mean is 9. So 10% Winsorization of these values means that the value 6 is reset to the value 9. In a similar manner, the largest observation, 43, is pulled down to the next largest value, 37. So 10% Winsorization of the values yields

$$37, 14, 26, 17, 21, 37, 25, 9, 9, 11.$$

The 10% Winsorized mean is just the average of the Winsorized values:

$$\bar{X}_w = \frac{37 + 14 + 26 + 17 + 21 + 37 + 25 + 9 + 9 + 11}{10} = 20.6.$$

■

> **EXAMPLE.** To compute a 20% Winsorized mean using the values in the previous example, first note that when computing the 20% trimmed mean, $g = [.2(10)] = 2$, so the two smallest values, 6 and 9, would be removed and the smallest value not trimmed is 11. Thus, Winsorizing the values means that the values 6 and 9 become 11. Similarly, when computing the 20% trimmed mean, the largest value not trimmed is 26, so Winsorizing means that the two largest values become 26. Consequently, 20% Winsorization of the data yields
>
> $$26, 14, 26, 17, 21, 26, 25, 11, 11, 11.$$
>
> The 20% Winsorized mean is $\bar{X}_w = 18.8$, the average of the Winsorized values. ■

Here is a general description of the Winsorized mean using a common notation. To compute a γ-Winsorized mean, first compute g as was done when computing a γ-trimmed mean. That is, g is γn rounded down to the nearest integer. The γ-Winsorized mean is

$$\bar{X}_w = \frac{1}{n}\{(g+1)X_{(g+1)} + X_{(g+2)} + \cdots + X_{(n-g-1)} + (g+1)X_{(n-g)}\}. \qquad (3.6)$$

3.2.7 S-PLUS Function winmean

S-PLUS does not have a built-in function for computing the Winsorized mean, so one has been provided in the library of S-PLUS functions written especially for this book. (These functions can be obtained as described in the Section 1.2.) The function has the form

```
winmean(x,tr=.2),
```

where tr now indicates the amount of Winsorizing, which defaults to .2. So if the values 2, 6, 8, 12, 23, 45, 56, 65, 72 are stored in the S-PLUS variable x, the command winmean(x) returns 31.78, which is the 20% Winsorized mean. The command winmean(x,0) returns the sample mean, 32.11.

3.2.8 M-Estimators

So-called *M-estimators* provide yet another class of measures of location that have practical value. To provide some intuitive sense of M-estimators, imagine a game where someone has written down five numbers and two contestants are asked to pick a number that is close to all five without knowing what the five numbers happen to be. Further, suppose that the first contestant picks the number 22 and that the five numbers written down are

$$46, 18, 36, 23, 9.$$

We can measure how close 22 is to the first value simply by taking the absolute value of their difference: $|46 - 22| = 24$. In a similar manner, the accuracy of the first contestant's guess compared to the second number is $|18 - 22| = 4$. To get an overall measure of accuracy for the first contestant's guess, we might use the sum of the absolute errors associated with each of the five values:

$$|46 - 22| + |18 - 22| + |36 - 22| + |23 - 22| + |9 - 22| = 56.$$

If the second contestant guessed 28, we can measure her overall accuracy in a similar manner:

$$|46 - 28| + |18 - 28| + |36 - 28| + |23 - 28| + |9 - 28| = 60.$$

So based on our criterion of the sum of the absolute differences, contestant 1 is generally more accurate.

Now consider what happens if instead of absolute values, we use squared differences to measure accuracy. So for the first contestant, the accuracy relative to the first number written down is $(46 - 22)^2 = 576$, and the overall accuracy is

$$(46 - 22)^2 + (18 - 22)^2 + (36 - 22)^2 + (23 - 22)^2 + (9 - 22)^2 = 958.$$

As for the second contestant, we get

$$(46 - 28)^2 + (18 - 28)^2 + (36 - 28)^2 + (23 - 28)^2 + (9 - 28)^2 = 874.$$

Now the overall accuracy of the second contestant is judged to be better, in contrast to the situation where absolute differences were used. That is, the choice of a winner can change depending on whether we use squared error or absolute error.

Generalizing, let c be any number and suppose we measure its closeness to the five values under consideration with

$$\sum |X_i - c| = |46 - c| + |18 - c| + |36 - c| + |23 - c| + |9 - c|.$$

A value for c that is closest based on this criterion is the median, $M = 23$. In our contest, if one of the contestants had picked the value 23, they could not be beat by the other contestant, provided we use the sum of the absolute differences to measure closeness. But if we use squared differences instead,

$$\sum (X_i - c)^2 = (46 - c)^2 + (18 - c)^2 + (36 - c)^2 + (23 - c)^2 + (9 - c)^2,$$

the optimal choice for c is the sample mean, $\bar{X} = 26.4$.

Generalizing even further, if for any n values $X_1, \ldots, X_n$ we want to choose c so that it minimizes the sum of squared errors,

$$\sum (X_i - c)^2 = (X_1 - c)^2 + (X_2 - c)^2 + \cdots + (X_n - c)^2, \qquad (3.7)$$

it can be shown that it must be the case that

$$(X_1 - c) + (X_2 - c) + \cdots + (X_n - c) = 0. \qquad (3.8)$$

From this last equation it can be seen that $c = \bar{X}$. That is, when we choose a measure of location based on minimizing the sum of the squared errors given by Equation (3.7), which is an example of what is called the *least squares principle*, this leads to using the sample mean. But if we measure how close c is to the n values using the sum of the absolute differences, the sample median M minimizes this sum.

We can state this result in a more formal manner as follows. The sign of a number is -1, 0, or 1, depending on whether the number is less than, equal to, or greater than zero. So the sign of -6 is -1 and 10 has a sign of 1. A common abbreviation for the sign of a number is simply sign(X). So for any constant c, sign$(X - c)$ is equal to -1, 0, or 1, depending on whether the value X is less than c, equal to c, or greater than c, respectively. If, for instance, $c = 10$, sign$(12 - c) = 1$. It can be shown that if we want to choose c so as to minimize

$$\sum |X_i - c| = |X_1 - c| + |X_2 - c| + \cdots + |X_n - c|, \tag{3.9}$$

then it must be that

$$\text{sign}(X_1 - c) + \text{sign}(X_2 - c) + \cdots + \text{sign}(X_n - c) = 0, \tag{3.10}$$

and the sample median satisfies this last equation.

Here is the point: There are infinitely many ways of measuring closeness that lead to reasonable measures of location. For example, one might measure closeness using the absolute difference between an observation and c raised to some power, say, a. (That is, use $|X - c|^a$.) Setting $a = 1$ leads to the median, as just explained, and $a = 2$ results in the mean. In 1844, the Cambridge mathematician R. L. Ellis pointed out that an even broader class of functions might be used. (See Hald, 1998, p. 496.) Ellis noted that for any function Ψ having the property $\Psi(-x) = -\Psi(x)$, we get a reasonable measure of location, provided the probability curve is symmetric, if we choose c so that it satisfies

$$\Psi(X_1 - c) + \Psi(X_2 - c) + \cdots + \Psi(X_n - c) = 0. \tag{3.11}$$

(For some choices for Ψ, reasonable measures of location require special treatment when distributions are skewed, as is explained in Section 3.5.) Measures of location based on this last equation are called *M-estimators*, a modern treatment of which was first made by Huber (1964). For example, if we take $\Psi(x) = x$, in which case $\Psi(X - c) = X - c$, we get the mean, and $\Psi(x) = \text{sign}(x)$ leads to the median. Said another way, it is arbitrary how we measure closeness, and because different measures of closeness lead to different measures of location, there is the obvious dilemma of deciding which measure of closeness to use. Chapter 4 will begin to address this problem. For now it is merely remarked that two additional choices for Ψ will be seen to have practical value. The first is *Huber's* Ψ, where $\Psi(x) = x$, provided $|x| < K$, with K some constant to be determined. (For reasons explained in Chapter 4, a good choice for general use is $K = 1.28$.) If $x < -K$, then $\Psi(x) = -K$; if $X > K$, then $\Psi(x) = K$. (A technical detail is being ignored at this point but is addressed in Section 3.5.) Said another way, Huber's Ψ is like the Ψ corresponding to least squares, provided an observation is not too far from zero. The constant K is chosen so as to deal with certain practical issues, but the details are best postponed for now.

Another well-studied choice for Ψ is the so-called *biweight*, where $\Psi(x) = x(1 - x^2)$ if $|x| \leq 1$, otherwise $\Psi(x) = 0$. In Section 3.5 we see that the biweight has a serious practical problem when estimating location, but it has practical value for some other goals that will be discussed. Figure 3.1 shows a graph of the four choices for Ψ just discussed.

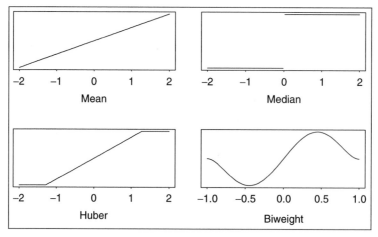

FIGURE 3.1   Examples of Ψ functions that have been considered in connection with M-estimators of location.

There remains the problem of how to compute an M-estimator of location once we have data. But before addressing this issue, we need some results in Section 3.3.

3.2.9 Other Measures of Location

The list of measures of location already covered is far from exhaustive, but it covers the measures that form the heart of this book. However, for completeness, some other classes of location estimators are mentioned. One consists of what are called *R-estimators*, which includes the so-called *Hodges–Lehmann estimator*, as a special case. To compute the Hodges–Lehmann estimator, first average every pair of observations. The median of all such averages is the Hodges–Lehmann estimator. This measure of location, as well as R-estimators in general, often have good properties when sampling from a perfectly symmetric distribution. But for asymmetric distributions, they can be quite unsatisfactory (e.g., Bickel & Lehmann, 1975; Huber, 1981, p. 65). Consequently, further details are not given here. (For more details about R-estimators, see Hettmansperger, 1984; Hettmansperger & McKean, 1998.) For a situation where the Hodges–Lehmann estimator can be unsatisfactory even when sampling from a symmetric distribution, see Morgenthaler and Tukey (1991, p. 15.)

L-estimators form yet another class of estimators that include the sample mean and trimmed means as special cases. L-estimators are like weighted means, only the weights are applied to the observed values after putting them in order. In formal terms, an L-estimator is

$$w_1 X_{(1)} + w_2 X_{(2)} + \cdots + w_n X_{(n)},$$

where, as in Section 3.2.3, $w_1, \ldots, w_n$ are constants chosen to achieve some desired goal.

For an extensive list of location estimators, and how they compare, see Andrews et al. (1972). Morgenthaler and Tukey (1991) describe yet another interesting class of estimators, but currently they can be difficult to compute.

3.3 Measures of Variation or Scale

As is evident, not all people are alike — they respond differently to similar conditions. If we ask 15 people to rate the job the president is doing on a 10-point scale, some might give a rating of 1 or 2, and others might give a rating of 9 or 10. There is *variation* among their ratings. In many ways, it is variation that provides the impetus for sophisticated statistical techniques.

For example, suppose all adults would give the same rating for the president — say, 7 — if they were polled. In statistical terms, the population mean is $\mu = 7$ because, as is evident, if all ratings are equal to 7, the average rating will be 7. This implies that we need sample only one adult in order to determine that the population mean is 7. That is, with $n = 1$ adult, the sample mean, $\bar{X}$, will be exactly equal to the population mean. Because in reality there is variation, in general the sample mean will not be equal to the population mean. If the population mean is $\mu = 7$ and we poll 15 adults, we might get a sample mean of $\bar{X} = 6.2$. If we poll another 15 adults, we might get $\bar{X} = 7.3$. We get different sample means because of the variation among the population of adults we want to study. One common goal in statistics is finding ways of taking into account how variation affects our ability to estimate the population mean with the sample mean. In a similar manner, there is the issue of how well the sample trimmed mean estimates the population trimmed mean.

To make progress, we need appropriate measures of variation, which are also called *measures of scale*. Like measures of location, many measures of scale have been proposed and studied.

3.3.1 Sample Variance

Imagine you sample 10 adults ($n = 10$) and ask each to rate the president; the ratings are:

$$3, 9, 10, 4, 7, 8, 9, 5, 7, 8.$$

The sample mean is $\bar{X} = 7$, and this is your estimate of the population mean, μ. The *sample variance* is

$$s^2 = \frac{\sum (X_i - \bar{X})^2}{n - 1}. \tag{3.12}$$

In words, the sample variance is computed by subtracting the sample mean from each observation and squaring. Then you add the results and divide by $n - 1$, the number of observations minus 1. For the data at hand, the calculations can be summarized as follows:

i	X_i	$X_i - \bar{X}$	$(X_i - \bar{X})^2$
1	3	−4	16
2	9	2	4
3	10	3	9

Continued

i	X_i	$X_i - \bar{X}$	$(X_i - \bar{X})^2$
4	4	−3	9
5	7	0	0
6	8	1	1
7	9	2	4
8	5	−2	4
9	7	0	0
10	8	1	1
$\sum$		0	48

The sum of the observations in the last column is $\sum (X_i - \bar{X})^2 = 48$. Then, because there are $n = 10$ subjects, the sample variance is

$$s^2 = \frac{48}{10 - 1} = 5.33.$$

The sample variance, s^2, is used to estimate the population variance σ^2, the variance we would get if only we could poll all adults. Under random sampling (which is formally described in Section 4.2), the sample variance gives us an increasingly more accurate estimate of the population variance as the sample size gets large. The square root of s^2, s, is called the *sample standard deviation* and estimates the population standard deviation, σ.

It is important to realize that a single unusual value can dominate the sample variance. This is one of several facts that wreaks havoc with standard statistical techniques. To provide a glimpse of problems to come, consider the values

$$8,8,8,8,8,8,8,8,8,8.$$

The sample variance is $s^2 = 0$, meaning there is no variation. If we increase the last value to 10, the sample variance is $s^2 = .36$. Increasing the last observation to 12, $s^2 = 1.45$, and increasing it to 14, $s^2 = 3.3$. The point is, even though there is no variation among the bulk of the observations, a single value can make the sample variance arbitrarily large. In modern terminology, the sample variance is not *resistant*, meaning roughly that a single unusual value can inflate the sample variance and give a misleading indication of how much variation there is among the bulk of the observations. Said more formally, the sample variance has a finite-sample breakdown point of only $1/n$. In some cases, this sensitivity to extreme values is desirable, but for many applied problems it is not, as will be seen.

3.3.2 The Interquartile Range

Another measure of scale or dispersion that is frequently used in applied work, particularly when the goal is to detect outliers, is called the *interquartile range*. For a population of individuals, let Q_u and Q_ℓ be the .75 and .25 quartiles, respectively. Q_u and Q_ℓ are called the *upper* and *lower quartiles*. So the probability that an observation is less than Q_ℓ is .25, the probability that an observation is less than Q_u is .75, and the probability that an observation is between Q_ℓ and Q_u is .5. The difference between the upper and lower quartiles, $Q_u - Q_\ell$, is the *population interquartile range* and reflects the variation of the middle portion of a distribution.

How should the quartiles be estimated based on data available to us? Many methods have been proposed and compared. (See Harrell & Davis, 1982; Dielman, Lowry, & Pfaffenberger, 1994; Parrish, 1990.) The choice of estimation method can depend in part on how the quartiles are to be used. In this book their main use is to detect unusually small or large values among a batch of numbers called *outliers*, in which case results in Cleveland (1985), Hoaglin and Iglewicz (1987), Hyndman and Fan (1996), Frigge, Hoaglin, and Iglewicz (1989) as well as Carling (2000) are relevant.

As usual, let $X_{(1)} \leq \cdots \leq X_{(n)}$ be the observations written in ascending order. Then estimates of the lower quartile typically have the form

$$q_1 = (1 - h)X_{(j)} + hX_{(j+1)}, \tag{3.13}$$

and the problem is determining appropriate choices for j and h. Among the eight choices considered by Frigge, Hoaglin, and Iglewicz (1989) when trying to detect outliers, the method based on the so-called *ideal fourth*, also known as the *machine fourth*, was found to be best, where j is the integer portion of $(n/4) + (5/12)$, meaning that j is $(n/4) + (5/12)$ rounded down to the nearest integer, and

$$h = \frac{n}{4} + \frac{5}{12} - j.$$

The estimate of the upper quartile is taken to be

$$q_2 = (1 - h)X_{(k)} + hX_{(k-1)}, \tag{3.14}$$

where $k = n - j + 1$, in which case the interquartile range is estimated with

$$\text{IQR} = q_2 - q_1. \tag{3.15}$$

EXAMPLE. Consider the values

$$-29.6, -20.9, -19.7, -15.4, -12.3, -8.0, -4.3, 0.8, 2.0, 6.2, 11.2, 25.0.$$

There are 12 observations ($n = 12$), so

$$\frac{n}{4} + \frac{5}{12} = 3.41667.$$

Rounding this last quantity down to the nearest integer gives $j = 3$, so $h = 3.416667 - 3 = .41667$. Because $X_{(3)} = -19.7$, the resulting estimate of the lower quartile is

$$q_1 = (1 - .41667)(-19.7) + .41667(-15.4) = -17.9.$$

In a similar manner, an estimate of the upper quartile is

$$q_2 = (1 - .41667)(6.2) + .41667(2) = 4.45,$$

so the estimate of the interquartile range is

$$\text{IQR} = 4.45 - (-17.9) = 22.35.$$

■

3.3.3 Winsorized Variance

When working with the trimmed mean, we will see that the so-called Winsorized variance plays an important role. To compute the Winsorized variance, simply Winsorize the observations as was done when computing the Winsorized mean in Section 3.2.6. The Winsorized variance is just the sample variance of the Winsorized values. Its finite-sample breakdown point is γ. So, for example, when computing a 20% Winsorized sample variance, more than 20% of the observations must be changed in order to make the sample Winsorized variance arbitrarily large.

3.3.4 S-PLUS Function winvar

Among the library of S-PLUS functions written for this book is

$$\text{winvar}(x, tr = .2),$$

which computes the Winsorized variance, where again tr represents the amount of Winsorizing and defaults to .2. So if the values

$$12, 45, 23, 79, 19, 92, 30, 58, 132$$

are stored in the S-PLUS variable x, winvar(x) returns the value 937.9, which is the 20% Winsorized variance. The command winvar(x,0) returns the sample variance, s^2, which is 1596.8. Typically the Winsorized variance will be smaller than the sample variance s^2 because Winsorizing pulls in extreme values.

3.3.5 Median Absolute Deviation

Another measure of dispersion, which plays an important role when trying to detect outliers (using a method described in Section 3.4.2) is the *median absolute deviation* (MAD) statistic. To compute it, first compute the sample median, M, subtract it from every observed value, and then take absolute values. In symbols, compute

$$|X_1 - M|, \dots, |X_n - M|.$$

The median of the n values just computed is the MAD. Its finite sample breakdown point is .5.

EXAMPLE. Again using the values

$$12, 45, 23, 79, 19, 92, 30, 58, 132,$$

the median is $M = 45$, so $|X_1 - M| = |12 - 45| = 33$ and $|X_2 - M| = 0$. Continuing in this manner for all nine values yields

$$33, 0, 22, 34, 26, 47, 15, 13, 87.$$

The MAD is the median of the nine values just computed: 26.   ■

There is a useful and commonly employed connection between the sample standard deviation, s, and the MAD. Recall that s is intended as an estimate of the population standard deviation, σ. In general, MAD does not estimate σ, but it can be shown that when sampling from a normal distribution, $MAD/.6745$ estimates σ as well. In Section 3.4.2 we will see that this suggests an approach to detecting outliers that plays an important role in data analysis. For convenience we set

$$MADN = \frac{MAD}{.6745}. \qquad (3.16)$$

Statisticians define MAD in the manner just described. S-PLUS has a built-in function called mad, but it computes MADN, not MAD. So if you wanted to compute MAD using S-PLUS, you would use the command .6745 * mad(x). In most applications, MADN is employed, so typically you would use the S-PLUS function mad without multiplying it by .6745.

3.3.6 Average Absolute Distance from the Median

There is a measure of dispersion closely related to MAD that is frequently employed. Rather than take the median of the values $|X_1 - M|, \ldots, |X_n - M|$, take the average of these values instead. That is, use

$$D = \frac{1}{n} \sum |X_i - M|. \qquad (3.17)$$

Despite using the median, D has a finite-sample breakdown point of only $1/n$. If, for example, we increase $X_{(n)}$, the largest of the X_i values, M does not change, but the difference between the largest value and the median becomes increasingly large, which in turn can make D arbitrarily large as well.

3.3.7 Biweight Midvariance and Percentage Bend Midvariance

There are many methods for measuring the variation among a batch of numbers, over 150 of which were compared by Lax (1985). There is little reason to list all of them here, but some additional methods should be mentioned: the *biweight midvariance* and the *percentage midvariance*. Recall that the Winsorized variance and MAD measure the variation of the middle portion of your data. In contrast, both the biweight and percentage bend midvariances make adjustments according to whether a value is flagged as being unusually large or small. The biweight midvariance empirically determines whether a value is unusually large or small using a slight modification of the outlier detection method described in Section 3.4.2. These values are discarded and the variation among the remaining values is computed. But the motivation for the remaining computational details is not remotely obvious without delving deeper into the theory of M-estimators, so further details regarding the derivation of these scale estimators are omitted. The percentage bend midvariance uses a different outlier detection rule and treats outliers in a different manner. Computational details are described in Box 3.1. Presumably readers will use S-PLUS to perform the computations, so detailed illustrations are omitted.

BOX 3.1 How to Compute the Percentage Bend Midvariance

and the Biweight Midvariance

Computing the Percentage Bend Midvariance
As usual, let $X_1, \ldots, X_n$ represent the observed values. Choose a value for the finite-sample breakdown point and call it β. A good choice for general use is .2. Set

$$m = [(1 - \beta)n + .5],$$

the value of $(1 - \beta)n + .5$ rounded down to the nearest integer. Let $W_i = |X_i - M|$, $i = 1, \ldots, n$, and let $W_{(1)} \leq \cdots \leq W_{(n)}$ be the W_i values written in ascending order. Set

$$\hat{\omega}_\beta = W_{(m)},$$

$$Y_i = \frac{X_i - M}{\hat{\omega}_\beta},$$

$$a_i = \begin{cases} 1, & \text{if } |Y_i| < 1 \\ 0, & \text{if } |Y_i| \geq 1, \end{cases}$$

in which case the estimated percentage bend midvariance is

$$\hat{\zeta}_{pb}^2 = \frac{n\hat{\omega}_\beta^2 \sum \{\Psi(Y_i)\}^2}{\left(\sum a_i\right)^2}, \tag{3.18}$$

where

$$\Psi(x) = \max[-1, \min(1, x)].$$

Computing the Biweight Midvariance
Set

$$Y_i = \frac{X_i - M}{9 \times MAD},$$

$$a_i = \begin{cases} 1, & \text{if } |Y_i| < 1 \\ 0, & \text{if } |Y_i| \geq 1, \end{cases} \tag{3.19}$$

$$\hat{\zeta}_{bimid} = \frac{\sqrt{n}\sqrt{\sum a_i(X_i - M)^2 \left(1 - Y_i^2\right)^4}}{\left|\sum a_i \left(1 - Y_i^2\right) \left(1 - 5Y_i^2\right)\right|}.$$

The biweight midvariance is $\hat{\zeta}_{bimid}^2$.

The percentage bend midvariance plays a role when searching for robust analogs of Pearson's correlation. Such measures can be useful when trying to detect an association between two variables, as we see in Chapter 13. Also, the most common strategy for comparing two groups of individuals is in terms of some measure of location, as indicated in Chapter 1. However, comparing measures of variation can be of interest, in which case both the biweight and percentage midvariances can be useful, for reasons best postponed until Chapter 8.

The finite-sample breakdown point of the percentage bend midvariance can be controlled through the choice of a constant labeled β in Box 3.1. Setting $\beta = .1$, the finite-sample breakdown point is approximately .1, and for $\beta = .2$ it is approximately .2. Currently it seems that $\beta = .2$ is a good choice for general use, based on criteria to be described. The finite-sample breakdown point of the biweight midvariance is .5.

3.3.8 S-PLUS Functions bivar and pbvar

The S-PLUS functions

$$\text{pbvar}(x, \text{beta} = .2) \qquad \text{and} \qquad \text{bivar}(x)$$

(written for this book) compute the percentage bend midvariance and the biweight midvariance, respectively. Storing the values

$$12, \ 45, \ 23, \ 79, \ 19, \ 92, \ 30, \ 58, \ 132$$

in the S-PLUS variable x, pbvar(x) returns the value 1527.75. If the largest value, 132, is increased to 1000, pbvar still returns the value 1527.75. If the two largest values (92 and 132) are increased to 1000, again pbvar returns the value 1527.75. With beta equal to .2, it essentially ignores the two largest and two smallest values for the observations used here.

A point that cannot be stressed too strongly is that when we discard outliers, this is not to say that they are uninteresting or uninformative. Outliers can be very interesting, but for some goals they do more harm than good. Again, this issue is discussed in detail after some more basic principles are covered.

For the original values used to illustrate pbvar, the S-PLUS function bivar returns 1489.4, a value very close to the percentage bend midvariance. But increasing the largest value (132) to 1000 means bivar now returns the value 904.7. Its value decreases because it did not consider the value 132 to be an outlier, but increasing it to 1000 means pbvar considers 1000 to be an outlier and subsequently ignores it. Increasing the value 92 to 1000 means bivar returns 739. Now bivar ignores the two largest values because it flags both as outliers.

It is not the magnitude of the biweight midvariance that will interest us in future chapters. Rather, the biweight midvariance plays a role when comparing groups of individuals, and it plays a role when studying associations among variables, as we see in Chapter 13.

3.4 Detecting Outliers

For reasons that will become clear, detecting outliers — unusually large or small values among a batch of numbers — can be very important. This section describes several strategies for accomplishing this goal. The first, which is based on the sample mean and sample variance, is frequently employed and is a rather natural strategy based on properties of the normal curve described in Chapter 2. Unfortunately, it can be highly unsatisfactory, for reasons illustrated in the next subsection. The other methods are designed to correct the problem associated with the first.

3.4.1 A Natural but Unsatisfactory Method for Detecting Outliers

Equation (2.13) described a method for detecting outliers assuming normality: Declare a value an outlier if it is more than two standard deviations from the mean. Here, however, consistent with a general approach to outlier detection suggested by Rousseeuw and van Zomeren (1990), a very slight modification of this rule is used: Declare a value an outlier if it is more than 2.24 standard deviations from the mean. So the value X is flagged an outlier if

$$\frac{|X - \mu|}{\sigma} > 2.24. \tag{3.20}$$

Under normality, the probability of declaring a value an outlier using Equation (3.20) is .025.

Section 2.7 described a problem with this outlier detection rule, but another aspect of this rule should be described and emphasized. Generally we do not know μ and σ, but they can be estimated from data using the sample mean and sample variance. This suggests the commonly used strategy of declaring X an outlier if

$$\frac{|X - \bar{X}|}{s} > 2.24. \tag{3.21}$$

EXAMPLE. Consider the values

$$2, 2, 2, 2, 2, 3, 3, 3, 3, 3, 4, 4, 4, 4, 4, 1000.$$

The sample mean is $\bar{X} = 65.94$, the sample variance is $s = 249.1$,

$$\frac{|1000 - 65.94|}{249.1} = 3.75,$$

3.75 is greater than 2.24, so the value 1000 is declared an outlier. As is evident, the value 1000 is certainly unusual, and in this case our outlier detection rule gives a reasonable result.   ■

EXAMPLE. Consider

$$2, 2, 2, 2, 2, 3, 3, 3, 3, 3, 4, 4, 4, 4, 4, 1000, 10,000.$$

These are the same values as in the last example, but with another outlier added. The value 10,000 is declared an outlier using Equation (3.21). Surely 1000 is unusual compared to the bulk of the observations, but it is not declared an outlier. The reason is that the two outliers inflate the sample mean and especially the sample standard deviation. Moreover, the influence of the outliers on s is so large, the value 1000 is not declared an outlier. In particular, $\bar{X} = 650.3$, $s = 2421.4$, so

$$\frac{|1000 - \bar{X}|}{s} = .14.$$

■

EXAMPLE. Now consider the values

$$2, 2, 3, 3, 3, 4, 4, 4, 100,000, 100,000.$$

It is left as an exercise to verify that the value 100,000 is not declared an outlier using Equation (3.21), yet surely it is unusually large.   ■

The last two examples illustrate the problem known as *masking*. Outliers inflate both the sample mean and the sample variance, which in turn can mask their presence when using Equation (3.21). What is needed is a rule for detecting outliers that is not itself affected by outliers. One way of accomplishing this goal is to switch to measures of location and scale that have a reasonably high breakdown point.

3.4.2 A Better Outlier Detection Rule

Here is a simple outlier detection rule that has received a great deal of attention. Declare X to be an outlier if

$$\frac{|X - M|}{\text{MAD}/.6745} > 2.24. \tag{3.22}$$

When sampling from a normal curve, Equation (3.22) mimics our rule based on Equation (3.21) because for this special case the sample median M estimates the population mean μ, and MAD/.6745 estimates the population standard deviation σ. An important advantage of the method is that it addresses the problem of masking by employing measures of location and scale both of which have a breakdown point of .5. That is, the method can handle a large number of outliers without making the problem of masking an issue.

The use of the value 2.24 in Equation (3.22) stems from Rousseeuw and van Zomeren (1990). It should be noted that Equation (3.22) is known as the

Hampel identifier, but Hampel used the value 3.5 rather than 2.24. (For some refinements, see Davies & Gather, 1993.)

EXAMPLE. Consider again the values

$$2, 2, 3, 3, 3, 4, 4, 4, 100,000, 100,000.$$

Using our rule based on the sample mean and sample variance, we saw that the two values equal to 100,000 are not declared outliers. It can be seen that $M = 3.5$, MADN = MAD/.6745 = .7413, and

$$\frac{100,000 - 3.5}{.7413} = 134,893.4.$$

So in contrast to the rule based on the mean and variance, 100,000 would now be declared an outlier.   ■

3.4.3 S-PLUS Function out

An S-PLUS function

$$\text{out}(x)$$

has been supplied that detects outliers using Equation (3.22). If the values from the last example are stored in the S-PLUS variable data, then part of the output from the command out(data) is

```
$out.val:
[1] 100000 100000

$out.id:
[1] 9 10
```

That is, there are two values declared outliers, both equal to 100,000, and they are the ninth and tenth observations stored in the S-PLUS variable data. (That is, the outliers are stored in data[9] and data[10].)

3.4.4 The Boxplot

Proposed by Tukey (1977), a *boxplot* is a commonly used graphical summary of data that provides yet another method for detecting outliers. The example boxplot shown in Figure 3.2 was created with the built-in S-PLUS command boxplot. As indicated, the ends of the rectangular box mark the lower and upper quartiles. That is, the box indicates where the middle half of the data lie. The horizontal line inside the box indicates the position of the median. The lines extending out from the box are called *whiskers*.

The boxplot declares the value X to be an outlier if

$$X < q_1 - 1.5(\text{IQR}) \tag{3.23}$$

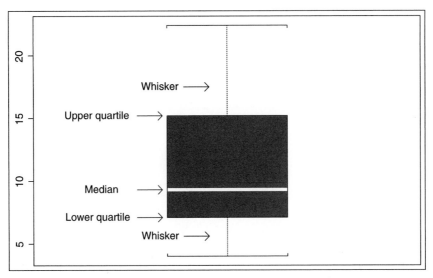

FIGURE 3.2   Example of a boxplot with no outliers.

or

$$X > q_2 + 1.5(\text{IQR}), \tag{3.24}$$

where IQR is the interquartile range defined in Section 3.3.2 and q_1 and q_2 are estimates of the lower and upper quartiles. To complicate matters, different software packages use different estimates of the quartiles. S-PLUS, for example, uses $j = n/4$ in Equation (3.13). If $n/4$ is an integer, S-PLUS sets $g = .5$, otherwise it uses $g = 0$. (A numerical summary of which points are declared outliers with the S-PLUS function boxplot can be determined with the S-PLUS command

$$\text{print}(\text{boxplot}(x, \text{plot}{=}F)\$\text{out}).$$

The S-PLUS command summary uses yet another method for estimating quartiles.)

   Figure 3.3 shows a boxplot with two outliers. The ends of the whiskers are called *adjacent values*. They are the smallest and largest values not declared outliers. Because the interquartile range has a finite sample breakdown point of .25, it takes more than 25% of the data to be outliers before the problem of masking occurs. A breakdown point of .25 seems to suffice in most situations, but exceptions can occur.

EXAMPLE.   For the data in Table 3.2, a boxplot declares all values greater than 13.5 to be outliers. This represents 11.4% of the data. In contrast, the rule in Section 3.4.2, based on the median and MAD, declares all values greater than or equal to 6 to be outliers which is 27.6% of the data. This suggests that masking might be a problem for the boxplot because the proportion of outliers using the rule in Section 3.4.2 exceeds .25, the breakdown point of the boxplot. If we use the sample mean and standard deviation to detect outliers, only the value 6,000 is declared an outlier.   ■

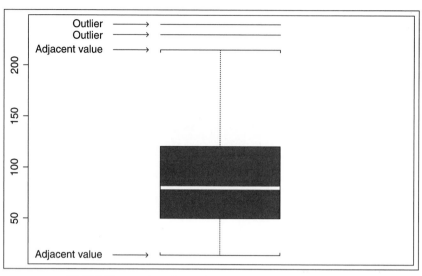

FIGURE 3.3   Boxplot with two values flagged as outliers.

3.4.5 A Modified Boxplot Rule for Detecting Outliers

A criticism of the traditional boxplot rule for detecting outliers is that the expected proportion of numbers that are declared outliers depends on the sample size; it is higher in situations where the sample size is small. To correct this problem, Carling (2000) uses the following rule: Declare the value X an outlier if

$$X > M + k\,\text{IQR} \tag{3.25}$$

or if

$$X < M - k\,\text{IQR}, \tag{3.26}$$

where

$$k = \frac{17.63n - 23.64}{7.74n - 3.71},$$

IQR is estimated with the ideal fourths, as described in Section 3.3.2, and, as usual, M is the sample median. So unlike standard boxplot rules, the median plays a role in determining whether a value is an outlier.

The choice between the outlier detection rule in Section 3.4.2 over the method just described is not completely straightforward. The rule used here is designed so that for normal distributions, the expected proportion of numbers declared outliers is .04. The choice .04 is arbitrary at some level and is clouded by the problem that the notion of outliers is vague. The extent to which this rate differs when using the method in Section 3.4.2 has not been studied. The method used here has a breakdown point of .25, in contrast to a breakdown point of .5 using the method in Section 3.4.2. As previously noted, generally a breakdown point of .25 suffices, but exceptions can occur. Moreover, for many purposes the method in Section 3.4.2, or some slight modification of it, has proven to have practical value in situations to be covered.

Finally, it is noted that there is a vast literature on detecting outliers. A few additional issues will be covered in subsequent chapters. Readers interested in a book-length treatment of this topic are referred to Barnett and Lewis (1994).

3.4.6 S-PLUS Function outbox

The S-PLUS function

$$outbox(x, mbox=F, gval=NA)$$

checks for outliers using one of two methods. With mbox=F, it uses the rule given by Equations (3.23) and (3.24), but unlike S-PLUS, quartiles are estimated with the ideal fourths. The argument gval corresponds to the constant 1.5 in Equations (3.23) and (3.24). Using gval=2, for example, causes this constant to be changed to 2. Setting mbox=T, outliers are detected using the method in the previous subsection.

> **EXAMPLE.** For the data in Table 3.2, values greater than or equal to 13 are declared outliers with the S-PLUS function outbox with mbox=T. So in this particular case, the values declared outliers are the same as those using the built-in S-PLUS function boxplot. But with mbox=F, 13 is no longer declared an outlier. Situations also arise where a value is declared an outlier with mbox=F but not when mbox=T. ■

3.5 Computing an M-Estimator of Location

Section 3.2.8 introduced the notion of an M-estimator based on Huber's Ψ, but no details were given on how it is computed. This is because we needed first to describe some measures of scale plus some outlier detection rules in order to address an important technical issue.

For any measure of location, it should be the case that if we multiply all observed values by some constant b, the measure of location should be multiplied by b as well. Such measures of location are said to be *scale equivariant*. For example, if the weights of five children are measured in pounds and found to be 65, 55, 72, 80, and 70, the sample mean is 68.4. So if we convert to kilograms by dividing each child's weight by 2.2, the sample mean becomes $68.4/2.2 = 31.09$. That is, the sample mean is scale equivariant. In a similar manner, the median in pounds is 70, and converting to kilograms the median becomes $70/2.2 = 31.8$. When defining an M-estimator as described by Equation (3.11), we do not get this property automatically when using Huber's Ψ or the biweight. However, there is a simple method for addressing this problem: Include a measure of scale in Equation (3.11). It can be shown that if we use Huber's Ψ plus a measure of scale that has a high finite-sample breakdown point, the resulting M-estimator will have the same breakdown point as the measure of scale. So if we use MAD as our measure of scale, which has a breakdown point of .5, the resulting M-estimator, still using Huber's Ψ, will be .5 as well. In this case, Equation (3.11) becomes

$$\Psi\left(\frac{X_1 - c}{\text{MAD}}\right) + \cdots + \Psi\left(\frac{X_n - c}{\text{MAD}}\right) = 0, \qquad (3.27)$$

where, as before, c is our measure of location.

An explicit equation for computing c once we have observations cannot be derived, but there are two simple and effective methods for dealing with this problem. One of these is to use a particular iterative technique for determining c that is easily applied on a computer. The second is to use a single step in this iterative method that inherits the positive features of M-estimators. The latter strategy is called a one-step M-estimator, and the resulting estimate of the measure of location is computed as follows. Let i_1 be the number of observations X_i for which $(X_i - M)/\text{MADN} < -K$, and let i_2 be the number of observations such that $(X_i - M)/\text{MADN} > K$, where typically $K = 1.28$ is used (for reasons that are difficult to explain until concepts in Chapter 4 are covered). The one-step M-estimator of location (based on Huber's Ψ) is

$$\hat{\mu}_{os} = \frac{K(\text{MADN})(i_2 - i_1) + \sum_{i=i_1+1}^{n-i_2} X_{(i)}}{n - i_1 - i_2}. \qquad (3.28)$$

This one-step M-estimator almost uses the following strategy: Determine which values are outliers using the method in Section 3.4.2, except that Equation (3.22) is replaced by

$$\frac{|X - M|}{\text{MAD}/.6745} > K. \qquad (3.29)$$

Next, remove the values flagged as outliers and average the values that remain. But for technical reasons, the one-step M-estimator makes an adjustment based on MADN, a measure of scale plus the number of outliers above and below the median.

EXAMPLE. Computing a one-step M-estimator (with $K = 1.28$) is illustrated with the following ($n = 19$) observations:

$$77 \quad 87 \quad 88 \quad 114 \quad 151 \quad 210 \quad 219 \quad 246 \quad 253 \quad 262$$

$$296 \quad 299 \quad 306 \quad 376 \quad 428 \quad 515 \quad 666 \quad 1310 \quad 2611.$$

It can be seen that $M = 262$ and that $\text{MADN} = \text{MAD}/.6745 = 169$. If for each observed value we subtract the median and divide by MADN we get

$$-1.09 \quad -1.04 \quad -1.035 \quad -0.88 \quad -0.66 \quad -0.31 \quad -0.25 \quad -0.095 \quad -0.05$$

$$0.00 \quad 0.20 \quad 0.22 \quad 0.26 \quad 0.67 \quad 0.98 \quad 1.50 \quad 2.39 \quad 6.2 \quad 13.90$$

Continued

EXAMPLE. (*Continued*) So there are four values larger than the median that are declared outliers: 515, 666, 1310, 2611. That is, $i_2 = 4$. No values less than the median are declared outliers, so $i_1 = 0$. The sum of the values not declared outliers is

$$77 + 87 + \cdots + 428 = 3411.$$

So the value of the one-step M-estimator is

$$\frac{1.28(169)(4 - 0) + 3411}{19 - 0 - 4} = 285.$$

■

3.5.1 S-PLUS Function onestep

The S-PLUS function

$$\text{onestep}(x, \text{bend}=1.28)$$

computes a one-step M-estimator with Huber's Ψ. The second argument, bend, corresponds to the constant K in Equation (3.28) and defaults to 1.28. If, for example, the data in Table 3.2 are stored in the S-PLUS variable sexm, onestep(sexm) returns the value 2.52.

3.5.2 A Modified One-Step M-Estimator

The one-step M-estimator has desirable theoretical properties, but when sample sizes are small, problems can arise when comparing groups using the methods covered in subsequent chapters. This section describes a simple modification of the one-step M-estimator given by Equation (3.28). The modification consists of dropping the term containing MADN. That is, use

$$\hat{\mu}_{\text{mom}} = \frac{\sum_{i=i_1+1}^{n-i_2} X_{(i)}}{n - i_1 - i_2} \tag{3.30}$$

as a measure of location, where now $K = 2.24$ is used to determine i_1 and i_2. In effect, use Equation (3.22) to detect outliers, discard any outliers that are found, and then use the mean of the remaining values. The finite-sample breakdown point is .5. This modified one-step M-estimator, MOM, is very similar to what are known as *skipped estimators*, which were studied by Andrews et al. (1972). (Skipped estimators use a boxplot rule to detect outliers rather than the median and MAD.) Initially, technical problems precluded skipped estimators from being used to compare groups of individuals, but recent advances have made MOM a viable measure of location. Note that MOM introduces a certain amount of flexibility versus using trimmed means. For example, MOM might discard zero observations. Moreover, if a distribution is heavy-tailed and highly skewed to the right, it might be desirable to trim more observations from the right tail versus the left, and MOM contains the possibility of doing this. (The relative merits of MOM versus M-estimators are discussed in subsequent chapters.)

It might help to summarize a fundamental difference among trimmed means, M-estimators, and MOM (or more generally the class of skipped estimators). Each represents a different approach to measuring location. Trimmed means discard a fixed proportion of large and small observations. MOM, and skipped estimators in general, empirically determine how many observations are to be trimmed and includes the possibility of different amounts of trimming in the tails as well as no trimming at all. M-estimators are based on how we measure the overall distance between some measure of location and the observations. Huber's measure of distance leads to the one-step M-estimator given by Equation (3.28), which has certain similarities to MOM, but unlike MOM, an adjustment is made based on a measure of scale when the amount of trimming in the left tail differs from the amount in the right.

3.5.3 S-PLUS Function mom

The S-PLUS function

$$mom(x, bend = 2.24)$$

computes the modified one-step M-estimator just described. As a brief illustration, consider again the data used in the last example. It can be seen that according to Equation (3.22), the three largest values (666, 1310, and 2611) are outliers. Discarding them and averaging the remaining values yields 245.4, which is the value returned by mom. So in contrast to the 20% trimmed mean, none of the lower values are discarded.

3.6 Histograms

Two additional graphical tools for summarizing data should be mentioned. One of these is the histogram; the other is a so-called *kernel density estimator*, which is described in the next section.

A histogram is illustrated with data from a heart transplant study conducted at Stanford University between October 1, 1967, and April 1, 1974. Of primary concern is whether a transplanted heart will be rejected by the recipient. With the goal of trying to address this issue, a so-called T5 mismatch score was developed by Dr. C. Bieber. It measures the degree of dissimilarity between the donor and the recipient tissue with respect to HL-A antigens. Scores less than 1 represent a good match, and scores greater than 1 a poor match. Of course, of particular interest is how well a T5 score predicts rejection, but this must wait for now. The T5 scores, written in ascending order, are shown in Table 3.3 and are taken from R. G. Miller (1976).

A histogram simply groups the data into categories and plots the corresponding frequencies. To illustrate the basic idea, we group the T5 values into eight categories: (1) values between −0.5 and 0.0, (2) values greater than 0.0 but less than or equal

TABLE 3.3   T5 Mismatch Scores from a Heart Transplant Study

0.00	0.12	0.16	0.19	0.33	0.36	0.38	0.46	0.47	0.60	0.61	0.61	0.66
0.67	0.68	0.69	0.75	0.77	0.81	0.81	0.82	0.87	0.87	0.87	0.91	0.96
0.97	0.98	0.98	1.02	1.06	1.08	1.08	1.11	1.12	1.12	1.13	1.20	1.20
1.32	1.33	1.35	1.38	1.38	1.41	1.44	1.46	1.51	1.58	1.62	1.66	1.68
1.68	1.70	1.78	1.82	1.89	1.93	1.94	2.05	2.09	2.16	2.25	2.76	3.05

TABLE 3.4   Frequencies and Relative Frequencies for Grouped T5 scores, $n = 65$

Test score (x)	Frequency	Relative frequency
−0.5–0.0	1	1/65 = .015
0.0–0.5	8	8/65 = .123
0.5–1.0	20	20/65 = .308
1.0–1.5	18	18/65 = .277
1.5–2.0	12	12/65 = .138
2.0–2.5	4	4/65 = .062
2.5–3.0	1	1/65 = .015
3.0–3.5	1	1/65 = .015

to 0.5, (3) values greater than 0.5 but less than or equal to 1.0, and so on. The frequency and relative frequency associated with each of these intervals is shown in Table 3.4. For example, there are eight T5 mismatch scores in the interval extending from 0.0 to 0.5, and the proportion of all scores belonging to this interval is .123.

Figure 3.4 shows the histogram for the T5 scores that was created with the built-in S-PLUS function hist. Notice that the base of the leftmost shaded rectangle extends from 0 to 0.5 and has a height of 9. This means that there are nine cases where a T5 score has a value between 0 and 0.5. The base of the next shaded rectangle extends from 0.5 to 1 and has a height of 20. This means that there are 20 T5 scores having a value between 0.5 and 1. The base of the next rectangle extends from 1 to 1.5 and has a height of 18, so there are 18 T5 scores between 1 and 1.5.

3.7 Kernel Density Estimators

As indicated in Section 2.6, probabilities associated with continuous variables are determined by the area under a curve called a *probability density function*. The equation for this curve is typically labeled $f(x)$. An example is Equation (2.10), which gives the equation for the probability density function of the normal distribution. For some purposes to be covered, it is useful to have an estimate of $f(x)$ (the equation for the probability density function) based on observations we make. A histogram provides a crude estimate, but for various reasons it can be unsatisfactory (e.g., Silverman, 1986, pp. 9–11).

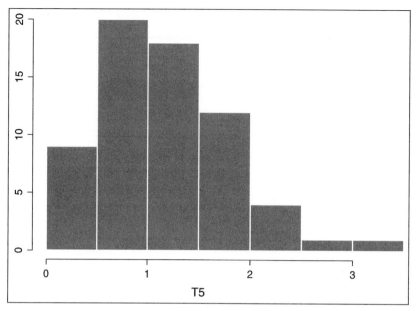

FIGURE 3.4   Example of a histogram.

Another seemingly natural but unsatisfactory approach to estimating the probability density function is to assume observations are sampled from a normal curve
and replace the population mean and variance (μ and σ^2) in Equation (2.10) with the
sample mean and variance ($\bar{X}$ and s^2). But this approach can be highly unsatisfactory,
even when sampling from a perfectly symmetric distribution. To understand why,
assume we are sampling from a mixed normal and that we happen to get an exact
estimate of the population mean and variance. The dotted curve in Figure 2.9 is the
probability density function being estimated, and the normal curve in Figure 2.9 is
the estimate of the mixed normal following the strategy just indicated. As is evident,
there is a substantial difference between the two curves, indicating that we get a poor
estimate.

For some purposes we get a much more effective estimate using what is called a
kernel density estimator. There are many variations of kernel density estimators (Silverman,
1986), but only one is used here. It is based on what is called *Rosenblatt's shifted histogram*
and employs results derived by D. W. Scott (1979) as well as Freedman and Diaconis
(1981). In particular, to estimate $f(x)$ for any x, set

$$h = \frac{1.2(\text{IQR})}{n^{1/5}},$$

where IQR is the interquartile range. Let A be the number of observations less than
or equal to $x + h$. Let B be the number of observations strictly less than $x - h$. Then
the estimate of $f(x)$ is

$$\hat{f}(x) = \frac{A - B}{2nh}.$$

A numerical illustration is not given because presumably readers will use the S-PLUS function described in the next subsection.

3.7.1 S-PLUS Function kdplot

The S-PLUS function kdplot estimates the probability density function $f(x)$ for a range of x values, using the kernel density estimator just described, and then plots the results. (The interquartile range is estimated with the S-PLUS built-in function summary.) It has the form

$$kdplot(data, rval=15),$$

where data is any S-PLUS variable containing data. By default, the function begins by estimating $f(x)$ at 15 values spread over the range of observed values. For example, if the data consist of values ranging between 2 and 36, kdplot picks 15 values between 2 and 36, estimates $f(x)$ at these points, and then plots the results. For very large sample sizes ($n \geq 500$) it might help to plot $f(x)$ at 20 or 25 values instead. This can be accomplished with the second argument. For example, rval=25 will plot an estimate of $f(x)$ at 25 x values evenly spaced between the smallest and largest of the observed values. For small and even moderate sample sizes, rval=15 or smaller gives a better (less ragged) estimate of the probability density function in most cases. For small sample sizes ($n < 40$), kdplot can be rather unrevealing. Figure 3.5 shows an estimate of the probability density function associated with the T5 mismatch scores in Table 3.3 based on the S-PLUS function kdplot (with the argument rval set equal to 10).

For another approach to estimating $f(x)$, readers might consider the built-in S-PLUS function density. The S-PLUS command plot(density(x)) plots an estimate of $f(x)$ using the data in the S-PLUS variable x.

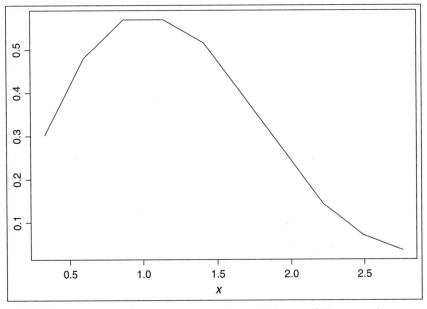

FIGURE 3.5   A kernel density estimate of the distribution of T5 mismatch scores.

TABLE 3.5   Word Identification Scores

58	58	58	58	58	64	64	68	72	72	72	75	75	77	77	79	80	
82	82	82	82	82	84	84	85	85	90	91	91	92	93	93	93	95	
95	95	95	95	95	95	95	98	98	99	101	101	101	102	102	102	102	
102	103	104	104	104	104	104	105	105	105	105	105	107	108	108	110	111	
112	114	119	122	122	125	125	125	127	129	129	132	134					

3.8 Stem-and-Leaf Displays

A stem-and-leaf display is another method of gaining some overall sense of what data are like. The method is illustrated with measures taken from a study aimed at understanding how children acquire reading skills. A portion of the study was based on a measure that reflects the ability of children to identify words.[1] Table 3.5 lists the observed scores in ascending order.

The construction of a stem-and-leaf display begins by separating each value into two components. The first is the *leaf*, which in this example is the number in the 1s position (the single digit just to the left of the decimal place). For example, the leaf corresponding to the value 58 is 8. The leaf for the value 64 is 4, and the leaf for 125 is 5. The digits to the left of the leaf are called the *stem*. Here the stem of 58 is 5, the number to the left of 8. Similarly, 64 has a stem of 6 and 125 has a stem of 12. We can display the results for all 81 children as follows:

Stems	Leaves
5	88888
6	448
7	22255779
8	0222224455
9	011233355555555889
10	1112222234444455555788
11	01249
12	22555799
13	24

There are five children who have the score 58, so there are five scores with a leaf of 8, and this is reflected by the five 8s displayed to the right of the stem 5, in the Leaves column. Two children got the score 64, and one child got the score 68. That is, for the stem 6, there are two leaves equal to 4 and one equal to 8, as indicated by the list of leaves in the display. Now look at the third row of numbers, where the stem is 7. The leaves listed are 2, 2, 2, 5, 5, 7, 7, and 9. This indicates that the value 72 occurred three times, the value 75 occurred two times, as did the value 77, and the value 79 occurred once. Notice that the display of the leaves gives us some indication of the

---

1 These data were supplied by L. Doi.

values that occur most frequently and which are relatively rare. Like the histogram, the stem-and-leaf display gives us an overall sense of what the values are like.

The choice of which digit is to be used as the leaf depends in part on which digit provides a useful graphical summary of the data. But details about how to address this problem are not covered here. Suffice it to say that algorithms have been proposed for deciding which digit should be used as the leaf and determining how many lines a stem-and-leaf display should have (e.g., Emerson & Hoaglin, 1983). Here we merely note that S-PLUS has a built-in function for computing a stem-and-leaf display that has the form

$$stem(x),$$

where, as usual, x now represents any S-PLUS variable containing data. For the T5 mismatch scores, the stem-and-leaf display created by S-PLUS is

```
Decimal point is at the colon

0 : z122344
0 : 556667777788889999
1 : 000001111111223334444
1 : 5566777788999
2 : 0122
2 : 8
3 : 0
```

The z in the first row stands for zero. This function also reports the median and the quartiles.

3.9 Exercises

1. For the observations

$$21, 36, 42, 24, 25, 36, 35, 49, 32$$

verify that the sample mean, trimmed mean, and median are $\bar{X} = 33.33$, $\bar{X}_t = 32.9$, and $M = 35$.

2. The largest observation in Exercise 1 is 49. If 49 is replaced by the value 200, verify that the sample mean is now $\bar{X} = 50.1$ but the trimmed mean and median are not changed. What does this illustrate about the resistance of the sample mean?

3. For the data in Exercise 1, what is the minimum number of observations that must be altered so that the 20% trimmed mean is greater than 1000?

4. Repeat the previous problem but use the median instead. What does this illustrate about the resistance of the mean, median, and trimmed mean?

5. For the observations

$$6, 3, 2, 7, 6, 5, 8, 9, 8, 11$$

verify that the sample mean, trimmed mean, and median are $\bar{X} = 6.5$, $\bar{X}_t = 6.7$, and $M = 6.5$.

6. A class of fourth-graders was asked to bring a pumpkin to school. Each of the 29 students counted the number of seeds in his or her pumpkin and the results were

   250, 220, 281, 247, 230, 209, 240, 160, 370, 274,
   210, 204, 243, 251, 190, 200, 130, 150, 177, 475,
   221, 350, 224, 163, 272, 236, 200, 171, 98.

   (These data were supplied by Mrs. Capps at the La Cañada Elementary School, La Cañada, CA.) Verify that the sample mean, trimmed mean, and median are $\bar{X} = 229.2$, $\bar{X}_t = 220.8$, and $M = 221$.

7. Suppose health inspectors rate sanitation conditions at restaurants on a five-point scale, where a 1 indicates poor conditions and a 5 is excellent. Based on a sample of restaurants in a large city, the frequencies are found to be $f_1 = 5$, $f_2 = 8$, $f_3 = 20$, $f_4 = 32$, and $f_5 = 23$. What is the sample size, n? Verify that the sample mean is $\bar{X} = 3.7$.

8. For the frequencies $f_1 = 12$, $f_2 = 18$, $f_3 = 15$, $f_4 = 10$, $f_5 = 8$, and $f_6 = 5$, verify that the sample mean is 3.

9. For the observations

   21, 36, 42, 24, 25, 36, 35, 49, 32

   verify that the sample variance and the sample Winsorized variance are $s^2 = 81$ and $s_w^2 = 51.4$, respectively.

10. In Exercise 9, what is your estimate of the standard deviation, σ?

11. In general, will the Winsorized sample variance, s_w^2, be less than the sample variance, s^2?

12. Among a sample of 25 subjects, what is the minimum number of subjects that must be altered to make the sample variance arbitrarily large?

13. Repeat Exercise 12 but for s_w^2 instead. Assume 20% Winsorization.

14. For the observations

   6, 3, 2, 7, 6, 5, 8, 9, 8, 11

   verify that the sample variance and Winsorized variance are 7.4 and 1.8, respectively.

15. For the data in Exercise 6, verify that the sample variance is 5584.9 and the Winsorized sample variance is 1375.6.

16. For the data in Exercise 14, determine the ideal fourths and the corresponding interquartile range.

17. For the data in Exercise 6, which values would be declared outliers using the rule based on MAD?

18. Referring to the description of the population variance, σ^2, devise a method of estimating σ^2 based on the frequencies, f_x.

19. Referring to Exercise 18, and using the data in Exercise 7, estimate σ, the population standard deviation.

20. For a sample of n subjects, the relative frequencies are $f_0/n = .2$, $f_1/n = .4$, $f_2/n = .2$, $f_3/n = .15$, and $f_4/n = .05$. The sample mean is 1.45. Using your answer to Exercise 18, verify that the sample variance is 1.25.

21. Snedecor and Cochran (1967) report results from an experiment on weight gain in rats as a function of source of protein and levels of protein. One of the groups was fed beef with a low amount of protein. The weight gains were

$$90, 76, 90, 64, 86, 51, 72, 90, 95, 78.$$

Verify that there are no outliers among these values when using a boxplot but that there is an outlier using Equation (3.22).

22. For the values 1, 2, 3, 4, 5, 6, 7, 8, 9, and 10 there are no outliers. If you increase 10 to 20, then 20 would be declared an outlier by a boxplot. If the value 9 is also increased to 20, would the boxplot find two outliers? If the value 8 is also increased to 20, would the boxplot find all three outliers?

23. Use the results of the last problem to come up with a general rule about how many outliers a boxplot can detect.

24. For the data in Table 3.3, verify that the value 3.05 is an outlier, based on a boxplot.

25. Figure 3.6 shows a boxplot of data from a reading study where the general goal was to find predictors of reading ability in children. What, approximately, are the lower and upper quartiles and the interquartile range? How large or small would an observation have to be in order to be declared an outlier?

26. Under normality, it can be shown that $(n - 1)s^2/\sigma^2$ has a chi-squared distribution with $n - 1$ degrees of freedom. If $n = 11$ and $\sigma^2 = 100$, use Table 3 in Appendix B to determine $P(s^2 > 159.8)$.

27. A researcher claims that for a certain population of adults, cholesterol levels have a normal distribution with $\sigma^2 = 10$. If you randomly sample 21 individuals, verify that $P(s^2 \leq 14.2) = .9$.

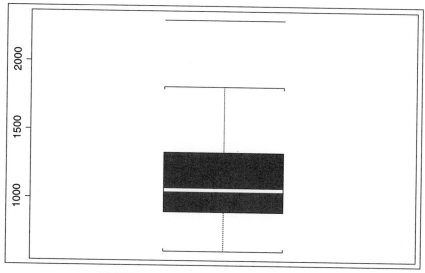

FIGURE 3.6   Boxplot of reading study data.

SAMPLING DISTRIBUTIONS AND CONFIDENCE INTERVALS

Understanding and appreciating modern statistical methods requires, among other things, a close look at the notion of a *sampling distribution*. We begin with the basics typically covered in an introductory course and then take up issues that explain why contemporary statistical techniques have practical value. This chapter also introduces the notion of a *confidence interval*, but important practical issues not typically covered in an introductory course are described.

4.1 Basics

To illustrate the notion of a *sampling distribution* in a concrete manner, imagine you are interested in the health risks associated with ozone in the atmosphere and that one of your concerns is how ozone affects weight gain in infants. Obviously you cannot experiment on human infants, so suppose you conduct an experiment where 22 rats are exposed to an ozone environment and their average weight gain is $\bar{X} = 11$ grams. Now imagine that another team of researchers repeats your experiment with a new sample of 22 rats. Of course their sample mean will probably differ from yours; they might get $\bar{X} = 16$. In a similar manner, a third team of researchers might get $\bar{X} = 9$. If infinitely many teams of researchers could repeat your experiment, then we would know what is called the *sampling distribution of the sample mean*. In particular, we would know the probability that the sample mean is less than 1, less than 10, or less than c for any constant c we might pick.

Being able to approximate the sampling distribution of the sample mean allows us to address a problem of fundamental importance. There are millions of rats that could have been used in your experiment. How well does the sample mean ($\bar{X}$) estimate the population mean (μ), the average weight gain you would observe if all living rats took part in this experiment? Based on your data, can you be reasonably certain that the population mean has a value close to the sample mean? Can you be reasonably certain that the population mean is less than 30 or less than 20? Can you be reasonably certain that it is greater than 1? If we can get a good approximation of the sampling distribution of the sample mean, we can answer these questions.

Sampling distributions also provide a perspective on how different location estimators compare to one another. That is, they provide both a graphical and a numerical summary of how the accuracy of the sample mean, for example, compares to the accuracy of the sample median, trimmed mean, or MOM estimate of location. Section 4.3 covers this important topic and provides results that begin to reveal the relative merits of the location estimators described in Section 3.2.

The conventional method for approximating the sampling distribution of the sample mean was derived by Laplace nearly 200 years ago. Laplace's strategy, which forms the foundation of many methods routinely used today, consists of several components, each of which must be understood if we want to appreciate the advantages of modern techniques. The immediate goal is to describe each of these components and then to indicate how they come together in an attempt to make inferences about the population mean. Then we will scrutinize these components and consider how we might get better results with modern technology.

Before proceeding, it might help to outline in a bit more detail where we are going. Continuing our illustration concerning weight gain in rats, imagine that we repeat an experiment infinitely many times and that for each experiment we sample 22 rats and compute the sample mean. Then there will be some average value for these infinitely many sample means. In formal terms, this average is written as $E(\bar{X})$, the expected value of the sample mean. Simultaneously, there will be variation among the sample means, and we can measure this variation in terms of the expected squared difference between the sample mean and the population mean. That is, we use a strategy similar to how the population variance was defined, only rather than use the average squared distance of a *single* observation from the mean, we now focus on the average squared distance between the sample mean and the population mean. In formal terms, we use

$$E(\bar{X} - \mu)^2,$$

which is called the *expected squared error* of the sample mean.

Notice that the expected squared error of the sample mean reflects how close the sample mean tends to be to the population mean. That is, it provides a crude indication of whether the sample mean tends to be an accurate estimate of μ. If the sample mean is always identical to the population mean, then the expected squared error of the sample mean is zero. But if situations arise where the sample mean provides a poor estimate of the population mean, the expected squared error of the sample mean will be large.

Now, our immediate goal is to approximate the distribution of the sample means without actually repeating our experiment infinitely many times. Recall from Chapter 2 that if we assume normality, probabilities are determined exactly once we know the mean and variance. So if we assume that a plot of infinitely many sample means would be a normal curve, then we would know the distribution of the sample mean if we could determine its mean ($E(\bar{X})$) and variance (VAR($\bar{X}$)). Laplace realized that these quantities can be determined if we assume random sampling, which is formally described in the next section.

A comment should be made about the normality assumption. We have seen indications that nonnormality is a practical concern. Under what conditions can normality be assumed when trying to approximate the plot of infinitely many sample means? There are two that are important here. The first is that if we (randomly) sample observations from a normal curve, then the sample mean will have a normal distribution as well. In the event sampling is not from a normal curve, Laplace appealed to his central limit theorem, which is formally introduced in Section 4.6. There are indeed situations where this latter strategy provides reasonably accurate results, but it can fail miserably, as we shall see.

4.2 Random Sampling

We need to be more formal about the notion of random sampling in order to understand why some strategies for dealing with nonnormality are theoretically unsound. *Random sampling* means that the observations available to us satisfy two conditions: (1) They are identically distributed, and (2) they are independent. (This is sometimes called *simple random sampling* to distinguish it from other types of sampling strategies not covered in this book.) Two measures are identically distributed if they have the same probability function. For continuous variables this means that for any constant c we might pick, the probability that an observation is less than c is the same regardless of which measure we use. So, for example, if X_1 is the first rat in our experiment on the effects of ozone on weight gain, there is a certain probability that this rat will gain 14 grams or less during the course of the experiment. Of course, there is some corresponding probability for the second rat, X_2. If these two probabilities are the same and in fact $P(X_1 < c) = P(X_2 < c)$ for any c we might pick, then X_1 and X_2 are said to be identically distributed. More generally, n observations, which we label $X_1, \ldots, X_n$, are said to be identically distributed if any two of them are identically distributed. Consequently, the expected value of each observation is the same, namely, μ, the population mean. That is,

$$E(X_1) = E(X_2) = \cdots = E(X_n) = \mu.$$

So a slight extension of results in Section 2.9 shows that if the observations are identically distributed, then the average value of their sum is just the sum of their average values. That is,

$$E(X_1 + \cdots + X_n) = \mu + \cdots + \mu = n\mu.$$

Using our rule for expected values when we multiply by a constant, if we multiply the sum in this last equation by $1/n$, we have that

$$E(\bar{X}) = \frac{n\mu}{n} = \mu. \tag{4.1}$$

In words, on average, the sample mean estimates the population mean when observations are identically distributed. When Equation (4.1) is satisfied, we say that the sample mean is an *unbiased* estimate of the population mean.

The variance of the sample mean refers to the variance of all sample means if we were to repeat a study infinitely many times. Said another way, it is the variance associated with the sampling distribution. The variance of any measure is defined as the expected squared distance from its mean (or average) value. For example, if we have a single observation X, its expected value is labeled μ, and the variance of X is the expected or average value of the squared distance between X and μ. In symbols, the variance of X is $E(X - \mu)^2$, as explained in Chapter 2. In a similar manner, the variance of the sample mean is the average squared difference between it and its mean. We just saw that the expected value of the sample mean is μ, so by definition its variance is $E(\bar{X} - \mu)^2$, which happens to be its expected squared error as well. It is common to write the variance of the sample mean as $\mathrm{VAR}(\bar{X})$ or $\sigma_{\bar{X}}^2$. So we have that

$$\mathrm{VAR}(\bar{X}) = E(\bar{X} - \mu)^2 = \sigma_{\bar{X}}^2.$$

To make a clear distinction between the variance of a single observation and the variance of the sample mean, the variance of the sample mean is often called the *squared standard error of the sample mean*.

Now consider the problem of getting an expression for the variance of the sample mean. The assumption that observations are identically distributed implies that they have identical variances. That is,

$$\mathrm{VAR}(X_1) = \mathrm{VAR}(X_2) = \cdots = \mathrm{VAR}(X_n) = \sigma^2.$$

As indicated by Equation (2.15), the variance of a sum of observations requires that we take into account their correlation. But if we assume independence, which is implied when we assume (simple) random sampling, each pair of observations has zero correlation, so the variance of the sum is just the sum of the variances. That is, under random sampling,

$$\mathrm{VAR}(X_1 + X_2 + \cdots + X_n) = n\sigma^2. \tag{4.2}$$

Finally, the rules for expected values can be used to show that as a result,

$$\mathrm{VAR}(\bar{X}) = \frac{\sigma^2}{n}. \tag{4.3}$$

Said another way, σ^2/n is the squared standard error of the sample mean, and $\sigma/\sqrt{n}$ is its standard error.

The results just given are so important, they are illustrated in another manner to make sure they are clear. To be concrete, imagine you have been asked to determine the attitude of the typical adult regarding the death penalty. You randomly sample $n = 100$ adults, and each gives one of three responses: the death penalty is not a deterrent, it makes a slight difference in some cases, or it is a strong deterrent. Suppose the responses are recorded as 1, 2, 3, respectively and that *unknown* to you, among the millions of adults you might interview, the probability function is

x:	1	2	3
$p(x)$:	.3	.5	.2

It can be seen that the population mean and variance are $\mu = 1.9$ and $\sigma^2 = .49$, respectively. If in your particular study 10 adults respond that it is not a deterrent,

TABLE 4.1 Illustration of a Sampling Distribution

Sample	X_1	X_2	X_3	...	X_{100}	$\bar{X}$	s^2
1	2	3	2	...	2	2.4	.43
2	1	3	2	...	1	2.2	.63
3	1	2	1	...	3	2.1	.57
4	3	1	3	...	1	2.3	.62
⋮	⋮	⋮	⋮	⋮	⋮	⋮	⋮

Average of $\bar{X}$ values is $\mu = 1.9$; i.e., $E(\bar{X}) = \mu$
Average of s^2 values is $\sigma^2 = .49$; i.e., $E(s^2) = \sigma^2$

$n = 100$ for each sample.

25 say it makes a slight difference, and the remaining 65 say it is a strong deterrent, then the sample mean is

$$\bar{X} = 1\frac{10}{100} + 2\frac{25}{100} + 3\frac{65}{100} = 2.55.$$

So there is a discrepancy between the sample mean and the population mean, as we would expect.

Suppose we repeat the survey of adults millions of times, and each time we interview $n = 100$ subjects. Again assume that the possible outcomes we observe are 1, 2, and 3. Table 4.1 summarizes what might be observed. Each row represents a replication of the study, and there are 100 columns, corresponding to the 100 subjects observed each time the study is replicated. The 100 subjects in the first row differ from the 100 subjects in the second row, which differ from the 100 subjects in third row, and so on. Then for the first column, among the millions of times we repeat the experiment, there is a certain proportion of times we would observe a 1 or a 2 or a 3, and we can think of these proportions as the corresponding probabilities. If the probability of a 1 is .3, then the proportion of 1's in the column headed by X_1 will be .3. Similarly, the column headed by X_2 has a certain proportion of 1's, 2's, and 3's, and again they can be thought of as probabilities. If the probability of a 2 is .5, then 50% of the values in this column will be 2. When we say that X_1 and X_2 are identically distributed, we mean that the proportion of 1's, 2's, and 3's in columns 1 and 2 are exactly the same. That is, it does not matter whether a subject is first or second when determining the probability of a 1 or a 2 or a 3. More generally, if all 100 observations are identically distributed, then any column we pick in Table 4.1 will have the same proportions of 1's, 2's, and 3's over the millions of subjects.

Now, an implication of having identically distributed random variables is that if we average all the sample means in the next to last column of Table 4.1, we would get μ, the population mean. That is, the average or expected value of all sample means we might observe based on n subjects is equal to μ, which in this case is 1.9. In symbols, $E(\bar{X}) = \mu$, as previously indicated. Moreover, the variance of these sample means is σ^2/n, which in this particular case is $.49/100 = .0049$, so the standard error of the sample mean is $\sqrt{.0049} = .07$.

But how do we determine the standard error of the sample mean in the more realistic situation where the probability function is not known? A natural guess is to estimate σ^2 with the sample variance, s^2, in which case $\text{VAR}(\bar{X})$ is estimated with s^2/n, and this is exactly what is done in practice.

To provide a bit more justification for estimating σ^2 with s^2, again consider the survey of adults regarding the death penalty. Imagine that for a sample of 100 adults we compute the sample variance, s^2. Repeating this process infinitely many times, each time sampling 100 adults and computing s^2, the sample variances we might get are shown in the last column of Table 4.1. It can be shown that under random sampling, the average of the resulting sample variances is σ^2, the population variance from which the observations were sampled. In symbols, $E(s^2) = \sigma^2$ for any sample size n we might use, which means that the sample variance is an *unbiased estimator* of σ^2, the population variance.[1] Because s^2 estimates σ^2, we estimate the standard error of the sample mean with $s/\sqrt{n}$.

EXAMPLE. Imagine you are a health professional interested in the effects of medication on the diastolic blood pressure of adult women. For a particular drug being investigated, you find that for $n = 9$ women, the sample mean is $\bar{X} = 85$ and the sample variance is $s^2 = 160.78$. An estimate of the squared standard error of the sample mean, assuming random sampling, is $s^2/n = 160.78/9 = 17.9$. An estimate of the standard error of the sample mean is $\sqrt{17.9} = 4.2$, the square root of s^2/n. ■

4.3 Approximating the Sampling Distribution of $\bar{X}$

Consider again the ozone experiment described at the beginning of this chapter. Suppose a claim is made that if all rats could be measured, we would find that weight gain is normally distributed, with $\mu = 14$ grams and a standard deviation of $\sigma = 6$. Under random sampling, this claim implies that the distribution of the sample mean has a mean of 14 as well; and because there are $n = 22$ rats, the variance of the sample mean is $6^2/22$. In symbols, $E(\bar{X}) = 14$ and $\text{VAR}(\bar{X}) = \sigma^2/22 = 6^2/22 = 36/22$. The sample mean based on the 22 rats in the experiment is $\bar{X} = 11$. Is it reasonable to expect a sample mean this low if the claimed values for the population mean (14) and standard deviation (6) are correct? In particular, what is the probability of observing $\bar{X} \leq 11$?

Recall from Chapter 2 that if X has a normal distribution, $P(X \leq c)$ can be determined by standardizing X. That is, subtract the mean and divide by the standard deviation, yielding $Z = (X - \mu)/\sigma$. If X is normal, Z has a standard normal distribution. That is, we solve the problem of determining $P(X \leq c)$ by transforming it into a problem involving the standard normal distribution. *Standardizing a random variable* is a recurrent theme in statistics, and it is useful when determining $P(\bar{X} \leq c)$ for any constant c.

---

1 Recall that the sample variance is given by $s^2 = \sum (X_i - \bar{X})^2/(n-1)$. If we divide by n rather than $n - 1$, it is no longer true that $E(s^2) = \sigma^2$.

Again consider the ozone experiment, where $n = 22$. Still assuming that the mean is $\mu = 14$ and the standard deviation is $\sigma = 6$, suppose we want to determine the probability that the sample mean will be less than 11. In symbols, we want to determine $P(\bar{X} \leq 11)$. When sampling from a normal distribution, $\bar{X}$ also has a normal distribution, so essentially the same technique described in Chapter 2 can be used to determine $P(\bar{X} \leq 11)$: We convert the problem into one involving a standard normal. This means that we *standardize the sample mean* by subtracting its mean and dividing by its standard deviation which is $\sigma/\sqrt{n}$. That is, we compute

$$Z = \frac{\bar{X} - \mu}{\sigma/\sqrt{n}}.$$

When sampling from a normal distribution, it can be mathematically verified that Z has a standard normal distribution. This means that

$$P(\bar{X} \leq 11) = P\left(Z \leq \frac{11 - \mu}{\sigma/\sqrt{n}}\right).$$

For the problem at hand,

$$\frac{11 - \mu}{\sigma/\sqrt{n}} = \frac{11 - 14}{6/\sqrt{22}} = -2.35,$$

and Table 1 in Appendix B tells us that the probability that a standard normal variable is less than -2.35 is .0094. More succinctly,

$$P(Z \leq -2.35) = .0094,$$

so the probability of getting a sample mean less than or equal to 11 is .0094. That is, if simultaneously the assumption of random sampling from a normal distribution is true, the population mean is $\mu = 14$, and the population standard deviation is $\sigma = 6$, then the probability of getting a sample mean less than or equal to 11 is .0094.

More generally, given c, some constant of interest, the probability that the sample mean is less than c is

$$P(\bar{X} < c) = P\left(\frac{\bar{X} - \mu}{\sigma/\sqrt{n}} < \frac{c - \mu}{\sigma/\sqrt{n}}\right)$$

$$= P\left(Z < \frac{c - \mu}{\sigma/\sqrt{n}}\right). \tag{4.4}$$

In other words, $P(\bar{X} < c)$ is equal to the probability that a standard normal random variable is less than $(c - \mu)/(\sigma/\sqrt{n})$.

EXAMPLE. To be sure Equation (4.4) is clear, the ozone example is repeated in a more concise manner. The claim is that $\mu = 14$ and $\sigma = 6$. With $n = 22$ rats, what is the probability that the sample mean will be less than or equal to $c = 11$ if this claim is true and it is assumed that we are randomly sampling from a normal distribution? To answer this question, compute

$$\frac{c - \mu}{\sigma/\sqrt{n}} = \frac{11 - 14}{6/\sqrt{22}} = -2.35.$$

Then according to Equation (4.4),

$$P(\bar{X} \leq 11) = P(Z \leq -2.35).$$

Referring to Table 1 in Appendix B, the probability that a standard normal random variable is less than or equal to -2.35 is .0094. That is, if the claim is correct,

$$P(\bar{X} \leq 11) = .0094.$$

◼

For the 22 rats in the ozone experiment, the sample mean is $\bar{X} = 11$. As just indicated, getting a sample mean less than or equal to 11 is unlikely if the claims $\mu = 14$ and $\sigma = 6$ are true. That is, the data suggest that perhaps these claims are false.

Analogous to results in Chapter 2, we can determine the probability that the sample mean is greater than some constant c, and we can determine the probability that the sample mean is between two numbers, say, a and b. The probability of getting a sample mean greater than c is

$$P(\bar{X} > c) = 1 - P\left(Z < \frac{c - \mu}{\sigma/\sqrt{n}}\right), \tag{4.5}$$

and the probability that the sample mean is between the numbers a and b is

$$P(a < \bar{X} < b) = P\left(Z < \frac{b - \mu}{\sigma/\sqrt{n}}\right) - P\left(Z < \frac{a - \mu}{\sigma/\sqrt{n}}\right). \tag{4.6}$$

EXAMPLE. A researcher claims that for college students taking a particular test of spatial ability, the scores have a normal distribution with mean 27 and variance 49. If this claim is correct, and you randomly sample 36 subjects, what is the probability that the sample mean will be greater than $c = 28$? Referring to Equation (4.5), first compute

$$\frac{c - \mu}{\sigma/\sqrt{n}} = \frac{28 - 27}{\sqrt{49/36}} = .857.$$

Continued

EXAMPLE. (*Continued*) Because $P(Z \leq .857) = .20$, Equation (4.5) tells us that

$$P(\bar{X} > 28) = 1 - P(Z \leq .857) = 1 - .20 = .80.$$

This means that if we randomly sample $n = 25$ subjects and the claims of the researcher are true, the probability of getting a sample mean greater than 28 is .8.

■

EXAMPLE. Suppose observations are randomly sampled from a normal distribution with $\mu = 5$ and $\sigma = 3$. If $n = 36$, what is the probability that the sample mean is between $a = 4$ and $b = 6$? To find out, compute

$$\frac{b - \mu}{\sigma/\sqrt{n}} = \frac{6 - 5}{3/\sqrt{36}} = 2.$$

Referring to Table 1 in Appendix B, $P(\bar{X} < 4) = P(Z < 2) = .9772$. Similarly,

$$\frac{a - \mu}{\sigma/\sqrt{n}} = \frac{4 - 5}{3/\sqrt{36}} = -2,$$

and $P(Z < -2) = .0228$. So according to Equation (4.6),

$$P(2 < \bar{X} < 4) = .9772 - .0228 = .9544.$$

This means that if $n = 36$ observations are randomly sampled from a normal distribution with mean $\mu = 5$ and standard deviation $\sigma = 3$, there is a .9544 probability of getting a sample mean between 4 and 6.

■

An important point is that as the number of observations increases, the sample mean will provide a better estimate of the population mean, μ. The sampling distribution of $\bar{X}$, under normality, provides an illustration of the sense in which this is true. Suppose we randomly sample a single observation ($n = 1$) from a standard normal distribution. This single observation provides an estimate of the population mean, and the standard normal distribution in Figure 4.1 graphically illustrates how close this observation will be to the mean, μ. Figure 4.1 also shows the distribution of $\bar{X}$ when $n = 16$. Notice that this distribution is more tightly centered around the mean. That is, $\bar{X}$ is more likely to be close to μ when $n = 16$ rather than 1. If we increase n to 25, the distribution of $\bar{X}$ would be even more tightly centered around the mean.

4.4 The Sample Mean versus MOM, the Median, Trimmed Mean, and M-Estimator

The notion of a sampling distribution generalizes to any of the location estimators considered in Chapter 3. For example, if we conduct a study and compute the median based on 20 observations and repeat the study billions of times, each time computing

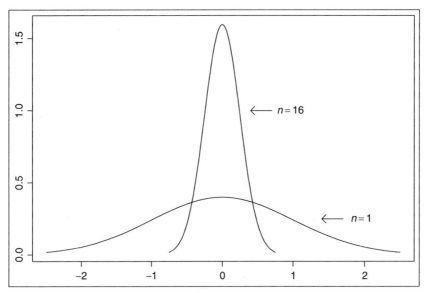

FIGURE 4.1   The distribution of $\bar{X}$, under normality, when $n = 1$ and $n = 16$.

a median based on 20 observations, we would know the sampling distribution of the median. That is, we would know the probability that the median is less than 4, less than 10, or less than any constant c that might be of interest. Temporarily assume sampling is from a normal distribution, and, for convenience only, suppose the population mean is zero and the variance is 1. Because sampling is from a symmetric distribution, the sample median, M, is a reasonable estimate of the population mean, μ. But of course the sample median will be in error, as was the case when using the sample mean.

Now, when sampling from a normal distribution, theory tells us that, on average, the sample mean will be a more accurate estimate of the population mean versus the sample median. To illustrate the extent to which this is true, 20 observations were generated on a computer from a standard normal distribution, the sample mean and median were computed, and this process was repeated 5000 times. Figure 4.2 shows a plot of the resulting means and medians. That is, Figure 4.2 shows an approximation of the sampling distributions of both the mean and median. Note that the plot of the means is more tightly centered around the value zero, the value both the mean and median are trying to estimate. This indicates that in general, the sample mean is more accurate.

But what about nonnormal distributions? How does the median compare to the mean in accuracy? To illustrate an important point, rather than sample from the standard normal distribution, we repeat the computer experiment used to create Figure 4.2, only now sampling is from the mixed normal shown in Figure 2.8. Recall that the mixed normal represents a small departure from normality in the sense described in Section 2.7.

Figure 4.3 shows the results of our computer experiment, and, in contrast to Figure 4.2, now the sample median is much more accurate, on average, relative to the mean. That is, the sample median is likely to be substantially closer to the true population mean than is the sample mean, $\bar{X}$. This illustrates a general result of considerable

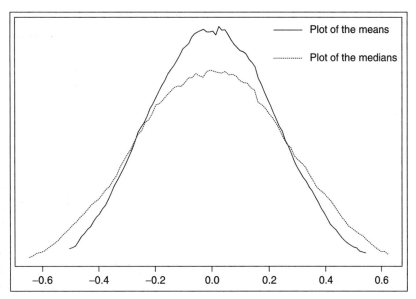

FIGURE 4.2 A plot of 5000 means and 5000 medians when sampling from a normal distribution.

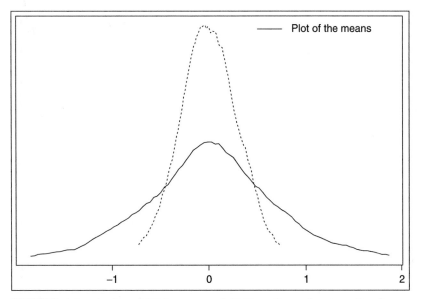

FIGURE 4.3 A plot of 5000 means and 5000 medians when sampling from a mixed normal distribution.

importance. Even for an extremely small departure from normality, the median can be a much more accurate estimate of the center of a symmetric distribution than the mean. But this is not a very compelling reason to routinely use the median, because Figure 4.2 illustrates that it can be substantially less accurate than the mean as well.

Now we consider the accuracy of the 20% trimmed mean and the M-estimator (based on Huber's Ψ) versus the mean. We repeat the process used to create

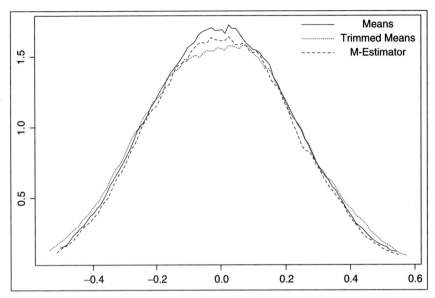

FIGURE 4.4   Plots of 5000 means, 20% trimmed means and M-estimators when sampling from a normal distribution.

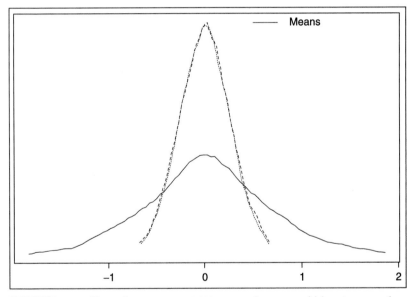

FIGURE 4.5   Plots of 5000 means, 20% trimmed means and M-estimators when sampling from a mixed normal distribution.

Figures 4.2 and 4.3, only the median is replaced by $\bar{X}_t$ and $\hat{\mu}_{os}$, where $\hat{\mu}_{os}$ is the one-step M-estimator given by Equation (3.28). Figure 4.4 shows the resulting plots when sampling is from a normal distribution. The mean is most accurate, but its improvement over the 20% trimmed mean and M-estimator is not very striking. If we use the MOM estimator given by Equation (3.30), the plot of the values reveals that it is slightly less accurate than the M-estimator used here. Figure 4.5 shows the sampling distributions when observations are sampled from a mixed normal instead. As is evident, the sample mean performs poorly relative to the other two estimators considered. Here, MOM gives nearly the same results as the 20% trimmed mean and M-estimator.

The results illustrated in Figures 4.4 and 4.5 are not surprising, in the sense that MOM, the 20% trimmed mean, and the M-estimator used here were designed to give nearly the same accuracy as the mean when sampling from a normal distribution. That is, theoretical results tell us how much we can trim without sacrificing too much accuracy under normality, and theory also suggests how to design an M-estimator so that again relatively good accuracy is obtained. Simultaneously, theory tells us that these estimators will continue to perform relatively well when sampling from a heavy-tailed distribution, such as the mixed normal.

To broaden our perspective on the relative merits of the mean, we now describe a general situation where the sample mean is optimal among all weighted means, described in Section 3.2.3. Without assuming normality, Gauss showed that under random sampling, among all the weighted means we might consider, the sample mean will have a smaller variance (or squared standard error) than any other weighted mean we might consider. (This result is a special case of the Gauss–Markov theorem, which is discussed in more detail when we take up regression.) So if we were to plot the sampling distribution of any weighted mean versus the sample mean, the sample mean would be more tightly centered around the population mean.

The result just described might seem to contradict our finding that the median, 20% trimmed mean, and one-step M-estimator can be substantially more accurate than the mean. There is no contradiction, however, because all three of these estimators do not belong to the class of weighted means. The median and trimmed mean, for example, involve more than weighting the observations — they require putting the observations in order.

It is easy to see why Gauss's result might suggest using the mean rather than the median or the trimmed mean. The 20% trimmed mean, for example, removes 40% of the observations. How could this possibly improve accuracy as opposed to giving some weight to all of the values? It is important to develop some intuition about this issue because despite graphical and mathematical arguments demonstrating that a 20% trimmed mean can be much more accurate than the sample mean, often there is reluctance to use any trimming at all because it seems counterintuitive. One way of understanding this issue is covered in Section 4.9. Here, an alternate but less technical explanation is given.

From the perspective about to be described, it is not surprising that a trimmed mean beats the mean, but it is rather amazing that the sample mean is optimal in any situation at all. To explain, suppose we sample 20 observations from a standard normal distribution. So the population mean and population 20% trimmed mean are zero. Now consider the smallest of the 20 values. It can be shown that, with probability .983, this value will be less than −0.9, and with probability .25 it will be less than −1.5. That is, with fairly high certainty, it will not be close to zero, the value we are trying to estimate. In a similar manner, the largest of the 20 observations has probability .983 of being greater than 0.9 and probability .25 of being greater than 1.5. Simultaneously, if we put the observations in ascending order, the probability that the two middle values do not differ by more than .5 from zero is .95. So a natural reaction is that extreme values should be given less weight in comparison to the observations in the middle. A criticism of this simple argument is that the smallest value will tend to be less than zero, the largest will be greater than zero, and their average value will be exactly zero, so it might seem that there is no harm in using them to estimate the population mean. However, the issue is how much these extreme values contribute to the variance of our estimator. When sampling from a normal distribution, we are better off on average using the sample mean, despite the fact that the extreme values are highly likely to be inaccurate. But as we move away from a normal distribution toward a heavier-tailed distribution, the sample mean becomes extremely inaccurate relative to the 20% trimmed mean.

There is, however, a practical concern about 20% trimming that should be stressed: Its standard error might be substantially higher versus the standard error of MOM. This can happen even when the number of outliers in both tails does not exceed 20%.

EXAMPLE. Consider the data

77 87 87 114 151 210 219 246 253 262
296 299 306 376 428 515 666 1310 2611,

which are from a study on self-awareness conducted by E. Dana (1990). The estimated standard error of the 20% trimmed mean is 56.1 (using the method described in Section 4.9.2). But the estimated standard error of MOM (using a method described in Chapter 7) is 37, which is substantially smaller. The outlier detection rule based on M and MAD [given by Equation (3.22)] flags the three largest values as outliers, and these are removed by MOM as well as the 20% trimmed mean. But the 20% trimmed also removes the three smallest values, which are not flagged as outliers.   ■

4.5 A Confidence Interval for the Population Mean

If you were a psychologist, you might be interested in a new method for treating depression and how it compares to a standard, commonly used technique. Assume that based on a widely used measure of effectiveness, the standard method has been applied thousands of times and found to have a mean effectiveness of $\mu = 48$. That is, the standard method has been used so many times, for all practical purposes we know that the population mean is 48. Suppose we estimate the effectiveness of the new method by trying it on $n = 25$ subjects and computing the sample mean. A crucial issue is whether the resulting sample mean is close to the population mean being estimated, the mean we would obtain if all depressed individuals were treated with the new method. Assume that for the experimental method we obtain a sample mean of $\bar{X} = 54$. This means that based on our experiment, the average effectiveness of the new method is estimated to be 54, which is larger than the average effectiveness of the standard method, suggesting that the new method is better for the typical individual.

But suppose that unknown to us, the average effectiveness of the new method is actually 46, meaning that on average the standard technique is better for treating depression. By chance we might get a sample mean of 54 and incorrectly conclude that the experimental method is best. What is needed is some way of determining whether $\bar{X} = 54$ makes it unlikely that the mean is 46 or some other value less than 48. If we can be reasonably certain that 54 is close to the actual population mean, and in particular we can rule out the possibility that the population mean is less than 48, we have evidence that, on average, the new treatment is more effective.

A major advantage of being able to determine the sampling distribution of $\bar{X}$ is that it allows us to address the issue of how well the sample mean $\bar{X}$ estimates the population mean μ. We can address this issue with what is called a confidence interval for μ. A *confidence interval* for μ is just a range of numbers that contains μ with some specified probability. If for the experimental method for treating depression you can be reasonably certain that the population mean is between 50 and 55, then there is evidence that its mean is greater than 48, the average effectiveness of the standard technique. That is, the new method appears to be more effective than the standard technique.

Suppose we want to use the observations to determine a range of values that contains μ with probability .95. From the previous section, if sampling is from a normal distribution, then $(\bar{X} - \mu)/(\sigma/\sqrt{n})$ has a standard normal distribution. Recall from Chapter 2 that the probability of a standard normal random variable being between -1.96 and 1.96 is .95. In symbols,

$$P\left(-1.96 \leq \frac{\bar{X} - \mu}{\sigma/\sqrt{n}} \leq 1.96\right) = .95.$$

We can rearrange terms in this last equation to show that

$$P\left(\bar{X} - 1.96\frac{\sigma}{\sqrt{n}} \leq \mu \leq \bar{X} + 1.96\frac{\sigma}{\sqrt{n}}\right) = .95.$$

This says that although the population mean, μ, is not known, there is a .95 probability that its value is between

$$\bar{X} - 1.96\frac{\sigma}{\sqrt{n}} \quad \text{and} \quad \bar{X} + 1.96\frac{\sigma}{\sqrt{n}}.$$

When sampling from a normal distribution,

$$\left(\bar{X} - 1.96\frac{\sigma}{\sqrt{n}}, \bar{X} + 1.96\frac{\sigma}{\sqrt{n}}\right) \tag{4.7}$$

is called a .95 confidence interval for μ. This means that if the experiment were repeated billions of times, and each time a confidence interval is computed using Equation (4.7), 95% of the resulting confidence intervals will contain μ if observations are randomly sampled from a normal distribution.

> **EXAMPLE.** In the example for treating depression, assume for illustrative purposes that the standard deviation is $\sigma = 9$ and sampling is from a normal distribution. Because the sample mean is $\bar{X} = 54$ and the sample size is $n = 25$, the .95 confidence interval for μ is
>
> $$\left(54 - 1.96\frac{9}{\sqrt{25}}, 54 + 1.96\frac{9}{\sqrt{25}}\right) = (50.5, 57.5).$$
>
> That is, based on the 25 subjects available to us, we can be reasonably certain that μ is somewhere between 50.5 and 57.5. ■

DEFINITION. The *probability coverage* of a confidence interval is the probability that the interval contains the parameter being estimated. The previous example described a confidence interval for the mean that has probability coverage .95.

A standard notation for the probability that a confidence interval does *not* contain the population mean, μ, is α. When computing a .95 confidence interval, $\alpha = 1 - .95 = .05$. For a .99 confidence interval, $\alpha = .01$. The quantity α is the probability of making a mistake. That is, if we perform an experiment with the goal of computing a .95 confidence interval, there is a .95 probability that the resulting interval contains the mean, but there is an $\alpha = 1 - .95 = .05$ probability that it does not.

The method of computing a confidence interval can be extended to any value of $1 - \alpha$ you might choose. The first step is to determine c such that the probability that a standard normal random variable lies between $-c$ and c is $1 - \alpha$. In symbols, determine c such that

$$P(-c \leq Z \leq c) = 1 - \alpha.$$

From Chapter 2, this means that you determine c such that

$$P(Z \leq c) = \frac{1 + (1 - \alpha)}{2}$$

$$= 1 - \frac{\alpha}{2}.$$

Put another way, c is the $1 - \alpha/2$ quantile of a standard normal distribution. For example, if you want to compute a $1 - \alpha = .95$ confidence interval, then

$$\frac{1 + (1 - \alpha)}{2} = \frac{1 + .95}{2} = .975,$$

and from Table 1 in Appendix B we know that

$$P(Z \leq 1.96) = .975,$$

so $c = 1.96$. For convenience, the values of c for $1 - \alpha = .9$, .95, and .99 are listed in Table 4.2.

TABLE 4.2   Common Choices for $1 - \alpha$ and c

$1 - \alpha$	c
.90	1.645
.95	1.96
.99	2.58

Once c is determined, a $1 - \alpha$ confidence interval for μ is

$$\left(\bar{X} - c\frac{\sigma}{\sqrt{n}}, \bar{X} + c\frac{\sigma}{\sqrt{n}} \right). \tag{4.8}$$

Equation (4.8) is a special case of a general technique developed by Laplace.

EXAMPLE. A college president claims that IQ scores at her institution are normally distributed with a mean of $\mu = 123$ and a standard deviation of $\sigma = 14$. Suppose you randomly sample $n = 20$ students and find that $\bar{X} = 110$. Does the $1 - \alpha = .95$ confidence interval for the mean support the claim that the average of all IQ scores at the college is $\mu = 123$? Because $1 - \alpha = .95$, $c = 1.96$, as just explained, so the .95 confidence interval is

$$\left(110 - 1.96\frac{14}{\sqrt{20}}, 110 + 1.96\frac{14}{\sqrt{20}} \right) = (103.9, 116.1).$$

The interval (103.9, 116.1) does not contain the value 123, suggesting that the president's claim might be false. Note that there is a .05 probability that the confidence interval will not contain the true population mean, so there is some possibility that the president's claim is correct. ∎

EXAMPLE. For 16 observations randomly sampled from a normal distribution, $\bar{X} = 32$ and $\sigma = 4$. To compute a .9 confidence interval (meaning that $1 - \alpha = .9$), first note from Table 4.2 that $c = 1.645$. So a .9 confidence interval for μ is

$$\left(32 - 1.645\frac{4}{\sqrt{16}}, 32 + 1.645\frac{4}{\sqrt{16}} \right) = (30.355, 33.645).$$

Although $\bar{X}$ is not, in general, equal to μ, the confidence interval provides some sense of how well $\bar{X}$ estimates the population mean. ∎

4.6 An Approach to Nonnormality: The Central Limit Theorem

We have seen how to compute a confidence interval for the mean when the standard deviation is known and sampling is from a normal distribution. Assuming observations

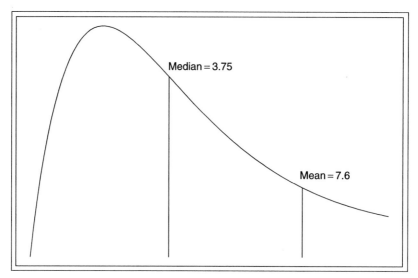

FIGURE 4.6   Example of a skewed, heavy-tailed distribution.

are randomly sampled from a normal distribution is convenient because the sampling distribution of $\bar{X}$ turns out to be a normal distribution as well. But how do we deal with nonnormal distributions? In the ozone experiment, how do we compute a confidence interval for μ if, unknown to us, observations are randomly sampled from the distribution shown in Figure 4.6, which is a reproduction of Figure 2.11.

Laplace's solution was to appeal to his central limit theorem, which says that under very general conditions, even if observations are randomly sampled from a nonnormal distribution, the sampling distribution of the sample mean will approach a normal distribution as the sample size gets large. In more practical terms, if n is sufficiently large, we can pretend that $\bar{X}$ has a normal distribution with mean μ and variance σ^2/n, in which case the method in the previous section can be employed. Of course this last statement is rather vague, in an important sense. How large must n be? Many books claim that $n = 25$ suffices and others claim that $n = 40$ is more than sufficient. These are not wild speculations, but now we know that much larger sample sizes are needed for many practical problems. The immediate goal is to provide some sense of why larger sample sizes are needed than once thought.

To begin, there is no theorem telling us when n is sufficiently large. The answer depends in part on the skewness of the distribution from which observations are sampled. (Boos & Hughes-Oliver, 2000, provide a recent summary of relevant details.) Generally, we must rely on empirical investigations, at least to some extent, in our quest to address this problem.

To illustrate how such empirical studies are done and why it might seem that $n = 40$ is sufficiently large for most purposes, suppose $n = 20$ observations are randomly sampled from the distribution in Figure 4.7, which is an example of a uniform distribution, and then we compute the sample mean. As is evident, the uniform distribution in Figure 4.7 does not remotely resemble a normal curve. If we generate 5000 sample means in this manner and plot the results, we get the curve shown in Figure 4.8. Also shown is the normal curve implied by the central limit

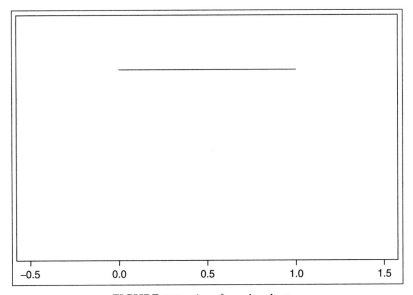

FIGURE 4.7 A uniform distribution.

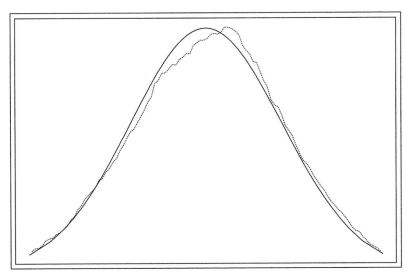

FIGURE 4.8 A plot of 5000 means based on observations sampled from a uniform distribution. The solid, symmetric curve is the plot of the means based on the central limit theorem.

theorem. As we see, the two curves are very similar, indicating that the central limit theorem is performing rather well in this particular case. That is, if we sample 20 observations from a uniform distribution, for all practical purposes we can assume the sample mean has a normal distribution.

Now consider the probability curve in Figure 4.9 (which is called an *exponential distribution*). Again this curve is nonnormal in an obvious way. If we repeat our computer experiment used to create Figure 4.8, but sampling from the distribution in Figure 4.9, then the plot of the resulting sample means appears as shown in Figure 4.10.

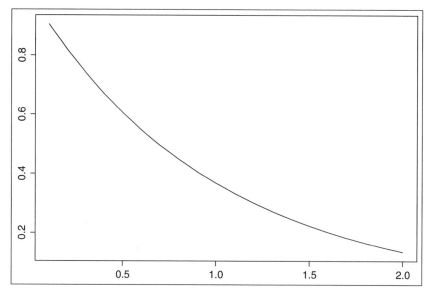

FIGURE 4.9   An exponential distribution which is relatively light-tailed.

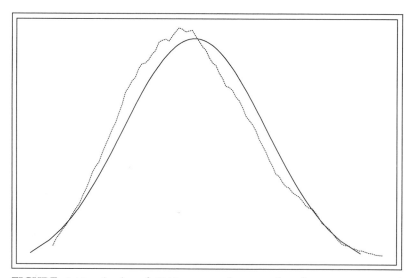

FIGURE 4.10   A plot of 5000 means when sampling from an exponential distribution.

Again we get fairly close agreement between the empirical distribution for the sample means and the theoretical distribution implied by the central limit theorem. These two illustrations are classic ways of demonstrating the central limit theorem, and the obvious speculation based on these results is that in general, with $n \geq 25$, we can assume the sample mean has a normal distribution. There are, however, two fundamental problems that have been overlooked. The first of these is illustrated here, and the second and even more serious problem is described in Section 4.7.

We repeat our computer experiment one more time, only now we sample 25 observations from the distribution shown in Figure 4.6. Figure 4.11 shows the resulting

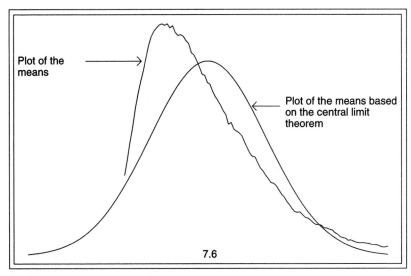

FIGURE 4.11   A plot of 5000 means when sampling from the distribution in Figure 4.6. The sample size for each mean is 25.

plot of the sample means. Now there is an obvious discrepancy between the plot of the sample means and the curve implied by the central limit theorem. Why do we get a different result from the two previous situations? The reason is that the curves in Figures 4.7 and 4.9 are distributions with relatively light tails. That is, outliers are fairly rare when sampling observations. In contrast, Figure 4.11 is obtained by sampling from a heavy-tailed distribution, where outliers are more common.

To provide some sense of how quickly matters improve as the sample size increases, we again sample observations from the distribution in Figure 4.6, but with a sample size of $n = 50$. Figure 4.12 shows the results. Note that the left tail of the plot of the means is lighter than what we would expect via the central limit theorem. Increasing n to 100, the central limit theorem gives a reasonable approximation of the actual distribution of the sample mean.

The illustrations just given might seem to suggest that as long as a distribution is not too heavy-tailed or if $n = 100$, an accurate confidence interval for μ can be computed using Equation (4.8). So a seemingly reasonable speculation is that if a boxplot indicates that there are no outliers, the actual probability coverage is reasonably close to the nominal level. Unfortunately, this strategy can fail, for reasons covered in Section 4.8. Moreover, even with $n = 100$, practical problems might occur. Briefly, when we take up the more realistic situation where σ is unknown and estimated with s, problems arise even when sampling from light-tailed distributions and $n = 100$.

Notice that when sampling from the distribution shown in Figure 4.6, although the population mean is in the extreme right portion of the distribution under considera-tion, the sample means become more tightly centered around the population mean, $\mu = 7.6$, as the sample size increases. So the sample mean is fulfilling its intended goal: As the sample size increases, it provides a better estimate of the population mean, which in this case is a quantity that happens to be in the tail of the distribution.

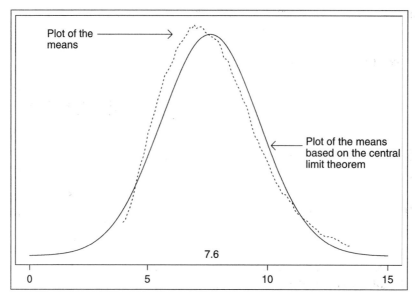

FIGURE 4.12   A plot of 5000 means when sampling from the distribution in Figure 4.6, only now the sample size for each mean is 50.

A criticism of the illustrations based on Figure 4.6 is that we are considering a hypothetical distribution. Perhaps in practice the central limit theorem will give satisfactory results even with $n = 25$. There are several reasons, however, for concluding that, in practice, problems can arise even when $n = 100$. Details are covered in Section 4.8.

One point should be stressed, because it plays an important role in subsequent chapters. Again consider the ozone experiment, and first assume that weight gain in rats has a normal distribution with variance $\sigma^2 = 1$. Then because the sample mean is $\bar{X} = 11$ and $n = 22$, the .95 confidence interval for μ is

$$\left(11 - 1.96\frac{1}{\sqrt{22}}, 11 + 1.96\frac{1}{\sqrt{22}}\right) = (10.58, 11.42).$$

The length of the confidence interval is just the difference between these two numbers, namely, $11.42 - 10.58 = .84$. When computing a .95 confidence interval, we want the length of the confidence interval to be as short as possible because a short confidence interval means we can be reasonably certain about what the value of μ happens to be. In the ozone experiment, we can be reasonably certain that the population mean is somewhere between 10.58 and 11.42, and this reflects how well $\bar{X}$ estimates μ.

Now suppose that sampling is from a mixed normal instead, which was discussed in Chapter 2. Although the mixed normal differs only slightly from the standard normal, the mixed normal has variance 10.9. Consequently, the .95 confidence interval for μ is

$$\left(11 - 1.96\frac{\sqrt{10.9}}{\sqrt{22}}, 11 + 1.96\frac{\sqrt{10.9}}{\sqrt{22}}\right) = (9.62, 12.38).$$

The length of this interval is $12.38 - 9.62 = 2.76$, more than three times longer than the situation where we sampled from a standard normal distribution instead. This illustrates that *small shifts away from normality, toward a heavy-tailed distribution, can drastically increase the length of a confidence interval.* Modern methods have been found for getting much shorter confidence intervals in situations where the length of the confidence interval based on Equation (4.8) is relatively long, some of which will be described.

4.7 Confidence Intervals when σ Is Unknown

The previous section described how to compute a confidence interval for μ when the standard deviation, σ, is known. However, typically σ is not known, so a practical concern is finding a reasonably satisfactory method for dealing with this issue. This section describes the classic method for addressing this problem, which was derived by William Gosset about a century ago. It is used routinely today, but unfortunately problems with nonnormality are exacerbated relative to the situation in Section 4.6, where σ is known.

Consider again the study of women's blood pressure, where the goal is to determine the average diastolic blood pressure of adult women taking a certain drug. Based on $n = 9$ women, we know that $\bar{X} = 85$, and $s^2 = 160.78$, but, as is usually the case, we do not know the population variance, σ^2. Although σ is not known, it can be estimated with s, the sample standard deviation, which in turn yields an estimate of the standard error of the sample mean, namely, $s/\sqrt{n} = 4.2$, as previously explained. If we assume that 4.2 is indeed an accurate estimate of $\sigma/\sqrt{n}$, then a reasonable suggestion is to assume that $\sigma/\sqrt{n} = 4.2$ when computing a confidence interval. In particular, a .95 confidence interval for the mean would be

$$(85 - 1.96(4.2), 85 + 1.96(4.2)) = (76.8, 93.2).$$

Prior to the year 1900, this was the strategy used, and based on a version of the central limit theorem, it turns out that this approach is reasonable if the sample size is sufficiently large, assuming random sampling. However, even when sampling from a normal distribution, a concern is that when the sample size is small, the population standard deviation, σ, might differ enough from its estimated value, s, to cause practical problems. Gosset realized that problems can arise and derived a solution assuming random sampling from a normal distribution. Gosset worked for a brewery and was not immediately allowed to publish his results, but eventually he was permitted to publish under the pseudonym Student.

Let

$$T = \frac{\bar{X} - \mu}{s/\sqrt{n}}. \tag{4.9}$$

The random variable T in Equation (4.9) is the same as Z used in Section 4.3, except σ has been replaced by s. Note that like $\bar{X}$ and Z, T has a distribution. That is, for any constant c we might pick, there is a certain probability that $T < c$ based on a random

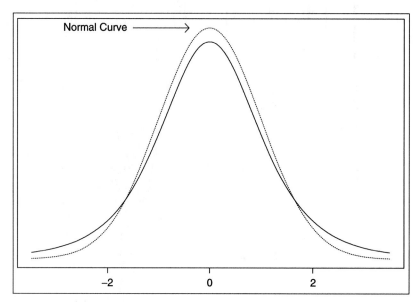

FIGURE 4.13   Student's T distribution with 4 degrees of freedom.

sample of n subjects. If the distribution of T can be determined, then a confidence interval for μ could be computed without knowing σ.

If we assume that observations are randomly sampled from a normal distribution, then the distribution of T, called *Student's T Distribution*, can be determined exactly. It turns out that the distribution depends on the sample size, n. By convention, the quantiles of the distribution are reported in terms of *degrees of freedom*: $\nu = n - 1$. Figure 4.13 shows Student's T distribution with $\nu = 4$ degrees of freedom. Note that the distribution is similar to a standard normal. In particular, it is symmetric about zero, so $E(T) = 0$. With infinite degrees of freedom, Student's T and the standard normal are identical.

Table 4 in Appendix B reports some quantiles of Student's T distribution. The first column gives the degrees of freedom. The next column, headed by $t_{.9}$, reports the .9 quantiles. For example, with $\nu = 1$, we see 3.078 under the column $t_{.9}$. This means that $P(T < 3.078) = .9$. That is, if we randomly sample two observations from a normal distribution, in which case $\nu = n - 1 = 1$, there is a .9 probability that the resulting value for T is less than 3.078. Similarly, if $\nu = 24$, then $P(T < 1.318) = .9$. The column headed by $t_{.99}$ lists the .99 quantiles. For example, if $\nu = 3$, we see 4.541 under the column headed $t_{.99}$, so the probability that T is less than 4.541 is .99. If $\nu = 40$, Table 2 indicates that $P(T < 2.423) = .99$. Many modern computer programs, such as Minitab and S-PLUS, contain functions that compute Student's T distribution for any $\nu \geq 1$. [In S-PLUS you can use the built-in function pt. For example, pt(1,5) will return the probability that T is less than 1 with $\nu = 5$ degrees of freedom.]

Similar to the situation when working with normal distributions,

$$P(T \geq c) = 1 - P(T \leq c), \qquad (4.10)$$

where c is any constant that might be of interest. For example, with $\nu = 4$, $P(T \leq 2.132) = .95$, as previously indicated, so $P(T \geq 2.132) = 1 - P(T \leq 2.132) = .05$.

To assist in learning how to use Table 4 in Appendix B, the following examples are presented.

EXAMPLE. Suppose you are involved in a study on the effects of alcohol on reaction times. Assuming normality, you randomly sample $n = 13$ observations and compute the sample mean and variance. To determine the probability that $T = (\bar{X} - \mu)/(s/\sqrt{n})$ is less than 2.179, first note that the degrees of freedom are $\nu = n - 1 = 13 - 1 = 12$. From Table 4 in Appendix B, looking at the row with $\nu = 12$, we see 2.179 in the column headed by $t_{.975}$, so $P(T < 2.179) = .975$.   ■

EXAMPLE. If $\nu = 30$ and $P(T > c) = .005$, what is c? Because Table 4 gives the probability that T is less than or equal to some constant, we must convert the present problem into one where Table 4 can be used. Based on Equation (4.10), if $P(T > c) = .005$, then

$$P(T \leq c) = 1 - P(T > c) = 1 - .005 = .995.$$

Looking at the column headed by $t_{.995}$ in Table 4, we see that with $\nu = 30$, $P(T < 2.75) = .995$, so $c = 2.75$.   ■

With Student's T distribution, we can compute a confidence interval for μ when σ is not known, assuming that observations are randomly sampled from a normal distribution. Recall that when σ is known, the $1 - \alpha$ confidence interval for μ is

$$\bar{X} \pm c\frac{\sigma}{\sqrt{n}} = \left(\bar{X} - c\frac{\sigma}{\sqrt{n}}, \bar{X} + c\frac{\sigma}{\sqrt{n}} \right),$$

where c is the $1 - \alpha/2$ quantile of a standard normal distribution and read from Table 1 in Appendix B. When σ is not known, this last equation becomes

$$\bar{X} \pm c\frac{s}{\sqrt{n}}, \tag{4.11}$$

where now c is the $1 - \alpha/2$ quantile of Student's T distribution with $n - 1$ degrees of freedom and read from Table 4 of Appendix B. If observations are randomly sampled from a normal distribution, then the probability coverage is exactly $1 - \alpha$. [The S-PLUS built-in function t.test computes a confidence interval using Equation (4.11).]

EXAMPLE. Returning to the ozone experiment, we compute a $1 - \alpha = .95$ confidence interval for μ. Because there are $n = 22$ rats, the degrees of freedom

Continued

EXAMPLE. (*Continued*) are $n - 1 = 22 - 1 = 21$. Because $1 - \alpha = .95, \alpha = .05$; so $\alpha/2 = .025$, and $1 - \alpha/2 = .975$. Referring to Table 4 in Appendix B, we see that the .975 quantile of Student's T distribution with 22 degrees of freedom is approximately $c = 2.08$. Because $\bar{X} = 11$ and $s = 19$, a .95 confidence interval is

$$11 \pm 2.08 \frac{19}{\sqrt{22}} = (2.6, 19.4).$$

That is, although both the population mean and variance are not known, we can be reasonably certain that the population mean, μ, is between 2.6 and 19.4, if the assumption of sampling from a normal distribution is true.   ■

EXAMPLE. Suppose you are interested in the reading abilities of fourth-graders. A new method for enhancing reading is being considered. You try the new method on 11 students and then administer a reading test yielding the scores

$$12, 20, 34, 45, 34, 36, 37, 50, 11, 32, 29.$$

For illustrative purposes, imagine that after years of using a standard method for teaching reading, the average scores on the reading test have been found to be $\mu = 25$. Someone claims that if the new teaching method is used, the population mean will remain 25. Assuming normality, we determine whether this claim is consistent with the .99 confidence interval for μ. That is, does the .99 confidence interval contain the value 25? It can be seen that the sample mean is $\bar{X} = 30.9$ and $s/\sqrt{11} = 3.7$. Because $n = 11$, the degrees of freedom are $\nu = 11 - 1 = 10$. Because $1 - \alpha = .99$, it can be seen that $1 - \alpha/2 = .995$, so, from Table 4 in Appendix B, $c = 3.169$. Consequently, the .99 confidence interval is

$$30.9 \pm 3.169(3.7) = (19.2, 42.6).$$

This interval contains the value 25, so the claim that $\mu = 25$ cannot be refuted based on the available data. Note, however, that the confidence interval also contains 35 and even 40. Although we cannot rule out the possibility that the mean is 25, there is some possibility that the new teaching method enhances reading by a substantial amount, but with only 11 subjects, the confidence interval is too long to resolve how effective the new method happens to be.   ■

4.8 Student's *T* and Nonnormality

In the next to last example dealing with weight gain in an ozone environment, it was shown that the .95 confidence interval for the mean is (2.6,19.4) *if* observations are randomly sampled from a normal distribution. But can we be reasonably certain that it contains μ if sampling is from a nonnormal distribution instead?

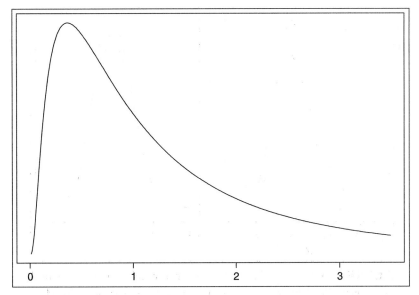

FIGURE 4.14 Example of what is called a lognormal distribution.

Unfortunately, if observations are randomly sampled from a skewed distribution, the actual probability coverage can be less than .7 (e.g., Wilcox, 1997a, p. 74). That is, the confidence interval is too short and there is a $1 - .7 = .3$ probability that it does not contain μ. To get .95 probability coverage, we need a longer interval. If we increase the sample size, the actual probability coverage will be closer to .95, as desired, but probability coverage might be unsatisfactory even with $n = 160$ (e.g., Westfall & Young, 1993, p. 40).

To elaborate a little, imagine that, unknown to us, observations have the distribution shown in Figure 4.14. This is an example of a lognormal distribution, which is skewed to the right and has relatively light tails, meaning that outliers can occur but the number of outliers is relatively low on average. (With $n = 20$, the expected number of outliers using the method in Section 3.4.5 is about 1.4. The median number of outliers is approximately 1.) The symmetric smooth curve in the left plot of Figure 4.15 shows Student's T distribution when $n = 20$ and sampling is from a normal distribution. The other curve shows a close approximation of the actual distribution of T when sampling from the distribution in Figure 4.14 instead. (The approximation is based on 5000 T-values generated on a computer.) Note that the actual distribution is skewed, not symmetric. Moreover, its mean is not zero but $-.5$, approximately. The right plot of Figure 4.15 shows the distribution of T when $n = 100$. There is closer agreement between the two distributions, but the tails of the distributions differ enough that practical problems arise.

It was noted that in the left plot of Figure 4.15, T has a mean of $-.5$. This might seem to be impossible because the numerator of T is $\bar{X} - \mu$, which has an expected value of zero. Under normality, T does indeed have a mean of zero, the proof of which is based on the result that under normality, $\bar{X}$ and s are independent. But for nonnormal distributions, $\bar{X}$ and s are dependent, and this makes it possible for T to

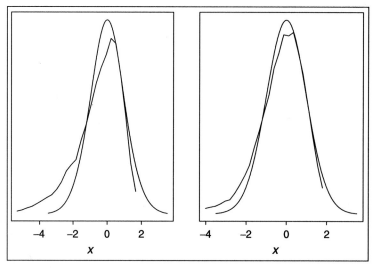

FIGURE 4.15   A plot of 5000 T values when sampling from a lognormal distribution. The left plot is based on $n = 20$ and the right plot is based on $n = 100$.

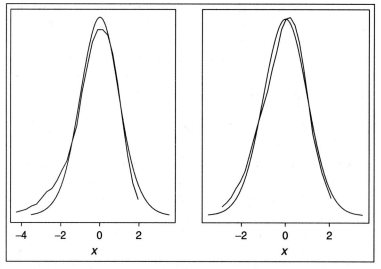

FIGURE 4.16   A plot of 5000 T values when sampling from an exponential distribution. The left plot is with $n = 20$ and the right plot is with $n = 100$.

have a mean that differs from zero. (Gosset was aware of this problem but did not have the tools and technology to study it to the degree he desired.)

It might be thought that the lognormal distribution (shown in Figure 4.14) represents an extreme case and therefore denigrates Student's T in an unfair manner. However, when working with Student's T distribution, other skewed, light-tailed distributions — where outliers are rare — also cause serious practical problems. Consider the (exponential) distribution shown in Figure 4.9. The left panel of Figure 4.16 shows the distribution of Student's T when $n = 20$. Again the left tails of the actual distribution differ noticeably from the distribution we get under normality. For example, under normality, the probability that T is less than -2 is $P(T \leq -2) = .03$, but when sampling from the exponential distribution, it is $.08$. The right panel shows the distribution when $n = 50$. Now, $P(T \leq -2) = .026$ when sampling from a normal distribution, but for the exponential distribution it is $.053$. In some situations this discrepancy is unsatisfactory. Increasing n to 100 gives good results.

Figure 4.16 illustrates another point worth stressing. We saw that when sampling observations from the distribution in Figure 4.9, the sampling distribution of the sample mean is in fairly close agreement with the distribution implied by the central limit theorem. We have just illustrated, however, that this does not imply that the actual distribution of T will be in close agreement with the distribution we get under normality.

When sampling from a skewed heavy-tailed distribution, the discrepancy between the actual distribution of T and the distribution obtained under normality becomes even more striking. If, for example, we sample $n = 20$ observations from the distribution shown in Figure 4.6, the distribution of T is as shown in Figure 4.17 and differs substantially from the distribution we get under normality, particularly in the left tail.

The illustrations so far are based on hypothetical distributions. Experience with actual data suggests that the problems just illustrated are real and, in at least some situations, these theoretical illustrations appear to underestimate problems with Student's T. To describe one reason for this remark, we consider data from a study on hangover symptoms reported by the sons of alcoholics. The 20 observed values were

$$1 \ 0 \ 3 \ 0 \ 3 \ 0 \ 15 \ 0 \ 6 \ 10 \ 1 \ 1 \ 0 \ 2 \ 24 \ 42 \ 0 \ 0 \ 0 \ 2.$$

(These data were supplied by M. Earleywine.) Figure 4.18 shows an approximation of the sampling distribution of T (which was obtained using methods covered

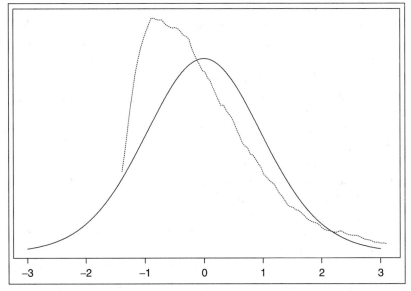

FIGURE 4.17　A plot of 5000 T values when sampling from a skewed, heavy-tailed distribution, $n = 20$.

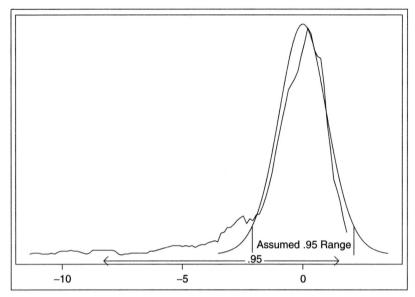

FIGURE 4.18   A plot of 5000 T values when resampling with replacement using data from a study dealing with sons of alcoholics. The smooth symmetric curve is the plot we get when sampling from a normal distribution.

in Chapter 7). Also shown is the distribution of T when sampling from a normal distribution instead. As is evident, there is a considerable discrepancy between the two distributions, particularly in the left tail. The practical implication is that when using T, the actual probability coverage, when computing a confidence interval based on Equation (4.11), might differ substantially from the nominal level.

It might be argued that this last example is somehow unusual or that with a slightly larger sample size, satisfactory probability coverage will be obtained. First note that if we were to repeat an experiment 5000 times, each time computing Student's T based on observations sampled from a normal distribution, a boxplot of the T values would be symmetric about the value zero. Now consider again the data in Table 3.2, where $n = 105$. We can use these observations to approximate the boxplot of T values we would get if this particular study were repeated 5,000 times. (The approximation is based on a method covered in Chapter 7.) Figure 4.19 shows the result. As is evident, there is extreme skewness, indicating that any confidence interval based on Student's T will be highly inaccurate.

An objection to this last illustration is that there is an extreme outlier among the data. Although the data are from an actual study, it might be argued that having such an extreme outlier is a highly rare event. So we repeat the last illustration, but with this extreme outlier removed. Figure 4.20 shows an approximation of the distribution of T. Again we see that there is a substantial difference as compared to the distribution implied by the central limit theorem.

Yet another possible objection to illustrations based on data from actual studies is that although we are trying to determine empirically the correct distribution for T, nevertheless we are approximating the correct distribution, and perhaps the method used to approximate the distribution is itself in error. That is, if were to take millions of

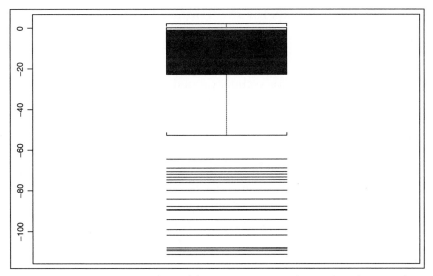

FIGURE 4.19 A boxplot of 5000 *T* values when resampling with replacement from the data in Table 3.2.

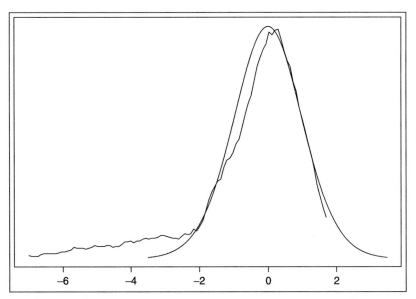

FIGURE 4.20 A plot of 5000 *T* values when resampling with replacement from the data in Table 3.2, but with the extreme outlier removed.

samples from the population under study, each time computing *T*, and if we were then to plot the results, perhaps this plot would better resemble a Student's *T* distribution. There is in fact reason to suspect that the approximation of the distribution of *T* used here is in error, but unfortunately all indications are that problems with *T* are being *underestimated*. For example, in Figure 4.20, it is highly likely that the actual distribution of *T* is more skewed and that the left tail extends even farther to the left than is indicated.

4.9 Confidence Intervals for the Trimmed Mean

There are at least three practical concerns with computing confidence intervals for the mean with Student's T. First, the probability coverage can be unsatisfactory, for reasons explained in the last section. Second, as noted in Chapter 2, there is the concern that when distributions are skewed, the population mean might provide a poor reflection of the typical subject under study. Third, slight departures from normality can greatly inflate the length of the confidence interval, regardless of whether sampling is from a skewed or a symmetric distribution. Theoretical results (e.g., Huber, 1981; Staudte and Sheather, 1990; Wilcox, 1993a) suggest a strategy that addresses all of these concerns: Switch to the Tukey–McLaughlin confidence interval for the population trimmed mean, μ_t (which is described in Section 4.9.3). But before describing this method, we first consider the more fundamental problem of how the standard error of the trimmed mean might be estimated.

4.9.1 Estimating VAR($\bar{X}_t$): A Natural but Incorrect Method

A seemingly natural method for computing a confidence when using a trimmed mean is to apply Student's T method to the values left after trimming. But this strategy is unsatisfactory, because the remaining observations are no longer identically distributed and they are not independent. Consequently, we are using an incorrect estimate of the standard error. We need to use a technique that addresses this problem, otherwise we run the risk of getting an increasingly inaccurate confidence interval as the sample size gets large.

When first encountered, the statement just made might seem counterintuitive. To be sure the problem just described is appreciated, we now elaborate. First we illustrate that if we trim observations, the remaining observations are dependent. Suppose $n = 5$ observations are randomly sampled from a standard normal distribution. We might get the values $X_1 = 1.5$, $X_2 = -1.2$, $X = 3.89$, $X_4 = .4$, and $X_5 = -.6$. Random sampling means that the observations are independent, as explained in Section 4.2. Now suppose we repeat this process 500 times, each time generating five observations but only recording the fourth and fifth values we observe. Figure 4.21 shows a scatterplot of the 500 pairs of points; this is the type of scatterplot we should observe if the observations are independent.

Next, suppose we generate five values from a normal distribution as was done before, but now we put the values in ascending order. As was done in Chapter 3, we label the ordered values $X_{(1)} \leq X_{(2)} \leq X_{(3)} \leq X_{(4)} \leq X_{(5)}$. If we observe $X_1 = 1.5$, $X_2 = -1.2$, $X_3 = .89$, $X_4 = .4$, and $X_5 = -.6$, then $X_{(1)} = -1.2$ is the smallest of the five values, $X_{(2)} = -.6$ is the second smallest, and so on. Now we repeat this process 500 times, each time randomly sampling five observations, but this time we record the two largest values. Figure 4.22 shows the resulting scatterplot for $X_{(4)}$ (the x-axis) versus $X_{(5)}$. There is a discernible pattern because they are dependent.

An important point is that we get dependence when sampling from any distribution, including normal distributions as a special case. If, for example, you are told that

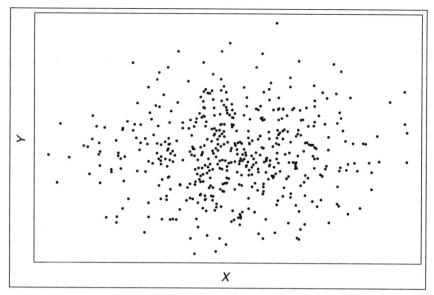

FIGURE 4.21   Five values were randomly sampled from a normal distribution and the fourth and fifth observations were recorded. Repeating this process 500 times yielded the pairs of points shown, which are independent.

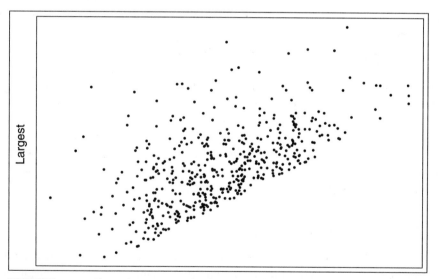

FIGURE 4.22   Five values were randomly sampled from a normal distribution, but now the two largest observations were recorded. Repeating this process 500 times yielded the pairs of points shown, illustrating that the two largest observations are dependent.

$X_{(4)}$ is .89, it follows that the largest value ($X_{(5)}$, still assuming $n = 5$) cannot be .8, .2, 0, or -1; it must be as large or larger than .89. Put another way, if we focus attention on the largest value, there is some probability — greater than zero — that its value is less than .89. In symbols, $P(X_{(5)} < .89) > 0$. But this probability is altered if you are told that $X_{(4)} = .89$; now it is exactly equal to zero. That is, it is impossible for the largest value to be less than .89 if the second-largest value is equal to .89. Said another way, if knowing the value of $X_{(4)}$ alters the range of possible values for $X_{(5)}$, then $X_{(4)}$ and $X_{(5)}$ are dependent. (See the end of Section 2.3.) This argument generalizes: Any two ordered values, say, $X_{(i)}$ and $X_{(j)}$, $i \neq j$, are dependent. Moreover, if we discard even one unusually large or small observation, the remaining observations are dependent.

The point of all this is that the method for determining the variance of the sample mean cannot be used to determine the variance of the trimmed mean. Recall that the derivation of the variance of the sample mean made use of the fact that under random sampling, all pairs of observations have zero correlation. But when we discard extreme observations, the remaining observations are dependent and have nonzero correlations, and this needs to be taken into account when trying to derive an expression for the variance of the trimmed mean. In symbols, we could determine the variance of the trimmed mean if we could determine

$$\mathrm{VAR}(X_{(g+1)} + \cdots + X_{(n-g)}),$$

where $X_{(g+1)} + \cdots + X_{(n-g)}$ is the numerator of the sample trimmed mean as given by Equation (3.5). The difficulty is that the variables in this last equation have nonzero correlations, so this variance is *not* equal to $\mathrm{VAR}(X_{(g+1)}) + \cdots + \mathrm{VAR}(X_{(n-g)})$, the sum of the individual variances. That is, after trimming, it is no longer true that the variance of the sum of the remaining observations is equal to the sum of the variances of the individual observations. (There is also the problem that the observations left after trimming do not have the same variance as the distribution from which they were sampled.)

We conclude this subsection with the following remark. Various studies suggest that outliers and heavy-tailed distributions are common in applied research. A tempting method for dealing with outliers is simply to discard any that are found and then to compute a confidence interval with Student's T using only the data that remain. This strategy is theoretically unsound, however, because, as just indicated, the observations not discarded are dependent, in which case the mathematical derivation of Student's T is no longer valid.

4.9.2 A Theoretically Correct Approach

During the 1960s, some mathematical techniques were developed that provide a convenient and useful method for estimating the variance of a trimmed mean. The theoretical details are described in Huber (1981), but here we merely describe the resulting method. In particular, the variance of the 20% trimmed mean can be estimated with

$$\frac{s_w^2}{.6^2 n}, \tag{4.12}$$

where s_w^2 is the Winsorized sample variance introduced in Chapter 3. The .6 in the denominator is related to the amount of trimming, which is assumed to be 20%.

More generally, for a γ-trimmed mean, the squared standard error is estimated with

$$\frac{s_w^2}{(1 - 2\gamma)^2 n'} \tag{4.13}$$

where now s_w^2 is the γ-Winsorized variance. For example, if 10% trimming is used instead, the .6 in Equation (4.12) is replaced by .8. The standard error of the trimmed mean is estimated with the square root of this last equation.

EXAMPLE. Again consider the weight-gain data for rats and assume 20% trimming is to be used. The sample Winsorized standard deviation can be computed as described in Chapter 3 and is $s_w = 3.927$. The trimmed mean is $\bar{X}_t = 23.27$. Because there are $n = 23$ rats, the estimated standard error of the sample trimmed is

$$\frac{3.927}{.6\sqrt{23}} = 1.4.$$

In contrast, the estimated standard error of the sample mean is $s/\sqrt{n} = 2.25$. Note that the ratio of these two values is $1.4/2.25 = .62$, and this is substantially less than 1 when viewed in light of techniques to be covered. Put another way, it might seem that the trimmed mean would have a larger standard error because only six of the ten observations are used to compute the trimmed mean. In fact the exact opposite is true, and this turns out to be important.   ■

EXAMPLE. The data in Table 4.3, from a study on self-awareness, reflect how long an individual could keep a portion of an apparatus in contact with a specified target. The trimmed mean is $\bar{X}_t = 283$ and its estimated standard error is 56.1. In contrast, the standard error of the sample mean is $s/\sqrt{n} = 136$, a value approximately 2.4 times larger than the sample standard error of the trimmed mean. This difference will be seen to be substantial.   ■

A practical issue is whether using a correct estimate of the standard error of the trimmed mean can make a difference in applied work versus the strategy of applying methods for means to the data left after trimming. That is, if we trim and ignore the

TABLE 4.3 Self-Awareness Data

77	87	88	114	151	210	219	246	253	262
296	299	306	376	428	515	666	1310	2611	

dependence among the remaining values, can this have any practical consequences? The answer is an unequivocal yes, as illustrated in the next example.

EXAMPLE. For the data in Table 3.2, the estimated standard error of the 20% trimmed mean is .532 using Equation (4.12). If we trim 20% and simply use the method for the sample mean on the remaining 63 values (meaning that we compute s using these 63 values only and then compute $s/\sqrt{63}$), we get 0.28, which is less than half of the value based on Equation (4.12). So we see that using a theoretically motivated estimate of the standard error of the trimmed mean, rather than using methods for the sample mean based on data not trimmed, is not an academic matter. The incorrect estimate can differ substantially from the estimate based on theory.   ■

4.9.3 A Confidence Interval for the Population Trimmed Mean

As was the case when working with the population mean, we want to know how well the sample trimmed mean, $\bar{X}_t$, estimates the population trimmed mean, μ_t. What is needed is a method for computing a $1 - \alpha$ confidence interval for μ_t. A solution was derived by Tukey and McLaughlin (1963) and is computed as follows. Let h be the number of observations left after trimming, as described in Chapter 3. Let c be the $1 - \alpha/2$ quantile of the Student's T distribution with $h - 1$ degrees of freedom and let s_w be the Winsorized sample standard deviation, which is also described in Chapter 3. A confidence interval for the γ-trimmed mean is

$$\left(\bar{X}_t - c \frac{s_w}{(1 - 2\gamma)\sqrt{n}}, \bar{X}_t + c \frac{s_w}{(1 - 2\gamma)\sqrt{n}} \right). \tag{4.14}$$

So for the special case of 20% trimming, a $1 - \alpha$ confidence interval is given by

$$\left(\bar{X}_t - c \frac{s_w}{.6\sqrt{n}}, \bar{X}_t + c \frac{s_w}{.6\sqrt{n}} \right). \tag{4.15}$$

In terms of probability coverage, we get reasonably accurate confidence intervals for a much broader range of nonnormal distributions versus confidence intervals for μ based on Student's T.

Section 4.4 provided one reason why a trimmed mean can be a more accurate estimate of the population mean when sampling from a symmetric distribution. The method for computing a confidence interval just described provides another perspective and explanation. We saw in Chapter 3 that generally the Winsorized standard deviation, s_w, can be substantially smaller than the standard deviation s. Consequently, a confidence interval based on a trimmed mean can be substantially shorter. However, when computing a confidence interval based on 20% trimming, for example, the estimate of the standard error of the trimmed mean is $s_w/(.6\sqrt{n})$. Because $s_w/.6$ can be greater than s, such as when sampling from a normal distribution, it is possible to get a shorter confidence interval using means. Generally, however, any improvement achieved with the mean is small, but substantial improvements based on a trimmed mean are often possible.

EXAMPLE. Suppose a test of open-mindedness is administered to $n = 10$ subjects, yielding the observations

$$5, 60, 43, 56, 32, 43, 47, 79, 39, 41.$$

We compute a .95 confidence interval for the 20% trimmed mean and compare the results to the confidence interval for the mean. With $n = 10$, the number of trimmed observations is four, as explained in Chapter 3. That is, the two largest and two smallest observations are removed, leaving $h = 6$ observations, and the average of the remaining observations is the trimmed mean, $\bar{X}_t = 44.8$. The mean using all 10 observations is $\bar{X} = 44.5$. This suggests that there might be little difference between the population mean, μ, and the population trimmed mean, μ_t. With $\nu = 6 - 1 = 5$ degrees of freedom, Table 4 in Appendix B indicates that the .975 quantile of Student's T distribution is $c = 2.57$. It can be seen that the Winsorized sample variance is $s_w^2 = 54.54$, so $s_w = \sqrt{54.54} = 7.385$, and the resulting confidence interval for the trimmed mean is

$$44.8 \pm 2.57 \frac{7.385}{.6\sqrt{10}} = (34.8, 54.8).$$

In contrast, the .95 confidence interval for the mean is $(30.7, 58.3)$. The ratio of the lengths of the confidence intervals is

$$\frac{54.8 - 34.8}{58.3 - 30.7} = .72.$$

That is, the length of the confidence interval based on the trimmed mean is substantially shorter.   ■

In the previous example, a boxplot of the data reveals that there is an outlier. This explains why the confidence interval for the mean is longer than the confidence interval for the trimmed mean: The outlier inflates the sample variance, s^2, but has no effect on the Winsorized sample variance, s_w^2. Yet another method for trying to salvage means is to check for outliers, and if none are found, compute a confidence interval for the mean. Recall, however, that even when sampling from a skewed light-tailed distribution, the distribution of T can differ substantially from the case where observations are normal. This means that even though no outliers are detected, when computing a .95 confidence interval for μ, the actual probability coverage could be substantially smaller than intended unless the sample size is reasonably large. When attention is turned to comparing multiple groups of subjects, this problem becomes exacerbated, as will be seen. Modern theoretical results tell us that trimmed means reduce this problem substantially.

TABLE 4.4   Average LSAT Scores for 15 Law Schools

545	555	558	572	575	576	578	580
594	605	635	651	653	661	666	

EXAMPLE. Table 4.4 shows the average LSAT scores for the 1973 entering classes of 15 American law schools. (LSAT is a national test for prospective lawyers.) The sample mean is $\bar{X} = 600.3$ with an estimated standard error of 10.8. The 20% trimmed mean is $\bar{X}_t = 596.2$ with an estimated standard error of 14.92. The .95 confidence interval for μ_t is (561.8, 630.6). In contrast, the .95 confidence interval for μ is (577.1, 623.4), assuming T does indeed have a Student's T distribution. Note that the length of the confidence interval for μ is smaller and, in fact, is a subset of the confidence interval for μ_t. This might suggest that the sample mean is preferable to the trimmed mean for this particular set of data, but closer examination suggests that this might not be true. The concern here is the claim that the confidence interval for the mean has probability coverage .95. If sampling is from a light-tailed, skewed distribution, the actual probability coverage for the sample mean can be substantially smaller than the nominal level. Figure 4.23 shows a boxplot of the data, indicating that the central portion of the data is skewed to the right. Moreover, there are no outliers, suggesting the possibility that sampling is from a relatively light-tailed distribution. Thus, the actual probability coverage of the confidence interval for the mean might be too low—a longer confidence interval might be needed to achieve .95 probability coverage. That is, an unfair comparison of the two confidence intervals has probably been made because they do not have the same probability coverage. If we were able to compute a .95 confidence interval for the mean, there is some possibility that it would be longer than the confidence interval for the trimmed mean. When sampling from a skewed, heavy-tailed distribution, problems with the mean can be exacerbated (as illustrated in Chapter 7).   ■

The confidence interval for the 20% trimmed mean given by Equation (4.15) assumes that

$$T_t = \frac{.6(\bar{X}_t - \mu_t)}{s_w/\sqrt{n}}$$

has a Student's T distribution with $h - 1 = n - 2g - 1$ degrees of freedom, where $g = [.2n]$ and $[.2n]$ is $.2n$ rounded down to the nearest integer. To graphically illustrate how nonnormality affects this assumption, we repeat the method used to create the left panel of Figure 4.15. That is, we sample $n = 20$ observations from the (lognormal) distribution in Figure 4.14, compute T_t, and repeat this 5000 times, yielding 5000 T_t values. The left panel of Figure 4.24 shows a plot of these T_t values versus Student's T with 11 degrees of freedom. To facilitate comparisons, the right panel shows a plot of 5000 T values (based on the mean and variance) versus Student's T distribution

FIGURE 4.23   A boxplot of the data in Table 4.4.

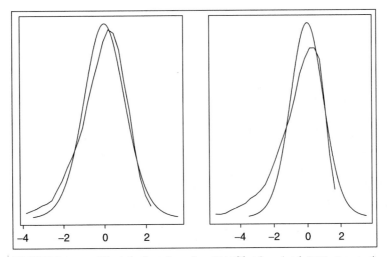

FIGURE 4.24   The left plot is based on 5000 T_t values (with 20% trimming) when sampling from a lognormal distribution. The right plot is based on 5000 T values (no trimming).

with 19 degrees of freedom. Student's T distribution gives a better approximation of the actual distribution of T_t versus T. Of particular importance is that the tails of the distribution are better approximated when using a trimmed mean. This indicates that more accurate probability coverage will be achieved using a 20% trimmed mean versus using the mean with no trimming. Generally, as the sample size increases, problems with nonnormality diminish more rapidly when using a trimmed mean versus using a mean. But switching to a trimmed mean does not eliminate all practical problems when sample sizes are small. Fortunately there are methods for getting even more accurate results, as we shall see.

4.9.4 S-PLUS Functions trimse and trimci

The S-PLUS function

$$\text{trimse}(x, tr=.2, alpha=.05)$$

has been supplied that computes an estimate of the standard error of the trimmed mean for the data stored in any S-PLUS variable x. The default amount of trimming (tr) is .2. The function

$$\text{trimci}(x, tr=.2, alpha=.05)$$

computes a $1 - \alpha$ confidence interval for μ_t. If the argument alpha is unspecified, $\alpha = .05$ is used.

EXAMPLE. If the data in Table 4.4 are stored in the S-PLUS variable blob, the command trimci(blob) returns a .95 confidence interval for the 20% trimmed mean: (561.8, 630.6). The command trimci(blob,tr=0,alpha=.01) returns a .99 confidence interval for the population mean using Student's T method described in Section 4.7.   ■

4.10 Transforming Data

For completeness, it should be remarked that simple transformations of data are often recommended for dealing with problems due to nonnormality. For example, a common recommendation is to replace all the values in a study with their logarithm and then to use methods based on means. Another common strategy when all observations are positive is to replace each value by its square root. These simple transformations can be useful for certain purposes, but for the situations covered in this book they are not always satisfactory. One problem is that they do not eliminate the deleterious effect of outliers (e.g., Rasmussen, 1989). Transformations can make a distribution appear to be reasonably normal, at least in some cases, but when using simple transformations it has been found that trimming is still beneficial (Doksum & Wong, 1983). For this reason, simple transformations of data are not discussed.

4.11 Confidence Interval for the Population Median

Although the median is a member of the class of trimmed means, the method for computing a confidence interval for the trimmed mean gives absurd results for the extreme case of the median. One way of dealing with this problem is to use a result derived by Laplace, which gives an expression for the variance of the median that depends in part on $f(x)$, the equation for the probability density function from which observations are randomly sampled. Details of this approach are covered in Wilcox (1997a) but not here. Instead we rely on the method outlined in Box 4.1, which was suggested by Hettmansperger and Sheather (1986). Results supporting the use of this method are reported by Sheather and McKean (1987) as well as Hall and Sheather (1988).

BOX 4.1 How to Compute a Confidence Interval for the Median

As indicated by Equation (2.9), for some p, $0 \leq p \leq 1$, the binomial probability function is

$$f(x) = \binom{n}{x} p^x (1 - p)^{n-x}, x = 0, \ldots, n.$$

For any integer k between 0 and $n/2$, let

$$\gamma_k = f(k) + f(k + 1) + \cdots + f(n - k)$$

when $p = .5$.
  Then

$$(X_{(k)}, X_{(n-k+1)}) \tag{4.16}$$

is a confidence interval for the median that has probability coverage exactly equal to γ_k.
  Because the binomial distribution is discrete, it is not possible, in general, to choose k so that the probability coverage is exactly equal to $1 - \alpha$. To get a $1 - \alpha$ confidence interval, first determine k such that $\gamma_{k+1} < 1 - \alpha < \gamma_k$. Next, compute

$$I = \frac{\gamma_k - 1 - \alpha}{\gamma_k - \gamma_{k+1}} \quad \text{and} \quad \lambda = \frac{(n - k)I}{k + (n - 2k)I}.$$

Then an approximate $1 - \alpha$ confidence interval is

$$(\lambda X_{(k+1)} + (1 - \lambda)X_{(k)}, \lambda X_{(n-k)} + (1 - \lambda)X_{(n-k+1)}). \tag{4.17}$$

4.11.1 S-PLUS Function sint

The S-PLUS function

$$\text{sint}(x, \text{alpha}=.05)$$

has been supplied for applying the method in Box 4.1. As usual, x is any S-PLUS variable containing data.

EXAMPLE. Staudte and Sheather (1990) illustrate the use of Equation (4.17) with data from a study on the lifetimes of EMT6 cells. The values are 10.4, 10.9, 8.8, 7.8, 9.5, 10.4, 8.4, 9.0, 22.2, 8.5, 9.1, 8.9, 10.5, 8.7, 10.4, 9.8, 7.7, 8.2, 10.3, 9.1. Storing these values in the S-PLUS variable blob, the command sint(blob) returns a .95 confidence interval of (8.72, 10.38). The command sint(blob,.01) returns a .99 confidence interval of (8.5, 10.4). ■

4.11.2 Estimating the Standard Error of the Sample Median

The method just described for computing a confidence interval for the median does not require an estimate of the standard error of M, the sample median. However, for situations to be covered, an explicit estimate will be needed. Many strategies have been proposed, comparisons of which can be found in Price and Bonett (2001). Here, the method derived by McKean and Schrader (1984) is described because it is very simple and currently appears to have practical value for problems addressed in subsequent chapters.

Compute

$$k = \frac{n+1}{2} - z_{.995}\sqrt{\frac{n}{4}},$$

where k is rounded to the nearest integer and $z_{.995}$ is the .995 quantile of a standard normal distribution. Put the observed values in ascending order, yielding $X_{(1)} \leq \cdots \leq X_{(n)}$. Then the McKean–Schrader estimate of the squared standard error of M is

$$\left(\frac{X_{(n-k+1)} - X_{(k)}}{2z_{.995}}\right)^2.$$

4.11.3 S-PLUS Function msmedse

The S-PLUS function

$$\text{msmedse}(x)$$

computes the estimated standard error of M given by the square root of the last equation.

4.12 A Remark About MOM and M-Estimators

How do we compute a confidence interval based on an M-estimator of location or the MOM estimator in Section 3.5.2? An expression for the standard error of an M-estimator has been derived and can be used to derive a confidence interval using methods similar to those described in this chapter. But for $n < 100$ it yields a reasonably accurate confidence interval only when sampling from a perfectly symmetric distribution. With n sufficiently large, an accurate confidence interval can be computed when sampling from a skewed distribution, but it remains unclear just how large n must be. As for MOM, no expression for the standard error has been derived. However, Chapter 7 describes an effective method for estimating its standard and computing accurate confidence intervals.

4.13 Confidence Intervals for the Probability of Success

Section 2.5 introduced the binomial probability function, where p represents the probability of success and x represents the number of successes among n randomly sampled observations. As was noted, the usual estimate of p is simply

$$\hat{p} = \frac{x}{n},$$

the proportion of successes among the n observations.

Results in Chapter 2 plus the central limit theorem suggest a simple method for computing a $1 - \alpha$ confidence interval for p:

$$\hat{p} \pm c\sqrt{\frac{p(1-p)}{n}},$$

where c is the $1 - \alpha/2$ quantile of a standard normal distribution. We do not know the value of the quantity under the radical, but it can be estimated with $\hat{p}$, in which case a simple $1 - \alpha$ confidence interval for p is

$$\hat{p} \pm c\sqrt{\frac{\hat{p}(1-\hat{p})}{n}}. \tag{4.18}$$

The resulting probability coverage will be reasonably close to $1 - \alpha$ if n is not too small and p is not too close to zero or 1. Just how large n must be depends on how close p is to zero or 1. An obvious concern is that we do not know p, so there is some difficulty in deciding whether n is sufficiently large. Numerous methods have been proposed for dealing with this issue. For the special cases $x = 0, 1, n - 1$, and n, Blyth (1986) suggests proceeding as follows:

- If $x = 0$, use

$$(0, 1 - \alpha^{1/n}).$$

- If $x = 1$, use

$$\left(1 - \left(1 - \frac{\alpha}{2}\right)^{1/n}, 1 - \left(\frac{\alpha}{2}\right)^{1/n}\right).$$

- If $x = n - 1$, use

$$\left(\left(\frac{\alpha}{2}\right)^{1/n}, \left(1 - \frac{\alpha}{2}\right)^{1/n}\right).$$

- If $x = n$, use

$$(\alpha^{1/n}, 1).$$

For all other situations, Blyth's comparisons of various methods suggest using Pratt's (1968) approximate confidence interval, which is computed as shown in Box 4.2.

BOX 4.2 Computing a $1 - \alpha$ Confidence Interval for p Based on x

Successes Among n Trials

Let c be the $1 - \alpha/2$ quantile of a standard normal distribution read from Table 1 in Appendix B. That is, if Z is a standard normal random variable, then $P(Z \leq c) = 1 - \alpha/2$. To determine c_U, the upper end of the confidence interval, compute

$$A = \left(\frac{x+1}{n-x}\right)^2$$

$$B = 81(x+1)(n-x) - 9n - 8$$

$$C = -3c\sqrt{9(x+1)(n-x)(9n+5-c^2)+n+1}$$

$$D = 81(x+1)^2 - 9(x+1)(2+c^2) + 1$$

$$E = 1 + A\left(\frac{B+C}{D}\right)^3$$

$$c_U = \frac{1}{E}.$$

To get the lower end of the confidence interval, c_L, compute

$$A = \left(\frac{x}{n-x-1}\right)^2$$

$$B = 81(x)(n-x-1) - 9n - 8$$

$$C = 3c\sqrt{9x(n-x-1)(9n+5-c^2)+n+1}$$

$$D = 81x^2 - 9x(2+c^2) + 1$$

$$E = 1 + A\left(\frac{B+C}{D}\right)^3$$

$$c_L = \frac{1}{E}.$$

An approximate $1 - \alpha$ confidence interval for p is

$$(c_L, c_U).$$

4.13.1 S-PLUS Function binomci

The S-PLUS function

binomci(x=sum(y), nn=length(y), y=NA, n=NA, alpha=0.05)

has been supplied to compute Pratt's approximate confidence interval for p. In the event $x = 0, 1, n - 1$, or n, Blyth's method is used instead. The first argument, x, is the number of successes and the second argument, nn, indicates the value of n, the number of observations. If the data are stored as a vector of 1's and 0's in some S-PLUS variable, where a 1 indicates a success and 0 a failure, use the third argument, y. (Generally, the fourth argument can be ignored.)

EXAMPLE. The command

$$binomci(5, 25)$$

returns $(0.07, 0.41)$ as a .95 confidence interval for p based on five successes among 25 observations. If the values 1, 1, 1, 0, 0, 1, 1, 0, 0, 0, 0, 1 are stored in the S-PLUS variable obs, the command

$$binomci(y=obs)$$

returns $(0.25, 0.79)$ as a .95 confidence interval for p.   ■

4.14 Exercises

1. Explain the meaning of a .95 confidence interval.
2. If you want to compute a .80, .92, or .98 confidence interval for μ when σ is known, and sampling is from a normal distribution, what values for c should you use in Equation (4.8)?
3. Assuming random sampling is from a normal distribution with standard deviation $\sigma = 5$, if you get a sample mean of $\bar{X} = 45$ based on $n = 25$ subjects, what is the .95 confidence interval for μ?
4. Repeat the previous example, but compute a .99 confidence interval instead.
5. A manufacturer claims that their light bulbs have an average life span of $\mu = 1200$ hours with a standard deviation of $\sigma = 25$. If you randomly test 36 light bulbs and find that their average lifespan is $\bar{X} = 1150$, does a .95 confidence interval for μ suggest that the claim $\mu = 1200$ is unreasonable?
6. Compute a .95 confidence interval for the mean in the following situations: (a) $n = 12$, $\sigma = 22$, $\bar{X} = 65$, (b) $n = 22$, $\sigma = 10$, $\bar{X} = 185$, (c) $n = 50$, $\sigma = 30$, $\bar{X} = 19$.
7. Describe the two components of a random sample.
8. If $n = 10$ observations are randomly sampled from a distribution with mean $\mu = 9$ and variance $\sigma^2 = 8$, what is the mean and variance of the sample mean?
9. Determine $E(\bar{X})$ and $\sigma_{\bar{X}}^2$ for a random sample of $n = 12$ observations from a discrete distribution with the following probability function:

x:	1	2	3	4
$p(x)$:	.2	.1	.5	.2

10. In Exercise 9, again suppose you sample $n = 12$ subjects and compute the sample mean. If you repeat this process 1000 times, each time using $n = 12$ subjects, and if you averaged the resulting 1000 sample means, approximately what would be the result? That is, approximate the average of the 1000 sample means.

11. Answer the same question posed in Exercise 10, except replace means with sample variances.

12. Estimate the variance and standard error of the sample mean for a random sample of $n = 8$ subjects from whom you get

$$2, 6, 10, 1, 15, 22, 11, 29.$$

13. If you randomly sample a single observation and get 32, what is the estimate of the population mean, μ? Can you get an estimate of the squared standard error? Explain, in terms of the squared standard error, why only a single observation is likely to be a less accurate estimate of μ versus a sample mean based on $n = 15$ subjects.

14. As part of a health study, a researcher wants to know the average daily intake of vitamin E for the typical adult. Suppose that for $n = 12$ adults, the intake is found to be

$$450, 12, 52, 80, 600, 93, 43, 59, 1000, 102, 98, 43.$$

Estimate the squared standard error of the sample mean.

15. In Exercise 14, verify that there are outliers. Based on results in Chapter 2, what are the effects of these outliers on the estimated squared standard error?

16. Estimate the variance and standard error of the sample mean when you randomly sample $n = 8$ subjects and get

$$2, 6, 10, 1, 15, 22, 11, 29.$$

17. In Exercise 16, if the observations are dependent, can you still estimate the standard error of the sample mean?

18. Section 2.7 described a mixed normal distribution that differs only slightly from a standard normal. Suppose we randomly sample $n = 25$ observations from a standard normal distribution. Then the squared standard error of the sample mean is 1/25. Referring back to Section 2.7, what is the squared standard error if sampling is from the mixed normal instead? What does this indicate about what might happen under slight departures from normality?

19. Explain why knowing the mean and squared standard error is not enough to determine the distribution of the sample mean. Relate your answer to results on nonnormality described in Section 2.7.

20. Suppose $n = 16$, $\sigma = 2$, and $\mu = 30$. Assume normality and determine (a) $P(\bar{X} < 29)$, (b) $P(\bar{X} > 30.5)$, (c) $P(29 < \bar{X} < 31)$.

21. Suppose $n = 25$, $\sigma = 5$, and $\mu = 5$. Assume normality and determine (a) $P(\bar{X} < 4)$, (b) $P(\bar{X} > 7)$, (c) $P(3 < \bar{X} < 7)$.

22. Someone claims that within a certain neighborhood, the average cost of a house is $\mu = \$100,000$ with a standard deviation of $\sigma = \$10,000$. Suppose that based on $n = 16$ homes, you find that the average cost of a house is

$\bar{X} = \$95,000$. Assuming normality, what is the probability of getting a sample mean this low or lower if the claims about the mean and standard deviation are true?

23. In Exercise 22, what is the probability of getting a sample mean between $97,500 and $102,500?

24. A company claims that the premiums paid by its clients for auto insurance have a normal distribution with mean $\mu = \$750$ and standard deviation $\sigma = \$100$. Assuming normality, what is the probability that for $n = 9$ randomly sampled clients, the sample mean will have a value between $700 and $800?

25. You sample 16 observations from a discrete distribution with mean $\mu = 36$ and variance $\sigma^2 = 25$. Use the central limit theorem to determine (a) $P(\bar{X} < 34)$, (b) $P(\bar{X} < 37)$, (c) $P(\bar{X} > 33)$, (d) $P(34 < \bar{X} < 37)$.

26. You sample 25 observations from a nonnormal distribution with mean $\mu = 25$ and variance $\sigma^2 = 9$. Use the central limit theorem to determine (a) $P(\bar{X} < 24)$, (b) $P(\bar{X} < 26)$, (c) $P(\bar{X} > 24)$, (d) $P(24 < \bar{X} < 26)$.

27. Describe a situation where Equation (4.11), used in conjunction with the central limit theorem, might yield a relatively long confidence interval.

28. Describe a type of continuous distribution where the central limit theorem gives good results with small sample sizes.

29. Compute a .95 confidence interval if (a) $n = 10$, $\bar{X} = 26$, $s = 9$, (b) $n = 18$, $\bar{X} = 132$, $s = 20$, (c) $n = 25$, $\bar{X} = 52$, $s = 12$.

30. Repeat Exercise 29, but compute a .99 confidence interval instead.

31. Table 4.3 reports data from a study on self-awareness. Compute a .95 confidence interval for the mean.

32. Rats are subjected to a drug that might affect aggression. Suppose that for a random sample of rats, measures of aggression are found to be

$$5, 12, 23, 24, 18, 9, 18, 11, 36, 15.$$

Compute a .95 confidence for the mean assuming the scores are from a normal distribution.

33. Describe in general terms how nonnormality can affect Student's T distribution.

34. When sampling from a light-tailed, skewed distribution, where outliers are rare, a small sample size is needed to get good probability coverage, via the central limit theorem, when the variance is known. How does this contrast with the situation where the variance is not known and confidence intervals are computed using Student's T distribution?

35. Compute a .95 confidence for the trimmed mean if (a) $n = 24$, $s_w^2 = 12$, $\bar{X}_t = 52$, (b) $n = 36$, $s_w^2 = 30$, $\bar{X}_t = 10$, (c) $n = 12$, $s_w^2 = 9$, $\bar{X}_t = 16$.

36. Repeat Exercise 35, but compute a .99 confidence interval instead.

37. Compute a .95 confidence interval for the 20% trimmed mean using the data in Table 4.3.

38. Compare the length of the confidence interval in Exercise 37 to the length of the confidence interval for the mean you got in Exercise 31. Comment on why they differ.

39. In a portion of a study of self-awareness, Dana observed the values

$$59, 106, 174, 207, 219, 237, 313, 365, 458, 497, 515,$$

$$529, 557, 615, 625, 645, 973, 1065, 3215.$$

Compare the lengths of the confidence intervals based on the mean and 20% trimmed mean. Why is the latter confidence interval shorter?

40. The ideal estimator of location would have a smaller standard error than any other estimator we might use. Explain why such an estimator does not exist.

41. Under normality, the sample mean has a smaller standard error than the trimmed mean or median. If observations are sampled from a distribution that appears to be normal, does this suggest that the mean should be preferred over the trimmed mean and median?

42. Chapter 3 reported data on the number of seeds in 29 pumpkins. The results were

$$250, 220, 281, 247, 230, 209, 240, 160, 370, 274, 210, 204, 243, 251, 190,$$
$$200, 130, 150, 177, 475, 221, 350, 224, 163, 272, 236, 200, 171, 98.$$

The 20% trimmed mean is $\bar{X}_t = 220.8$ and the mean is $\bar{X} = 229.2$. Verify that the .95 confidence interval for μ is (200.7, 257.6) and that for the trimmed mean, μ_t, it is (196.7, 244.9).

43. In Exercise 42, the length of the confidence interval for μ is $257.6 - 200.7 = 56.9$ and the length based on the trimmed mean is $244.9 - 196.7 = 48.2$. Comment on why the length of the confidence interval for the trimmed mean is shorter.

44. If the mean and trimmed mean are nearly identical, it might be thought that it makes little difference which measure of location is used. Based on your answer to Exercise 43, why might it make a difference?

45. For the past 16 presidential elections in the United States, the incumbent party won or lost the election depending on whether the Washington Redskins, American football team, won their last game just prior to the election. That is, there has been perfect agreement between the two events during the last 16 elections. Verify that according to Blyth's method, a .99 confidence for the probability of agreement is (.75, 1).

46. An ABC news program reported that a standard method for rendering patients unconscious led patients to wake up during surgery. These individuals were not only aware of their plight, they suffered from nightmares later on. Some physicians tried monitoring brain function during surgery to avoid this problem, the strategy being to give patients more medication if they showed signs of regaining consciousness, and they found that among 200,000 trials, no patients woke during surgery. However, administrators concerned about cost argued that with only 200,000 trials, the probability of waking up using the new method could not be accurately estimated. Verify that a .95 confidence interval for p, the probability of waking up, is (0, .000015).

5

HYPOTHESIS TESTING

In applied research, it is common to make some speculation about the population mean and then to try to assess whether this speculation is reasonable based on the data that are available. Roughly, if the likelihood of an observed value for the sample mean is small based on an assumed value for the population mean, then perhaps the assumed value of the population mean is incorrect. Another possibility is that the sample mean does not accurately reflect the population mean. That is, the speculation about the population mean is correct, but by chance the sample mean differs substantially from the population mean.

As an example, imagine a researcher who claims that on a test of open-mindedness, the population mean (μ) for adult men is 50. Suppose you randomly sample $n = 10$ adult males, give them the test for open-mindedness, and get the scores

$$25, 60, 43, 56, 32, 43, 47, 59, 39, 41.$$

The sample mean is $\bar{X} = 44.5$. Does this make the claim $\mu = 50$ unreasonable? Do the data support the claim? If in reality $\mu = 50$, what is the probability that you will get a sample mean less than 45?

Chapter 4 touched on how you might decide whether the claim $\mu = 50$ is reasonably consistent with the 10 open-mindedness scores just given. If the .95 confidence interval for μ happens to be (40, 48), then this interval does not contain 50, suggesting that the claim $\mu = 50$ is not reasonable. If the .95 confidence interval is (46, 52), this interval contains 50, which suggests that the claim $\mu = 50$ should not be ruled out. The purpose of this chapter is to expand on the topic of making decisions about whether some claim about the population mean or some other parameter of interest is consistent with data. As usual, we begin by describing basic concepts and techniques typically covered in an introductory statistics course. Then we describe modern insights into when and why the standard method based on the mean might be highly unsatisfactory and how these practical problems might be addressed.

5.1 The Basics of Hypothesis Testing

We continue the illustration described in the introduction to this chapter, but for convenience we first consider a situation where it is claimed that the population mean is greater than or equal to 50. A typical way of writing this claim more succinctly is

$$H_0 : \mu \geq 50,$$

where the notation H_0 is read "H naught." This last expression is an example of what is called a *null hypothesis*. A null hypothesis is just a statement — some speculation — about some characteristic of a distribution. In the example, the null hypothesis is a speculation about the population mean, but it could just as easily be some speculation about the population median or trimmed mean. If someone claims that the mean is greater than or equal to 60, then our null hypothesis would be written as $H_0 : \mu \geq 60$. If there is some reason to speculate that $\mu \leq 20$, and the goal is to see whether this speculation is consistent with observations we make, then the null hypothesis is $H_0 : \mu \leq 20$.

The goal of statistical hypothesis testing is to find a decision rule about whether the null hypothesis is true, or should be ruled out, based on observations we make. When the null hypothesis is rejected, this means you decide that the corresponding alternative hypothesis is accepted. For example, if the null hypothesis is $H_0 : \mu \geq 50$, the *alternative hypothesis* is typically written as

$$H_1 : \mu < 50,$$

and if you reject H_0, you in effect accept H_1. That is, you conclude that the mean is less than 50 based on the data available in your study.

Suppose we sample some adult men, measure their open-mindedness, and find that the resulting sample mean is $\bar{X} = 61$. Thus, our estimate of the population mean μ is 61, which is consistent with the null hypothesis $H_0 : \mu \geq 50$, so there is no empirical evidence to doubt the claim that the population mean is greater than or equal to 50. For the data given at the beginning of this chapter, $\bar{X} = 44.5$. That is, μ is estimated to be less than 50, which suggests that the null hypothesis is false and should be rejected. But if it were true that $\mu = 50$, then there is some possibility of observing a sample mean less than or equal to 44.5. That is, if we reject the null hypothesis and conclude that μ is less than 50 based on this observed sample mean, there is some possibility that our decision is in error.

DEFINITION. A *Type I error* refers to a particular type of mistake, namely, rejecting the null hypothesis when in fact it is correct. A common notation for the probability of a Type I error is α, which is often referred to as the *level of significance*.

We can avoid a Type I error by never rejecting the null hypothesis. In this case, $\alpha = 0$, meaning that the probability of erroneously rejecting the null hypothesis is zero. But a problem with this rule is that it is impossible to discover situations where indeed the null hypothesis is false. If in our illustration $\mu = 46$, then the null hypothesis is false and we want a method that will detect this. That is, we need a rule that allows the possibility of rejecting, but simultaneously we want to control the probability of a Type I error.

A natural strategy is to try to determine how small the sample mean must be before we reject the hypothesis that μ is greater than or equal to 50. But rather than work with $\bar{X}$, it is more convenient to work with

$$Z = \frac{\bar{X} - 50}{\sigma/\sqrt{n}},$$

where for the moment we assume the population standard deviation (σ) is known. Using Z is convenient because when sampling from a normal distribution, it provides a simple method for controlling the probability of a Type I error.

For illustrative purposes, temporarily assume $\sigma = 12$ and $n = 10$. So if $\bar{X} = 50$, then $Z = 0$. If $\bar{X} = 49$, then $Z = -0.26$; and if $\bar{X} = 48$, then $Z = -0.53$. That is, as the sample mean decreases and moves further away from the null hypothesis, Z decreases as well. But *if* the null hypothesis is true and in fact $\mu = 50$, then the assumption that observations are sampled from a normal distribution implies that Z has a standard normal distribution. This in turn provides a simple way of controlling the probability of making a Type I error. For example, if the null hypothesis is true and $\mu = 50$, then from Table 1 in Appendix B we see that $P(Z \leq -1.645) = .05$. So if we reject when $Z \leq -1.645$, the probability of a Type I error is $\alpha = .05$.

EXAMPLE. For the hypothesis $H_0 : \mu \geq 50$, imagine that, based on $n = 10$ subjects, with $\sigma = 12$, you find that the sample mean is $\bar{X} = 48$. Then, as previously indicated, $Z = -0.53$. If you want the probability of a Type I error to be .05, then you should reject the null hypothesis only if Z is less than or equal to -1.645. Because -0.53 is greater than -1.645, you fail to reject. That is, the sample mean is less than 50, but you do not have convincing evidence for ruling out the possibility that the population mean is greater than or equal to 50. This does *not* mean, however, that it is reasonable to accept H_0 and conclude that $\mu \geq 50$. (This issue is elaborated upon in Section 5.2.)   ▪

Figure 5.1 illustrates the decision rule just described. If the null hypothesis is true, and in particular $\mu = 50$, then Z has a standard normal distribution as shown in Figure 5.1, in which case the probability that Z is less than -1.645 is the area of the shaded region, which is .05. In summary, if you assume the null hypothesis is true, are willing to have a Type I error probability of .05, and $Z \leq -1.645$, then this suggests that the assumption $\mu \geq 50$ is not reasonably consistent with empirical evidence and should be rejected.

Recall from Chapter 2 that $P(Z \leq -1.96) = .025$. This means that if we reject $H_0 : \mu \geq 50$ when $Z \leq -1.96$, then the probability of a Type I error is .025 if it happens to be the case that $\mu = 50$. In a similar manner, $P(Z \leq -2.58) = .005$, and if we reject when $Z \leq -2.58$, then the probability of a Type I error is .005. A *critical value* is the value used to determine whether the null hypothesis should be rejected. If it is desired to have a Type I error probability of .05 when testing $H_0 : \mu \geq 50$, the critical value is -1.645, meaning that you reject if $Z \leq -1.645$ (assuming normality and that σ is known). The set of all Z values such that $Z \leq -1.645$ is called the *critical region*; it corresponds to the shaded region in Figure 5.1. If you want the probability of a Type I error to be $\alpha = .025$, then the critical value is -1.96 and the critical region

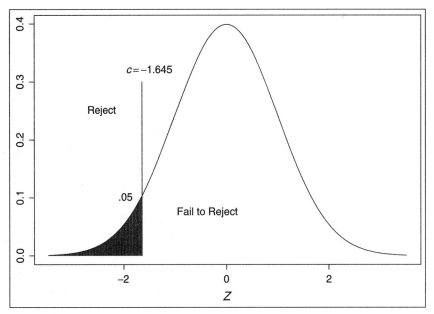

FIGURE 5.1 Graphical depiction of a decision rule. Here you reject when $Z \leq -1.645$ and fail to reject otherwise. The shaded region corresponds to the probability of a Type I error, which is .05.

consists of all Z values less than or equal to -1.96. If $\alpha = .005$, then the critical value is -2.58 and the critical region is the set of all Z values less than -2.58.

EXAMPLE. Continuing the illustration regarding open-mindedness, where the goal is to test $H_0 : \mu \geq 50$, suppose we want the probability of a Type I error to be .025, in which case the critical value is -1.96, the .025 quantile of a standard normal distribution. By assumption, $\sigma = 12$, there are $n = 10$ subjects, and, for the data reported at the beginning of this chapter, $\bar{X} = 44.5$. Therefore,

$$Z = \frac{44.5 - 50}{12/\sqrt{10}} = -1.45.$$

Because -1.45 is greater than the critical value, -1.96, you fail to reject the null hypothesis. If instead you are willing to have a Type I error probability of .1, the critical value is -1.28, and because $Z = -1.45$ is less than -1.28, you reject. That is, you conclude that the mean is less than 50, and the probability of making a mistake, rejecting when you should not have rejected, is .1. ■

EXAMPLE. Rather than test $H_0 : \mu \geq 50$, imagine that the goal is to test $H_0 : \mu \geq 60$. The calculations are exactly the same as before, except that 50 is

Continued

> **EXAMPLE.** (*Continued*) replaced by 60 when computing Z. So now
>
> $$Z = \frac{44.5 - 60}{12/\sqrt{10}} = -4.08.$$
>
> If you test at the .005 level, the critical value is -2.58, -4.08 is less than -2.58, so you reject. That is, you conclude that the sample mean is not consistent with the assumption that $\mu \geq 60$ and that μ is less than 60. Again, there is the possibility of incorrectly rejecting when in fact $\mu \geq 60$ is true, and by design the probability of making this mistake is $\alpha = .005$ when $\mu = 60$.   ■

There are two variations of the hypothesis-testing method just described. To illustrate the first, imagine that you work in the research and development department of a company that helps students train for the SAT examination. After years of experience, it is found that the typical student attending the training course gets an SAT mathematics score of $\mu = 580$ and the standard deviation is $\sigma = 50$. You suspect that the training course could be improved and you want to empirically determine whether this is true. You try the new method on $n = 20$ students and get a sample mean of $\bar{X} = 610$. For illustrative purposes, assume that the standard deviation is again 50. We need to consider carefully how the null hypothesis should be stated. You have evidence that the new training method is better for the typical student because the estimate of the population mean is 610, which is greater than 580. But you need to convince management, so you assume that the new method is actually worse, with the goal of determining whether this assumption can be ruled out based on $\bar{X} = 610$. That is, you decide to test the hypothesis $H_0 : \mu \leq 580$, so if you reject, there is empirical evidence suggesting that it is unreasonable to believe that the new method is not beneficial for the typical student.

Testing $H_0 : \mu \leq 580$ is like the mirror image of testing $H_0 : \mu \geq 580$. If you get a sample mean of $\bar{X} = 550$, your estimate of μ is 550, and this is consistent with the hypothesis that μ is less than or equal to 580. That is, you would not reject. In the illustration, $\bar{X} = 610$, which suggests that the null hypothesis might be false. But if $\mu = 580$, getting a mean of 610 or larger could happen by chance. The issue is whether 610 is large enough to rule out the possibility that, beyond a reasonable doubt, $\mu \leq 580$. To find out, you compute Z as before, only now you reject if Z is sufficiently large. If you reject when $Z \geq 1.645$, the probability of a Type I error is $\alpha = P(Z \geq 1.645) = .05$. In the illustration,

$$Z = \frac{\bar{X} - \mu}{\sigma/\sqrt{n}} = \frac{610 - 580}{50/\sqrt{20}} = 2.68.$$

Because 2.68 is greater than 1.645, you reject and conclude that the mean is greater than 580. That is, you have empirical evidence to present to management that the new training method offers an advantage over the conventional approach, and there is a .05 probability that you made a mistake.

EXAMPLE. We repeat the illustration just given, only now imagine you want the probability of a Type I error to be .005 instead. From Table 1 in Appendix B, we see that $P(Z \geq 2.58) = .005$. This means that if you reject $H_0 : \mu \leq 580$ when Z is greater than or equal to 2.58, the probability of a Type I error is $\alpha = .005$. As already indicated, $Z = 2.68$, and because this exceeds 2.58, you again reject and conclude that the mean is greater than 580.   ■

5.1.1 p-Value (Significance Level)

There is an alternative way of describing hypothesis testing that is frequently employed and therefore important to understand. It is based on what is called the *significance level*, or *p-value*, an idea that appears to have been proposed first by Deming (1943), which is just the probability of a Type I error if the observed value of Z is used as a critical value.[1] If you reject when the p-value is less than or equal to .05, then the probability of a Type I error is .05, assuming normality. If you reject when the p-value is less than or equal to .01, then the probability of a Type I error is .01.

EXAMPLE. Again consider the open-mindedness example where we want to test $H_0 : \mu \geq 50$ and $\sigma = 12$. Imagine that you randomly sample $n = 10$ subjects and compute the sample mean, $\bar{X}$. If, for example, you get $\bar{X} = 48$, then

$$Z = \frac{48 - 50}{12/\sqrt{10}} = -0.53.$$

The p-value is just the probability of a Type I error if you reject when Z is less than or equal to -0.53. This probability is

$$P(Z \leq -0.53) = .298.$$

If you want the probability of a Type I error to be no greater than .05, then you would not reject, because .298 is greater than .05. Put another way, if you reject when your test statistic, Z, is less than or equal to -0.53, the probability of a Type I error is .298.   ■

The idea behind the p-value (or significance level) is that it gives you more information about the α level at which the null hypothesis would be rejected. If you are told that you reject with $\alpha = .05$, and nothing else, this leaves open the issue of whether you would also reject with $\alpha = .01$. If you are told that the p-value is .024, say, then you know that you would reject with $\alpha = .05$ but not $\alpha = .01$. If the p-value is .003, then in particular you would reject with $\alpha = .05$, $\alpha = .01$, and even $\alpha = .005$.

---

1 Level of significance refers to α, the Type I error probability specified by the investigator. Consequently, some authorities prefer the term *p-value* over the expression *significance level* as it is used here.

In case it helps, the p-value can be described in a slightly different fashion. Again consider the null hypothesis $H_0 : \mu \geq 50$ with $n = 10$ and $\sigma = 10$. Given $\bar{X}$, let

$$p = P\left(Z \leq \frac{\bar{X} - 50}{\sigma/\sqrt{n}}\right). \qquad (5.1)$$

The quantity p is called the *significance level*, or *p-value*, associated with the null hypothesis $H_0 : \mu \geq 50$. In the example, $\bar{X} = 48$,

$$p = P\left(Z \leq \frac{48 - 50}{12/\sqrt{10}}\right)$$

$$= P(Z \leq -0.53)$$

$$= .298,$$

so, as in the preceding example, the p-value is .298.

Next, consider a situation where the null hypothesis is that the mean is less than or equal to some specified value. To be concrete, consider $H_0 : \mu \leq 580$. Then given $\bar{X}$, the significance level is now

$$p = P\left(Z \geq \frac{\bar{X} - 580}{\sigma/\sqrt{n}}\right).$$

If $\bar{X} = 590$, $\sigma = 60$, and $n = 20$, then the significance level is

$$p = P\left(Z \geq \frac{590 - 580}{60/\sqrt{20}}\right)$$

$$= P(Z \geq 0.745)$$

$$= .228$$

This means that when $\bar{X} = 590$, $Z = 0.745$ and that if you reject when $Z \geq 0.745$, the probability of a Type I error is .228.

5.1.2 A Two-Sided Test: Testing for Exact Equality

One other variation of hypothesis testing needs to be described: testing the hypothesis that the mean is exactly equal to some specified value. Returning to the example regarding open-mindedness, suppose it is claimed that the average score of all adult men is exactly 50, as opposed to being greater than or equal to 50. Then the null hypothesis is

$$H_0 : \mu = 50.$$

If the sample mean is exactly equal to 50, you would not reject, because this is consistent with H_0. If $\bar{X} > 50$, then the larger the sample mean happens to be, the more doubt there is that $\mu = 50$. Similarly, if $\bar{X} < 50$, then the smaller the sample mean,

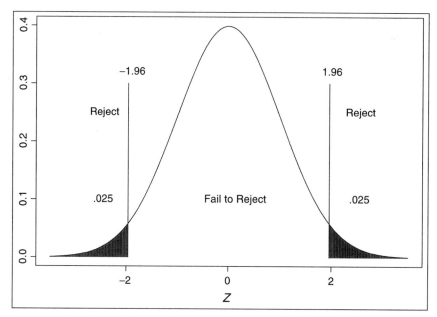

FIGURE 5.2   Critical region for a two-sided test such that $P(\text{Type I error}) = .05$.

the more doubt there is that $\mu = 50$. That is, now it is reasonable to reject H_0 if $\bar{X}$ is either too large or too small. An equivalent way of saying this is that you should reject if

$$Z = \frac{\bar{X} - 50}{\sigma/\sqrt{n}}$$

is too large or too small.

Suppose you reject H_0 if either $Z \leq -1.96$ or $Z \geq 1.96$. A more succinct way of describing this decision rule is that you reject if the absolute value of Z is greater than or equal to 1.96. In symbols, reject if $|Z| \geq 1.96$. If the null hypothesis is true and sampling is from a normal distribution, then Z has a standard normal distribution, so the probability of rejecting is

$$P(Z \leq -1.96) + P(Z \geq 1.96) = .025 + .025 = .05,$$

which is the total area of the two shaded regions in Figure 5.2.

EXAMPLE. Imagine a list of 55 minor malformations babies might have at birth. For illustrative purposes, it is assumed that the average number of malformations is 15 and the population standard deviation is $\sigma = 6$. For babies born to diabetic women, is the average number different from 15? That is, can you reject the hypothesis $H_0 : \mu = 15$? To find out, you sample $n = 16$ babies having diabetic mothers, count the number of malformations for each, and find

Continued

EXAMPLE. (*Continued*) that the average number of malformations is $\bar{X} = 19$. Then

$$Z = \frac{19 - 15}{6/\sqrt{16}} = 2.67.$$

If the goal is to have the probability of a Type I error equal to .05, then the critical values are -1.96 and 1.96, for the reasons just given. Because 2.67 is greater than 1.96, you reject the null hypothesis and conclude that the average number of malformations is greater than 15.   ■

The significance level, or p-value, can be determined when testing for exact equality, but you must take into account that the critical region consists of both tails of the standard normal distribution.

EXAMPLE. Continuing the last example, where $Z = 2.67$, if you had decided to reject the null hypothesis if $Z \leq -2.67$ or if $Z \geq 2.67$, then the probability of a Type I error is

$$P(Z \leq -2.67) + P(Z \geq 2.67) = .0038 + .0038 = .0076.$$

This means that the significance level is 0.0076.   ■

5.1.3 Criticisms of Two-Sided Hypothesis Testing and p-Values

Testing for exact equality has met with some criticism, on the grounds that exact equality is impossible. If, for example, one tests $H_0 : \mu = 50$, the argument is that surely μ differs from 50 at some decimal place, meaning that the null hypothesis is false and will be rejected with a sufficiently large sample size. A related criticism is that because H_0 is surely false, the p-value (or significance level) is meaningless.

Assuming that these criticisms have merit, there are at least two ways one might address them. One is to reformulate the goal (cf. Shaffer, 1974). Rather than test $H_0 : \mu = 50$, for example, suppose the goal is to determine whether μ is less than or greater than 50. Further assume that when addressing this goal we make one of three decisions: (1) if $Z \leq -1.96$, then decide $\mu < 50$, (2) if $Z \geq 1.96$, then decide $\mu > 50$, and (3) if $-1.96 < Z < 1.96$, then make no decision about whether μ is less than or greater than 50. Then there is some probability of making an incorrect decision, and, still assuming normality, the maximum probability of deciding $\mu < 50$ when in fact $\mu \geq 50$ is .025. Similarly, the maximum probability of deciding $\mu > 50$ when in fact $\mu \leq 50$ is .025. So in this context, α, or the p-value, tells us something about how certain we can be that μ is less than or greater than 50. However, the p-value tells us nothing about the degree to which μ differs from the hypothesized value. (This last issue is discussed in more detail in Chapter 8.)

A second approach is to rely exclusively on a confidence interval for μ. A confidence interval not only tells us whether we should reject H_0 and conclude that μ is less than or greater than some hypothesized value, it also provides information about the degree to which the population mean differs from the hypothesized value. For example, if the .95 confidence interval is $(12, 19)$, we would reject $H_0 : \mu = 10$ because this interval does not contain 10. In general, if we compute a $1 - \alpha$ confidence interval for μ, and if we reject whenever this confidence interval does not contain the hypothesized value for μ, the probability of a Type I error is α. In the example, the confidence interval tells us more. In addition to rejecting H_0, we can be reasonably certain that the population mean exceeds 10 (the hypothesized value) by at least 2.

5.1.4 Summary and Generalization

The basics of hypothesis testing, assuming normality and that σ is known, can be summarized in the following manner. Let μ_0 (read "mu naught") be some specified constant. The goal is to make some inference about how the population mean, μ, compares to μ_0. For the hypothesis $H_0 : \mu \leq 50$, $\mu_0 = 50$. For $H_0 : \mu = 15$, $\mu_0 = 15$; and for $H_0 : \mu \geq 580$, $\mu_0 = 580$. Furthermore, you want the probability of a Type I error to be α. Once the sample mean has been determined, compute

$$Z = \frac{\bar{X} - \mu_0}{\sigma/\sqrt{n}}.$$

Case 1. $H_0 : \mu \geq \mu_0$. Reject H_0 if $Z \leq c$, the α quantile of a standard normal distribution.

Case 2. $H_0 : \mu \leq \mu_0$. Reject H_0 if $Z \geq c$, the $1 - \alpha$ quantile of a standard normal distribution.

Case 3. $H_0 : \mu = \mu_0$. Reject H_0 if $Z \geq c$ or if $Z \leq -c$, where now c is the $1 - \alpha/2$ quantile of a standard normal distribution. Equivalently, reject if $|Z| \geq c$.

The hypotheses $H_0 : \mu \geq \mu_0$ and $H_0 : \mu \leq \mu_0$ are called *one-sided hypotheses*. In contrast, $H_0 : \mu = \mu_0$ is called a *two-sided hypothesis*.

5.1.5 A Property of p-Values

p-Values have a property that is rarely discussed in applied statistics books, but it might provide some sense of why certain modern techniques (described in subsequent chapters) are reasonable. Imagine that a the null hypothesis is true and that an experiment is repeated infinitely many times. Here it is assumed that the probability of a Type I error can be controlled exactly for any α we might pick. This will be true, for example, when using Z as described in Section 5.1.4 and sampling is from a normal distribution. Further imagine that each time the experiment is performed, the p-value (or significance level) is computed. If the infinitely many p-values were plotted, we would get the uniform distribution (described in Exercise 10 of Chapter 2). That is, all p-values are equally likely and are centered around .5 (e.g., Sackrowitz & Samuel-Cahn, 1999).

In particular, the probability of getting a p-value less than or equal to .025 or greater than or equal to .975 is exactly .05. More generally, when sampling from a nonnormal distribution, if a method for testing some hypothesis controls the probability of a Type I error for a sufficiently large sample size (meaning that the central limit theorem applies), then the distribution of p converges to a uniform distribution as the sample size increases.

5.2 Power and Type II Errors

After years of production, a manufacturer of batteries for automobiles finds that on average, their batteries last 42.3 months with a standard deviation of $\sigma = 4$. A new manufacturing process is being contemplated and one goal is to determine whether the batteries have a longer life on average. Ten batteries are produced by the new method and their average life is found to be 43.4 months. For illustrative purposes, assume that the standard deviation is again $\sigma = 4$. Based on these $n = 10$ test batteries, it is estimated that the average life of all the batteries produced using the new manufacturing method is greater than 42.3 (the average associated with the standard manufacturing method), in which case the new manufacturing process has practical value. To add support to this speculation, it is decided to test $H_0 : \mu \leq 42.3$ versus $H_1 : \mu > 42.3$, where μ is the population mean using the new method.

The idea is to determine whether $\bar{X}$ is sufficiently larger than 42.3 to rule out the possibility that $\mu \leq 42.3$. That is, the goal is to determine whether the new method is no better and possibly worse on average. If H_0 is rejected, there is empirical evidence that the new method should be adopted. As explained in the previous section, you test this hypothesis by computing $Z = (43.4 - 42.3)/(4/\sqrt{10}) = .87$. If you want the probability of a Type I error to be $\alpha = .01$, the critical value is 2.33, because $P(Z \leq 2.33) = .99$. In the present context, a Type I error is concluding that the new method is better on average when in reality it is not. Because .87 is less than 2.33, you fail to reject. Does this imply that you should accept the alternative hypothesis that μ is less than 42.3? In other words, should you conclude that the average battery lasts less than 42.3 months under the new manufacturing method?

Suppose that if the null hypothesis is not rejected, you conclude that the null hypothesis is true and that the population mean is less than 42.3. Then there are four possible outcomes, which are summarized in Table 5.1. The first possible outcome is

TABLE 5.1   Four Possible Outcomes When Testing Hypotheses

Decision	Reality	
	H_0 **true**	H_0 **false**
H_0 true	Correct decision	Type II error (probability β)
H_0 false	Type I error (probability α)	Correct decision (power)

that the null hypothesis is true and you correctly decide not to reject. The second possible outcome is that the null hypothesis is false but you fail to reject and therefore make a mistake. That is, your decision that $\mu \leq 42.3$ is incorrect — in reality the mean is greater than 42.3. The third possible outcome is that the null hypothesis is true but you make a mistake and reject. This is a Type I error, already discussed in Section 5.1. The fourth possible outcome is that in reality $\mu > 42.3$ and you correctly detect this by rejecting H_0.

This section is concerned with the error depicted by the upper right portion of Table 5.1. That is, the null hypothesis is false but you failed to reject. If, for example, the actual average life of a battery under the new manufacturing method is $\mu = 44$, the correct conclusion is that $\mu > 42.3$. The practical problem is that even if in reality $\mu = 44$, by chance you might get $\bar{X} = 41$, suggesting that the hypothesis $H_0 : \mu \leq 42.3$ should be accepted. And even if $\bar{X} > 42.3$, it might be that the sample mean is not large enough to reject even though in reality H_0 is false. Failing to reject when you should reject is called a *Type II error*.

DEFINITION.   A *Type II error* is failing to reject a null hypothesis when it should be rejected. The probability of a Type II error is often labeled β.

DEFINITION.   *Power* is the probability of rejecting H_0 when in fact it is false. In symbols, power is $1 - \beta$, which is 1 minus the probability of a Type II error. In the illustration, if the new manufacturing method is actually better, meaning that μ is greater than 42.3, and the probability of rejecting $H_0 : \mu \leq 42.3$ is .8, say, this means that power is $1 - \beta = .8$, and the probability of a Type II error is $\beta = .2$.

Power and the probability of making a Type II error are of great practical concern. In the illustration, if $\mu = 44$, the manufacturer has found a better manufacturing method, and clearly it is in their interest to discover this. What is needed is a method for ascertaining power, meaning the probability of correctly determining that the new method is better when in fact $\mu > 42.3$. If power is high but the company fails to detect an improvement over the standard method of production, the new method can be discarded. That is, there is empirical evidence that H_0 is true and the new method has no practical value. However, if power is low, meaning that there is a low probability of discovering that the new method produces longer-lasting batteries even when the new method is in fact better, then simply failing to reject does not provide convincing empirical evidence that H_0 is true.

In the present context, power depends on four quantities: σ, α, n, and the value of $\mu - \mu_0$, where μ is the unknown mean of the new manufacturing method. Although μ is not known, you can address power by considering values of μ that are judged to be interesting and important in a given situation. In the illustration, suppose you want to adopt the new manufacturing method if $\mu = 44$. That is, the average life of a battery using the standard method is 42.3 and you want to be reasonably certain of adopting the new method if the average life is now 44. In the more formal terminology of hypothesis testing, you want to test $H_0 : \mu \leq 42.3$, and if $\mu = 44$, you want power to be reasonably close to 1. What is needed is a convenient way of assessing power given α, σ, μ, μ_0, and the sample size, n, you plan to use.

BOX 5.1 How to Compute Power, σ Known

Goal

Assuming normality, compute power when testing $H_0 : \mu < \mu_0$ or $H_0 : \mu > \mu_0$ or $H_0 : \mu = \mu_0$ given:

1. n, the sample size
2. σ, the standard deviation
3. α, the probability of a Type I error
4. some specified value for μ
5. μ_0, the hypothesized value.

Case 1

$H_0 : \mu < \mu_0$. Determine the critical value c as described in Section 5.1. (The critical value is the $1 - \alpha$ quantile of a standard normal distribution.) Then power, the probability of rejecting the null hypothesis, is

$$1 - \beta = P\left(Z \geq c - \frac{\sqrt{n}(\mu - \mu_0)}{\sigma}\right).$$

In words, power is equal to the probability that a standard normal random variable is greater than or equal to

$$c - \frac{\sqrt{n}(\mu - \mu_0)}{\sigma}.$$

Case 2

$H_0 : \mu > \mu_0$. Determine the critical value c, which is now the α quantile of a standard normal distribution. Then power is

$$1 - \beta = P\left(Z \leq c - \frac{\sqrt{n}(\mu - \mu_0)}{\sigma}\right).$$

Case 3

$H_0 : \mu = \mu_0$. Now c is the $1 - \alpha/2$ quantile of a standard normal distribution. Power is

$$1 - \beta = P\left(Z \leq -c - \frac{\sqrt{n}(\mu - \mu_0)}{\sigma}\right) + P\left(Z \geq c - \frac{\sqrt{n}(\mu - \mu_0)}{\sigma}\right).$$

Box 5.1 summarizes how to compute power given n, σ, μ, μ_0, and α. Continuing the illustration where $H_0 : \mu \leq 42.3$, suppose $\alpha = .05$, $n = 10$, and you want to determine how much power there is when $\mu = 44$. Because $\alpha = .05$, the critical value is $c = 1.645$. Referring to Box 5.1, power is

$$1 - \beta = P\left(Z \geq c - \frac{\sqrt{n}(\mu - \mu_0)}{\sigma}\right)$$

$$= P\left(Z \geq 1.645 - \frac{\sqrt{10}(44 - 42.3)}{4}\right)$$

$$= P(Z \geq .30)$$

$$= .38.$$

This says that if battery life has a normal distribution and, unknown to us, the actual average life of a battery under the new manufacturing method is $\mu = 44$, then the probability of rejecting the hypothesis that the mean is less than 42.3 is .38. That is, for this situation where we should reject and conclude that the new manufacturing method is better on average, there is a .38 probability of making the correct decision that the null hypothesis is false. Consequently, the probability of committing a Type II error and failing to reject even though the null hypothesis is false is $1 - .38 = .62$.

Figure 5.3 graphically illustrates power when testing $H_0 : \mu \leq 42.3$ with $\alpha = .05$ and $\mu = 46$. It can be seen that power is $1 - \beta = .9$, so the probability of a Type II error is $\beta = .1$. The left normal distribution is the distribution of Z when the null hypothesis is true; it is standard normal and you reject if $Z \geq 1.645$, as already discussed. When the null hypothesis is false and in fact $\mu = 46$, Z still has a normal distribution, but its mean is no longer zero — it is larger, as indicated by Figure 5.3. That is, the right distribution reflects the actual distribution of Z when $\mu = 46$. Power is the area under the right (nonnull) curve and to the right of the critical value ($c = 1.645$). The area of the shaded region represents the probability of a Type II error, which is .1.

Notice that we do not know the actual value of μ, the average life of batteries manufactured with the new method. To deal with this issue, we must ask ourselves a series of questions: What if $\mu = 44$ or 45 or 46, and so on? By computing power for each of these situations, we get some idea about the probability of rejecting when in fact the null hypothesis is false. Figure 5.4 graphs power as μ increases. Notice that the more the mean μ exceeds the hypothesized value of 42.3, the higher the power. This is, of course, a property we want. The larger the difference between the hypothesized value and the actual value of μ, the more likely we are to reject and correctly conclude that μ is greater than 42.3.

5.2.1 Understanding How n, α, and σ Are Related to Power

Power is a function of three fundamental components of any study: the sample size, n, the Type I error probability you pick, α, and the population standard deviation, σ. As already explained, power plays a crucial role in applied work, so it is important to understand how each of these quantities is related to power.

First consider how the sample size, n, affects power. If the null hypothesis is false, we want the probability of rejecting to go up as the sample size increases. That is,

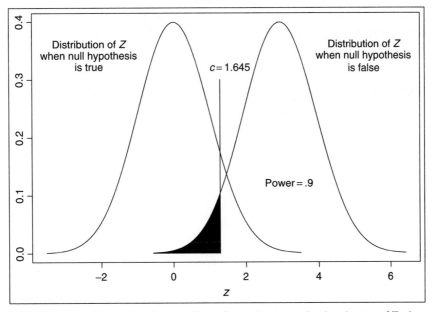

FIGURE 5.3   Illustration of power. The left distribution is the distribution of Z when the null hypothesis is true. The right distribution is the distribution of Z when the null hypothesis is false. The area of the shaded region, which is the area under the right (nonnull) distribution and to the left of 1.645 is .1; it is equal to the probability of a Type II error. So power is the area under the nonnull distribution and to the right of 1.645.

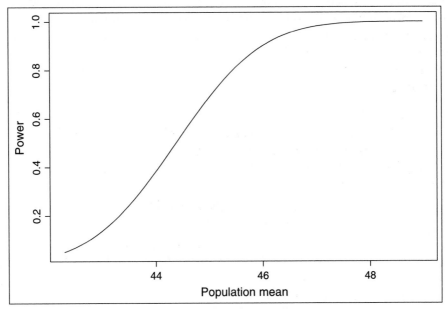

FIGURE 5.4   Power curve for Z when the hypothesized value for the population mean (μ_0) is 42.3. As μ increases, meaning that we are moving away from the hypothesized value for μ, power increases.

as the sample size, n, gets large, we should have an increasingly higher probability of making a correct decision about H_0 when it is false. Examining the expressions for power in Box 5.1 reveals that is exactly what happens.

EXAMPLE. Consider the battery example once again, where we want to test $H_0 : \mu \leq 42.3$, and suppose we want to know how much the power is when $\mu = 44$, but this time we consider three sample sizes: 10, 20, and 30. For $n = 10$ we have already seen that power is .38. If $n = 20$, then power is

$$1 - \beta = P\left(Z \geq c - \frac{\sqrt{n}(\mu - \mu_0)}{\sigma}\right)$$

$$= P\left(Z \geq 1.645 - \frac{\sqrt{20}(44 - 42.3)}{4}\right)$$

$$= P(Z \geq -0.256)$$

$$= .60$$

Increasing n to 30, it can be seen that power is now $1 - \beta = .75$, meaning that your probability of making a correct decision and rejecting H_0 when it is false is now .75. ∎

This example illustrates how the sample size might be determined in applied work. First determine the difference between μ and μ_0 that is important in a given situation. In the battery illustration, it might be decided that if $\mu - \mu_0 = 44 - 42.3 = 1.7$, we want to be reasonably certain of rejecting the null hypothesis and deciding that the new manufacturing method is better on average. Next, compute power for some value of n. If the power is judged to be sufficiently large, use this sample size in your study. If not, increase the sample size.

Your choice for α, the probability of a Type I error you are willing to allow, also affects power.

EXAMPLE. For the battery example with $n = 30$, consider three choices for α: .05, .025, and .01. For $\alpha = .05$ we have already seen that power is .75 when testing $H_0 : \mu < 42.3$ and $\mu = 44$. For $\alpha = .025$, the critical value is now $c = 1.96$, so power is

$$1 - \beta = P\left(Z \geq c - \frac{\sqrt{n}(\mu - \mu_0)}{\sigma}\right)$$

$$= P\left(Z \geq 1.96 - \frac{\sqrt{20}(44 - 42.3)}{4}\right)$$

$$= P(Z \geq .059)$$

$$= .47.$$

Continued

> **EXAMPLE.** (*Continued*) If instead you use $\alpha = .01$, the critical value is now $c = 2.33$ and power can be seen to be .33. This illustrates that if we adjust the critical value so that the probability of a Type I error goes down, power goes down as well. Put another way, the more careful you are not to commit a Type I error by choosing α close to zero, the more likely you are to commit a Type II error if the null hypothesis happens to be false.   ■

Finally, the population standard deviation, σ, is also related to power. The larger σ happens to be, the lower the power will be given α, n, and a value for $\mu - \mu_0$.

> **EXAMPLE.** For the battery example, consider $\alpha = .05$, $\mu - \mu_0 = 1.7$, and $n = 30$, with $\sigma = 4$. Then power is .75, as previously explained. But if $\sigma = 8$, power is now .31. If $\sigma = 12$, power is only .19.   ■

The results just described on how n, α, and σ are related to power can be summarized as follows:

- As the sample size, n, gets large, power goes up, so the probability of a Type II error goes down.
- As α goes down, in which case the probability of a Type I error goes down, power goes down and the probability of a Type II error goes up.
- As the standard deviation, σ, goes up, with n, α, and $\mu - \mu_0$ fixed, power goes down.

Notice that once you have chosen an outcome variable of interest (X) and the population of individuals you want to study, there are two types of factors that affect power. The first type consists of factors that are under your control: n, the sample size, and α, the probability of a Type I error you are willing to allow. By increasing n or α, you increase power. The population standard deviation also affects power, but it is not under your control, it merely reflects a state of nature. (In some situations the variance of X can be influenced based on how an outcome variable is designed or constructed.) However, understanding how σ affects power is important in applied work, because it plays a role in choosing an accurate hypothesis-testing method, as will be seen.

5.3 Testing Hypotheses About the Mean When σ Is Not Known

Next we describe the classic method for testing hypotheses about the population mean when the population standard deviation is not known. Then we describe recent insights into why this technique has several practical problems.

When σ is known, we can test hypotheses about the population mean if we can determine the distribution of

$$Z = \frac{\bar{X} - \mu_0}{\sigma/\sqrt{n}}.$$

When σ is not known, we estimate σ with s, the sample standard deviation, and we can test hypotheses if the distribution of

$$T = \frac{\bar{X} - \mu_0}{s/\sqrt{n}} \tag{5.2}$$

can be determined. As indicated in Chapter 4, the distribution of T can be determined when sampling from a normal distribution. For this special case, hypotheses can be tested as described in Section 5.1.4, except that critical values are read from Table 4 in Appendix B with the degrees of freedom set to $\nu = n - 1$. The details can be summarized as follows.

GOAL: Test hypotheses regarding how the population mean, μ, compares to a specified constant, μ_0. The probability of a Type I error is to be α.

ASSUMPTIONS: Random sampling and normality.

DECISION RULES:

- For $H_0 : \mu \geq \mu_0$, reject if $T \leq c$, where c is the α quantile of Student's T distribution with $\nu = n - 1$ degrees of freedom and T is given by Equation (5.2).
- For $H_0 : \mu \leq \mu_0$, reject if $T \geq c$, where c is now the $1 - \alpha$ quantile of Student's T distribution with $\nu = n - 1$ degrees of freedom.
- For $H_0 : \mu = \mu_0$, reject if $T \geq c$ or $T \leq -c$, where c is now the $1 - \alpha/2$ quantile of Student's T distribution with $\nu = n - 1$ degrees of freedom. Equivalently, reject if $|T| \geq c$.

EXAMPLE. For the measures of open-mindedness given at the beginning of this chapter, test the hypothesis $H_0 : \mu \geq 50$ with $\alpha = .05$. The sample standard deviation is $s = 11.4$ and the sample mean is $\bar{X} = 44.5$. Because $n = 10$, the degrees of freedom are $\nu = n - 1 = 9$ and

$$T = \frac{\bar{X} - \mu_0}{s/\sqrt{n}} = \frac{44.5 - 50}{11.4/\sqrt{10}} = -1.5.$$

Referring to Table 4 in Appendix B, $P(T \leq -1.83) = .05$, so the critical value is -1.83. This means that if we reject when T is less than or equal to -1.83, the probability of a Type I error will be $.05$, assuming normality. Because the observed value of T is -1.5, which is greater than the critical value, you fail to reject. In other words, the sample mean is not sufficiently smaller than 50 to be reasonably certain that the speculation $\mu \geq 50$ is false. ■

As you can see, the steps you follow when σ is not known mirror the steps you use to test hypotheses when σ is known.

EXAMPLE. Suppose you observe the values

$$12, 20, 34, 45, 34, 36, 37, 50, 11, 32, 29$$

and the goal is to test $H_0 : \mu = 25$ such that the probability of a Type I error is $\alpha = .05$. Here, $n = 11$, $\mu_0 = 25$, and it can be seen that $\bar{X} = 33.24$, $s/\sqrt{11} = 3.7$, so

$$T = \frac{\bar{X} - \mu}{s/\sqrt{n}} = \frac{33.24 - 25}{3.7} = 2.23.$$

The null hypothesis is that the population mean is exactly equal to 25. So the critical value is the $1 - \alpha/2 = .975$ quantile of Student's T distribution with degrees of freedom $\nu = 11 - 1 = 10$. Table 4 in Appendix B indicates that

$$P(T \leq 2.28) = .975,$$

so our decision rule is to reject H_0 if the value of T is greater than or equal to 2.28 or less than or equal to -2.28. Because the absolute value of T is less than 2.28, you fail to reject. ■

5.4 Controlling Power and Determining *n*

Problems of fundamental importance are determining what sample size to use and finding methods that ensure power will be reasonably close to 1. Two approaches are described in this section, both of which assume random sampling from a normal distribution. The first is based on choosing *n* prior to collecting any data. The second is used after data are available and is aimed at determining whether *n* was sufficiently large to ensure that power is reasonably high. One fundamental difference between the two methods is how they measure the extent to which the null hypothesis is false.

5.4.1 Choosing *n* Prior to Collecting Data

First consider how one might choose *n* prior to collecting data so that power is reasonably close to 1. To begin, we need a measure of the difference between the hypothesized value for the mean (μ_0) and its true value (μ). One possibility is

$$\delta = \mu - \mu_0, \tag{5.3}$$

which is consistent with how we discussed power in Section 5.2 when σ is known. However, when using T, it is impossible to control power given some value for δ without first obtaining data, because when using T, power depends on the unknown variance (Dantzig, 1940). The standard method for dealing with this problem is to replace δ with

$$\Delta = \frac{\mu - \mu_0}{\sigma}. \tag{5.4}$$

So if $\Delta = 1$, for example, the difference between the means is one standard deviation. That is, $\mu - \mu_0 = \sigma$. If $\Delta = .5$, the difference between the mean is half a standard deviation. (That is, $\mu - \mu_0 = .5\sigma$.) We saw in Chapter 2 that for normal distributions, σ has a convenient probabilistic interpretation, but Section 2.7 illustrated that for even a small departure from normality, this interpretation breaks down. This causes practical problems when using Δ, but we temporarily ignore this issue. (These practical problems are discussed in detail in Chapter 8.) Here it is merely remarked that under normality, power can be determined for any choice of n, Δ, and α. Rather than describe the details, we merely provide an S-PLUS function that performs the computations.

5.4.2 S-PLUS Function pow1

The S-PLUS function

$$\text{pow1}(n, \text{Del}, \text{alpha})$$

(written for this book) computes power when performing a one-sided test, where the argument Del is Δ and alpha is α. For example, if you want to determine how much power you have when testing $H_0 : \mu \geq 15$ with $n = 10$, $\Delta = -.3$, and $\alpha = .05$, the S-PLUS command pow1(10, −.3, .05) returns the value .219. Increasing n to 30, power is now .479. With $n = 100$, power is .9. So in this particular case, $n = 10$ is inadequate if $\Delta = -.3$ is judged to be a difference that is important to detect. To ensure high power requires a sample size of around 100. In a similar manner, if the goal is to test $H_0 : \mu \leq 15$ and now $\Delta = .3$, then pow1(10, .3, .05) again returns the value .219. (The function assumes that if Δ is positive, you are testing $H_0 : \mu \leq \mu_0$, and that if Δ is negative, you are testing $H_0 : \mu \geq \mu_0$.)

This S-PLUS function can handle two-sided tests in a simple manner. You simply divide α by 2. In the previous illustration, if instead you want to test $H_0 : \mu = 15$ at the .05 level, the command pow1(30, .3, .025) returns the value .35, indicating that power is .35. (The same result is returned if the argument Del is $-.3$.) If this amount of power is judged to be too small, simply increase n until a more satisfactory power level is obtained.

5.4.3 Stein's Method

Assuming normality, Stein (1945) derived a method that indicates whether the sample size n is sufficiently large to achieve some specified amount of power. In contrast to the method in Section 5.4.1, it is used after data are collected and it is based on $\delta = \mu - \mu_0$ rather than Δ. Said another way, if you fail to reject some null hypothesis, Stein's method helps you decide whether this is because power is low due to too small an n. Stein's method does even more; it indicates how many additional observations are needed to achieve some specified amount of power.

For convenience, assume that a one-sided test is to be performed. Also assume that n observations have been randomly sampled from some normal distribution, yielding a sample variance s^2. If the goal is to ensure that power is at least $1 - \beta$, then compute

$$d = \left(\frac{\delta}{t_{1-\beta} - t_\alpha} \right)^2,$$

where $t_{1-\beta}$ and t_α are, respectively, the $1 - \beta$ and α quantiles of Student's T distribution with $\nu = n - 1$ degrees of freedom. For example, if $n = 10$ and you want power to be $1 - \beta = .9$, then $t_{1-\beta} = t_{.9} = 1.383$, which can be read from Table 4 in Appendix B. Then the number of required observations is

$$N = \max\left(n, \ \left[\frac{s^2}{d} \right] + 1 \right), \qquad (5.5)$$

where the notation $[s^2/d]$ means you compute s^2/d and round down to the nearest integer and max refers to the larger of the two numbers inside the parentheses. Continuing the example where $n = 10$ and $1 - \beta = .9$, if $s = 21.4$, $\delta = 20$, and $\alpha = .01$, then $\nu = 9$,

$$d = \left(\frac{20}{1.383 - (-2.82)} \right)^2 = 22.6,$$

so

$$N = \max\left(10, \ \left[\frac{(21.4)^2}{22.6} \right] + 1 \right) = \max(10, 21) = 21.$$

If $N = n$, the sample size is adequate, but in the illustration $N - n = 21 - 10 = 11$. That is, 11 additional observations are needed to achieve the desired amount of power.

A two-sided test ($H_0 : \mu = \mu_0$) is handled in a similar manner. The only difference is that α is replaced by $\alpha/2$. So if in the last example we wanted the Type I error probability to be .02 when testing a two-sided test, then again $N = 21$. If we want the Type I error probability to be .05, then $t_{\alpha/2} = t_{.025} = -2.26$, so if again we want power to be .9 when $\delta = \mu - \mu_0 = 20$,

$$d = \left(\frac{20}{1.383 - (-2.26)} \right)^2 = 30.14,$$

so we need a total of

$$N = \max\left(10, \ \left[\frac{(21.4)^2}{30.14} \right] + 1 \right) = 16$$

observations.

Stein also indicated how to test the null hypothesis if the additional $(N - n)$ observations can be obtained. But rather than simply perform Student's T on all N values, Stein used instead

$$T_s = \frac{\sqrt{n}(\hat{\mu} - \mu_0)}{s},\tag{5.6}$$

where $\hat{\mu}$ is the mean of all N observations. You test hypotheses by treating T_s as having a Student's T distribution with $\nu = n - 1$ degrees of freedom. That is, you test hypotheses as described in Section 5.3 but with T replaced by T_s. For example, you reject $H_0 : \mu \leq \mu_0$ if $T_s \geq c$, where c is the $1 - \alpha$ quantile of Student's T distribution with $\nu = n - 1$ degrees of freedom. A two-sided confidence interval for the population mean is given by

$$\hat{\mu} \pm c\frac{s}{\sqrt{n}},\tag{5.7}$$

where c is now the $1 - \alpha/2$ quantile of Student's T distribution with $n - 1$ degrees of freedom. For the special case where $N = n$ (meaning that the original sample size was sufficient for your power needs), $T_s = T$ and you are simply using Student's T test, but with the added knowledge that the sample size meets your power requirements. What is unusual about Stein's method is that if $N > n$, it uses the sample variance of the original n observations — not the sample variance of all N observations. Also, the degrees of freedom remain $n - 1$ rather than the seemingly more natural $N - 1$. By proceeding in this manner, Stein showed that power will be at least $1 - \beta$ for whatever value of $1 - \beta$ you pick. (Simply performing Student's T on all N values, when $N > n$, results in certain technical problems that are described by Stein, 1945, but not here. For a survey of related methods, see Hewett & Spurrier, 1983.)

A popular alternative to Stein's method when trying to assess power when a nonsignificant result is obtained is based on what it called *observed power*. The approach assumes that the observed difference between the sample mean and its hypothesized value is indeed equal to the true difference $(\mu - \mu_0)$; it also assumes that $s^2 = \sigma^2$, and then based on these assumptions one computes power. Hoenig and Heisey (2001) illustrate that this approach is generally unsatisfactory.

5.4.4 S-PLUS Functions stein1 and stein2

The S-PLUS function

stein1 (x,del,alpha=.05,pow=.8,oneside=F)

returns N, the sample size needed to achieve power given by the argument pow (which defaults to .8), given some value for δ (which is the argument del) and α. The function assumes that a two-sided test is to be performed. For a one-sided test, set the argument oneside to T for true.

The S-PLUS function

$$\text{stein2}(x1, x2, mu0 = 0, alpha = .05)$$

tests the hypothesis $H_0 : \mu = \mu_0$ using Stein's method, assuming that the initial n observations are stored in x1 and that the additional $N - n_1$ observations are stored in x2. The argument mu0 is the hypothesized value, μ_0, which defaults to 0.

EXAMPLE. The last example in Section 5.3 used Student's T to test $H_0 : \mu = 25$ with $\alpha = .05$, based on the data

$$12, 20, 34, 45, 34, 36, 37, 50, 11, 32, 29.$$

A nonsignificant result was obtained. If we want power to be .9 when $\mu = 28$, in which case $\delta = 28 - 25 = 3$, was the sample size sufficiently large? Storing these data in the S-PLUS variable y, the command

$$\text{stein1}(y, 3, pow = .9)$$

returns the value 220. That is, we need $N = 220$ observations to achieve this much power. Since we have only 11 observations, $220 - 11 = 209$ additional observations are needed. For a one-sided test, $N = 94$. ■

5.5 Practical Problems with Student's *T*

Student's T deals with the common situation where σ is not known, but it assumes observations are randomly sampled from a normal distribution. Because distributions are never exactly normal, it is important to understand how nonnormality affects conclusions based on T. A version of the central limit theorem tells us that as n gets large, the distribution of T becomes more like Student's T distribution, and in fact its distribution approaches a standard normal. That is, if the sample size is large enough and observations are randomly sampled, then violating the normality assumption is not a serious concern. Conventional wisdom is that assuming T has a Student's T distribution with $v = n - 1$ degrees of freedom provides reasonably accurate results with n fairly small, and surely accurate results are obtained with $n = 100$. However, in recent years, much more sophisticated methods have been derived for understanding how nonnormality affects T, and serious concerns have been discovered, two of which are described here.

The first is that *very* small departures from normality can drastically reduce power. The main reason is that even small departures from normality can inflate the population variance; this in turn inflates the standard error of the sample mean, so power can be relatively low. As indicated in Section 5.2.1, as σ gets large, power goes down when using Z to test hypotheses, and the same is true when using T. (One of the earliest results indicating theoretical concerns about this problem can be found in Bahadur & Savage, 1956.)

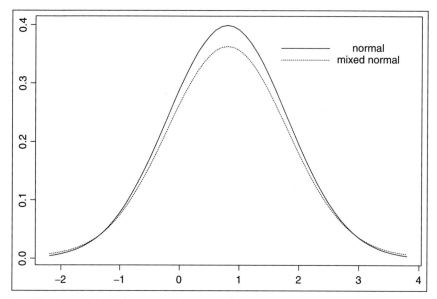

FIGURE 5.5   Small departures from normality can greatly reduce power when using Student's T. If we sample 20 observations from the normal distribution shown here, which has mean .8, and we test $H_0 : \mu = 0$ at the .05 level, power is .93. But if we sample from the mixed normal instead, power is only .39.

EXAMPLE. As an illustration, first consider using Student's T to test $H_0 : \mu = 0$ with $\alpha = .05$ when sampling from the normal distribution shown in Figure 5.5, which has mean .8 and standard deviation $\sigma = 1$. It can be shown that with $n = 20$, power is .93. That is, there is a 93% chance of correctly rejecting the null hypothesis that the mean is $\mu = 0$. Now suppose that sampling is from the other distribution shown in Figure 5.5. This is the mixed normal distribution described in Chapter 2, but with mean .8. This distribution is very similar to the normal distribution, in the sense described in Chapter 2, yet power is now .39. This demonstrates that if you test hypotheses with Student's T or any method based on the sample mean, small departures from normality can result in a substantial decrease in your ability to detect situations where the null hypothesis is false.   ■

The second problem is that nonnormality can affect your ability to control the probability of a Type I error or control the probability coverage when computing a confidence interval. First consider situations where sampling is from a perfectly symmetric distribution and imagine you want the probability of a Type I error to be .05. When sampling is from a normal distribution, you can accomplish your goal with Student's T test, as already demonstrated. However, if you happen to be sampling observations from a mixed normal instead, the actual probability of a Type I error is only .022 with $n = 20$. This might seem desirable because the probability of incorrectly rejecting when the null hypothesis is true is less than the nominal level of .05. However, recall that the smaller α happens to be, the lower your power. This means that the probability of correctly rejecting the null hypothesis when it is false might be low, contrary to what you want, because you are inadvertently testing at the .022 level. Increasing the sample size to 100, now the actual probability of a Type I error is .042, but low power remains a possible concern because sampling is from a heavy-tailed distribution.

If observations are sampled from a skewed distribution, the actual probability of a Type I error can be substantially higher or lower than .05. With $n = 12$, there are situations where the actual probability of a Type I error is .42 when testing at the .05 level, and it can be as low as .001 when testing a one-sided test at the .025 level (e.g., Wilcox, 1997a, p. 74). As another illustration, imagine you are interested in how response times are affected by alcohol and that, *unknown to you*, response times have the skewed (lognormal) distribution shown in Figure 4.14, which has mean $\mu = 1.65$. Further imagine that you want to test the hypothesis $H_0 : \mu \geq 1.65$ with $\alpha = .05$. That is, unknown to you, the null hypothesis happens to be true, so you should not reject. With $n = 20$ observations, the actual probability of a Type I error is .14. That is, your intention was to have a 5% chance of rejecting in the event the null hypothesis is true, but in reality there is a 14% chance of rejecting by mistake. Increasing the sample size to $n = 160$, the actual probability of Type I error is now .11. That is, control over the probability of a Type I error improves as the sample size gets large, in accordance with the central limit theorem, but at a rather slow rate. Even with $n = 160$, the actual probability of rejecting might be more than twice as large as intended. The seriousness of a Type I error will depend on the situation, but at least in some circumstances the discrepancy just described would be deemed unsatisfactory.

The lognormal distribution used in the previous paragraph is relatively light-tailed. When sampling from a skewed, heavy-tailed distribution, Student's T can deteriorate even more. Consider the distribution in Figure 5.6, which has a mean of .0833. With $n = 20$ and $\alpha = .05$, the actual probability of a Type I error is .20. Increasing n to 100, the actual probability of a Type I error drops to only .19. It is getting closer to the nominal level, in accordance with the central limit theorem, but at a very slow rate. (For theoretical results indicating drastic sensitivity to nonnormality, see Basu & DasGupta, 1995.)

When you use Student's T test under the assumption that sampling is from a normal distribution, you are assuming that T has a symmetric distribution about zero. But as pointed out in Section 4.8, the actual distribution of T can be asymmetric with a mean that differs from zero. For example, when sampling from the (lognormal) distribution in Figure 4.14, with $n = 20$, the distribution of T is skewed, with a mean of -0.5, the result being that the probability of a Type I error is not equal to the value you want. In fact, Student's T test is *biased*, meaning that the probability of rejecting is not minimized when the null hypothesis is true. That is, situations arise where there is a higher probability of rejecting when the null hypothesis is true than in a situation where the null hypothesis is false. (Generally, any hypothesis-testing method is said to be unbiased if the probability of rejecting is minimized when the null hypothesis is true. Otherwise it is biased.)

To illustrate the possible effect of skewness on the power curve of Student's T, suppose we sample 20 observations from the (lognormal) distribution shown in

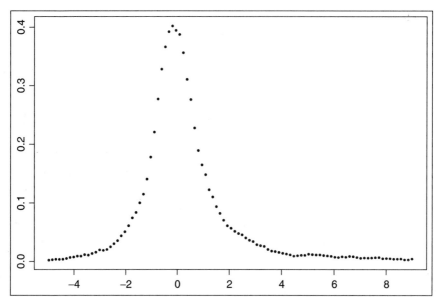

FIGURE 5.6   Example of a skewed, heavy-tailed distribution used to illustrate the effects of nonnormality on Student's T. This distribution has a mean of .8033.

Figure 4.14, which has a population mean approximately equal to 1.649. So if we test $H_0 : \mu = 1.649$ with $\alpha = .05$, the intention is to have the probability of a Type I error equal to .05, but in reality it is approximately .14. Now suppose we add δ to every observation. In effect, we shift the distribution in Figure 4.14 so that its mean is now $1.649 + \delta$. If, for example, we use $\delta = .3$, we have in effect increased the population mean from 1.649 to 1.949. So now we should reject $H_0 : \mu = 1.649$, but the actual probability of rejecting is .049. That is, we have described a situation where the null hypothesis is false, yet we are less likely to reject than in the situation where the null hypothesis is true. If we set $\delta = .6$ so that now $\mu = 2.249$ and we again test $H_0 : \mu = 1.649$, the probability of rejecting is .131, approximately the same probability of rejecting when the null hypothesis is true. As we increase δ even more, power continues to increase as well. Figure 5.7 shows the power curve of Student's T for δ ranging between 0 and 1.

Section 5.2.1 summarized factors that are related to power when using Z. Now we summarize factors that influence power for the more common situation where T is used to make inferences about the population mean. These features include the same features listed in Section 5.2.1 plus some additional features related to nonnormality.

- As the sample size, n, gets large, power goes up, so the probability of a Type II error goes down.
- As α goes down, in which case the probability of a Type I error goes down, power goes down and the probability of a Type II error goes up.
- As the standard deviation, σ, goes up, with n, α, and $\mu - \mu_0$ fixed, power goes down.

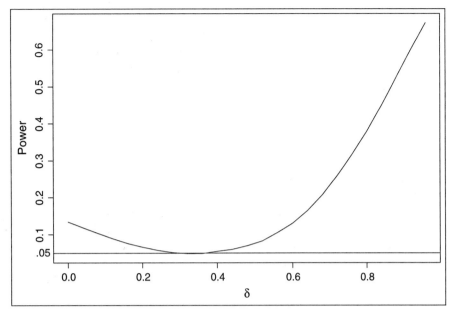

FIGURE 5.7 Power curve of Student's T when sampling from a lognormal distribution, $n = 20$, $\alpha = .05$. The null hypothesis corresponds to $\delta = 0$. Ideally the power curve should be strictly increasing as δ gets large.

- Small departures from normality can inflate the standard error of the sample mean ($\sigma/\sqrt{n}$), which in turn can substantially reduce power.
- Student's T can be biased due to skewness. That is, power might be low (relative to other inferential methods you might use) because as μ moves away from the hypothesized value, the probability of rejecting can actually decrease. Practical problems arise even when sampling from a distribution where outliers are rare.

5.6 Hypothesis Testing Based on a Trimmed Mean

An argument for testing hypotheses based on the mean is that under normality, the sample mean has a smaller standard error than any other measure of location we might use. This means that no other hypothesis-testing method will have more power than the method based on Student's T. However, this argument is not very compelling, because arbitrarily small departures from normality can result in extremely low power relative to other methods you might use, and of course there is the concern about getting accurate confidence intervals and good control over the probability of a Type I error. Currently, there seem to be two general strategies for dealing with these problems that are relatively effective. The first is to switch to a robust measure of location, and the trimmed mean is particularly appealing based on recently published studies. The second is to switch to a rank-based method, some of which are described in Chapter 15.

Here, attention is focused on the 20% trimmed mean. The method for computing a confidence interval for the trimmed mean described in Chapter 4 is easily extended to the problem of testing some hypothesis about μ_t, the population trimmed mean. The process is the same as Student's T test, only you adjust the degrees of freedom, you replace the sample mean with the trimmed mean, you replace the sample variance with the Winsorized variance, and for technical reasons you multiply by .6. In symbols, your test statistic is now

$$T_t = \frac{.6(\bar{X}_t - \mu_0)}{s_w/\sqrt{n}},\tag{5.8}$$

where again μ_0 is some specified value of interest and s_w is the 20% Winsorized standard deviation. Then reject $H_0 : \mu_t = \mu_0$ if $T_t \leq -c$ or $T_t \geq c$, where c is now the $1 - \alpha/2$ quantile of Student's T distribution with $n - 2g - 1$ degrees of freedom and g is the number of observations trimmed from each tail, as described and illustrated in Chapter 3. (The total number of trimmed observations is $2g$, so $n - 2g$ is the number of observations left after trimming.)

More generally, when using a γ-trimmed mean,

$$T_t = \frac{(1 - 2\gamma)(\bar{X}_t - \mu_0)}{s_w/\sqrt{n}},\tag{5.9}$$

where s_w is now the γ-Winsorized standard deviation. (In the previous paragraph, $\gamma = .2$.) The degrees of freedom are $\nu = n - 2g - 1$, where $g = [\gamma n]$ and $[\gamma n]$ is γn rounded down to the nearest integer.

As for the one-sided hypothesis $H_0 : \mu \geq \mu_0$, reject if $T_t \leq c$, where c is now the α quantile of Student's T distribution with $n - 2g - 1$ degrees of freedom. The hypothesis $H_0 : \mu \leq \mu_0$ is rejected if $T_t \geq c$, where c is now the $1 - \alpha$ quantile of Student's T distribution with $n - 2g - 1$ degrees of freedom.

5.6.1 S-PLUS Function trimci

The S-PLUS function

$$\text{trimci}(x, \text{tr} = .2, \text{alpha} = .05, \text{nv} = 0),$$

introduced in Chapter 4, also tests hypotheses about the population trimmed mean. By default it tests $H_0 : \mu_t = 0$. To test $H_0 : \mu_t = 2$, set the argument nv equal to 2. In addition to a confidence interval, the function returns the p-value.

EXAMPLE. Doksum and Sievers (1976) report data on weight gain among rats. One group was the control and the other was exposed to an ozone environment. (The data are given in Table 8.6.) For illustrative purposes, attention is focused on the control group, and we consider the claim that the typical weight

Continued

EXAMPLE. (*Continued*) gain is 26.4. If we test the hypothesis $H_0 : \mu = 26.4$ with Student's T, we get

$$T = \frac{\bar{X} - \mu_0}{s/\sqrt{n}} = \frac{22.4 - 26.4}{10.77/\sqrt{23}} = -1.8.$$

With $\nu = 23 - 1 = 22$ degrees of freedom, and $\alpha = .05$, the critical value is $c = 2.07$. Because $|T| = 1.8$ is less than 2.07, we fail to reject. In contrast, the 20% trimmed mean is $\bar{X}_t = 23.3$, $s_w = 3.9$, and for $H_0 : \mu_t = 26.4$ we see that

$$T_t = \frac{.6(\bar{X}_t - \mu_0)}{s_w/\sqrt{n}} = \frac{.6(23.3 - 26.4)}{3.9/\sqrt{23}} = -2.3.$$

Because there are 23 rats, $g = 4$, so the number of trimmed observations is $2g = 8$, the degrees of freedom are $\nu = 23 - 8 - 1 = 14$, and the critical value is $c = 2.14$. Because $|T| = |-2.3| = 2.3$ is greater than the critical value, we reject the hypothesis that the trimmed mean is 26.4. Thus, although you cannot rule out the possibility that the population mean for all rats is 26.4 if our Type I error probability is to be .05, it is unlikely that the population trimmed mean has this value. ■

In the last example, the sample mean exceeds the hypothesized value by more than the sample trimmed mean. The difference between the hypothesized value of 26.4 and the mean is $22.4 - 26.4 = -4$. The difference between the hypothesized value and the trimmed mean is $23.3 - 26.4 = -3.1$, yet you reject with the trimmed mean but not with the mean. The reason is that the standard error of the trimmed mean is smaller than the standard error of the mean. This illustrates one of the practical advantages of using a trimmed mean. Situations often arise where the trimmed mean has a substantially smaller standard error, and this can translate into a substantial gain in power.

Once again it is stressed that for skewed distributions, population means and trimmed means are generally not equal. So, for example, the null hypothesis $H_0 : \mu = 26.4$ is not necessarily the same as $H_0 : \mu_t = 26.4$. In the context of hypothesis testing, an argument for the trimmed mean is that good control over the probability of a Type I error can be achieved in situations where Student's T gives poor results. Trimmed means often have a smaller standard error than the mean which can result in substantially higher power. If there is some reason for preferring the mean to the trimmed mean in a particular study, Student's T might be unsatisfactory unless the sample size is very large. Just how large n must be depends on the *unknown* distribution from which observations were sampled. In some cases even a sample size of 300 is unsatisfactory. A small sample size will suffice in some instances, but an effective method for establishing whether this is the case, simply by examining your data, has not been found.

In this book, only two-sided trimming is considered. If a distribution is skewed to the right, for example, a natural reaction is to trim large observations but not small ones. An explanation can now be given as to why one-sided trimming is not recommended. In terms of Type I errors and probability coverage, you get more accurate results if two-sided trimming is used. There is nothing obvious or intuitive about this result, but all of the studies cited by Wilcox (1997a) support this view.

Thanks to the central limit theorem, we know that when working with means, problems with Student's T diminish as the sample size increases. Theory and simulations indicate that when using a 20% trimmed mean instead, problems diminish much more rapidly. That is, smaller sample sizes are required to get good control over the probability of a Type I error; but with very small sample sizes, practical problems persist. (Methods covered in Chapter 7 provide a basis for dealing with very small sample sizes in an effective manner.)

Finally, no mention has been made about how to determine whether power is adequate based on the sample size used when making inferences about a trimmed mean. Theoretical results suggest how an analog of Stein's method might be derived; a reasonable speculation is that the method should perform well when sample sizes are small. But this issue has not yet been investigated. An alternative solution has been found, but we will need some tools covered in Chapter 7 before describing it.

5.7 Exercises

1. Given that $\bar{X} = 78$, $\sigma^2 = 25$, $n = 10$, and $\alpha = .05$, test $H_0 : \mu > 80$, assuming observations are randomly sampled from a normal distribution. Also, draw the standard normal distribution indicating where Z and the critical value are located.

2. Repeat Exercise 1, but test $H_0 : \mu = 80$.

3. For Exercise 2, compute a .95 confidence interval and verify that this interval is consistent with your decision about whether to reject the null hypothesis.

4. For Exercise 1, determine the p-value.

5. For Exercise 2, determine the p-value.

6. Given that $\bar{X} = 120$, $\sigma = 5$, $n = 49$, and $\alpha = .05$, test $H_0 : \mu > 130$, assuming observations are randomly sampled from a normal distribution.

7. Repeat Exercise 6, but test $H_0 : \mu = 130$.

8. For Exercise 7, compute a .95 confidence interval and compare the result with your decision about whether to reject H_0.

9. If $\bar{X} = 23$ and $\alpha = .025$, can you make a decision about whether to reject $H_0 : \mu < 25$ without knowing σ?

10. An electronics firm mass-produces a component for which there is a standard measure of quality. Based on testing vast numbers of these components, the company has found that the average quality is $\mu = 232$ with $\sigma = 4$. However, in recent years the quality has not been checked, so management asks you to check their claim with the goal of being reasonably certain that an average quality of less than 232 can be ruled out. That is, assume the quality is poor and in fact less than 232, with the goal of empirically establishing that this assumption is unlikely. You get $\bar{X} = 240$ based on a sample $n = 25$ components, and you want the probability of a Type I error to be .01. State the null hypothesis, and perform the appropriate test assuming normality and $\sigma = 4$.

11. An antipollution device for cars is claimed to have an average effectiveness of exactly 546. Based on a test of 20 such devices you find that $\bar{X} = 565$. Assuming normality and that $\sigma = 40$, would you rule out the claim with a Type I error probability of .05?

12. Comment on the relative merits of using a .95 confidence interval for addressing the effectiveness of the antipollution device in Exercise 11.

13. For $n = 25$, $\alpha = .01$, $\sigma = 5$, and $H_0 : \mu \geq 60$, verify that power is .95 when $\mu = 56$.

14. For $n = 36$, $\alpha = .025$, $\sigma = 8$, and $H_0 : \mu \leq 100$, verify that power is .61 when $\mu = 103$.

15. For $n = 49$, $\alpha = .05$, $\sigma = 10$, and $H_0 : \mu = 50$, verify that power is approximately .56 when $\mu = 47$.

16. A manufacturer of medication for migraine headaches knows that their product can cause liver damage if taken too often. Imagine that by a standard measuring process, the average liver damage is $\mu = 48$. A modification of their product is being contemplated, and, based on $n = 10$ trials, it is found that $\bar{X} = 46$. Assuming $\sigma = 5$, they test $H_0 : \mu \geq 48$, the idea being that if they reject, there is convincing evidence that the average amount of liver damage is less than 48. Then

$$Z = \frac{46 - 48}{5/\sqrt{10}} = -1.3.$$

With $\alpha = .05$, the critical value is -1.645, so they do not reject, because Z is not less than the critical value. What might be wrong with accepting H_0 and concluding that the modification results in an average amount of liver damage greater than or equal to 48?

17. For Exercise 16, verify that power is .35 if $\mu = 46$.

18. Exercise 17 indicates that power is relatively low with only $n = 10$ observations. Imagine that you want power to be at least .8. One way of getting more power is to increase the sample size, n. Verify that for sample sizes of 20, 30, and 40, power is .56, .71, and .81, respectively.

19. For Exercise 18, rather than increase the sample size, what else might you do to increase power? What is a negative consequence of using this strategy?

20. Test the hypothesis $H_0 : \mu = 42$ with $\alpha = .05$ and $n = 25$ given the following values for $\bar{X}$ and s: (a) $\bar{X} = 44$, $s = 10$, (b) $\bar{X} = 43$, $s = 10$, (c) $\bar{X} = 43$, $s = 2$.

21. For part b of Exercise 20, you fail to reject, but you reject for the situation in part c. What does this illustrate about power?

22. Test the hypothesis $H_0 : \mu < 42$ with $\alpha = .05$ and $n = 16$ given the following values for $\bar{X}$ and s: (a) $\bar{X} = 44$, $s = 10$, (b) $\bar{X} = 43$, $s = 10$, (c) $\bar{X} = 43$, $s = 2$.

23. Repeat Exercise 22, except test $H_0 : \mu > 42$.

24. A company claims that on average, when exposed to their toothpaste, 45% of all bacteria related to gingivitis are killed. You run ten tests and find that the percentages of bacteria killed among these tests are 38, 44, 62, 72, 43, 40, 43, 42, 39, 41. The mean and standard deviation of these values are $\bar{X} = 46.4$ and $s = 11.27$. Assuming normality, test the hypothesis that the average percentage is 45 with $\alpha = .05$.

25. A portion of a study by Wechsler (1958) reports that for 100 males taking the Wechsler Adult Intelligence Scale (WAIS), the sample mean and variance on picture completion are $\bar{X} = 9.79$ and $s = 2.72$. Test the hypothesis $H_0 : \mu \geq 10.5$ with $\alpha = .025$.

26. Assuming 20% trimming, test the hypothesis $H_0 : \mu_t = 42$ with $\alpha = .05$ and $n = 20$ given the following values for $\bar{X}_t$ and s_w: (a) $\bar{X}_t = 44$, $s_w = 9$, (b) $\bar{X}_t = 43$, $s_w = 9$, (c) $\bar{X}_t = 43$, $s_w = 3$.

27. Repeat Exercise 26, except test the hypothesis $H_0 : \mu_t < 42$ with $\alpha = .05$ and $n = 16$.

28. For the data in Exercise 24, the trimmed mean is $\bar{X}_t = 42.17$ with a Winsorized standard deviation of $s_w = 1.73$. Test the hypothesis that the population trimmed mean is 45 with $\alpha = .05$.

29. A standard measure of aggression in 7-year-old children has been found to have a 20% trimmed mean of 4.8 based on years of experience. A psychologist wants to know whether the trimmed mean for children with divorced parents differs from 4.8. Suppose $\bar{X}_t = 5.1$ with $s_w = 7$ based on $n = 25$. Test the hypothesis that the population trimmed mean is exactly 4.8 with $\alpha = .01$.

6

LEAST SQUARES REGRESSION AND PEARSON'S CORRELATION

Two common goals are determining whether and how two variables are related. This chapter describes the basics of the two most frequently used approaches to these problems. The first is based on what is called *least squares regression*. The second employs an estimate of Pearson's correlation (ρ). It will be seen that these methods are satisfactory for some purposes, but they are completely inadequate for others. In terms of hypothesis testing, the techniques in this chapter inherit the problems associated with means, and new problems are introduced. There are methods for dealing with these difficulties, but the details are given in subsequent chapters. The main goal here is to summarize some basic principles and to motivate the use of more modern techniques.

6.1 Fitting a Straight Line to Data: The Least Squares Principle

When describing the association between two variables, certainly the most common strategy is to assume that the association is linear. There are practical situations where the assumption of a linear association is unreasonable, but this issue is ignored for the moment.

We illustrate the basic strategy for fitting a straight line to a scatterplot of points using data from a classic study in astronomy. Is there some pattern to how the galaxies in the universe are moving relative to one another? Edwin Hubble collected data on two measures relevant to this issue in the hope of gaining some insight into how the universe was formed. He measured the distance of 24 galaxies from earth plus their recession velocity. His measurements, published in 1929, are shown in Table 6.1, where X is a galaxy's distance from earth in megaparsecs and Y is its speed in kilometers per second. (One parsec is 3.26 light-years.) For example, the first galaxy is .032 megaparsecs from earth and moving away from earth at the rate of 170 kilometers per second. The third galaxy is .214 megaparsecs from earth and approaching earth at the rate of 130 kilometers per second.

TABLE 6.1   Hubble's Data on the Distance and Recession Velocity of 24 Galaxies

Distance (X):	0.032	0.034	0.214	0.263	0.275	0.275	0.450	0.500	0.500	0.630	0.800	0.900
	0.900	0.900	0.900	1.000	1.100	1.100	1.400	1.700	2.000	2.000	2.000	2.000
Velocity (Y):	170	290	−130	−70	−185	−220	200	290	270	200	300	−30
	650	150	500	920	450	500	500	960	500	850	800	1090

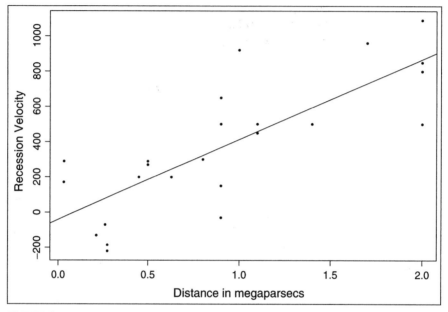

FIGURE 6.1   Scatterplot of Hubble's data on the recession velocity of galaxies. The straight line has slope 454 and intercept −40.8.

A common notation for representing n pairs of points is

$$(X_1, Y_1), \ldots, (X_n, Y_n).$$

So in Table 6.1, $(X_1, Y_1) = (.032, 170)$, where the subscript 1 indicates the first pair of observations. Here $n = 24$ and $(X_{24}, Y_{24}) = (2, 1090)$.

To fit a straight line to Hubble's data, we need a criterion for judging how well any line fits a scatterplot of the points. Given a criterion, we can then search for the line that is optimal. More precisely, consider lines having the form

$$\hat{Y} = b_0 + b_1 X, \tag{6.1}$$

where b_1 and b_0 are, respectively, the unknown slope and intercept that are to be determined from the data and the notation $\hat{Y}$ is used to make a distinction between the predicted value of Y based on X, which is $\hat{Y}$, and the Y we observe. Figure 6.1 shows a scatterplot of Hubble's data together with the line $\hat{Y} = 454X - 40.8$ (so $b_1 = 454$ and $b_0 = -40.8$). For the first galaxy in Table 6.1, which is $X_1 = .032$ megaparsecs from earth, its predicted recession velocity is $\hat{Y}_1 = 454(.032) - 40.8 = -26.3$ kilometers per second. But from Table 6.1 we see that its actual recession velocity is $Y_1 = 170$,

TABLE 6.2   Fitted Values and Residuals for the Data in Table 6.1

Observation number	Y_i	$\hat{Y}_i$	r_i
1	170	−26.3	196.3
2	290	−25.3	315.3
3	−130	56.4	−186.4
4	−70	78.7	−148.7
5	−185	84.1	−269.1
6	−220	84.1	−304.1
7	200	163.6	36.4
8	290	186.3	103.7
9	270	186.3	83.7
10	200	245.3	−45.3
11	300	322.5	−22.5
12	−30	368.0	−398.0
13	650	282	367.0
14	150	368.0	−217.0
15	500	368.0	132.0
16	920	413.4	506.6
17	450	458.8	−8.8
18	500	458.8	41.2
19	500	595.0	−95.0
20	960	731.3	228.7
21	500	867.5	−367.5
22	850	867.5	−17.5
23	800	867.5	−67.5
24	1090	867.5	222.5

so for the first galaxy having $(X_1, Y_1) = (.032, 170)$, there is a discrepancy between the actual and predicted velocity of $r_1 = Y_1 - \hat{Y}_1 = 170 + 26.3 = 196.3$. More generally, for the ith pair of observations, there is a discrepancy between the observed and predicted Y values, given by

$$r_i = Y_i - \hat{Y}_i, \tag{6.2}$$

where r_i ($i = 1, \ldots, n$) is called the ith *residual*. Table 6.2 shows the $\hat{Y}$ values and residuals for the data in Table 6.1.

For any slope (b_1) and intercept (b_0) we might choose, one way of judging the overall fit of the resulting line to a scatterplot of points is to use the sum of the squared residuals:

$$\sum r_i^2 = \sum (Y_i - b_1 X_i - b_0)^2. \tag{6.3}$$

If we choose the slope and intercept so as to minimize the sum of squared residuals, we are using what is called the *least squares principle*. Without making any assumptions about the distribution of X or Y, least squares leads to taking the slope to be

$$b_1 = \frac{\sum (X_i - \bar{X})(Y_i - \bar{Y})}{\sum (X_i - \bar{X})^2} \tag{6.4}$$

and the intercept is

$$b_0 = \bar{Y} - b_1 \bar{X}. \tag{6.5}$$

That is, if we use the regression equation

$$\hat{Y} = b_1 X + b_0,$$

with b_1 and b_0 given by Equations (6.4) and (6.5), we minimize the sum of squared residuals. The straight line in Figure 6.1 is the least squares regression line for Hubble's data.

Chapter 3 noted that when summarizing a single measure, the least squares principle leads to the sample mean, which has a finite-sample breakdown point of only $1/n$. That is, only one unusual point can render the sample mean meaningless. And even when sampling from a perfectly symmetric distribution, in which case the population mean provides a reasonable measure of location, the sample mean can be a relatively inaccurate estimate of μ. These problems extend to the situation at hand, and there are new ways practical problems can arise.

To begin to explain the practical problems associated with least squares regression, we first note that the estimate of the slope can be written as a weighted mean of the Y values, with the weights depending on the X values. In particular, the least squares estimate of the slope can be written as

$$b_1 = \sum w_i Y_i,$$

where

$$w_i = \frac{X_i - \bar{X}}{(n-1)s_x^2}$$

and s_x^2 is the sample variance of the X values. But we saw in Chapter 3 that a weighted mean has a finite-sample breakdown point of only $1/n$, so the finite-sample breakdown point of the least squares estimate of the slope is $1/n$ as well. In particular, a single unusual X value can have an inordinate influence on the estimate of the slope, as can a single unusual Y value. This suggests checking whether any of the X values are outliers, doing the same for the Y values; if no outliers are found, assume that there are no influential points that result in a poor fit to the bulk of the points. Unfortunately, this relatively simple strategy can be inadequate, for at least two reasons. First, there are situations where an outlying X value is beneficial. (Details are covered in Section 6.3.1.) Second, a small number of unusual points can greatly influence the least squares estimate, giving a distorted view of how the bulk of the points are associated, even though none of the corresponding X or Y values is declared outliers using the methods of Section 3.4. When addressing this latter problem, what is needed is some method that takes into account the overall structure of the points, but for now we merely illustrate the problem.

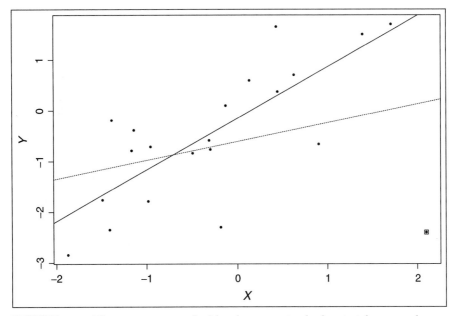

FIGURE 6.2   The two points marked by the square in the lower right corner have a substantial impact of the least squares regression line. Ignoring these two points, the least squares regression is given by the solid line. Including them, the least squares regression line is given by the dashed line. Moreover, none of the X or Y values are declared outliers using the methods in Chapter 3.

EXAMPLE. To illustrate the last problem, Figure 6.2 shows 20 points that were generated on a computer, where both X and Y are normal and the points are centered around the line $Y = X$. So the true slope is 1. The solid straight line passing through the bulk of the points in Figure 6.2 is this least squares estimate of the regression line and has slope $b_1 = 1.01$. Then two additional points were added at $X = 2.1$ and $Y = -2.4$ and are marked by the square in the lower right corner of Figure 6.2. Among all 22 X values, none is declared an outlier by any of the methods in Section 3.4, and the same is true for the Y values. Yet these two additional points are clearly unusual relative to the other 20, and they have a substantial influence on the least squares estimate of the slope. Now the estimated slope is .37, and the resulting least squares regression line is represented by the dashed line in Figure 6.2. Note that the estimated intercept changes substantially as well.   ■

EXAMPLE. Figure 6.3 shows the surface temperature (X) of 47 stars versus their light intensity. The solid line is the least squares regression line. As is evident, the regression line does a poor job of summarizing the association

Continued

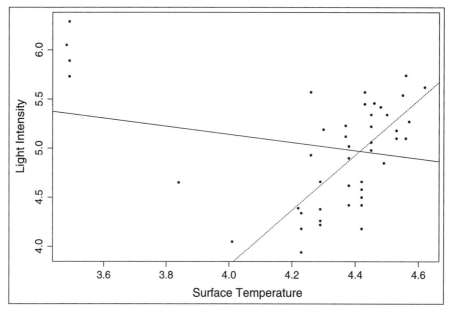

FIGURE 6.3   The solid line is the least squares regression line using all of the star data. When ignoring outliers among the X values, the least squares regression line is given by the dashed line.

EXAMPLE. (*Continued*) between the two variables under study. In this particular case it is clear that the four points in the upper left portion of Figure 6.3 are unusual. Moreover, the S-PLUS function outbox (described in Section 3.4.6) indicates that X values less than or equal to 3.84 are outliers. If we simply exclude all points with X values declared outliers and apply least squares regression to the points that remain, we get the dashed line shown in Figure 6.3. This provides a better summary for the bulk of the points, but even then the least squares regression line does not seem quite satisfactory.   ■

One point should be stressed. In the last example we simply restricted the range of X values to get a better fit to the bulk of the points under study. A similar strategy might be used when dealing with unusual Y values. But when restricting the range of Y values, caution must be used. The reason is that when testing hypotheses about the slope and intercept, simply eliminating points with unusual Y values leads to technical problems that are described in Section 6.3.3. Special techniques for dealing with unusual Y values are required, and they are described in Chapter 13.

6.2 The Standard Least Squares Model

This section describes the standard least squares model, which is important to understand for at least two general reasons. First, it describes conditions under which the least squares estimate of the slope and intercept [given by Equations (6.4) and (6.5)] have the smallest standard error among a large class of estimators. Said another way, understanding this model is important when trying to understand when and why the least squares estimate of the slope and intercept can be unsatisfactory. Second, a special case of this model is used when computing confidence intervals and testing hypotheses, and understanding the standard model helps explain some of the reasons conventional hypothesis-testing methods can be highly inaccurate.

To be concrete, we describe the model using a study conducted by G. Margolin and A. Medina, where the goal was to examine how children's information processing is related to a history of exposure to marital aggression. Results for two of their measures are shown in Table 6.3. The first, labeled X, is a measure of marital aggression that reflects physical, verbal, and emotional aggression during the last year; Y is a child's score on a recall test. If aggression in the home (X) has a relatively low value, what would we expect a child to score on the recall-test (Y)? If the measure of aggression is high, now what would we expect the recall-test score to be?

TABLE 6.3 Measures of Marital Aggression and Recall-Test Scores

| Family | Aggression | Test score | Family | Aggression | Test score |
i	X_i	Y_i	i	X_i	Y_i
1	3	0	25	34	2
2	104	5	26	14	0
3	50	0	27	9	4
4	9	0	28	28	0
5	68	0	29	7	4
6	29	6	30	11	6
7	74	0	31	21	4
8	11	1	32	30	4
9	18	1	33	26	1
10	39	2	34	2	6
11	0	17	35	11	6
12	56	0	36	12	13
13	54	3	37	6	3
14	77	6	38	3	1
15	14	4	39	3	0
16	32	2	40	47	3
17	34	4	41	19	1
18	13	2	42	2	6
19	96	0	43	25	1
20	84	0	44	37	0
21	5	13	57	11	2
22	4	9	46	14	11
23	18	1	47	0	3
24	76	4			

Among the millions of homes in the world, temporarily consider all homes that have an aggression measure of 50 ($X = 50$). Among these homes there will be some average value on the recall test (Y). In more formal terms, this average is $E(Y|X = 50)$, the mean of Y given that $X = 50$. The standard regression model assumes that the population mean of Y, given X, is $\beta_0 + \beta_1 X$, where β_1 and β_0 are unknown parameters that we want to estimate based on observations available to us. [Least squares regression estimates β_1 and β_0 with b_1 and b_0, respectively, given by Equations (6.4) and (6.5).] One problem is that in general, still restricting attention to those homes with $X = 50$, the Y values will differ from one another. In particular, they will not always be equal to the population mean of the Y values. Of course typically this will be true for any value of X we might pick. That is, for a randomly sampled pair of observations, (X, Y), ordinarily there will be a discrepancy between Y and its (conditional) mean given X. In formal terms, this discrepancy is $e = Y - \beta_1 X - \beta_0$. Rearranging the terms of this last equation we get the *standard regression model*:

$$Y = \beta_0 + \beta_1 X + e, \qquad (6.6)$$

where e is the so-called error term, which is assumed to have a mean of zero. That is, $E(e) = 0$ is assumed, which implies that given X, $E(Y) = \beta_0 + \beta_1 X$.

This regression model is said to be *homoscedastic* if the variance of the error term does not depend on X. In our example, this means, for example, that the variance of recall-test scores (Y), given that aggression in the home (X) is 50, is equal to the variance of the recall-test scores given that aggression in the home is 75. More generally, for any value of X we might pick, homoscedasticity means that the conditional variance of Y, given X, does not change with X—it is some constant value, as illustrated in Figure 6.4.

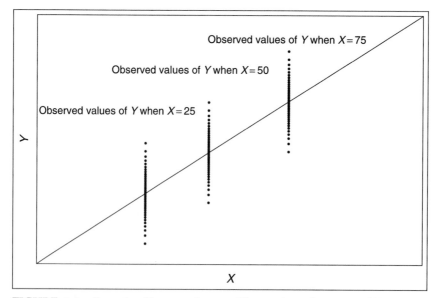

FIGURE 6.4   Example of homoscedasticity. The conditional variance of Y, given X, does not change with X.

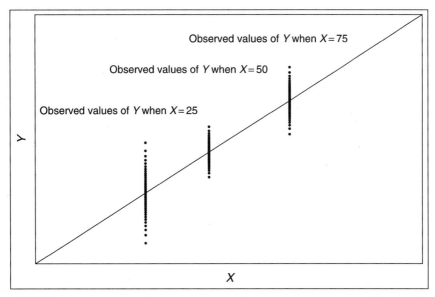

FIGURE 6.5 Example of heteroscedasticity. The conditional variance of Y, given X, changes with X.

A *heteroscedastic* regression model refers to a situation where this conditional variance changes with X, such as illustrated in Figure 6.5, where the variance of Y is smaller when $X = 50$ versus the other two X values shown.

Notice that b_1 has a sampling distribution. That is, if we were to repeat a study infinitely many times with each replication based on n randomly sampled pair of observations, we would get a collection of estimated slopes that generally differ from one another, and of course the same is true for the intercept. For the regression model given by Equation (6.6), it can be shown that

$$E(b_1) = \beta_1 \qquad \text{and} \qquad E(b_0) = \beta_0.$$

That is, b_1 and b_0 are unbiased estimates of the slope and intercept, respectively, roughly meaning that on average, they give a correct estimate of the true slope and intercept. The squared standard error of b_1 refers to the variance of the b_1 values obtained when repeating an experiment infinitely many times. If we can find an unbiased estimator with a smaller standard error, then generally it will be a more accurate estimate of the true slope, β_1. Our immediate goal is to describe conditions under which the least squares estimates of the slope and intercept have the smallest possible standard error.

For the homoscedastic regression model just described, among all the weighted means of the Y values we might consider for estimating the slope and intercept, the least squares estimate given by Equations (6.4) and (6.5) has the smallest standard error. This result is a special case of what is known as the *Gauss–Markov* theorem. This theorem also yields the weighted mean with the smallest standard error in the heteroscedastic case. In particular, Gauss showed that if the conditional variance of the Y values, given that $X = X_i$, is σ_i^2, say, the optimal estimates of the slope and intercept are the values b_1 and b_0 that minimize

$$\sum w_i(Y_i - b_1X_i - b_0)^2, \qquad (6.7)$$

where $w_i = 1/\sigma_i^2$. In our illustration, for example, we see from Table 6.3 that the first aggression score is $X_1 = 3$, the second is $X_2 = 104$, and so on. If we knew σ_1^2, the variance of recall-test scores (Y) given that the aggression measure is $X_1 = 3$, and more generally if we knew σ_i^2, the conditional variance of Y given that $X = X_i$, we could compute the optimal weighted mean estimate of the slope and intercept, which are the values b_1 and b_0 that minimize Equation (6.7). In the homoscedastic case, all of the w_i values in Equation (6.7) have a common value and the b_1 and b_0 values that minimize this equation are the same values that minimize Equation (6.3). Determining the slope and intercept with Equation (6.7) is an example of what is called *weighted least squares*, and using Equation (6.3) is called *ordinary least squares*.

A problem is that typically the σ_i^2 are not known, so a common strategy is to assume homoscedasticity and simply estimate the slope and intercept with ordinary least squares [using Equations (6.4) and (6.5)], the optimal least squares estimates in the homoscedastic case. A practical issue is whether knowing the σ_i^2 would result in a substantially more accurate estimate of the slope and intercept. If the answer is no, heteroscedasticity is not a concern; but if such situations arise, methods for dealing with heteroscedasticity become important.

EXAMPLE. Consider a situation where both X and the error term (e) are standard normal and the slope is $\beta_1 = 1$. Further assume that the standard deviation of Y, given X, is 1 if $|X| < 1$, otherwise it is $|X| + 1$. So, for example, if $X = .5$, Y has variance 1; but if $X = 1.5$, Y has standard deviation 2.5. Figure 6.6 shows a scatterplot of points generated in this fashion. Figure 6.7 shows the sampling distribution of the weighted and ordinary least squares estimates of the slope. The standard error of the ordinary least squares estimator is .52, versus .37 using the weighted least squares estimator instead. As is evident, the weighted least squares estimate tends to be much closer to the correct value for the slope. ■

EXAMPLE. Consider the same situation as in the last example, except that the conditional standard deviation of Y, given X, is X^2. Now the standard error of the ordinary least squares estimator is more than 10 times larger than the standard error of the optimal weighted least squares estimator. So we see that even under normality, the ordinary least squares estimate of the slope and intercept can be highly inaccurate relative to the optimal weighted least squares approach. Nonnormality can make the least squares estimator perform even more poorly. ■

One way of trying to improve upon ordinary least squares is to attempt to estimate the variance of Y given X. If we could do this in a reasonably accurate manner for

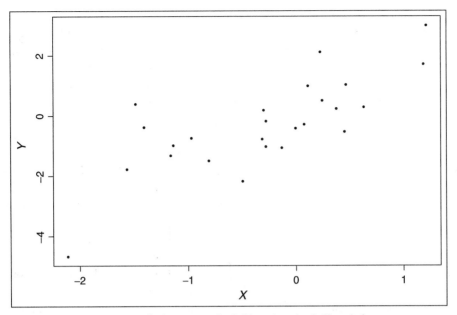

FIGURE 6.6   Example of what points look like when both X and the error term are standard normal and the variance of Y is 1 if $|X| < 1$ and $|X| + 1$ if not.

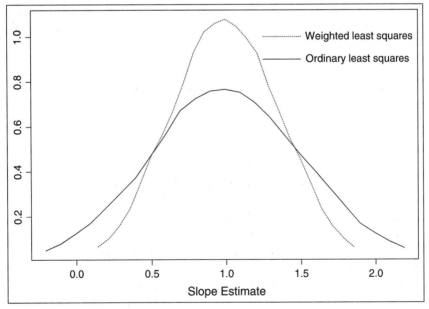

FIGURE 6.7   Comparison of the sampling distributions of ordinary least squares estimator versus the optimal weighted least squares estimator. There is heteroscedasticity, which is why the weighted least squares estimator tends to be closer to the correct value of 1.

every X value we observe, weighted least squares could be employed. In the aggression study, for example, the first home had an aggression score of 3, so we need to estimate the variance of the recall-test scores for all homes having an aggression score of 3. In a similar manner, the second home had an aggression score of 104, so we need to estimate the variance of the recall-test scores for all homes having an aggression score of 104. In more formal terms, $X_2 = 104$, and to apply weighted least squares we need to know σ_2^2, the variance of Y given that $X = X_2$. The third home has an aggression score of $X_3 = 50$, so we need to know σ_3^2 as well, which is the variance of the recall-test scores when the aggression measure is 50. If we had a reasonably large number of homes with an aggression score of 50, we could simply compute the sample variance of the corresponding recall-test scores to get an estimate of σ_3^2. The practical problem is that we have only one home with an aggression score of 50, so this method cannot be used.

There are at least two general strategies for dealing with heteroscedasticity when the goal is to find an estimator of the slope and intercept that has a relatively small standard error. The first uses what are called *smoothers* to estimate the σ_i^2 values (e.g., Müller, 1988, p. 153; M. Cohen, Dalal, & Tukey, 1993; Wilcox, 1996a). We do not describe these methods here because either they ignore certain problems caused by outliers, or there is no known method that performs reasonably well when testing hypotheses or computing confidence intervals and the sample sizes are small or even moderately large. Instead we rely on the second general strategy, which uses one of the robust regression methods covered in Chapter 13. So the main point here is that heteroscedasticity matters in terms of getting an estimator of the slope and intercept that has a relatively small standard error, but we postpone how to deal with this problem for now.

6.2.1 Comments About Linearity and Homoscedasticity

Caution must be exercised when assuming that a regression line is straight. Consider, for example, the aggression data in Table 6.3, where Y is a recall-test score. If we fit a straight line using the least squares principle, we find that $b_1 = -0.0405$ and $b_0 = 4.581$. Figure 6.8 shows a scatterplot of the 47 pairs of observations along with the least squares regression line used to predict test scores. If, for instance, the measure of marital aggression is $X = 20$, the fitted value for the score on the recall test is $\hat{Y} = -0.0405(20) + 4.581 = 3.77$. Similarly, if $X = 40$, the estimate is $\hat{Y} = 2.961$. The estimated recall-test score corresponding to $X = 40$ is less than it is when $X = 20$ because the slope is negative. Generally, the regression equation $Y = -0.0405X + 4.581$ suggests that the higher the measure of aggression, the lower the score on the recall test. Is there some possibility that this is a misleading representation of the data?

Now consider the right portion of Figure 6.8, consisting of the 21 aggression scores greater than or equal to 25. For these 21 values, if we estimate the regression line, ignoring the other 26 values, we find that

$$\hat{Y} = 0.002285X + 1.93.$$

Note that the slope is slightly larger than zero. Now the fitted value for $X = 25$ is $\hat{Y} = 1.98$, and for $X = 40$ it is $\hat{Y} = 2.02$. In contrast, for the aggression scores

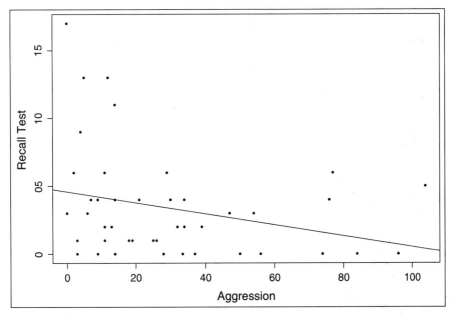

FIGURE 6.8 Scatterplot and least squares regression line for the aggression data in Table 6.3.

less than 25, ignoring the scores greater than 25, the regression equation is

$$\hat{Y} = -0.194X + 6.33.$$

The left panel of Figure 6.9 shows a scatterplot of the aggression data, with squares around the points having aggression scores greater than 25. Also shown is the least squares regression line based on these rightmost points, ignoring the points not marked with a square. The right panel shows the least squares regression line based on the remaining 26 pairs of observations, which are now marked with squares. As is evident, these two plots suggest something quite different from the regression line based on all of the data. In particular, one possibility is that there is a negative association for low aggression scores, but as the aggression scores increase, there seems to be little or no association at all. This means that, for the range of aggression scores available, perhaps there is a nonlinear association between aggression and test scores. Another possibility is that with more data, a straight regression line would prove to be adequate.

In recent years, better and more sophisticated methods have been developed for studying, describing, and detecting nonlinear associations. The only goal here is to illustrate a situation where nonlinearity might be an issue. Some very useful exploratory methods for studying nonlinearity can be found in Hastie and Tibshirani (1990). (A portion of these methods will be covered in Chapter 13.) There are also methods that can be used to test the hypothesis that an association between two random variables is linear.

A natural strategy when dealing with curvature is to add a quadratic term. That is, use a prediction rule having the form $\hat{Y} = \beta_0 + \beta_1 X + \beta_2 X^2$, or, more generally, include a term with X raised to some power. This might suffice in some situations,

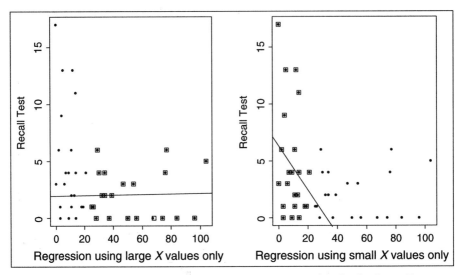

FIGURE 6.9   The left panel shows the least squares regression line for the data in Figure 6.8 using only the points marked with a square. The right panel shows the least squares regression line using only the data ignored in the right panel. As is evident, the left portion of the data gives a substantially different impression of how aggression and recall test scores are associated.

but experience with modern methods suggests that another type of nonlinearity is common: A straight line with a nonzero slope gives a reasonable prediction rule over some range of X values, but for X values outside this range there is little or no association at all. That is, the slope is zero. The aggression data in Table 6.3 are one example where this might be true.

Yet another interesting feature of the aggression data is the possibility is that there is heteroscedasticity. Perhaps for fairly low X values, the variance of Y is large relative to higher X values. It might even be the case that X and Y are dependent but that this dependence is due primarily to how the variance of Y, as opposed to the mean, is related to X. In any event, the three largest Y values, which appear in the upper left portion of Figure 6.8, have a substantial impact on the least squares regression line. If we ignore these three points, the least squares slope goes from $-.041$ to $-.017$.

6.3 Hypothesis Testing and Confidence Intervals

So far, attention has been focused on fitting a straight line to data without making any inferences about what the slope and intercept might be if you could sample all subjects of interest. That is, β_1 and β_0 represent the slope and intercept if all subjects could be sampled, and there is interest in determining which values for these parameters can be ruled out based on the data at hand. One of the most common goals is to test $H_0 : \beta_1 = 0$. If $\beta_1 = 0$, knowing X provides no help in our attempts to estimate Y using a straight regression line. In the aggression study, for example, the regression equation becomes $\hat{Y} = 3.4$, the sample mean of all the Y values.

TABLE 6.4   Selling Price of Homes (divided by 1000) versus Size (in square feet)

Home i	Size (X_i)	Price (Y_i)	Home i	Size (X_i)	Price (Y_i)
1	2359	510	15	3883	859
2	3397	690	16	1937	435
3	1232	365	17	2565	555
4	2608	592	18	2722	525
5	4870	1125	19	4231	805
6	4225	850	20	1488	369
7	1390	363	21	4261	930
8	2028	559	22	1613	375
9	3700	860	23	2746	670
10	2949	695	24	1550	290
11	688	182	25	3000	715
12	3147	860	26	1743	365
13	4000	1050	27	2388	610
14	4180	675	28	4522	1290

This says that regardless of what the aggression score X might be, you estimate the score on the recall test to be 3.4. In general, if a regression equation is assumed to have the form given by Equation (6.1) and if $\beta_1 = 0$, then $\hat{Y} = \bar{Y}$. Thus, rejecting $H_0 : \beta_1 = 0$ says that knowing X provides some help in estimating Y, assuming a linear equation having the form $\hat{Y} = \beta_0 + \beta_1 X$ is to be used.

To begin to understand the standard method for testing hypotheses, consider the data in Table 6.4, which shows the selling price (divided by 1000) of homes in a particular suburb of Los Angeles during the month of May 1998. Also shown is the size of the home, in square feet. Given that you are interested in buying a home with 2000 square feet, what would you expect to pay? What would you expect to pay for a house having 1500 square feet?

If we assume that the mean selling price of a home (Y), given its square feet (X), is

$$E(Y) = \beta_0 + \beta_1 X,$$

then the least squares estimates of the slope and intercept are, respectively,

$$b_1 = .215 \quad \text{and} \quad b_0 = 38.192.$$

So, for example, the estimated cost of a house with 2000 square feet (in thousands of dollars) is

$$.00215(2000) + .38192 = 468.7921,$$

or \$468,792. But the data in Table 6.4 do not represent all homes sold in this area. If all homes were included, we would know β_1 and β_0. How can we compute confidence intervals or test hypotheses about these two parameters?

The standard solution is to assume random sampling, homoscedasticity, and that the distribution of the error term (e) in the regression model [given by Equation (6.6)] has a normal distribution. This last assumption means that the selling price of homes, given that they have 1400 square feet, for example, is normally distributed. Similarly, the selling price of homes, among all homes having 2000 square feet, is normally distributed. More generally, the distribution of Y, given X, is assumed to be normal.

As previously explained, homoscedasticity means that the conditional variance of Y, given X, does not change with X. Let σ^2 represent this common variance. In formal terms, $\sigma^2 = \text{VAR}(Y|X) = \text{VAR}(e)$, which does not depend on X. The standard estimate of σ^2 is

$$\hat{\sigma}^2 = \frac{1}{n-2} \sum r_i^2,$$

the sum of the squared residuals divided by $n-2$.

EXAMPLE. For the housing data in Table 6.4, the sum of squared residuals is $\sum r_i^2 = 26.99$. There are $n = 14$ homes, so, assuming homoscedasticity, the estimate of σ^2 is

$$\hat{\sigma}^2 = \frac{26.99}{14-2} = 2.25.$$

■

6.3.1 Conventional Hypothesis Testing and Confidence Intervals

Under the assumptions that the regression model is true and that there is random sampling, homoscedasticity, and normality, confidence intervals for the slope and intercept can be computed and hypotheses can be tested. In particular, a $1 - \alpha$ confidence interval for the slope, β_1, is

$$b_1 \pm t \sqrt{\frac{\hat{\sigma}^2}{\sum (X_i - \bar{X})^2}}, \tag{6.8}$$

where t is the $1 - \alpha/2$ quantile of Student's T distribution with $\nu = n - 2$ degrees of freedom. (The value of t is read from Table 4 in Appendix B.) The quantity

$$\sqrt{\frac{\hat{\sigma}^2}{\sum (X_i - \bar{X})^2}}$$

is the *estimated standard error* of b_1. As for the intercept, β_0, a $1 - \alpha$ confidence interval is given by

$$b_0 \pm t \sqrt{\frac{\hat{\sigma}^2 \sum X_i^2}{n \sum (X_i - \bar{X})^2}}. \tag{6.9}$$

The quantity

$$\sqrt{\frac{\hat{\sigma}^2 \sum X_i^2}{n \sum (X_i - \bar{X})^2}}$$

is the *estimated standard error* of b_0. For the common goal of testing $H_0 : \beta_1 = 0$, the hypothesis that the slope is zero, you reject if the confidence interval does not contain zero. Alternatively, you reject if

$$|T| \geq t,$$

where again t is the $1 - \alpha/2$ quantile of Student's T distribution with $\nu = n - 2$ degrees of freedom and

$$T = b_1 \sqrt{\frac{\sum (X_i - \bar{X})^2}{\hat{\sigma}^2}}. \tag{6.10}$$

For convenience, it is noted that the hypothesis-testing methods just described can be applied in S-PLUS with the command

```
summary(lm(y~x))
```

where x is any S-PLUS variable containing the predictor values and y is any S-PLUS variable containing the outcome values.

EXAMPLE. Using the aggression data in Table 6.3, we test the hypothesis $H_0 : \beta_1 = 0$ with the goal that the probability of a Type I error be .05, assuming normality and that the error term is homoscedastic. Because $\alpha = .05$, $1 - \alpha/2 = .975$. There are $n = 47$ pairs of observations, so the degrees of freedom are $\nu = 47 - 2 = 45$, and the critical value is $c = 2.01$. The least squares estimate of the slope is $b_1 = -0.0405$, and it can be seen that $\sum (X_1 - \bar{X})^2 = 34659.74$ and $\hat{\sigma}^2 = 14.15$, so the test statistic [given by Equation (6.10)] is

$$T = -0.0405 \sqrt{\frac{34659.74}{14.5}} = -1.98.$$

Because $|T| = 1.98 < 2.01$, fail to reject. ■

In regression, any outlier among the X values is called a *leverage point*. Notice that a single leverage point can inflate $\sum (X_i - \bar{X})^2$, which is just the numerator of the sample variance for the X values. But $\sum (X_i - \bar{X})^2$ appears in the denominator of the expression for the standard error of b_1, so a single leverage point can cause the standard error of b_1 to be smaller compared to a situation where no leverage points occurred. In practical terms, we can get shorter confidence intervals and more power when there are leverage points, but caution must be exercised because leverage points can result in a poor fit to the bulk of the data.

6.3.2 Violating Assumptions

Currently, the hypothesis-testing method described in the previous subsection is routinely employed. Unfortunately, violating the assumptions of the method can cause serious problems.

Consider a situation where the normality assumption is valid but there is heteroscedasticity. For the housing data, for example, imagine that the variation among the selling prices differs depending on how many square feet a house happens to have. For instance, the variation among houses having 1500 square feet might differ from the variation among homes having 2000 square feet. Then the standard method for testing hypotheses about the slope, given by Equation (6.10), might provide poor control over the probability of a Type I error and poor probability coverage (e.g., Long & Ervin, 2000; Wilcox, 1996b). If the distributions are not normal, the situation gets worse. In some cases, the actual probability of a Type I error can exceed .5 when testing at the $\alpha = .05$ level! Perhaps an even more serious concern is that violating the homoscedasticity assumption might result in a substantial loss in power.

The homoscedasticity assumption is valid when X and Y are independent. (Independence implies homoscedasticity, but $\beta_1 = 0$, for example, does not necessarily mean that there is homoscedasticity.) Practical problems arise when X and Y are dependent because now there is no particular reason to assume homoscedasticity; and if there is heteroscedasticity; the wrong standard error is being used to compute confidence intervals and test hypotheses. If we could determine how $VAR(Y|X)$ changes with X, a correct estimate of the standard error could be employed, but currently it seems that alternate strategies for dealing with heteroscedasticity (covered in subsequent chapters) are more effective.

There are methods for testing the assumption that there is homoscedasticity (see, for example, Lyon & Tsai, 1996). But given some data, it is unknown how to tell whether any of these tests have enough power to detect situations where there is enough heteroscedasticity to cause practical problems with standard inferential methods. Even if there is homoscedasticity, nonnormality remains a serious concern. Currently, a more effective approach appears to be to switch to some method that allows heteroscedasticity.

Long and Ervin (2000) compare three simple methods for computing confidence intervals when there is heteroscedasticity. One of these they recommend for general use. However, in situations where leverage points are likely and simultaneously the error term has a normal or light-tailed distribution, their recommended method can be unsatisfactory. A more effective method is described in Chapter 7; so no details are given here about the method recommended by Long and Ervin.

6.3.3 Restricting the Range of Y

The derivation of the inferential methods in Section 6.3.1 treats the X values as constants and the Y values as random variables. That is, the methods are derived by conditioning on the X values, which makes it a fairly simple matter to derive an estimate of the standard error of the least squares estimate of the slope and intercept. For example, we saw that

$$b_1 = \sum w_i Y_i,$$

where

$$w_i = \frac{X_i - \bar{X}}{(n-1)s_x^2}.$$

Treating the X values as constants, and using the rules of expected values in Chapter 2, it can be shown that the squared standard error of b_1 is given by

$$\frac{\sigma^2}{\sum (X_i - \bar{X})^2}, \tag{6.11}$$

as noted in Section 6.3.1. A practical implication is that if we restrict the range of X values, no technical problems arise when trying to estimate the standard error of b_1; we simply use Equation (6.11) on the points that remain (with n reduced to the number of points remaining). But if we restrict the range of Y values by eliminating outliers, the methods in Section 6.3.1 are no longer valid, even under normality and homoscedasticity. We saw in Section 4.9.1 that if we eliminate extreme values and compute the mean using the data that remain, the standard error of this mean should not be estimated with the sample variance based on the data that remain. A similar problem arises here. If we eliminate extreme Y values, the remaining Y values are no longer independent. So if we use least squares to estimate the slope based on the pairs of points not eliminated, estimating the standard error of the slope becomes a nontrivial problem — the dependence among the Y values must be taken into account.

6.3.4 Standardized Regression

Popular statistical software reports what is called a *standardized regression coefficient*. This simply means that rather than compute the least squares estimator using the raw data, the observations are first converted to Z scores. For the aggression data in Table 6.3, for example, it can be seen that the test scores (Y) have mean $\bar{Y} = 3.4$ and standard deviation $s_y = 3.88$. The first test score is $Y_1 = 0$, and its Z-score equivalent is

$$Z = \frac{0 - 3.4}{3.88} = -0.88.$$

Of course, the remaining Y values can be converted to a Z score in a similar manner. In symbols, for the ith observation, you compute

$$Z_{yi} = \frac{Y_i - \bar{Y}}{s_y}.$$

Next, convert all of the aggression values to Z scores. That is, for the ith pair of observations, compute

$$Z_{xi} = \frac{X_i - \bar{X}}{s_x}.$$

For the aggression data, $\bar{X} = 28.5$ and $s_x = 27.45$. For example, the first entry in Table 6.3 has $X_1 = 3$, so $Z_{x1} = -0.93$. Next, you determine the least squares estimate of the slope using the transformed X and Y values just computed. The resulting estimate of the slope will be labeled b_z. The resulting estimate of the intercept is always zero, so the regression equation takes the form

$$\hat{Z}_y = b_z Z_x.$$

For the aggression data, it can be seen that $b_z = -0.29$, so $\hat{Z}_y = -0.29(Z_x)$.

The standardized regression coefficient, b_z, can be computed in another manner. First compute the least squares estimate of the slope using the original data yielding b_1. Then

$$b_z = b_1 \frac{s_x}{s_y}.$$

EXAMPLE. For the aggression data in Table 6.3, the sample standard deviations of the X and Y values are $s_y = 3.88$ and $s_x = 27.45$. As previously indicated, the least squares estimate of the slope is $b_1 = -0.0405$. The standardized slope is just

$$b_z = -0.0405 \frac{27.45}{3.88} = -0.29. \qquad ■$$

One reason standardized regression has some appeal is that it attempts to provide perspective on the magnitude of a predicted value for Y. Recall from Chapter 2 that for normal distributions, the value of $Z = (X - \mu)/\sigma$ has a convenient probabilistic interpretation under normality. For example, half the observations fall below a Z score of zero. A Z score of 1 indicates we are 1 standard deviation above the mean, and about 84% of all observations are to below this point when observations have a normal distribution. (From Table 1 in Appendix B, $Z = 1$ is the .84

quantile.) Similarly, $Z = 2$ refers to a point 2 standard deviations above the mean, and approximately 98% of all observations are below this value. Thus, for normal distributions, Z scores give you some sense of how large or small a value happens to be. Standardized regression attempts to tell us, for example, how a change of 1 standard deviation in X is related to changes in Y, again measured in standard deviations.

EXAMPLE. For the aggression data, assume normality and suppose we want to interpret the standardized regression estimate of the recall test when the measure of aggression is 1 standard deviation above or below the mean. One standard deviation above the mean of the aggression scores, X, corresponds to $Z_x = 1$, so, as previously indicated,

$$\hat{Z}_y = (-0.29)1 = -0.29.$$

For a standard normal distribution, the probability of being less than -0.29 is approximately .39, and this provides a perspective on how the recall test is related to the measure of aggression. In a similar manner, 1 standard deviation below the mean of the aggression scores corresponds to $Z_x = -1$, so now

$$\hat{Z}_y = (-0.29)(-1) = 0.29,$$

and there is approximately a .61 probability that a standard normal random variable is less than .29. ■

For nonnormal distributions, situations arise where Z scores can be interpreted in much the same way as when distributions are normal. But based on results in Chapters 2 and 3, there are two points to keep in mind when using Z scores. The first is that they can give a misleading representation of the sample of observations being studied. The second is that interpretation problems can arise even with an arbitrarily large sample size and a very small departure from normality.

EXAMPLE. The last example illustrated how to interpret $\hat{Z}_y$ assuming normality, but now we take a closer look at the data to see whether this interpretation might be misleading. Table 6.5 shows all 47 Z_y scores for the recall-test values written in ascending order. We see that 24 of the 47 values are below $-.29$, so the proportion below -0.29 is $24/47 = .51$. This means that based on the available data, your estimate is that there is .51 probability of having a Z_y score less than -0.29. Put another way, a Z score of -0.29 corresponds, approximately, to the median. In contrast, for normal distributions, $Z_y = 0$ is the median and the probability of getting a Z_y score less than -0.29 is approximately .39.

Continued

TABLE 6.5   *Z Scores for the Recall-Test Scores in Table 6.3 (in ascending order)*

−0.88	−0.88	−0.88	−0.88	−0.88	−0.88	−0.88	−0.88	−0.88	−0.88	−0.88	−0.88
−0.62	−0.62	−0.62	−0.62	−0.62	−0.62	−0.62	−0.36	−0.36	−0.36	−0.36	−0.36
−0.10	−0.10	−0.10	−0.10	0.15	0.15	0.15	0.15	0.15	0.15	0.15	0.41
0.67	0.67	0.67	0.67	0.67	0.67	1.44	1.96	2.47	2.47	3.51	

EXAMPLE. (*Continued*) Thus, there is some discrepancy between the empirical estimate of the probability of getting a Z_y score less than −0.29 versus the probability you get assuming normality. The main point here is that switching to standardized regression does not necessarily provide a perspective that is readily interpretable.   ■

A criticism of this last example is that the estimated probability of getting a Z score less than −0.29 is based on only 47 observations. Perhaps with a larger sample size, the estimated probability would be reasonably close to 0.39, the value associated with a normal distribution. However, results in Chapter 2 indicate that even with a large sample size, there can be a considerable difference, so caution is recommended when interpreting standardized regression equations.

6.4   Pearson's Correlation

Chapter 2 introduced Pearson's correlation, ρ. This section takes up the problem of estimating ρ based on data available to us, plus the issue of interpreting what this estimate tells us about the association between the two measures under study.

Recall from Chapter 2 that

$$\rho = \frac{\sigma_{xy}}{\sigma_x \sigma_y},$$

where σ_{xy} is the (population) covariance between X and Y (as defined in Section 2.8). Given n pairs of observations

$$(X_1, Y_1), \dots, (X_n, Y_n),$$

the covariance is typically estimated with

$$s_{xy} = \frac{1}{n-1} \sum (X_i - \bar{X})(Y_i - \bar{Y}).$$

The usual estimate of ρ is

$$r = \frac{s_{xy}}{s_x s_y}, \tag{6.12}$$

where s_x and s_y are the standard deviations of the X and Y values, respectively.

It can be shown that $-1 \leq r \leq 1$. That is, the estimate of Pearson's correlation, based on r, always lies between -1 and 1, as does ρ. If all of the points lie on a straight line with a positive slope, $r = 1$. If all of the points lie on a straight line with a negative slope, $r = -1$.

The estimate of the slope of the least squares regression line (b_1) is related to the estimate of Pearson's correlation in the following manner:

$$b_1 = r\frac{s_y}{s_x}. \tag{6.13}$$

So if $r > 0$, the least squares regression line has a positive slope; if $r < 0$, the slope is negative. In standardized regression (as discussed in Section 6.3.4), the slope is equal to the correlation between X and Y. That is, the least squares regression line between

$$Z_x = \frac{X - \bar{X}}{s_x} \qquad \text{and} \qquad Z_y = \frac{Y - \bar{Y}}{s_y}$$

is

$$\hat{Z}_y = rZ_x.$$

6.4.1 Five Features of Data That Affect the Magnitude of r

Interpreting r is complicated by the fact that various features of the data under study affect its magnitude. Five such features are described here, and a sixth is briefly indicated at the end of this section.

Assuming that there is a linear association between X and Y, the first feature is the distance of the points from the line around that they are centered. That is, the magnitude of the residuals is associated with the magnitude of r. The left panel of Figure 6.10 shows a scatterplot of points with $r = .92$. The right panel shows another scatterplot of points that are centered around the same line as in the left panel, only they are farther from the line around which they are centered. Now $r = .42$.

A second feature that affects the magnitude of r is the magnitude of the slope around which the points are centered (e.g., Barrett, 1974; Loh, 1987). Figure 6.11 shows the same points as in the left panel of Figure 6.10, only rotated so that the slope around which they are centered has been decreased from 1 to .5. This causes the correlation to drop from .92 to .83. If we continue to rotate the points until they are centered around the x-axis, $r = 0$.

A third feature of data that affects r is outliers. This is not surprising, because we already know that the least squares regression line has a breakdown point of only $1/n$, and we have seen how r is related to the least squares estimate of the slope [as indicated by Equation (6.13)]. For the star data in Figure 6.3, $r = -.21$, which is consistent with the negative slope associated with the least squares regression line. But we have already seen that for the bulk of the points, there is a positive association. Generally, a single unusual value can cause r to be close to zero even when the remaining points

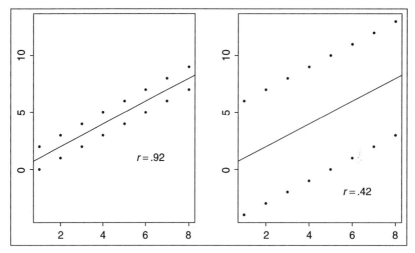

FIGURE 6.10 Illustration showing that the magnitude of the residuals affects Pearson's correlation.

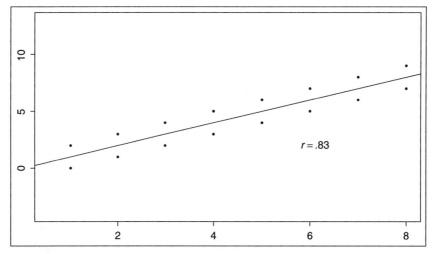

FIGURE 6.11 Illustration showing that Pearson's correlation is related to the magnitude of the slope of the line around which points are clustered.

are centered around a line having a nonzero slope, and one outlier can cause $|r|$ to be fairly large even when there is no association among the remaining points.

Moreover, when sampling points in situations where outliers are likely to occur, even small departures from normality can greatly affect the population correlation ρ, and r can be affected as well no matter how large the sample size might be. To provide some indication of why, the left panel of Figure 6.12 shows the distribution between X and Y when both X and Y are normal and $\rho = .8$. In the right panel, again X and Y are normal, but now $\rho = .2$. So under normality, decreasing ρ from .8 to .2 has a very noticeable effect on the joint distribution of X and Y. Now look at Figure 6.13. It looks similar to the left panel of Figure 6.12, where $\rho = .8$, but now $\rho = .2$. In Figure 6.13 X is again normal, but Y has the mixed normal distribution (described

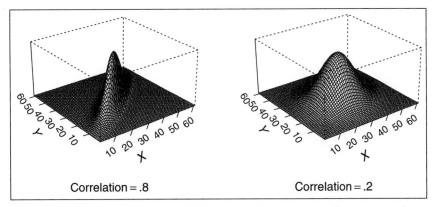

FIGURE 6.12 When both X and Y are normal, increasing ρ from .2 to .8 has a noticeable effect on the bivariate distribution of X and Y.

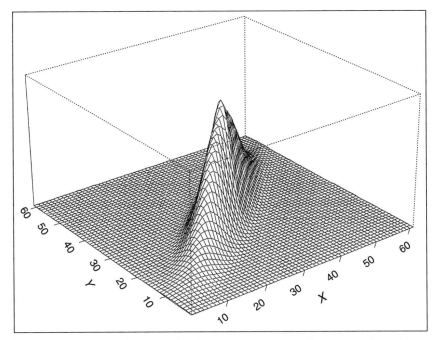

FIGURE 6.13 Two bivariate distributions can appear to be very similar yet have substantially different correlations. Shown is a bivariate distribution with $\rho = .2$, but the graph is very similar to the left panel of Figure 6.12, where $\rho = .8$.

in Section 2.7). This demonstrates that a very small change in any distribution can have a very large impact on ρ. Also, no matter how large the sample size might be, a slight departure from normality can drastically affect r.

A fourth feature that affects the magnitude of r is any restriction in range among the X (or Y) values. To complicate matters, restricting the range of X can increase or decrease r. For example, the left panel of Figure 6.14 shows a scatterplot of points for which $r = .98$. When we eliminate the points with $|X| > 1$, leaving the points shown in the right panel of Figure 6.14, $r = .79$.

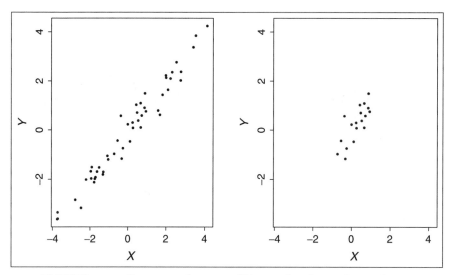

FIGURE 6.14   Restricting the range of X can reduce Pearson's correlation.

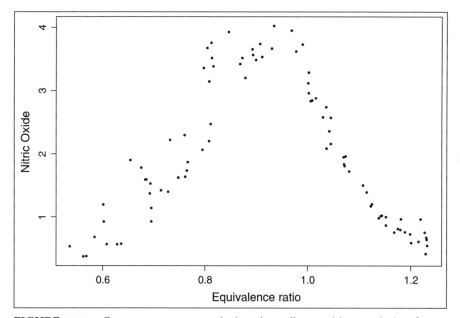

FIGURE 6.15   Curvature occurs in applied work, as illustrated here with data from a study of how concentrations of nitric oxides in engine exhaust are related to its equivalence ratio. Here, Pearson's correlation is only $-.1$ despite a rather strong association.

The star data in Figure 6.3 illustrate that restricting the range of X (or Y) can increase r as well. If we eliminate all points having $X \leq 4.1$, r increases from $-.21$ to $.65$.

A fifth feature that affects r is curvature. Figure 6.15 shows a scatterplot of points relating the concentration of nitric oxides in engine exhaust versus its equivalence ratio, a measure of the richness of the air–ethanol mix. There is a rather obvious association, but the correlation is $r = -.1$, a value relatively close to zero. As another example, if X is standard normal and $Y = X^2$, there is an exact association between X and Y, but $\rho = 0$.

In summary, the following features of data influence the magnitude of Pearson's correlation:

- The slope of the line around which points are clustered
- The magnitude of the residuals
- Outliers
- Restricting the range of the X values, which can cause r to go up or down
- Curvature

A point worth stressing is that although independence implies that $\rho = 0$, $\rho = 0$ does not necessarily imply independence. In fact there are various ways in which X and Y can be dependent, yet ρ is exactly zero. For example, if X and e are independent and $Y = |X|e$, then X and Y are dependent because there is heteroscedasticity, yet $\rho = 0$. More generally, if there is heteroscedasticity and the least squares slope is zero, then $\rho = 0$ as well. As another example, suppose U, V, and W are independent standard normal random variables. Then it can be shown that $X = U/W^2$ and $Y = V/W^2$ are dependent (roughly because both X and Y have the same denominator), yet they have correlation $\rho = 0$.

We conclude this section by noting that the foregoing list of factors that affect the magnitude of r is not exhaustive. Yet another feature of data that affects the magnitude of r is the reliability of the measures under study (e.g., Lord & Novick, 1968), but the details go beyond the scope of this book.

6.5   Testing $H_0 : \rho = 0$

Next we describe the classic test of

$$H_0 : \rho = 0. \tag{6.14}$$

If we can reject this hypothesis, then by implication X and Y are dependent.

If we assume that X and Y are independent and if at least one of these two variables is normal, then

$$T = r\sqrt{\frac{n-2}{1-r^2}} \tag{6.15}$$

has a Student's T distribution with $\nu = n - 2$ degrees of freedom (Muirhead, 1982, p. 146; also see Hogg & Craig, 1970, pp. 339–341). So the decision rule is to reject H_0 if $|T| \geq t$, where t is the $1 - \alpha/2$ quantile of Student's T distribution with $n - 2$ degrees of freedom.

EXAMPLE. For the data in Table 6.3, $n = 47$, $r = -0.286$, so $\nu = 45$ and

$$T = -0.286\sqrt{\frac{45}{1 - (-0.286)^2}} = -2.$$

Continued

> **EXAMPLE.** (*Continued*) With $\alpha = .05$, the critical value is $t = 2.01$; because $|-2| < 2.01$, we fail to reject. That is, we are unable to conclude that the aggression scores and recall-test scores are dependent with $\alpha = .05$. ■

Caution must be exercised when interpreting the implications associated with rejecting the hypothesis of a zero correlation with T. Although it is clear that T given by Equation (6.15) is designed to be sensitive to r, homoscedasticity plays a crucial role in the derivation of this test. When in fact there is heteroscedasticity, the derivation of T is no longer valid and can result in some unexpected properties. For instance, it is possible to have $\rho = 0$, yet the probability of rejecting $H_0 : \rho = 0$ *increases* as the sample size gets large.

> **EXAMPLE.** Figure 6.16 shows a scatterplot of 40 points generated on a computer, where both X and Y have normal distributions and $\mu_x = 0$. In this particular case, Y has variance 1 unless $|X| > .5$, in which case Y has standard deviation $|X|$. So X and Y are dependent, but $\rho = 0$. For this situation, when testing at the $\alpha = .05$ level, the actual probability of rejecting H_0 with T is .098 with $n = 20$. For $n = 40$ it is .125, and for $n = 200$ it is .159. The probability of rejecting is *increasing* with n even though $\rho = 0$. When we reject, a correct conclusion is that X and Y are dependent, but it would be incorrect to conclude that $\rho \neq 0$. ■

Some experts might criticize this last example on the grounds that it would be highly unusual to encounter a situation where $\rho = 0$ and there is heteroscedasticity. That is, perhaps we are describing a problem that is theoretically possible but unlikely ever to be encountered. Even if we agree with this argument, the more salient issue is that employing the wrong standard error can lead to highly erroneous results. Generally, conventional methods use correct estimates of standard errors when variables are independent but incorrect estimates when they are dependent, and in the latter case this can lead to poor power and highly inaccurate confidence intervals.

A common alternative to T when making inferences about ρ is to employ what is known as Fisher's *r-to-z transformation*, but we provide no details here because of results in Duncan and Layard (1973). Briefly, Fisher's method requires normality. For nonnormal distributions there are general conditions where the method does not converge to the correct answer as the sample size increases. That is, the method violates the basic principle that the accuracy of our results should increase as n gets large.

6.5.1 The Coefficient of Determination

A positive feature of r is that it provides a useful characterization of how well the least squares regression line summarizes the association between two variables. To explain,

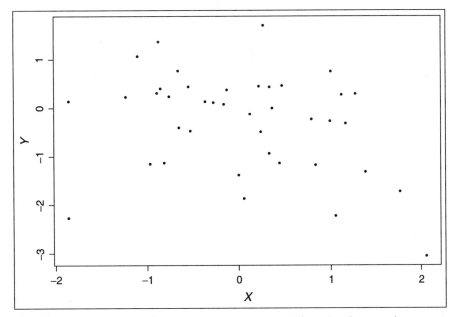

FIGURE 6.16   Example of 40 points that were generated from distributions where $\rho = 0$ and there is heteroscedasticity. Student's T test of $H_0 : \rho = 0$ is unsatisfactory in terms of making inferences about ρ, but it does detect the dependence between X and Y.

imagine that we ignore X in our attempts to predict Y and simply use $\hat{Y} = \bar{Y}$. Then we can measure the accuracy of our prediction rule with

$$\sum (Y_i - \bar{Y})^2,$$

the sum of the squared discrepancies between the Y values we observe and the predicted values, $\bar{Y}$. (This is the method for measuring accuracy already discussed in Section 3.2.8.) Notice that this sum is the numerator of the sample variance of the Y values. If instead we use the least squares regression line $\hat{Y} = \beta_0 + \beta_1 X_i$ to predict Y, then an overall measure of the accuracy of our prediction rule is

$$\sum (Y_i - \hat{Y}_i)^2,$$

as already explained. The difference between these two sums measures the extent to which using $\hat{Y}$ improves upon using $\hat{Y}$. In symbols, this difference is

$$\sum (Y_i - \bar{Y})^2 - \sum (Y_i - \hat{Y}_i)^2.$$

Finally, if we divide this difference by $\sum (Y_i - \bar{Y})^2$, we get a measure of how much $\hat{Y}$ improves upon $\bar{Y}$ relative to $\bar{Y}$. It can be seen that this measure is just r^2, which is called the *coefficient of determination*. In symbols, the coefficient of determination is,

$$r^2 = \frac{\sum (Y_i - \bar{Y})^2 - \sum (Y_i - \hat{Y}_i)^2}{\sum (Y_i - \bar{Y})^2}. \tag{6.16}$$

We have already seen that even a single outlier can have a substantial impact on both r and the least squares regression line. So when the coefficient of determination (r^2) is fairly large, this does not necessarily mean that the least squares regression line accurately reflects the association among the bulk of the points under study. It simply reflects the extent to which the least squares regression line improves upon the prediction rule $\hat{Y} = \bar{Y}$, the point being that both rules might be very ineffective.

6.5.2 Establishing Independence

Establishing that two measures are independent is a much more difficult goal than showing that they are dependent. If we reject the hypothesis that Pearson's correlation is zero, we can be reasonably certain that the two measures are dependent even though this tells us virtually nothing about what the dependence is like. But if we fail to reject, this is not remotely convincing evidence that we have independence. The basic problem is that the test of $H_0 : \rho = 0$ given by Equation (6.15) may not be sensitive to the type of association that exists between the variables under study. There is a rather lengthy list of alternative methods for detecting dependence that attempt to address this problem. (See, for example, Kallenberg & Ledwina, 1999, plus the references they cite.) A few alternative techniques are described in subsequent chapters.

6.6 Concluding Remarks

The purpose of this chapter was to introduce basic concepts and to describe standard hypothesis-testing methods associated with least squares regression and Pearson's correlation. Another goal was to provide some indication of what might go wrong with these standard methods. Some contemporary techniques for addressing these problems are covered in subsequent chapters. It is stressed, however, that regression is a vast topic and that not all methods and issues are discussed in this book. For more about regression, see Li (1985), Montgomery and Peck (1992), Staudte and Sheather (1990), Hampel, Ronchetti, Rousseeuw, and Stahel (1986), Huber (1981), Rousseeuw and Leroy (1987), Belsley, Kuh, and Welsch (1980), Cook and Weisberg (1992), Carroll and Ruppert (1988), Hettmansperger (1984), Hettmansperger and McKean (1998), and Wilcox (1997a).

6.7 Exercises

1. For the following pairs of points, verify that the least squares regression line is $\hat{Y} = 1.8X - 8.5$.

X:	5,	8,	9,	7,	14
Y:	3,	1,	6,	7,	19

2. Compute the residuals using the results from Exercise 1. Verify that if you square and sum the residuals, you get 47, rounding to the nearest integer.

3. Verify that for the data in Exercise 1, if you use $\hat{Y} = 2X - 9$, the sum of the squared residuals is larger than 47. Why would you expect a value greater than 47?

4. Suppose that based on $n = 25$ values, $s_x^2 = 12$, $s_y^2 = 25$, and $r = .6$. What is the slope of least squares regression?

5. Verify that for the data in Table 6.3, the least squares regression line is $\hat{Y} = -0.0405X + 4.581$.

6. The following table reports breast cancer rates plus levels of solar radiation (in calories per day) for various cities in the United States. Fit a least squares regression to the data with the goal of predicting cancer rates and comment on what this line suggests.

City	Rate	Daily calories	City	Rate	Daily calories
New York	32.75	300	Chicago	30.75	275
Pittsburgh	28.00	280	Seattle	27.25	270
Boston	30.75	305	Cleveland	31.00	335
Columbus	29.00	340	Indianapolis	26.50	342
New Orleans	27.00	348	Nashville	23.50	354
Washington, DC	31.20	357	Salt Lake City	22.70	394
Omaha	27.00	380	San Diego	25.80	383
Atlanta	27.00	397	Los Angeles	27.80	450
Miami	23.50	453	Fort Worth	21.50	446
Tampa	21.00	456	Albuquerque	22.50	513
Las Vegas	21.50	510	Honolulu	20.60	520
El Paso	22.80	535	Phoenix	21.00	520

7. For the following data, compute the least squares regression line for predicting gpa given SAT.

SAT:	500	530	590	660	610	700	570	640
gpa:	2.3	3.1	2.6	3.0	2.4	3.3	2.6	3.5

8. For the data in the Exercise 7, verify that the coefficient of determination is .36 and interpret what this tells you.

9. For the following data, compute the least squares regression line for predicting Y from X.

X:	40	41	42	43	44	45	46
Y:	1.62	1.63	1.90	2.64	2.05	2.13	1.94

10. In Exercise 6, what would be the least squares estimate of the cancer rate given a solar radiation of 600? Indicate why this estimate might be unreasonable.

11. Maximal oxygen uptake (mou) is a measure of an individual's physical fitness. You want to know how mou is related to how fast someone can run a mile. Suppose you randomly sample six athletes and get

mou (milliliters/kilogram):	63.3	60.1	53.6	58.8	67.5	62.5
time (seconds):	241.5	249.8	246.1	232.4	237.2	238.4

Compute the correlation. Can you be reasonably certain about whether it is positive or negative with $\alpha = .05$?

12. Verify that for the following pairs of points, the least squares regression line has a slope of zero. Plot the points and comment on the assumption that the regression line is straight.

X:	1	2	3	4	5	6
Y:	1	4	7	7	4	1

13. Repeat Exercise 12; but for the points

X:	1	2	3	4	5	6
Y:	4	5	6	7	8	2

14. Vitamin A is required for good health. You conduct a study and find that as vitamin A intake decreases, there is a linear association with bad health. However, one bite of polar bear liver results in death because it contains a high concentration of vitamin A. Comment on what this illustrates in the context of regression.

15. Sockett et al. (1987) report data related to patterns of residual insulin secretion in children. A portion of the study was concerned with whether age can be used to predict the logarithm of C-peptide concentrations at diagnosis. The observed values are

Age (X):	5.2	8.8	10.5	10.6	10.4	1.8	12.7	15.6	5.8	1.9
	2.2	4.8	7.9	5.2	0.9	11.8	7.9	1.5	10.6	8.5
	11.1	12.8	11.3	1.0	14.5	11.9	8.1	13.8	15.5	9.8
	11.0	12.4	11.1	5.1	4.8	4.2	6.9	13.2	9.9	12.5
	13.2	8.9	10.8							

Continued

C-peptide (Y):	4.8	4.1	5.2	5.5	5.0	3.4	3.4	4.9	5.6
	3.7	3.9	4.5	4.8	4.9	3.0	4.6	4.8	5.5
	4.5	5.3	4.7	6.6	5.1	3.9	5.7	5.1	5.2
	3.7	4.9	4.8	4.4	5.2	5.1	4.6	3.9	5.1
	5.1	6.0	4.9	4.1	4.6	4.9	5.1		

Replace the C-peptide values with their (natural) logarithms. For example, the value 4.8 would be replaced by $\log(4.8) = 1.5686$. Create a scatterplot for these data and consider whether a linear rule for predicting Y with X is reasonable. Also verify that $r = .4$ and that you reject $H_0 : \rho = 0$ with $\alpha = .05$.

16. For the data in Exercise 15, verify that a least squares regression line using only X values (age) less than 7 yields $b_1 = 0.247$ and $b_0 = 3.51$. Verify that when using only the X values greater than 7 you get $b_1 = .009$ and $b_0 = 4.8$. What does this suggest about using a linear rule for all of the data?

17. The housing data in Table 6.4 are from a suburb of Los Angeles where even a small empty lot would cost at least \$200,000 (and probably much more) at the time the data were collected. Verify that based on the least squares regression line for these data, if we estimate the cost of an empty lot by setting the square feet of a house to $X = 0$, we get 38,192. What does this suggest about estimating Y using an X value outside the range of observed X values?

18. For the data in Table 6.4, the sizes of the corresponding lots are:

18,200 12,900 10,060 14,500 76,670 22,800 10,880
10,880 23,090 10,875 3,498 42,689 17,790 38,330
18,460 17,000 15,710 14,180 19,840 9,150 40,511
9,060 15,038 5,807 16,000 3,173 24,000 16,600.

Verify that the least squares regression line for estimating the selling price, based on the size of the lot, is $\hat{Y} = 11X + 436{,}834$.

19. Imagine two scatterplots where in each scatterplot the points are clustered around a line having slope .3. If for the first scatterplot $r = .8$, does this mean that points are more tightly clustered around the line versus the other scatterplot, where $r = .6$?

20. You measure stress (X) and performance (Y) on some task and get

X:	18	20	35	16	12
Y:	36	29	48	64	18

Verify that you do not reject $H_0 : \beta_1 = 0$ using $\alpha = .05$. Is this result consistent with what you get when testing $H_0 : \rho = 0$? Why would it be incorrect to conclude that X and Y are independent?

21. Suppose you observe

X:	12.2,	41,	5.4,	13,	22.6,	35.9,	7.2,	5.2,	55,	2.4,	6.8,	29.6,	58.7,
Y:	1.8,	7.8,	0.9,	2.6,	4.1,	6.4,	1.3,	0.9,	9.1,	0.7,	1.5,	4.7,	8.2

TABLE 6.6   Reading Data

X:	34	49	49	44	66	48	49	39	54	57	39	65	43	43	44	42
	71	40	41	38	42	77	40	38	43	42	36	55	57	57	41	66
	69	38	49	51	45	141	133	76	44	40	56	50	75	44	181	45
	61	15	23	42	61	146	144	89	71	83	49	43	68	57	60	56
	63	136	49	57	64	43	71	38	74	84	75	64	48			
Y:	129	107	91	110	104	101	105	125	82	92	104	134	105	95	101	104
	105	122	98	104	95	93	105	132	98	112	95	102	72	103	102	102
	80	125	93	105	79	125	102	91	58	104	58	129	58	90	108	95
	85	84	77	85	82	82	111	58	99	77	102	82	95	95	82	72
	93	114	108	95	72	95	68	119	84	75	75	122	127			

Verify that the .95 confidence interval for the slope is $(0.14, 0.17)$. Would you reject $H_0 : \beta_1 = 0$? Based on this confidence interval only, can you be reasonably certain that, generally, as X increases, Y increases as well?

22. The data in Table 6.6 are from a study, conducted by L. Doi, where the goal is to understand how well certain measures predict reading ability in children. Verify that the .95 confidence interval for the slope is $(-0.16, .12)$ based on Equation (6.8).

7

BASIC BOOTSTRAP METHODS

This chapter covers the basics of a modern statistical tool called the *bootstrap*. There is a rather large collection of bootstrap methods, most of which will not be described here. For book-length descriptions of these techniques, the reader is referred to Efron and Tibshirani (1993), Chernick (1999), Davison and Hinkley (1997), Hall and Hall (1995), Lunneborg (2000), Mooney and Duval (1993), and Shao and Tu (1995). The goal here is to introduce and illustrate some of the more basic versions that have considerable practical value in applied work when computing confidence intervals or testing hypotheses.

7.1 The Percentile Method

We begin with what is called the *percentile bootstrap method*. It is stressed that this technique does not perform well when the goal is to make inferences about the population mean based on the sample mean, unless the sample size is very large. However, it has considerable practical value for a wide range of other problems, even with very small sample sizes, and modifications of the method have practical value as well. Although the percentile bootstrap is not recommended when working with the sample mean, it is perhaps easiest to explain in terms of the sample mean, so we start with this special case.

Imagine we want to compute a .95 confidence interval for the population mean, μ. The strategy behind the percentile bootstrap method is to estimate the .025 and .975 quantiles of the sampling distribution of the sample mean and then to use this estimate as a .95 confidence interval for μ. For example, if we estimated that $P(\bar{X} \leq 5) = .025$ and $P(\bar{X} \leq 26) = .975$, then 5 and 26 are the .025 and .975 quantiles, respectively, and (5, 26) would be a .95 confidence interval for μ. Of course, the practical problem is that we do not know the sampling distribution of $\bar{X}$, let alone what the .025 and .975 quantiles might be. If we assume that $\bar{X}$ has a normal distribution, which is reasonable with a sufficiently large sample size, the .025 and .975 quantiles are easily estimated: You simply estimate μ and σ^2 with the sample mean and sample variance, and, assuming these estimates are fairly accurate, the method in Section 4.3 can be used.

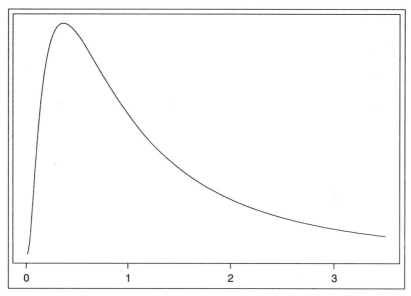

FIGURE 7.1   Example of a skewed, light-tailed distribution. This particular curve belongs to the family of lognormal distributions.

The percentile bootstrap method uses another estimate of the sampling distribution of the sample mean that makes no assumptions about what this distribution might be. In particular, it does not assume the distribution is normal. To understand the basic strategy, first we review a description of the sampling distribution given in Chapter 4. If we were to repeat an experiment infinitely many times, each time computing $\bar{X}$ based on n observations, we would know $P(\bar{X} \leq c)$, the probability that the sample mean is less than or equal to c, for any constant c we might choose. Said another way, if we knew the distribution from which observations were sampled, we could use a computer to get a very accurate estimate of the sampling distribution of $\bar{X}$.

For example, imagine we want to determine the sampling distribution of $\bar{X}$ when n observations are randomly sampled from the distribution shown in Figure 7.1. This is an example of a lognormal distribution, and observations can be generated from it using standard software. (In S-PLUS, the function rlnorm accomplishes this goal.) So if we want to know the sampling distribution of $\bar{X}$ when $n = 20$, say, simply generate 20 observations from the lognormal distribution and compute $\bar{X}$. If we repeat this process many times we will have an excellent approximation of what the sampling distribution happens to be. Figure 7.2 shows a plot of 2000 sample means generated in this manner.

Of course, the practical problem is that we do not know the distribution from which observations are sampled. However, we can use the data available to us to estimate this distribution, and this is the basic idea behind all bootstrap techniques. In the simplest case, the strategy is to use the observed relative frequencies as estimates of the probabilities associated with possible values we might observe, and then we simply use a computer to generate observations based on these probabilities. By repeatedly generating many samples of size n, the resulting sample means provide an estimate of the sampling distribution of $\bar{X}$, and the middle 95% of these generated sample means provide an approximate .95 confidence interval for μ.

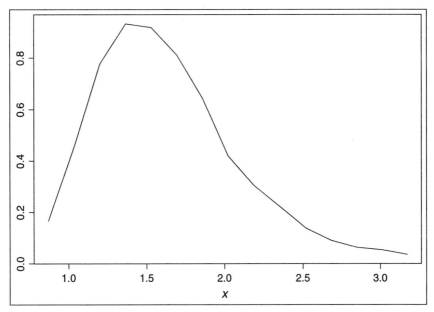

FIGURE 7.2 Plot of 2000 sample means, each mean based on $n = 20$ observations generated from the distribution in Figure 7.1.

To illustrate the idea, imagine we observe the following 10 values:

$$1, 4, 2, 19, 4, 12, 29, 4, 9, 16.$$

The value 1 occurred only once, so we estimate that the value 1 has probability 1/10. Similarly, the value 4 occurred three times, so we estimate its probability to be 3/10. The value 3 occurred zero times, so we estimate its probability to be 0. In the notation of Chapter 3, if the number of times we observe the value x is f_x, $p(x)$ is estimated with f_x/n.

A *bootstrap sample* is obtained by randomly sampling, *with replacement*, observations from the observed values. In our illustration, if we randomly sample a single observation from the values listed in the previous paragraph, we might get the value 12. The probability of getting 12 is 1/10. Or we might get the value 4, and the probability of getting a 4 is 3/10. If we randomly sample a second value from among all 10 values, we might again get 12, or we might get 9 or any of the 10 values from our original sample. If we randomly sample n observations in this manner, we get what is called a *bootstrap sample of size n*. In our example we might get

$$2, 9, 16, 2, 4, 12, 4, 29, 16, 19.$$

The mean of this bootstrap sample is $\bar{X}^* = 11.3$. This is in contrast to the sample mean of our original observations, which is $\bar{X} = 10$. If we were to obtain a second bootstrap sample, the mean of these bootstrap values will typically differ from the first bootstrap sample mean; it might be $\bar{X}^* = 9.6$.

Now imagine that we repeat this process B times, yielding B bootstrap sample means. If B is reasonably large, we get a collection of bootstrap sample means that yields an approximation of the sampling distribution of the sample mean. As an

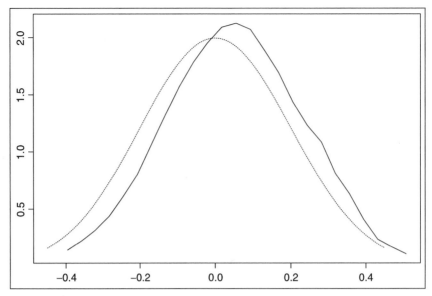

FIGURE 7.3   The solid line is a bootstrap approximation of the sampling distribution of $\bar{X}$ based on 25 values generated from a normal curve. The dashed line shows the exact sampling distribution of $\bar{X}$ under random sampling.

illustration, 25 observations were randomly sampled from a standard normal distribution and then 1000 bootstrap samples were generated in the manner just described. For each of these 1000 bootstrap samples, the sample mean was computed; a plot of these bootstrap means is shown in Figure 7.3. Also shown is the exact sampling distribution of $\bar{X}$. In this particular case, the middle 95% of the 1000 bootstrap means extend from $-.32$ to $.41$, and this interval contains 0. That is, $(-.32, .41)$ is an approximate $.95$ confidence interval for μ and it happens to contain the population mean. This is an example of what is called a *percentile bootstrap confidence interval*, because the strategy is to estimate percentiles of the sampling distribution. However, for the situation at hand, the actual probability that the bootstrap confidence interval will contain μ is less than $.95$. And for nonnormal distributions this method for computing a confidence interval for μ can be quite unsatisfactory. On the positive side, there are some nonnormal distributions for which the resulting confidence interval is more accurate than Student's T method given in Chapters 4 and 5. So progress has been made, but more needs to be done.

A more general and more formal description of the percentile bootstrap method will be helpful. Let $X_1, \ldots, X_n$ represent a random sample of observations, and let $X_1^*, \ldots, X_n^*$ represent a bootstrap sample of size n that is obtained by randomly sampling, with replacement, n values from $X_1, \ldots, X_n$. The sample mean of this bootstrap sample is just

$$\bar{X}^* = \frac{1}{n} \sum X_i^*.$$

Now suppose we repeat the process of generating a bootstrap sample mean B times, and we label these sample means $\bar{X}_1^*, \ldots, \bar{X}_B^*$. Then an approximate $1 - \alpha$ confidence interval for μ is

$$\left(\bar{X}^*_{(\ell+1)}, \bar{X}^*_{(u)}\right), \tag{7.1}$$

where $\bar{X}^*_{(1)} \leq \cdots \leq \bar{X}^*_{(B)}$ are the B bootstrap means written in ascending order, $\ell = \alpha B/2$, rounded to the nearest integer, and $u = B - \ell$. So if $B = 20$ and $1 - \alpha = .8$, ℓ is 2 and u is 18. For the special case $\alpha = .05$, $\ell = .025B$ (still rounding to the nearest integer), and $\bar{X}^*_{(\ell+1)}$ and $\bar{X}^*_{(u)}$ estimate the .025 and .975 quantiles of the distribution of $\bar{X}$, respectively. In general, $\bar{X}^*_{(\ell+1)}$ and $\bar{X}^*_{(u)}$ contain the middle $(1 - \alpha)$ percent of the B bootstrap sample means.

One final point might help before ending this section. An attempt has been made to provide some intuitive sense of how the percentile bootstrap is applied. But from a theoretical point of view, more needs to be done to justify this technique. Here it is merely noted that a formal justification has been derived (e.g., Hall, 1988a, 1988b; Liu and Singh, 1997). Although complete theoretical details cannot be given here, a crude description of the theoretical underpinnings might help. To this end, imagine you want to test the hypothesis $H_0 : \mu = 12$ and that the null hypothesis is true. When you generate a bootstrap sample, there will be some probability that the bootstrap sample mean will be less than 12. For convenience, label this probability p^*. That is, $p^* = P(\bar{X}^* < 12)$. Of course, p^* is not known, but it can be estimated with $\hat{p}^*$, the proportion of the B bootstrap sample means that is less than 12. Note that associated with every random sample is some p^* value. That is, repeating a study infinitely many times will produce infinitely many p^* values. With a sufficiently large sample size, p^* will have a uniform distribution when the null hypothesis is true and in fact is like a significance level or p-value. (Under random sampling, if the null hypothesis is true, the significance level of Student's T test also converges to a uniform distribution as the sample size increases.) Moreover, with B sufficiently large, $\hat{p}^*$ will provide a reasonably accurate estimate of p^*, and $2\hat{p}^*$ is the estimated significance level (or p-value) when testing a two-sided hypothesis. This also leads to the confidence interval given by Equation (7.1).

7.1.1 A Bootstrap Estimate of Standard Errors

Situations arise where expressions for the standard error of some estimator is not known or it takes on a rather complicated form. Examples are M-estimators and MOM. So if it is desired to estimate their standard errors, it would be convenient to have a relatively simple method for accomplishing this goal. The percentile bootstrap method, just described, is one way of tackling this problem.

Let $\hat{\mu}$ be any measure of location and let $\hat{\mu}^*$ be its value based on a bootstrap sample. Let $\hat{\mu}^*_b$ ($b = 1, \ldots, B$) be B bootstrap estimates of the measure of location. Then an estimate of the squared standard error of $\hat{\mu}$ is

$$S^2 = \frac{1}{B-1} \sum_{b=1}^{B} (\hat{\mu}^*_b - \bar{\mu}^*)^2,$$

where $\bar{\mu}^* = \sum_{b=1}^{B} \hat{\mu}^*_b / B$.

7.1.2 S-PLUS Function bootse

The S-PLUS function

$$bootse(x,nboot=1000,est=median)$$

computes S, the bootstrap estimate of the standard error of the measure of location specified by the argument est. By default, the median is used, and the argument nboot corresponds to B, which defaults to 1000.

7.2 The Bootstrap-*t* Interval

This section describes what is called the bootstrap-*t* (or the percentile-*t*) method. The basic idea is that if we knew the distribution of

$$T = \frac{\bar{X} - \mu}{s/\sqrt{n}},$$ (7.2)

a confidence interval for the population mean could be computed. As explained in Chapter 4, the conventional strategy is to assume normality or to assume that the sample size is sufficiently large, in which case T has a Student's T distribution. But we have already seen that confidence intervals and control over the probability of a Type I error can be unsatisfactory with $n = 160$ when sampling from a skewed, light-tailed distribution. And sample sizes greater than 300 can be required when sampling from a skewed, heavy-tailed distribution instead. A better approximation of the distribution of T is needed.

The bootstrap strategy for estimating the distribution of T begins in the same manner used in the percentile method: Obtain a bootstrap sample of size n. As in the previous section, we let $X_1,\ldots,X_n$ represent the original observations and $X_1^*,\ldots,X_n^*$ represent a bootstrap sample of size n that is obtained by randomly sampling, with replacement, n values from $X_1,\ldots,X_n$. Let $\bar{X}^*$ and s^* be the mean and standard deviation based on this bootstrap sample. That is,

$$\bar{X}^* = \frac{1}{n} \sum X_i^*$$

and

$$s^* = \sqrt{\frac{1}{n-1} \sum (X_i^* - \bar{X}^*)^2}.$$

Also let

$$T^* = \frac{\bar{X}^* - \bar{X}}{s^*/\sqrt{n}}.$$ (7.3)

Notice that when obtaining a bootstrap sample, we know the mean of the distribution from which the bootstrap sample was obtained. It is $\bar{X}$. So in the bootstrap world, $\bar{X}$ plays the role of μ, and $\bar{X}^*$ plays the role of $\bar{X}$.

If we repeat the foregoing process B times, yielding B T^* values, we obtain an approximation of the sampling distribution of T, and in particular we have an estimate of its .025 and .975 quantiles. The estimate of these quantiles is based on the middle 95% of the T^* values. In more formal terms, if we let $T^*_{(1)} \leq T^*_{(2)} \leq \cdots \leq T^*_{(B)}$ be the B bootstrap T^* values written in ascending order, and we let $\ell = .025B$, rounded to the nearest integer, and $u = B - \ell$, an estimate of the .025 and .975 quantiles of the distribution of T is $T^*_{(\ell+1)}$ and $T^*_{(u)}$. The resulting .95 confidence interval for μ is

$$\left(\bar{X} - T^*_{(u)} \frac{s}{\sqrt{n}}, \bar{X} - T^*_{(\ell+1)} \frac{s}{\sqrt{n}} \right). \tag{7.4}$$

(In this last equation, $T^*_{(\ell+1)}$ is negative, which is why it is subtracted, not added, from $\bar{X}$. Also, it might seem that $T^*_{(u)}$ should be used to compute the upper end of the confidence interval, not the lower end, but it can be shown that this is not the case.)

HYPOTHESIS TESTING. To test $H_0 : \mu = \mu_0$, compute

$$T = \frac{\bar{X} - \mu_0}{s/\sqrt{n}}$$

and reject if

$$T \leq T^*_{(\ell+1)},$$

or if

$$T \geq T^*_{(u)}.$$

EXAMPLE. Forty observations were generated from a standard normal distribution, and then the bootstrap-*t* method was used to approximate the distribution of T with $B = 1000$. A plot of the 1000 bootstrap T^* values is shown in Figure 7.4. The smooth symmetric curve is the correct distribution (a Student's T distribution with $\nu = 39$). In this particular case, the bootstrap estimate of the distribution of T is fairly accurate. The bootstrap estimates of the .025 and .975 quantiles are $-T^*_{(u)} = -2.059$ and $-T^*_{(\ell)} = 2.116$. The correct answers are -2.022 and 2.022, respectively. ■

Both theoretical and simulation studies indicate that generally, the bootstrap-*t* performs better than the percentile bootstrap or Student's T when computing a confidence interval or testing some hypothesis about μ. There are exceptions, such as when sampling from a normal distribution, but to avoid poor probability coverage, the bootstrap-*t* method is preferable to Student's T or the percentile bootstrap. (However, when working with robust measures of location, we will see that typically the percentile bootstrap is preferable to the bootstrap-*t*.)

From a theoretical point of view, the improvements achieved by the bootstrap-*t* method over Student's T are not surprising. To roughly explain why, note that when computing a $1 - \alpha$ confidence interval with Student's T, there will be some discrepancy between the actual probability coverage and the value for $1 - \alpha$ that you

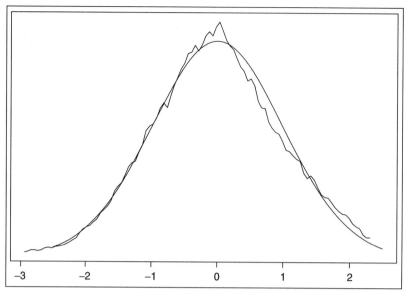

FIGURE 7.4   A plot of 1000 bootstrap T^* values. These T^* values are attempting to approximate the smooth, symmetric curve.

have picked. *When the sample size is large*, mathematicians are able to characterize the rate at which this discrepancy goes to zero; it is $1/\sqrt{n}$. When using the bootstrap-t interval instead, the rate this discrepancy goes to zero is now $1/n$. The discrepancy goes to zero faster using the bootstrap-t, suggesting that it will have better probability coverage and better control over the probability of a Type I error. This mathematical result is encouraging, but the theoretical tools being used tell us only what happens when sample sizes are large. There are known situations where these tools are highly misleading when sample sizes are small — say, less than 150 — but simulation studies aimed at assessing performance when sample sizes are small again indicate that the bootstrap-t is preferable to the percentile bootstrap or Student's T (e.g., Westfall & Young, 1993).

But despite the theoretical appeal of the bootstrap-t method when trying to find an accurate confidence interval for the mean, and even though it improves upon Student's T in certain situations, the method can be unsatisfactory. For example, if we sample 20 observations from the mixed normal shown in Figure 2.8, and we compute a .95 bootstrap-t confidence interval with $B = 1000$, the actual probability coverage is only .9. Put another way, if we reject $H_0 : \mu = \mu_0$ if the .95 bootstrap-t confidence interval does not contain μ_0, the actual probability of a Type I error will not be .05 as intended, but close to .1. Theory tells us that as both n and B get large, if we compute a $1 - \alpha$ confidence interval with the bootstrap-t method, the actual probability coverage will converge to $1 - \alpha$. For the situation at hand, simply increasing B, with n fixed, does not improve matters very much. Increasing n to 100, the actual probability of a Type I error (still testing at the .05 level) is .09. The seriousness of a Type I error will vary from one situation to the next, but some authorities would argue that when testing some hypothesis with $\alpha = .05$, usually the actual probability of a Type I error should not exceed .075 and should not drop below .025 (e.g., Bradley, 1978). One argument for being dissatisfied with an actual Type I error probability of .075 is that if a researcher believes that a Type I error probability of .075 is acceptable, she would have set $\alpha = .075$ in the first place to achieve higher power.

If we sample observations from a skewed heavy-tailed distribution, such as the one shown in Figure 5.6, and then we apply the bootstrap-*t* method at the $\alpha = .05$ level with $n = 20$, the actual probability of a Type I error is .198. This is not much better than using Student's T, where the actual Type I error probability is .202. Increasing n to 100 it drops to .168 using the bootstrap-*t* method. Student's T is even less satisfactory: The actual Type I error probability drops to only .190. So both methods are improving as the sample size gets large, but at a rather slow rate. Even with $n = 300$ the actual Type I error probability remains above .15 when using the bootstrap-*t*, and it is worse using Student's T.

We saw in Chapter 5 that Student's T is biased: When testing $H_0 : \mu = \mu_0$, the probability of rejecting is not minimized when $\mu = \mu_0$. (In practical terms, the probability of rejecting might be higher when H_0 is true versus certain situations where it is false.) The bootstrap-*t* method reduces this problem but does not eliminate it.

Chapter 5 pointed out that arbitrarily small departures from normality can destroy power when using Student's T to make inferences about the population mean. Switching to the bootstrap-*t* method, or any other bootstrap method, does not address this problem.

7.2.1 Symmetric Confidence Intervals

A variation of the bootstrap-*t* method should be mentioned that can be used when testing a two-sided hypothesis only. Rather than use T^* as defined by Equation (7.3), use

$$T^* = \frac{|\bar{X}^* - \bar{X}|}{s^*/\sqrt{n}},\tag{7.5}$$

and reject $H_0 : \mu = \mu_0$ if $|T| \geq T^*_{(c)}$, where $c = (1 - \alpha)B$ rounded to the nearest integer and again $T^*_{(1)} \leq \cdots \leq T^*_{(B)}$ are the B bootstrap T^* values written in ascending order. An approximate $1 - \alpha$ confidence interval for μ is now given by

$$\bar{X} \pm T^*_{(c)} \frac{s}{\sqrt{n}}.\tag{7.6}$$

This is called a *symmetric* two-sided confidence interval, meaning that the same quantity $(T^*_{(c)}s/\sqrt{n})$ is added and subtracted from the mean when computing a confidence interval. In contrast is the confidence interval given by Equation (7.4), which is called an *equal-tailed* confidence interval. With large sample sizes, the symmetric two-sided confidence interval enjoys some theoretical advantages over the equal-tailed confidence interval (Hall, 1988a, 1988b). The main point here is that when sample sizes are small, probability coverage and control over the probability of a Type I error can again be unsatisfactory. In some cases the actual probability coverage of these two methods differs very little, but exceptions arise. For example, when sampling from the mixed normal ($n = 20$) and testing at the .05 level, the actual Type I error probability using the symmetric confidence interval [given by Equation (7.6)] has probability coverage .014, compared to .10 when using the equal-tailed method [given by equation (7.4)]. So in this particular case, the symmetric confidence interval does a better job of avoiding a Type I error that is substantially higher than the nominal level. But there are situations where the symmetric confidence interval is less satisfactory than the equal-tailed method. Moreover, even when the equal-tailed method has a Type I error probability substantially higher than the nominal α level, switching to the symmetric confidence interval can make matters worse. In practical terms, given some data, it is difficult knowing which of these two methods should be preferred. With a large sample size, currently it seems that it makes little practical difference.

7.3 A Modified Percentile Method for Least Squares Regression and Pearson's Correlation

Both the percentile and bootstrap-t methods have been considered when computing a .95 confidence interval for the slope of a least squares regression line — and both have been found to be unsatisfactory when sample sizes are small or even moderately large. For example, there are known situations where the percentile method requires $n = 250$ to get reasonably accurate probability coverage. However, in a comparison of several methods, Wilcox (1996b) found a slight modification of the percentile method that performs reasonably well over a relatively broad class of nonnormal distributions, even when $n = 20$. Moreover, unlike the conventional method for computing a confidence interval for the slope, the method performs well when there is heteroscedasticity.

To begin, we first note that in regression, two general methods have been considered for generating a bootstrap sample. The first is based on resampling values from the residuals, but no details are given because the method deals poorly with heteroscedasticity (e.g., Wu, 1986). The second method is much more flexible (for general reasons detailed by Efron and Tibshirani, 1993, pp. 113–115). It allows heteroscedasticity, so we focus on this method here. In particular, we obtain a bootstrap sample simply by resampling, *with replacement*, n pairs of values from the original n pairs of values used to compute the least squares estimate of the slope and intercept. In symbols, if we observe $(X_1, Y_1), \ldots, (X_n, Y_n)$, a bootstrap sample is obtained by resampling n pairs of these points, with each pair of points having probability $1/n$ of being resampled. If, for example, we observe

$$(6, 2), (12, 22), (10, 18), (18, 24), (16, 29),$$

a bootstrap sample might be

$$(10, 18), (16, 29), (10, 18), (6, 2), (6, 2).$$

That is, there are $n = 5$ pairs of points, so with each resample there is a 1/5 probability that the first pair of points selected will be (6, 2), and this is true for the other four pairs of values as well. A common notation for a bootstrap sample obtained in this manner is $(X_1^*, Y_1^*), \ldots, (X_n^*, Y_n^*)$. In our illustration, $(X_1^*, Y_1^*) = (10, 18)$ and $(X_2^*, Y_2^*) = (16, 29)$.

The least squares estimate of the slope and intercept based on this bootstrap sample is represented by b_1^* and b_0^*, respectively.

The basic percentile bootstrap method described in Section 7.1 extends to the situation at hand in a simple manner. To compute a .95 confidence interval for the slope, first repeat the process of generating a bootstrap sample of size n B times, yielding B bootstrap estimates of the slope, which we label $b_{11}^*, \ldots, b_{1B}^*$. Then an approximate .95 confidence interval for the slope is given by the middle 95% of these bootstrap estimates. In symbols, we write these B bootstrap estimates of the slope in ascending order as $b_{1(1)}^* \leq b_{1(2)}^* \leq \cdots \leq b_{1(B)}^*$. Letting $\ell = .025B$ and setting $u = B - \ell$ then rounding ℓ and u to the nearest integer, an approximate .95 confidence interval for the slope is

$$\left(b_{1(\ell+1)}^*, b_{1(u)}^*\right).$$

Although the probability coverage of the confidence interval just given can differ substantially from .95 when n is less than 250, it has a property of considerable practical value: Given n, the actual probability coverage is fairly stable over a relatively wide range of distributions, even when there is a fairly large degree of heteroscedasticity and the sample size is small. This suggests a method for getting a reasonably accurate confidence interval: Adjust the confidence interval so that the actual probability coverage is close to .95 when sampling from a normal distribution and there is homoscedasticity. Then use this adjusted confidence interval for nonnormal distributions or when there is heteroscedasticity. So we adjust the percentile bootstrap method when computing a .95 confidence interval, based on the least squares regression estimator, in the following manner. Take $B = 599$, and for each bootstrap sample compute the least squares estimate of the slope. Next, put these 599 values in ascending order yielding $b_{1(1)}^* \leq \cdots \leq b_{1(599)}^*$. The .95 confidence interval is

$$\left(b_{1(a)}^*, b_{1(c)}^*\right), \tag{7.7}$$

where for $n < 40$, $a = 7$ and $c = 593$; for $40 \leq n < 80$, $a = 8$ and $c = 592$; for $80 \leq n < 180$, $a = 11$ and $c = 588$; for $180 \leq n < 250$, $a = 14$ and $c = 585$; while for $n \geq 250$, $a = 15$ and $c = 584$. Said another way, these choices for a and c stem from Gosset's strategy for dealing with small sample sizes: Assume normality for a given sample size determine the (critical) value so that the probability of a Type I error is α, and then hope that these values continue to give good results under nonnormality. This strategy performs relatively well here (Wilcox, 1996b), but it does not perform very well for other problems, such as when computing a confidence interval for the mean. Confidence intervals based on Equation (7.7) will be called the *modified percentile* bootstrap method.

HYPOTHESIS TESTING. Reject $H_0 : \beta_1 = 0$ if the confidence interval for the slope, given by Equation (7.7), does not contain zero.

A confidence interval for the intercept can be computed in a similar manner. You simply replace b_1, the least squares estimate of the slope, with b_0, the estimate of the intercept in the description of the modified bootstrap method just given.

Although situations arise where we get more accurate probability coverage than the conventional method covered in Chapter 6, practical problems still occur. That is, we get accurate probability coverage when computing a confidence interval for the slope under a relatively broad range of situations, but the modified percentile bootstrap method is less successful when dealing with the intercept. How to improve upon the modified bootstrap method when dealing with the intercept remains unknown.

There are two practical points to keep in mind when comparing the bootstrap confidence interval for the slope just described to the conventional method in Chapter 6. First, often the bootstrap method yields a longer confidence interval because its probability coverage is generally much closer to the nominal .95 level — the actual probability coverage of the conventional method is often much smaller than .95. In some cases the actual probability coverage drops below .5! That is, the conventional method often gives a shorter confidence interval because it is not nearly as accurate as the modified percentile bootstrap method. Second, despite having longer confidence intervals, situations arise where the bootstrap method rejects $H_0 : \beta_1 = 0$ and the conventional method does not.

> **EXAMPLE.** Using the aggression data in Table 6.3, it was already illustrated that the hypothesis $H_0 : \beta_1 = 0$ is not rejected with $\alpha = .05$ using the conventional Student's T test given by Equation (6.10). Using the modified percentile bootstrap method, the .95 confidence interval for the slope is $(-0.105, -0.002)$, this interval does not contain zero, so you reject. The .95 confidence interval based on Student's T [using Equation (6.8)] is $(-0.08, 0.0002)$.   ■

> **EXAMPLE.** For the selling price of homes in Table 6.4, the .95 confidence interval using the bootstrap method is $(.166, .265)$ versus $(.180, .250)$ using Student's T. Student's T gives a shorter confidence interval, but it might be substantially less accurate because it is sensitive to violations of the assumptions of normality and homoscedasticity.   ■

7.3.1 S-PLUS Function lsfitci

The S-PLUS function

$$lsfitci(x,y)$$

computes a modified .95 percentile bootstrap confidence interval for the slope and intercept of a least squares regression line. Here x is a vector of predictor values and y is a corresponding vector of outcome values.

7.3.2 Testing for Zero Correlation

The modified percentile bootstrap method just described performs relatively well when the goal is to test the hypothesis of a zero correlation (Wilcox & Muska, 2001).

You proceed exactly as already described in this section, except for every bootstrap sample you compute Pearson's correlation r rather than the least squares estimate of the slope. So now we have B bootstrap values for r, which, when written in ascending order, we label $r^*_{(1)} \leq \cdots \leq r^*_{(B)}$. Then a .95 confidence interval for ρ is

$$\left(r^*_{(a)}, r^*_{(c)} \right),$$

where again for $n < 40$, $a = 7$ and $c = 593$; for $40 \leq n < 80$, $a = 8$ and $c = 592$; for $80 \leq n < 180$, $a = 11$ and $c = 588$; for $180 \leq n < 250$, $a = 14$ and $c = 585$; while for $n \geq 250$, $a = 15$ and $c = 584$. As usual, if this interval does not contain zero, reject $H_0 : \rho = 0$.

We saw in Chapter 6 that heteroscedasticity causes Student's T test of $H_0 : \rho = 0$ to have undesirable properties. All indications are that the modified percentile bootstrap eliminates these problems. When $\rho \neq 0$, the actual probability coverage remains fairly close to the .95 level provided ρ is not too large. But if, for example, $\rho = .8$, the actual probability coverage of the modified percentile bootstrap method can be unsatisfactory in some situations (Wilcox & Muska, 2001). There is no known method for correcting this problem.

7.3.3 S-PLUS Function corb

The S-PLUS function

corb(x,y)

computes a .95 confidence interval for ρ using the modified percentile bootstrap method. Again, x and y are S-PLUS variables containing vectors of observations.

EXAMPLE. For the aggression data in Table 6.3, Student's T test fails to reject the hypothesis that the correlation is zero. Using the modified percentile bootstrap method instead, the S-PLUS function corb returns a .95 confidence interval of $(-0.54, -0.01)$. So now we reject $H_0 : \rho = 0$ (because the confidence interval does not contain zero), and we conclude that these two variables are dependent.   ■

7.4 More About the Population Mean

For many situations encountered in statistics, it is now possible to compute reasonably accurate confidence intervals even under fairly extreme departures from standard assumptions. But making accurate inferences about the population mean remains one of the more difficult problems. In terms of avoiding Type I errors greater than the nominal level, Student's T is satisfactory when sampling from a perfectly symmetric distribution. But for skewed distributions, it can be quite unsatisfactory, even with a sample size of 300. Yet another strategy is to use the modified bootstrap method introduced in Section 7.3. To provide some sense of how the modified bootstrap

TABLE 7.1   Actual Type I Error Probabilities for Four Methods
Based on the Mean, $\alpha = .05$

	Dist.	Method			
		BT	SB	MP	T
$n = 20$	N	.054	.051	.041	.050
	LN	.078	.093	.096	.140
	MN	.100	.014	.050	.022
	SH	.198	.171	.190	.202
$n = 100$	N	.048	.038	.049	.050
	LN	.058	.058	.063	.072
	MN	.092	.018	.054	.041
	SH	.168	.173	.177	.190

N = normal; LN = lognormal; MN = mixed normal; SH = skewed,
heavy-tailed; BT = equal-tailed, bootstrap-t; SB = symmetric bootstrap-t;
MP = modified percentile bootstrap; T = Student's T.

performs, Table 7.1 shows the actual probability of a Type I error when testing $H_0 : \mu = \mu_0$ with $\alpha = .05$ and $n = 20$ and 100. In Table 7.1, BT indicates the equal-tailed bootstrap-t method [given by Equation (7.4)], SB is the symmetric bootstrap-t method [given by Equation (7.6)], MP is the modified bootstrap method described in Section 7.3 in conjunction with least squares regression, and T indicates Student's T. The distributions considered here are normal (N), lognormal (LN), which is shown in Figure 7.1 and represents a distribution that is skewed with relatively light tails, mixed normal (MN), and a skewed, heavy-tailed distribution (SH), which is shown in Figure 5.6. So, for example, when sampling from a lognormal distribution and testing at the .05 level with $n = 20$, the actual probability of a Type I error with Student's T is .14.

Notice that for distribution LN, the equal-tailed bootstrap-t method is substantially better than Student's T; this has been one of the reasons the equal-tailed bootstrap-t method has been recommended. In this particular case, the equal-tailed bootstrap-t also beats the symmetric bootstrap-t as well as the modified percentile method, which is more accurate than the percentile method described in Section 7.1. However, for the mixed normal, the equal-tailed bootstrap-t method is the least satisfactory. In this particular case, if we switch to the symmetric bootstrap-t method, the Type I error probability is .014 with $n = 20$ and .018 with $n = 100$. But when sampling from a lognormal distribution, the actual Type I error probability is .093, and with $n = 100$ there is little difference between the two bootstrap-t methods for this special case. For the mixed normal, the symmetric bootstrap-t has an actual Type I error probability well below the nominal .05 level, suggesting that in this particular case its power might be low relative to the modified percentile bootstrap method. Unfortunately, all of these methods are highly unsatisfactory when sampling from a skewed, heavy-tailed distribution, and the situation improves very slowly as the sample size increases. So we see that in situations where Student's T is unsatisfactory in terms of Type I errors, we can get improved results with some type of bootstrap method. But the choice of which bootstrap method to use depends on the situation, and all four methods considered here can be unsatisfactory, even with $n = 100$, if sampling happens to be from a skewed, heavy-tailed distribution. One could check whether a distribution appears to be skewed and heavy-tailed, but an effective diagnostic tool that detects situations where these four methods fail to control the probability of a Type I error has not been established.

The lognormal distribution is a relatively light-tailed distribution. We have just seen that as we move toward a skewed distribution, where outliers are more common, all four methods in Table 7.1 begin to break down. Sutton (1993) proposed a bootstrap method that improves upon a method for handling skewed distributions proposed by Johnson (1978), but Sutton's method deals with skewed distributions for a certain special case only. In particular, if it is known that a distribution is skewed to the right and the goal is to test $H_0 : \mu \leq \mu_0$, the method can be employed, but the method is not designed to handle $H_0 : \mu \geq \mu_0$. If the distribution is skewed to the left, now you can test $H_0 : \mu \geq \mu_0$ but not the other. More recently, Chen (1995) proposed a modification that avoids the bootstrap, but it too is based on the same restrictions. That is, if a distribution is skewed to the right, you can test $H_0 : \mu \leq \mu_0$, but not $H_0 : \mu \geq \mu_0$. Chen's method appears to perform well, provided the distribution is not too heavy-tailed. If the distribution is heavy-tailed, its control over the probability of a Type I error becomes unsatisfactory. Currently, no method has been found that provides accurate inferences about μ when sampling from a skewed, heavy-tailed distribution unless the sample size is very large. The only certainty is that in some situations, even $n = 300$ is not large enough.

7.5 Inferences About a Trimmed Mean

The Tukey–McLaughlin method for making inferences about the trimmed mean (covered in Sections 4.9 and 5.6) reduces the problems associated with Student's T. Generally, as we increase the amount of trimming, the problem of low power under very small departures from normality is reduced, and we get improved control over the probability of a Type I error. But problems with controlling the Type I error probability, or probability coverage when computing a confidence interval, persist. Combining trimmed means with an appropriate bootstrap method reduces these problems considerably. In fact, with 20% trimming, good control over the probability of a Type I error can be achieved under fairly extreme departures from normality, even with $n = 11$.

7.5.1 Using the Percentile Method

The percentile bootstrap is applied using a simple modification of the method described in Section 7.1: Simply replace the sample mean by the trimmed mean. So if we generate a bootstrap sample of size n and compute the trimmed mean, $\bar{X}_t^*$, and if we repeat this B times, yielding $\bar{X}_{t1}^*, \ldots, \bar{X}_{tB}^*$, then an approximate $1 - \alpha$ confidence interval for the population trimmed mean is

$$\left(\bar{X}_{t(\ell+1)}^*, \bar{X}_{t(u)}^* \right), \tag{7.8}$$

where $\bar{X}_{t(1)}^* \leq \cdots \leq \bar{X}_{t(B)}^*$ are the B bootstrap trimmed means written in ascending order, $\ell = \alpha B/2$, rounded to the nearest integer, and $u = B - \ell$. As usual, reject $H_0 : \mu_t = \mu_0$ if the confidence interval for the trimmed mean [Equation (7.8) in this particular case] does not contain the hypothesized value, μ_0.

The performance of the percentile bootstrap improves in terms of Type I errors and probability coverage as we increase the amount of trimming and becomes fairly accurate with at least 20% trimming. Moreover, with 20% trimming, the modified percentile bootstrap considered in Section 7.3 is no longer needed and performs in an unsatisfactory manner. But the minimum amount of trimming needed to justify the percentile bootstrap is not known. The only rule currently available is that with a minimum of 20% trimming, accurate results can be obtained even with very small sample sizes. So in particular, the percentile bootstrap method performs well when working with the median. Perhaps the percentile bootstrap continues to perform well with 15% trimming, or even 10% trimming, but this has not been established.

7.5.2 Singh's Modification

Imagine 10 observations with two extreme outliers. For example, suppose we observe

$$2, 3, 6, 3, 9, 12, 15, 7, 200, 300.$$

With 20% trimming, the two outliers have no influence on $\bar{X}_t$. Notice, however, that when we generate a bootstrap sample, by chance we might get three outliers. That is, the number of outliers in a bootstrap sample might exceed the finite-sample breakdown point of the trimmed mean even though this is not the case for the original observations. The result is that the bootstrap trimmed mean becomes inflated, and this can lead to a relatively long confidence interval when using the percentile bootstrap method.

Singh (1998) showed that from a theoretical point of view, we can address this problem by first Winsorizing the data before we take bootstrap samples. The only restriction imposed by theory is that the amount of Winsorizing must be less than or equal to the amount of trimming. So in our example, if we plan to use a 20% trimmed mean, we are allowed to Winsorize our values by 20%, in which case the 10 values in our example become

$$6, 6, 6, 3, 9, 12, 15, 7, 15, 15.$$

Now we generate bootstrap samples as before, but we resample with replacement from the Winsorized values rather than from the original observations. After generating B bootstrap trimmed means, we apply the percentile bootstrap method in the usual manner. For example, the middle 95% of the bootstrap trimmed means provide a .95 confidence interval.

Unfortunately, if we Winsorize as much as we trim, it seems that probability coverage based on a percentile bootstrap method can be unsatisfactory, at least when the sample size is small (Wilcox, 2001d). However, if when we use 20% trimming we Winsorize by 10%, good probability coverage is obtained.

7.5.3 Using the Bootstrap-*t* Method

The bootstrap-*t* method can be applied with a trimmed mean in the following manner. Generate a bootstrap sample of size n and compute the trimmed mean and Winsorized standard deviation, which we label $\bar{X}_t^*$ and s_w^*, respectively. Next, compute

$$T_t^* = \frac{(1 - 2\gamma)(\bar{X}_t^* - \bar{X}_t)}{s_w^*/\sqrt{n}}, \tag{7.9}$$

where, as usual, γ is the amount of trimming, which we usually take to be .2. Repeating this process B times yields B T_t^* values. Writing these B values in ascending order we get $T_{t(1)}^* \le T_{t(2)}^* \le \cdots \le T_{t(B)}^*$. Letting $\ell = .025B$, rounded to the nearest integer, and $u = B - \ell$, an estimate of the .025 and .975 quantiles of the distribution of T_t is $T_{(\ell+1)}^*$ and $T_{(u)}^*$. The resulting .95 confidence interval for μ_t (the population trimmed mean) is

$$\left(\bar{X}_t - T_{t(u)}^* \frac{s_w}{(1 - 2\gamma)\sqrt{n}}, \bar{X}_t - T_{t(\ell+1)}^* \frac{s_w}{(1 - 2\gamma)\sqrt{n}} \right). \tag{7.10}$$

HYPOTHESIS TESTING. As for testing $H_0 : \mu_t = \mu_0$, compute

$$T_t = \frac{(1 - 2\gamma)(\bar{X}_t - \mu_0)}{s_w/\sqrt{n}}$$

and reject if

$$T_t \le T_{t(\ell+1)}^*,$$

or if

$$T_t \ge T_{t(u)}^*.$$

That is, we use the same method employed when making inferences about the mean, except we replace the sample mean with the trimmed mean and we replace the sample standard deviation s with $s_w/(1 - 2\gamma)$.

The symmetric bootstrap-*t* method can be used as well when testing a two-sided hypothesis. Now we use

$$T_t^* = \frac{|(1 - 2\gamma)(\bar{X}_t^* - \bar{X}_t)|}{s_w^*/\sqrt{n}}. \tag{7.11}$$

and reject H_0 if $|T_t| > T_{t(c)}^*$, where $c = (1 - \alpha)B$ rounded to the nearest integer. An approximate $1 - \alpha$ confidence interval for μ_t is

$$\bar{X}_t \pm T_{t(c)}^* \frac{s_w}{(1 - 2\gamma)\sqrt{n}}. \tag{7.12}$$

Table 7.1 reported the actual probability of a Type I error when using means with one of four methods. None of the methods was satisfactory for all four distributions

TABLE 7.2　Actual Type I Error Probabilities Using 20% Trimmed Means, $\alpha = .05$

	Dist.	Method			
		BT	SB	P	TM
$n = 20$	N	.067	.052	.063	.042
	LN	.049	.050	.066	.068
	MN	.022	.019	.053	.015
	SH	.014	.018	.066	.020

N = normal; LN = lognormal; MN = mixed normal; SH = skewed, heavy-tailed; BT = equal-tailed, bootstrap-t; SB = symmetric bootstrap-t; P = percentile bootstrap; TM = Tukey–McLaughlin.

considered, even after increasing the sample size to 300. Table 7.2 shows the actual probability of a Type I error when using 20% trimmed means instead. Notice that the percentile bootstrap method is the most stable; the actual probability of a Type I error ranges between .053 and .066. The other three methods do a reasonable job of avoiding Type I error probabilities above the nominal .05 level. But they can have actual Type I error probabilities well below the nominal level, which is an indication that their power might be less than when using the percentile method instead. With the caveat that no method is best in all situations, the percentile bootstrap with a 20% trimmed mean is a good candidate for general use.

In the previous subsection we noted that we can Winsorize our data before we take bootstrap samples. Theory allows us to do the same when working with the bootstrap-t method. But when sample sizes are small, probability coverage can be poor. Apparently with a sufficiently large sample size this problem becomes negligible, but just how large the sample size must be remains unknown.

7.5.4　S-PLUS Functions trimpb and trimcibt

The S-PLUS function

$$\text{trimpb}(x,\text{tr}=.2,\text{alpha}=.05,\text{nboot}=2000,\text{WIN}=F,\text{win}=.1)$$

(written for this book) computes a confidence interval for a trimmed mean using the percentile bootstrap method. The argument tr indicates the amount of trimming and defaults to .2 if not specified. As usual, alpha is α and defaults to .05. It appears that $B = 500$ suffices, in terms of achieving accurate probability coverage with 20% trimming. But to be safe, B (nboot) defaults to 2000. (An argument for using $B = 2000$ can be made along the lines used by Booth & Sarker, 1998.) The argument WIN indicates whether the values should be Winsorized before bootstrap samples are taken. By default WIN=F, for false, meaning that Winsorizing will not be done. If WIN=T, the amount of Winsorizing is given by the argument win, which defaults to .1.

The S-PLUS function

$$\text{trimcibt}(x, \text{tr} = 0.2, \text{alpha} = 0.05, \text{nboot} = 599, \text{side} = F)$$

computes a bootstrap-t confidence interval. The argument side indicates whether an equal-tailed or a symmetric confidence interval is to be computed. As indicated, side defaults to F, meaning that an equal-tailed confidence interval [given by Equation (7.10)] will be computed. Using side=T results in a symmetric confidence interval [given by Equation (7.12)].

EXAMPLE. Table 3.2 reported data on the desired number of sexual partners among 105 college males. As previously indicated, these data are highly skewed, with a relatively large number of outliers, and this can have a deleterious effect on many methods for computing a confidence interval and testing hypotheses. If we compute the Tukey–McLaughlin .95 confidence interval for the 20% trimmed mean [using Equation (4.41)], we get (1.62, 3.75). Using the S-PLUS function trimcibt with side=F yields an equal-tailed .95 confidence interval of (1.28, 3.61). With side=T it is (1.51, 3.61). Using the percentile bootstrap method, the S-PLUS function trimpb returns (1.86, 3.95). So in this particular case, the lengths of the confidence intervals do not vary that much among the methods used, but the intervals are centered around different values, which in some cases might affect any conclusions made.   ■

In summary, all indications are that the percentile bootstrap is more stable (with at least 20% trimming) than the bootstrap-t method. That is, the actual Type I error probability tends to be closer to the nominal level. And it has the added advantage of more power, at least in some situations, compared to any other method we might choose. However, subsequent chapters will describe situations where the bootstrap-t method outperforms the percentile method. And there are additional situations where the percentile bootstrap is best. So both methods are important to know.

7.6 Estimating Power When Testing Hypotheses About a Trimmed Mean

As when working with means, if we test some hypothesis about a trimmed mean and fail to reject, this might be because the null hypothesis is true, or perhaps power is too low to detect a meaningful difference. If we can estimate how much power we have based on the same data used to test some hypothesis, we are better able to discern which reason accounts for a nonsignificant result. We saw in Section 5.4.3 how, given some data, power can be controlled using Stein's method. A natural strategy is to use some analog of this method when working with trimmed means, but this approach has not been investigated as yet. There are several alternative methods one might employ, but most have proven to be unsatisfactory when sample sizes are small or even moderately large (Wilcox & Keselman, 2002). This section outlines the method that currently performs best when estimating power. It is a special case of another method covered in Chapter 8. So for brevity, we merely describe what the method attempts to do and then provide some software for implementing it. Readers interested in computational details can refer to Chapter 8.

The method in this section is designed specifically for the case where the percentile bootstrap method is used to test some hypothesis about a 20% trimmed mean with $\alpha = .05$. There are two goals: (1) Compute an (unbiased) estimate of power for some given value of $\delta = \mu_t - \mu_0$, and (2) provide a conservative estimate of power meaning a (one-sided) confidence interval for how much power we have. Roughly, the method estimates the standard error of the trimmed mean and then, given δ, provides an estimate of how much power we have. A possible concern, however, is that this estimate might underestimate power. Based on data, we might estimate power to be .7, but in reality it might be .5 or it might be as low as .4. So the method also computes a (lower) .95 confidence interval for the actual amount of power using a percentile bootstrap technique. Briefly, for every bootstrap sample, the standard error of the trimmed mean is estimated, which yields an estimate of power corresponding to whatever δ value is of interest. Repeating this process B times yields B estimates of power, which, when put into ascending order, we label $\hat{\xi}_{(1)} \leq \cdots \leq \hat{\xi}_{(B)}$. Then a conservative estimate of power is $\hat{\xi}_{(a)}$, where $a = .05B$ rounded to the nearest integer. So if $\hat{\xi}_{(a)} = .4$, say, we estimate that with probability .95, power is at least .4. If $\hat{\xi}_{(a)} = .6$, we estimate that with probability .95, power is at least .6.

7.6.1 S-PLUS Functions powt1est and powt1an

The S-PLUS function

$$\text{powt1est}(x, \text{delta}=0, \text{ci}=F, \text{nboot}=800)$$

returns an estimate of how much power there is for some value of δ. As usual, x now represents any S-PLUS variable containing data. The argument ci defaults to F (for false), meaning that no confidence interval for power is computed. If ci=T is used, a percentile bootstrap is used to get a conservative estimate of power. As usual, nboot indicates how many bootstrap samples are used. (That is, nboot corresponds to B.)

EXAMPLE. Consider the values

$$12, 20, 34, 45, 34, 36, 37, 50, 11, 32, 29.$$

Using the S-PLUS function trimpb, we get a .95 confidence interval for the 20% trimmed mean of $(22.86, 38.71)$. So we would not reject $H_0 : \mu_t = 32$. To gain perspective on why, we estimate how much power we have when $\delta = \mu_t - \mu_0 = 2$. The S-PLUS command powt1est(x,2,T) returns

```
$est.power:
[1] 0.09033971

$ci:
[1] 0.06400682
```

meaning that the estimated power is only .09 and that with probability .95 power is at least .06. So power is estimated to be inadequate, suggesting that H_0 should not be accepted. (Using Stein's method for means, we see that 368 observations are needed to get power equal to .8 with $\delta = 2$.) ■

The S-PLUS function

$$powt1an(x, ci = F, plotit = T, nboot = 800)$$

provides a power analysis without having to specify a value for δ. Rather, the function chooses a range of δ values so that power will be between .05 and .9, approximately. Then it estimates how much power there is for each δ value that it selects and plots the results. That is, the function estimates the power curve associated with the percentile bootstrap method of testing hypotheses with a 20% trimmed mean. If the argument plotit is set to F (for false), no plot is generated; the function merely reports the δ values it generates and the corresponding power. The function also reports a lower estimate of power if ci=T is used. That is, with probability .95, power is at least as high as this lower estimate.

EXAMPLE. Figure 7.5 shows the plot created by powt1an(x) using the data from the last example. This plot indicates that power is reasonably high with $\delta = 12$, but for $\delta = 6$ power is only about .3.

Figure 7.6 shows the plot generated by the command powt1an(x,ci=T). The upper, solid line is the same as in Figure 7.5 and represents the estimated power. The lower, dashed line indicates a conservative estimate of power. So we see that for $\delta = 12$, it is estimated that power exceeds .8, and we can be approximately 95% certain that power exceeds .4. That is, it appears that power is adequate for $\delta = 12$, but there is some possibility that it is not — perhaps power is adequate only for δ values much greater than 12. ■

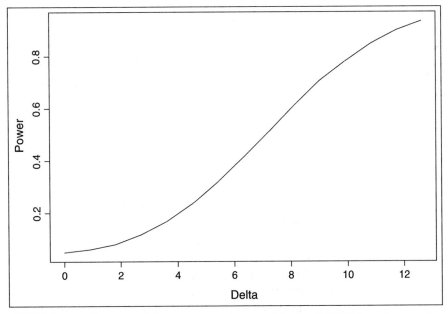

FIGURE 7.5   Estimate of the power curve returned by the S-PLUS function powt1an based on the data in Section 7.6.1.

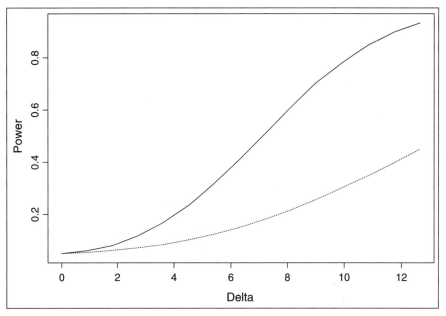

FIGURE 7.6   Same as Figure 7.5 but with lower confidence interval for power, indicated by the dashed lines.

7.7 Inferences Based on MOM and M-Estimators

Currently it appears that the best method for making inferences about an M-estimator of location is the percentile bootstrap technique. For very small sample sizes, probability coverage and control over the probability of a Type I error are a bit more stable using a 20% trimmed mean, but with a sample size of at least 20 the percentile bootstrap appears to perform reasonably well with an M-estimator. So to compute a confidence interval for μ_{os}, the measure of location being estimated by $\hat{\mu}_{os}$ [the one-step M-estimator given by Equation (3.28)], generate bootstrap samples as before until you have B values for $\hat{\mu}_{os}$. As usual, put these B values in ascending order, yielding $\hat{\mu}^*_{os(1)} \leq \cdots \leq \hat{\mu}^*_{os(B)}$. Then a $1 - \alpha$ confidence interval for μ_{os} is

$$\left(\hat{\mu}^*_{os(\ell+1)}, \hat{\mu}^*_{os(u)}\right) \tag{7.13}$$

where $\ell = \alpha B/2$, rounded to the nearest integer, and $u = B - \ell$. For $\alpha = .05$, all indications are that in terms of probability coverage, $B = 399$ suffices and that using $B = 599$ offers no practical advantage.

As for the MOM estimator, simply replace $\hat{\mu}_{os}$ with $\hat{\mu}_{mom}$, which is given by Equation (3.30). All indications are that confidence intervals based on MOM (with $B = 500$) provide accurate probability coverage and good control over the probability of a Type I error, even in situations where control over the probability of a Type I error using an M-estimator is unsatisfactory. In particular, currently it seems that good results are obtained even with $n = 11$.

HYPOTHESIS TESTING.   Reject $H_0 : \mu_{os} = \mu_0$ if the bootstrap confidence interval given by Equation (7.13) does not contain μ_0. A similar strategy is used when the one-step M-estimator is replaced by MOM.

7.7.1 S-PLUS Function mestci

The S-PLUS function

$$\text{mestci}(x,\text{alpha}=.05,\text{nboot}=399)$$

computes a confidence interval based on an M-estimator, where x is an S-PLUS variable containing data, alpha is α, which defaults to .05, and nboot is B, the number of bootstrap samples to be used, which defaults to 399. (This function contains two additional arguments, the details of which can be found in Wilcox, 1997a, p. 85.)

7.7.2 S-PLUS Functions momci and onesampb

The S-PLUS function

$$\text{momci}(x,\text{alpha}=.05,\text{nboot}=500)$$

computes a confidence interval based on the measure of location MOM described in Section 3.5.2. Unlike the S-PLUS function mestci, the default number of bootstrap samples (nboot) is 500. The S-PLUS function

$$\text{onesampb}(x,\text{est}=\text{mom},\text{alpha}=.05,\text{nboot}=500,\ldots)$$

can also be used to make inferences based on MOM; but unlike momci, any other measure of location or scale can be used through the argument est. For example, using est=pbvar would compute a confidence interval for the percentage bend midvariance.

EXAMPLE. For 15 law schools, the undergraduate GPA of entering students, in 1973, was

  3.39 3.30 2.81 3.03 3.44 3.07 3.00 3.43 3.36 3.13 3.12 2.74 2.76 2.88 2.96.

The .95 confidence interval returned by mestci is (2.95, 3.28). So among all law schools, it is estimated that the typical GPA of entering students is between 2.95 and 3.28. Using the MOM estimator instead (the S-PLUS function momci), the .95 confidence interval is (2.88, 3.35). ■

7.8 Detecting Nonlinear Associations

A possible concern about any method for testing the hypothesis that Pearson's correlation (ρ) is zero is that it might routinely miss an association that is not linear. The method described in this section, which allows heteroscedasticity, is one possible way of dealing with this problem. The theoretical justification for the method stems from Stute, Manteiga, and Quindimil (1998). Here, a very slight modification of their method is used that performs better in terms of Type I errors (in simulations) when the sample size is small.

The basic idea is to test the hypothesis that the regression line between two variables is both flat and straight. So the method is designed to be sensitive to any curved association between two variables; in the event the regression line is straight, the method is designed to detect situations where the slope differs from zero. Said another way, in the context of the regression model in Section 6.2, the method can be used to detect situations where the expected value of Y, given X, is not a constant. That is, the mean of Y, given X, changes with X in some unspecified manner that may be nonlinear. This is in contrast to standard regression methods, where it is assumed that the mean of Y, given X, is given by some straight line: $\beta_1 X + \beta_0$. Because the method in this section is designed to be sensitive to a broader range of associations than any test of $H_0 : \rho = 0$, a natural speculation is that it is more likely to detect an association, and experience suggests that this is indeed the case.

As in Section 7.3, let $(X_1, Y_1), \ldots, (X_n, Y_n)$ be a random sample of n pairs of points. The test statistic is computed as follows. Consider the jth pair of points: (X_j, Y_j). For any j we might pick $(1 \leq j \leq n)$, some of the X values will be less than or equal to X_j. For these X values, compute the sum of the difference between the corresponding Y values and $\bar{Y}$. We represent this sum, divided by the square root of the sample size, by R_j. Said more formally, if we fix j and set $I_i = 1$ if $X_i \leq X_j$, otherwise $I_i = 0$, then

$$R_j = \frac{1}{\sqrt{n}} \sum I_i (Y_i - \bar{Y})$$
$$= \frac{1}{\sqrt{n}} \sum I_i r_i, \tag{7.14}$$

where now

$$r_i = Y_i - \bar{Y}.$$

The test statistic is the maximum absolute value of all the R_j values. That is, the test statistic is

$$D = \max |R_j|, \tag{7.15}$$

where max means that D is equal to the largest of the $|R_j|$ values.

It is noted that the method just described is heteroscedastic. That is, unlike Student's T test of $H_0 : \rho = 0$, this wild bootstrap method is not sensitive to changes in the variation of Y (the conditional variance of Y given X) as X increases.

EXAMPLE. We illustrate the computation of D with the data in Table 7.3. These data are from a study conducted about 200 years ago with the goal of determining whether the earth bulges at the equator, as predicted by Newton, or whether it bulges at the poles. The issue was addressed by measuring latitude at

Continued

TABLE 7.3   Data on Meridian Arcs

Place	Transformed latitude (X)	Arc length (Y)	$r_i = Y_i - \bar{Y}$	R
Quito	0.0000	56,751	−301.6	−134.88
Cape of Good Hope	0.2987	57,037	−15.6	−141.86
Rome	0.4648	56,979	−73.6	−174.77
Paris	0.5762	57,074	21.4	−165.20
Lapland	0.8386	57,422	369.4	0.00

EXAMPLE. (*Continued*) various points on the earth and trying to determine how latitude is related to a measure of arc length. To compute R_1, we note that $X_1 = 0.0$ and there are no other X values less than or equal to 0.0, so

$$R_1 = \frac{r_1}{\sqrt{5}} = \frac{-301.6}{\sqrt{5}} = -134.88.$$

As for R_2, $X_2 = 0.2987$, there are two X values less than or equal to 0.2987 (namely, X_1 and X_2); the corresponding r_i values are -301.6 and -15.6. So

$$R_2 = \frac{1}{\sqrt{5}}(-301.6 - 15.6) = -141.86.$$

The remaining R values are computed in a similar manner and are shown in the last column of Table 7.3. The largest absolute value in this column is 174.77, so $D = 174.77$. ■

To determine how large D must be to reject the null hypothesis, a so-called *wild bootstrap* is used. (The other types of bootstrap methods already covered are known to be theoretically unsound for the problem at hand.) Let

$$r_i = Y_i - \bar{Y}$$

be the ith residual corresponding to the regression line $\hat{Y} = \bar{Y}$. Generate n observations from a uniform distribution (which is shown in Figure 4.7) and label the results $U_1, \ldots, U_n$. (This can be done in S-PLUS with the built-in function runif.) So each U has a value between 0 and 1, and all values between 0 and 1 are equally likely. Next, for every value of i ($i = 1, \ldots, n$) set

$$V_i = \sqrt{12}(U_i - .5),$$
$$r_i^* = r_i V_i,$$
$$Y_i^* = \bar{Y} + r_i^*.$$

Then based on the n pairs of points $(X_1, Y_1^*), \ldots, (X_n, Y_n^*)$, compute the test statistic as described in the previous paragraph and label it D^*. Repeat this process B times, and label the resulting (bootstrap) test statistics $D_1^*, \ldots, D_B^*$. Finally, put these B values in ascending order, which we label $D^*_{(1)} \leq \cdots \leq D^*_{(B)}$. Then the critical value is $D^*_{(u)}$, where $u = (1 - \alpha)B$ rounded to the nearest integer. That is, reject if

$$D \geq D^*_{(u)}.$$

There is a variation of the method that should be mentioned where the test statistic D is replaced by

$$W = \frac{1}{n} \left(R_1^2 + \cdots + R_n^2\right). \tag{7.16}$$

The critical value is determined in a similar manner as before. First, generate a wild bootstrap sample and compute W, yielding W^*. Repeating this B times, you reject if

$$W \geq W^*_{(u)},$$

where again $u = (1 - \alpha)B$ rounded to the nearest integer, and $W^*_{(1)} \leq \cdots \leq W^*_{(B)}$ are the B W^* values written in ascending order. The test statistic D is called the *Kolmogorov–Smirnov* test statistic, and W is called the *Cramér–von Mises* test statistic. The choice between these two test statistics is not clear-cut. For the situation at hand, currently it seems that there is little separating them in terms of controlling Type I errors. But for situations where there are multiple predictors, as described in Chapter 14, the Cramér–von Mises test statistic seems to have an advantage (Wilcox & Muska, 2001).

The method just described can be used with $\bar{Y}$ replaced by a trimmed mean, $\bar{Y}_t$. So now $r_i = Y_i - \bar{Y}_t$ and $Y_i^* = \bar{Y}_t + r_i^*$. Otherwise the computations are exactly the same as before. We have already seen that replacing the sample mean with a 20% trimmed mean can make a substantial difference in power and control over the probability of a Type I error. Here, however, all indications are that the improvement in Type I error control is negligible. As for power, often it makes little difference whether a mean or trimmed mean is used, but situations do arise where you reject with a mean but not a trimmed mean, and the reverse happens as well.

By design, the method in this section is not sensitive to heteroscedasticity. That is, if the regression line is horizontal and straight, the probability of a Type I error will be approximately α, even when there is heteroscedasticity. In contrast, Student's T test of $H_0: \rho = 0$, covered in Section 6.5, is sensitive to heteroscedasticity. This would seem to suggest that in some instances, Student's T test might have more power, but experience suggests that in practice the method in this section is more likely to detect an association.

7.8.1 S-PLUS Function indt

The S-PLUS function

indt(x, y, tr = 0.2, nboot = 500, alpha = 0.05, flag = 1)

(written for this book) performs the test of independence just described. As usual, x and y are S-PLUS variables containing data, tr indicates the amount of trimming, and nboot is B. The argument flag indicates which test statistic will be used:

- flag=1 means the Kolmogorov–Smirnov test statistic, D, is used.
- flag=2 means the Cramér–von Mises test statistic, W, is used.
- flag=3 both test statistics are computed.

EXAMPLE. For the aggression data in Table 6.3, Pearson's correlation is $r = -0.286$ and we fail to reject $H_0 : \rho = 0$ at the $\alpha = .05$ level. So we fail to conclude that there is an association between marital aggression and recall-test scores among children. (The significance level is .051, so we nearly reject.) The output from indt(x,y,flag=3) is

```
$dstat:
[1] 8.002455

$wstat:
[1] 37.31641

$critd:
[1] 6.639198

$critw:
[1] 24.36772
```

So $D = 8.002455$, the critical value is 6.639198, and because D exceeds the critical value, reject and conclude that marital aggression and recall-test scores are dependent. Similarly, the Cramér–von Mises test statistic is $W = 37.31641$; it exceeds the critical value 24.36772, so again we reject. So we have empirical evidence that aggression in the home is associated with recall-test scores among children living in the home, but the function indt tells us nothing about what this association might be like. ■

7.9 Exercises

1. For the following 10 bootstrap sample means, what would be an appropriate .8 confidence interval for the population mean?

    7.6, 8.1, 9.6, 10.2, 10.7, 12.3, 13.4, 13.9, 14.6, 15.2.

2. Rats are subjected to a drug that might affect aggression. Suppose that for a random sample of rats, measures of aggression are found to be

    5, 12, 23, 24, 18, 9, 18, 11, 36, 15.

    Verify that the equal-tailed .95 confidence interval for the mean returned by the S-PLUS function trimcibt is (11.8, 25.9). Compare this confidence interval to the confidence interval you got for Exercise 32 in Chapter 4.

3. For the data in Exercise 2, verify that the .95 confidence interval for the mean returned by the S-PLUS function trimpb is (12.3, 22.4).

4. Referring to the previous two Exercises, which confidence interval is more likely to have probability coverage at least .95?

5. For the data in Exercise 2, verify that the equal-tailed .95 confidence interval for the population 20% trimmed mean using trimcibt is (9.1, 25.7).

6. For the data in Exercise 2, verify that the .95 confidence interval for the population 20% trimmed mean using trimpb is (11.2, 22.0).

7. Which of the two confidence intervals given in the last two exercises is likely to have probability coverage closer to .95?

8. For the following observations, verify that the .95 confidence interval based on a one-step M-estimator returned by the S-PLUS function mestci is (7, 21.8).

$$2, 4, 6, 7, 8, 9, 7, 10, 12, 15, 8, 9, 13, 19, 5, 2, 100, 200, 300, 400$$

9. For the data in the previous exercise, verify that the .95 confidence interval for the 20% trimmed mean returned by trimpb is (7.25, 64). Why would you expect this confidence interval to be substantially longer than the confidence interval based on a one-step M-estimator?

10. Use trimpb on the data used in the previous two exercises, but this time Winsorize the data first by setting the argument WIN to T. Verify that the .95 confidence interval for the trimmed mean is now (7.33, 57.7). Why do you think this confidence interval is shorter than the confidence interval in the last exercise?

11. Repeat the last exercise, only now Winsorize 20% by setting the argument win to .2. Verify that the .95 confidence interval is now (7.67, 14). Why is this confidence interval so much shorter than the confidence interval in the last exercise?

12. For the confidence interval obtained in the last exercise, what practical problem might have occurred regarding probability coverage?

13. For the data in Exercise 8, compute a .95 confidence interval for the median using the S-PLUS function sint, described in Section 4.11.1. What does this suggest about using a median versus a one-step M-estimator?

14. Exercise 22 in Chapter 6 reports that the .95 confidence interval for the slope of the least squares regression line, based on the data in Table 6.6, is $(-0.16, .12)$. Using the S-PLUS function lsfitci, described in Section 7.3.1, verify that the .95 confidence interval based on the modified bootstrap method is $(-0.27, 0.11)$.

15. Again using the data in Table 6.6, verify that you do not reject $H_0 : \rho = 0$ using the S-PLUS function corb.

16. Using the data in Table 6.6, verify that when using the method in Section 7.8 based on a mean (meaning that you set tr=0 when using the S-PLUS function indt), both the Kolmogorov–Smirnov Cramér–von Mises test statistics reject the hypothesis that these two variables are independent.

17. The previous exercise indicates that the variables in Table 6.6 are dependent, but the results in Exercises 14 and 15 failed to detect any association. Describe a possible reason for the discrepant results.

18. Create a scatterplot of the data in Table 6.6, and note that six points having X values greater than 125 are visibly separated from the bulk of the observations.

Now compute a .95 confidence interval for the slope of the least squares regression line with the six points having X values greater than 125 eliminated. The resulting confidence interval is $(-0.84, -0.17)$, so you reject the hypothesis of a zero slope and conclude that the two variables are dependent. Comment on this result in the context of factors that affect the magnitude of Pearson's correlation, which are described in Section 6.4.

19. Restricting the range as was done in the previous exercise, verify that the S-PLUS function corb returns a .95 confidence interval of $(-0.63, -0.19)$. Compare this to the result in Exercise 15.

20. Using the data in Exercise 6 of Chapter 6, verify that the S-PLUS function lsfitci returns a .95 confidence interval for the slope (when predicting cancer rates given solar radiation) of $(-0.049, -0.25)$. The conventional .95 confidence interval based on Equation (6.8) is $(-0.047, -0.24)$. What does this suggest about the conventional method?

21. For Hubble's data on the recession velocity of galaxies, shown in Table 6.1, verify that the modified bootstrap method yields a .95 confidence interval of $(310.1, 630.1)$ for the slope.

COMPARING TWO
INDEPENDENT GROUPS

One of the most common goals in applied research is comparing two independent variables or groups. For example if one group of individuals receives an experimental drug for treating migraine headaches and a different, independent group of individuals receives a placebo, and we measure the effectiveness of a drug using some standard technique, how might we compare the outcomes corresponding to the two groups? How does the reading ability of children who watch 30 hours or more of television per week compare to children who watch 10 hours or less? How does the birth weight of newborns among mothers who smoke compare to the birth weight among mothers who do not smoke? In general terms, if we have two independent variables, how might we compare these two measures?

In this book, attention is focused on four interrelated and overlapping methods one might use to compare two independent groups or variables:

- Compare the groups using some measure of location, such as the mean, trimmed mean, or median. In particular, we might test the hypothesis that the measures of location are identical, or we might compute a confidence interval to get some sense of how much they differ.
- Test the hypothesis that the two groups have identical distributions. Identical distributions means that for any constant c we might pick, the probability that a randomly sampled observation is less than c is the same for both groups.
- Determine the probability that a randomly sampled observation from the first group will be less than a randomly sampled observation from the second. If the groups do not differ, this probability will be .5.
- Compare variances or some other measure of scale.

Each approach has its advantages and disadvantages. Each provides a different and useful perspective, and no single approach is optimal in all situations. A general goal is to explain the relative merits of each of the four strategies just listed, plus the practical advantages associated with the methods based on a specific strategy, so that applied researchers can make an informed decision as to which approach and which method might be used in a given situation.

The emphasis in this chapter is on the first strategy, but some methods covered here, including the best-known method for comparing means, are related to the second strategy as well. The fourth strategy is taken up in Section 8.10 (and the third strategy is discussed in Chapter 15). Section 8.12 describes how Pearson's correlation and regression slopes can be compared.

8.1 Student's T

We begin with the classic and most commonly used method for comparing two independent groups: Student's T test. The goal is to test

$$H_0 : \mu_1 = \mu_2, \tag{8.1}$$

the hypothesis that the two groups have identical means. That is, the goal is to determine whether the typical individual in the first group differs from the typical individual in the second. If we can find a method for computing a confidence interval for $\mu_1 - \mu_2$, we can get some sense of the degree to which the typical individual in the first group differs from the typical individual in the second. This assumes, of course, that the population mean provides a reasonable measure of what is typical. We have already seen that this assumption is dubious in some situations, but we ignore this issue for the moment.

Exact control over the probability of a Type I error can be had under the following assumptions:

- Sampling is random.
- Sampling is from normal distributions.
- The two groups have equal variances; that is, $\sigma_1^2 = \sigma_2^2$, where σ_1^2 and σ_2^2 are the variances corresponding to the groups having means μ_1 and μ_2, respectively.

The last assumption is called *homoscedasticity*. If the variances differ ($\sigma_1^2 \neq \sigma_2^2$), we say that there is *heteroscedasticity*.

Before describing how to test the hypothesis of equal means, first consider how we might estimate the assumed common variance. For convenience, let σ_p^2 represent the common variance and let s_1^2 and s_2^2 be the sample variances corresponding to the two groups. Also let n_1 and n_2 represent the corresponding sample sizes. The typical estimate of the assumed common variance is

$$s_p^2 = \frac{(n_1 - 1)s_1^2 + (n_2 - 1)s_2^2}{n_1 + n_2 - 2}, \tag{8.2}$$

where the subscript p is used to indicate that the sample variances are being pooled. Because s_1^2 and s_2^2 are assumed to estimate the same quantity, σ_p^2, a natural strategy for combining them into a single estimate is to average them, and this is exactly what is done when the sample sizes are equal. But with unequal sample sizes a slightly different strategy is used, as indicated by Equation (8.2). (A weighted average is used instead.)

Now consider the problem of testing the null hypothesis of equal means. Simultaneously, we want a confidence interval for the difference between the population means, $\mu_1 - \mu_2$. Under the assumptions already stated, the probability of a Type I error will be exactly α if we reject H_0 when

$$|T| \geq t, \tag{8.3}$$

where

$$T = \frac{\bar{X}_1 - \bar{X}_2}{\sqrt{s_p^2 \left(\frac{1}{n_1} + \frac{1}{n_2} \right)}}, \tag{8.4}$$

and t is the $1 - \alpha/2$ quantile of Student's T distribution with $\nu = n_1 + n_2 - 2$ degrees of freedom. An exact $1 - \alpha$ confidence interval for the difference between the population means is

$$(\bar{X}_1 - \bar{X}_2) \pm t \sqrt{s_p^2 \left(\frac{1}{n_1} + \frac{1}{n_2} \right)}. \tag{8.5}$$

EXAMPLE. Salk (1973) conducted a study where the general goal was to examine the soothing effects of a mother's heartbeat on her newborn infant. Infants were placed in a nursery immediately after birth, and they remained there for four days except when being fed by their mothers. The infants were divided into two groups. One group was continuously exposed to the sound of an adult's heartbeat; the other group was not. Salk measured, among other things, the weight change of the babies from birth to the fourth day. Table 8.1 reports the weight change for the babies weighing at least 3,510 grams at birth. The estimate of the assumed common variance is

$$s_p^2 = \frac{(20 - 1)(60.1^2) + (36 - 1)(88.4^2)}{20 + 36 - 2} = 6335.9.$$

So

$$T = \frac{18 - (-52.1)}{\sqrt{6335.9 \left(\frac{1}{20} + \frac{1}{36} \right)}} = \frac{70.1}{22.2} = 3.2.$$

The degrees of freedom are $\nu = 20 + 36 - 2 = 54$. If we want the Type I error probability to be $\alpha = .05$, then, from Table 4 in Appendix B, $t = 2.01$. Because $|T| \geq 2.01$, reject H_0 and conclude that the means differ. That is, we conclude that among all newborns we might measure, the average weight gain would be higher among babies exposed to the sound of a heartbeat compared to those who are not exposed. By design, the probability that our conclusion is in error is .05, assuming normality and homoscedasticity. The .95 confidence interval

Continued

TABLE 8.1   Weight Gain (in grams) for Large Babies

Subject	Gain	Subject	Gain	Subject	Gain	Subject	Gain	Subject	Gain	Subject	Gain
Group 1 (heartbeat)*				Group 2 (no heartbeat)†							
1	190	11	10	1	140	11	−25	21	−50	31	−130
2	80	12	10	2	100	12	−25	22	−50	32	−155
3	80	13	0	3	100	13	−25	23	−60	33	−155
4	75	14	0	4	70	14	−30	24	−75	34	−180
5	50	15	−10	5	25	15	−30	25	−75	35	−240
6	40	16	−25	6	20	16	−30	26	−85	36	−290
7	30	17	−30	7	10	17	−45	27	−85		
8	20	18	−45	8	0	18	−45	28	−100		
9	20	19	−60	9	−10	19	−45	29	−110		
10	10	20	−85	10	−10	20	−50	30	−130		

*$n_1 = 20$, $\bar{X}_1 = 18.0$, $s_1 = 60.1$, $s_1/\sqrt{n_1} = 13$.
†$n_2 = 36$, $\bar{X}_2 = -52.1$, $s_2 = 88.4$, $s_2/\sqrt{n_2} = 15$.

EXAMPLE. (*Continued*) for $\mu_1 - \mu_2$, the difference between the means, is

$$[18 - (-52.1)] \pm 2.01 \sqrt{6335.9 \left(\frac{1}{20} + \frac{1}{36} \right)} = (25.5, 114.7).$$

This interval does not contain zero, and it indicates that the difference between the means is likely to be at least 25.5, so again you would reject the hypothesis of equal means.   ■

8.2 Relative Merits of Student's *T*

We begin by describing some positive features of Student's *T*. If distributions are nonnormal, but otherwise *identical*, Student's *T* performs reasonably well in terms of controlling Type I errors. This result is somewhat expected based on features of the one-sample Student's *T* covered in Chapters 4 and 5. To get a rough idea of why, we first note that for any two independent variables having identical distributions, their difference will have a perfectly symmetric distribution about zero, even when the distributions are skewed. For example, suppose that in Salk's study, the first group of infants has weight gains that follow the (lognormal) distribution shown in Figure 4.14, and the second group has the same distribution. If we randomly sample an observation from the first group and do the same for the second group, then the distribution of the difference will be symmetric about zero. (That is, repeating this process billions of times, a plot of the resulting differences will be symmetric about zero.) More generally, if we sample *n* observations from each of two identical distributions, the difference between the sample means will have a symmetric distribution.

Note that when two distributions are identical, then not only are their means equal, but their variances are also equal.

Chapters 4 and 5 noted that in the one-sample case, problems with controlling Type I errors—ensuring that the actual Type I error probability does not exceed the nominal level—arise when sampling from a skewed distribution. For symmetric distributions, this problem is of little concern. So for the two-sample problem considered here, the expectation is that if there is absolutely no difference between the two groups, implying that they have not only equal means but equal variances and the same skewness, then the actual Type I error probability will not exceed by very much the α value you specify. All indications are that this argument is correct, but some simple extensions of this argument lead to erroneous conclusions. In particular, it might seem that generally, if we sample from perfectly symmetric distributions, probability coverage and control of Type I error probabilities will be satisfactory, but we will see that this is not always the case—even under normality. In particular, perfectly symmetric distributions with unequal variances can create practical problems. Nevertheless, all indications are that generally, when comparing identical distributions (so that in particular the hypothesis of equal means is true and the variances are equal), Type I error probabilities will not exceed the nominal level by very much, and for this special case power is not an issue.

Student's T begins having practical problems when distributions differ in some manner. If sampling is from normal distributions with equal sample sizes but unequal variances, Student's T continues to perform reasonably well in terms of Type I errors (Ramsey, 1980). But when sampling from nonnormal distributions, this is no longer the case. And even under normality there are problems when sample sizes are unequal. Basically, Type I error control, power and probability coverage can be very poor. In fact, when sample sizes are unequal, Cressie and Whitford (1986) describe general conditions under which Student's T does not even converge to the correct answer as the sample sizes get large.

A reasonable suggestion for salvaging the assumption that the variances are equal is to test it. That is, test $H_0 : \sigma_1^2 = \sigma_2^2$; if a nonsignificant result is obtained, proceed with Student's T. But this strategy has been found to be unsatisfactory, even under normality (Markowski & Markowski, 1990; Moser, Stevens & Watts, 1989; Wilcox, Charlin & Thompson, 1986). Nonnormality makes this strategy even less satisfactory. A basic problem is that tests of the hypothesis of equal variances do not always have enough power to detect situations where the assumption should be discarded.

Another general problem is that Student's T is designed to be sensitive to the differences between the means, but in reality it is sensitive to the myriad ways the distributions might differ, such as unequal skewnesses. Said in a more formal manner, if $F_1(x)$ is the probability that an observation from the first group is less than or equal to x, and $F_2(x)$ is the probability that an observation from the second group is less than or equal to x, then Student's T provides a reasonably satisfactory test of

$$H_0 : F_1(x) = F_2(x), \qquad \text{for any } x \tag{8.6}$$

in terms of controlling the probability of a Type I error. That is, the hypothesis is that for any constant x we pick, the probability of getting an observation less than or equal to x is the same for both groups being compared.

As a method for testing the hypothesis of equal population means, one might defend Student's T in the following manner. If we reject, we can be reasonably certain that the distributions differ. And if the distributions differ, some authorities would argue that by implication, the means are not equal. That is, they would argue that in practice we never encounter situations where distributions differ in shape but have equal means. But as soon as we agree that the distributions differ, there is the possibility that the actual probability coverage of the confidence interval for the difference between the means, given by Equation (8.5), differs substantially from the $1-\alpha$ value you have specified. A crude rule is that the more the distributions differ, particularly in terms of skewness, the more inaccurate the confidence interval [given by Equation (8.5)] might be. Outliers and heteroscedasticity can contribute to this problem. Said another way, Student's T can be used to establish that the means differ by implication, but it can be very inaccurate in terms of indicating the magnitude of this difference.

One more problem with Student's T is that situations arise where it is biased. That is, there is a higher probability of rejecting when the means are equal compared to some situations where the means differ. If the goal is to find a method that is sensitive to differences between the means, surely this property is undesirable.

Yet another concern about Student's T is that for a variety of reasons, its power can be very low relative to many other methods one might use. One reason is that very slight departures from normality can inflate the variances. Consider, for example, the two normal distributions shown in the left panel of Figure 8.1. Both have a variance of 1, and the means differ by 1. Using Student's T with $\alpha = .05$, power is .96 with $n_1 = n_2 = 25$. But if we sample from the two distributions shown in the right panel of Figure 8.1, power is only .28. One reason power is low is that when sampling from a heavy-tailed distribution, the actual probability of a Type I error can be substantially lower than the specified α value. For example, if you use Student's T with $\alpha = .05$, the actual probability of a Type I error can drop below .01. If an adjustment could be made so that the actual probability of a Type I error is indeed .05, power would be better, but it would still be low relative to alternative methods that are less sensitive to outliers. And even when outliers are not a concern, having unequal variances or even different degrees of skewness can result in relatively poor power as well.

Finally, we note that when using Student's T, even a single outlier in only one group can result in a rather undesirable property. The following example illustrates the problem.

EXAMPLE. Consider the following values:

Group 1:	4	5	6	7	8	9	10	11	12	13
Group 2:	1	2	3	4	5	6	7	8	9	10

The corresponding sample means are $\bar{X}_1 = 8.5$ and $\bar{X}_2 = 5.5$ and $T = 2.22$. With $\alpha = .05$, the critical value is $T = 2.1$, so Student's T would reject the hypothesis of equal means and conclude that the first group has a larger

Continued

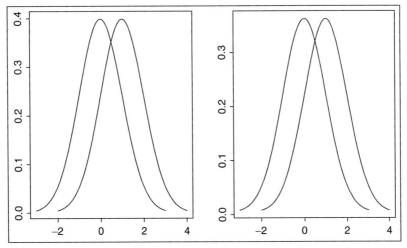

FIGURE 8.1 When sampling from the distributions on the left, power is .96 when using Student's T with $\alpha = .05$ and $n_1 = n_2 = 25$. But for the distributions on the right, power is only .28. This illustrates that very small departures from normality can destroy power when using means.

EXAMPLE. (*Continued*) population mean than the second (because the first group has the larger sample mean). Now, if we increase the largest observation in the first group from 13 to 23, the sample mean increases to $\bar{X}_1 = 9.5$. So the difference between $\bar{X}_1$ and $\bar{X}_2$ has increased from 3 to 4, and this would seem to suggest that we have stronger evidence that the population means differ and in fact the first group has the larger population mean. However, increasing the largest observation in the first group also inflates the corresponding sample variance, s_1^2. In particular, s_1^2 increases from 9.17 to 29.17. The result is that T *decreases* to $T = 2.04$ and we no longer reject. That is, increasing the largest observation has more of an effect on the sample variance than on the sample mean, in the sense that now we are no longer able to conclude that the population means differ. Increasing the largest observation in the first group to 33, the sample mean increases to 10.5, the difference between the two sample means increases to 5, and now $T = 1.79$. So again we do not reject and in fact our test statistic is getting smaller! This illustration provides another perspective on how outliers can mask differences between population means. ■

8.3 Welch's Heteroscedastic Method for Means

A step toward addressing the negative features of Student's T, still assuming that the goal is to compare means, is to switch to a method that allows unequal variances. Many techniques have been proposed, none of which is completely satisfactory. We describe only one such method here. One reason it was chosen is because it is a special case of a more general technique that gives more satisfactory results. That is, our goal is to build our way up to a method that performs relatively well over a broad range of situations.

Proposed by Welch (1938), we test the hypothesis of equal means ($H_0 : \mu_1 = \mu_2$) as follows. For the jth group ($j = 1, 2$) let

$$q_j = \frac{s_j^2}{n_j}, \tag{8.7}$$

where, as usual, s_1^2 and s_2^2 are the sample variances corresponding to the two groups being compared. That is, q_1 and q_2 are the usual estimates of the squared standard errors of the sample means, $\bar{X}_1$ and $\bar{X}_2$, respectively. We can get a confidence interval having probability coverage exactly equal to $1 - \alpha$ if we can determine the distribution of

$$W = \frac{(\bar{X}_1 - \bar{X}_2) - (\mu_1 - \mu_2)}{\sqrt{\frac{s_1^2}{n_1} + \frac{s_2^2}{n_2}}}. \tag{8.8}$$

Welch's strategy was to approximate the distribution of W using a Student's T distribution with degrees of freedom determined by the sample variances and sample sizes. In particular, the degrees of freedom are estimated with

$$\hat{\nu} = \frac{(q_1 + q_2)^2}{\frac{q_1^2}{n_1 - 1} + \frac{q_2^2}{n_2 - 1}}. \tag{8.9}$$

The $1 - \alpha$ confidence interval for $\mu_1 - \mu_2$ is

$$(\bar{X}_1 - \bar{X}_2) \pm t \sqrt{\frac{s_1^2}{n_1} + \frac{s_2^2}{n_2}}, \tag{8.10}$$

where t is the $1 - \alpha/2$ quantile of Student's T distribution and is again read from Table 4 in Appendix B.

HYPOTHESIS TESTING. When the hypothesis of equal means is true (so $\mu_1 - \mu_2 = 0$), Equation (8.8) reduces to

$$W = \frac{\bar{X}_1 - \bar{X}_2}{\sqrt{\frac{s_1^2}{n_1} + \frac{s_2^2}{n_2}}} \tag{8.11}$$

and you reject $H_0 : \mu_1 = \mu_2$ if

$$|W| \geq t, \tag{8.12}$$

where again t is the $1 - \alpha/2$ quantile of Student's T distribution with $\hat{\nu}$ degrees of freedom.

One-sided tests are performed in a similar manner. To test $H_0 : \mu_1 \leq \mu_2$, reject if $W \geq t$, where now t is the $1 - \alpha$ quantile of Student's T distribution with $\hat{v}$ degrees of freedom. As for $H_0 : \mu_1 \geq \mu_2$, reject if $W \leq t$, where t is the α quantile of Student's T.

EXAMPLE. For Salk's data in Table 8.1, the value of the test statistic given by Equation (8.11) is

$$W = \frac{18 - (-52.1)}{\sqrt{\frac{60.1^2}{20} + \frac{88.4^2}{36}}}$$

$$= \frac{70.1}{19.9}$$

$$= 3.52.$$

To compute the estimated degrees of freedom, first compute

$$q_1 = \frac{60.1^2}{20} = 180.6,$$

$$q_2 = \frac{88.4^2}{36} = 217,$$

in which case

$$\hat{v} = \frac{(180.6 + 217)^2}{\frac{180.6^2}{19} + \frac{217^2}{35}} \approx 52.$$

So if the Type I error probability is to be .05, then for a two-sided test ($H_0 : \mu_1 = \mu_2$), $t = 2.01$, approximately, and we reject because $|W| = 3.52 > 2.01$. That is, the empirical evidence indicates that infants exposed to the sounds of a mother's heart beat gain more weight, on average. The .95 confidence interval for the difference between the means can be seen to be (30, 110). In contrast, Student's T method yields a .95 confidence interval of (25.5, 114.7). So Welch's procedure indicates that the difference between the means of the two groups is at least 30, but Student's T leads to the conclusion that the difference between the means is at least 25.5. Generally, Welch's method provides more accurate confidence intervals than Student's T, so for the problem at hand, results based on Welch's method should be used. Also, Welch's method can provide a shorter confidence interval. There are situations where Student's T is more accurate than Welch's method, but in general the improvement is modest at best, while the improvement of Welch's method over Student's T can be substantial. ■

8.3.1 Nonnormality and Welch's Method

One reason Welch's method improves upon Student's T is that under random sampling, Welch's method satisfies the basic requirement of converging to the correct answer as the sample sizes get large when randomly sampling from nonnormal distributions, even when the sample sizes are unequal. That is, as the sample sizes increase, the actual probability coverage for the difference between the means will converge to $1 - \alpha$ when using Equation (8.10). For Student's T, this is true only when the sample sizes are equal. When sampling from normal distributions, Welch's method does a better job of handling unequal variances. This can translate into more power as well as shorter and more accurate confidence intervals. But unfortunately, nonnormality can be devastating in terms of probability coverage (e.g., Algina, Oshima & Lin, 1994) and especially power. Moreover, Welch's method can be biased: The probability of rejecting can be higher when the means are equal compared to situations where they are not. When the two groups being compared have identical distributions, Welch's method performs well in terms of controlling the probability of a Type I error, and because for this special case the population means are equal, power is not an issue. But when distributions differ in some manner, such as having different skewnesses or even unequal variances, Welch's method can have poor properties under nonnormality. So although Welch's method is an improvement over Student's T, more needs to be done.

8.4 Comparing Groups with Individual Confidence Intervals: An Example of What Not to Do

It might seem that one could compare population means simply by computing a confidence interval for μ_1 using Equation (4.8), computing a confidence interval for μ_2, and then checking whether these two confidence intervals overlap. If they do not overlap, it might seem reasonable to reject the hypothesis of equal population means. So, for example, if we compute a .95 confidence interval for the first mean, yielding (5, 8), and for the second mean we get a .95 confidence interval of (9, 13), it might appear reasonable to reject $H_0: \mu_1 = \mu_2$. However, this strategy violates a basic principle. Because there seems to be increasing interest in this approach, particularly when using error bars (defined shortly) to summarize data, some comments seem in order. (Also see Schenker & Gentleman, 2001; cf. Tryon, 2001.)

First we note that a general way of standardizing any variable, call it Y, is to subtract its mean and then divide this difference by the standard error of Y. In symbols, we standardize Y by transforming it to

$$\frac{Y - E(Y)}{SE(Y)}, \tag{8.13}$$

where $SE(Y)$ indicates the standard error of Y. That is, $SE(Y)$ is the square root of the variance of Y. For example, when working with a single sample mean, in which case $Y = \bar{X}$, we know that the standard error of the sample mean is $\sigma/\sqrt{n}$, $E(\bar{X}) = \mu$, so we standardize the sample mean by transforming it to

$$\frac{\bar{X} - \mu}{\sigma/\sqrt{n}}.$$

In practice we do not know σ, so we estimate it with s, the standard deviation. Then this last equation becomes

$$T = \frac{\bar{X} - \mu}{s/\sqrt{n}},$$

and a version of the central limit theorem tells us that with n sufficiently large, T is approximately standard normal. That is, inferences about the population mean can be made assuming T has a normal distribution, and this satisfies the basic requirement of converging to the correct answer as the sample size gets large.

When working with two independent groups, the standard error of the difference between the sample means is

$$\sqrt{\frac{\sigma_1^2}{n_1} + \frac{\sigma_2^2}{n_2}}. \tag{8.14}$$

This expression for the standard error is being estimated by Welch's method, which is part of the reason why Welch's method converges to the correct answer as the sample sizes get large. That is, a correct expression for the standard error is being used to derive Welch's method, and this is a fundamental component of any method when standardizing is being done. In contrast, under general conditions, Student's T does not converge to the correct answer when the sample sizes differ, and the reason is roughly because it does not use an estimate of the correct standard error.

To keep the discussion simple, we focus on what are called error bars. *Error bars*, often used in graphical summaries of data, are simply vertical lines plotted above and below the sample means. In symbols, error bars for the first means correspond to the values

$$\bar{X}_1 - \frac{s_1}{\sqrt{n_1}} \qquad \text{and} \qquad \bar{X}_1 + \frac{s_1}{\sqrt{n_1}},$$

so they are simply a type of confidence interval. As for the second group, error bars correspond to the values

$$\bar{X}_2 - \frac{s_2}{\sqrt{n_2}} \qquad \text{and} \qquad \bar{X}_2 + \frac{s_2}{\sqrt{n_2}}.$$

Now consider the strategy of deciding that the population means differ if the intervals based on these error bars do not overlap. In symbols, reject the null hypothesis of equal means if

$$\bar{X}_1 + \frac{s_1}{\sqrt{n_1}} < \bar{X}_2 - \frac{s_2}{\sqrt{n_2}}$$

or if

$$\bar{X}_1 - \frac{s_1}{\sqrt{n_1}} > \bar{X}_2 + \frac{s_2}{\sqrt{n_2}}.$$

Here is a fundamental problem with this strategy. Rearranging terms, we are rejecting if

$$\frac{|\bar{X}_1 - \bar{X}_2|}{\frac{s_1}{\sqrt{n_1}} + \frac{s_2}{\sqrt{n_2}}} \geq 1. \tag{8.15}$$

The denominator,

$$\frac{s_1}{\sqrt{n_1}} + \frac{s_2}{\sqrt{n_2}},$$

violates a basic principle — it does not estimate a correct expression for the standard error of the difference between the sample means, which is given by Equation (8.14). Consequently, the left side of Equation (8.15) does not converge to a standard normal distribution and statements about Type I error probabilities are at best difficult to make. A correct estimate of the standard error is

$$\sqrt{\frac{s_1^2}{n_1} + \frac{s_2^2}{n_2}},$$

which is used by Welch's method and which can differ substantially from the incorrect estimate used in Equation (8.15). So regardless of how large the sample sizes might be, using error bars to make decisions about whether groups differ can be highly inaccurate, even under normality. More generally, if we reject the hypothesis of equal means when confidence intervals for the individual means do not overlap, the wrong standard error is being used, so any statements about the probability of a Type I error are difficult to make. Yet another problem, for reasons explained in Chapter 4, is that under nonnormality the probability coverage for the individual population means can differ substantially from the nominal level, depending on the type of distributions from which we sample. (There is a variation of the method just described where error bars are computed assuming homoscedasticity, but again basic principles are being violated.)

8.5 A Bootstrap Method for Comparing Means

This section describes how we might improve upon Welch's method for comparing means. Similar to Chapter 7, we can compute accurate confidence intervals for the difference between means or test hypotheses about the equality of means if we can determine the distribution of W given by Equation (8.8). A bootstrap approximation of this distribution is obtained as follows. Generate a bootstrap sample of size n_1 from the first group and label the resulting sample mean and standard deviation $\bar{X}_1^*$ and s_1^*, respectively. Do the same for the second group and label the bootstrap sample mean

TABLE 8.2   Data Generated from a Mixed Normal (Group 1) and a Standard Normal (Group 2)

Group 1:	3.73624506	2.10039320	−3.56878819	−0.26418493	−0.27892175	
	0.87825842	−0.70582571	−1.26678127	−0.30248530	0.02255344	
	14.76303893	−0.78143390	−0.60139147	−4.46978177	1.56778991	
	−1.14150660	−0.20423655	−1.87554928	−1.62752834	0.26619836	
Group 2:	−1.1404168	−0.2123789	−1.7810069	−1.2613917	−0.3241972	1.4550603
	−0.5686717	−1.7919242	−0.6138459	−0.1386593	−1.5451134	−0.8853377
	0.3590016	0.4739528	−0.2557869			

and standard deviation $\bar{X}_2^*$ and s_2^*. Let

$$W^* = \frac{(\bar{X}_1^* - \bar{X}_2^*) - (\bar{X}_1 - \bar{X}_2)}{\sqrt{\frac{(s_1^*)^2}{n_1} + \frac{(s_2^*)^2}{n_2}}}.\qquad(8.16)$$

Repeating this process B times, yields $W_1^*, \ldots, W_B^*$. Next, put these B values in ascending order, which we label $W_{(1)}^* \leq \cdots \leq W_{(B)}^*$ in our usual way. Let $\ell = \alpha B/2$, rounded to the nearest integer, and $u = B - \ell$. Then an approximate $1 - \alpha$ confidence interval for the difference between the means $(\mu_1 - \mu_2)$ is

$$\left((\bar{X}_1 - \bar{X}_2) - W_{(u)}^* \sqrt{\frac{(s_1^*)^2}{n_1} + \frac{(s_2^*)^2}{n_2}}, \ (\bar{X}_1 - \bar{X}_2) - W_{(\ell+1)}^* \sqrt{\frac{(s_1^*)^2}{n_1} + \frac{(s_2^*)^2}{n_2}} \right).$$

$$(8.17)$$

> **EXAMPLE.** We illustrate that the confidence interval based on Welch's method can differ substantially from the confidence interval based on the bootstrap-t method just described. The data for group 1 in Table 8.2 were generated from a mixed normal distribution (using S-PLUS) and the data for group 2 were generated from a standard normal distribution. So both groups have population means equal to 0. Applying Welch's method, the .95 confidence interval for the difference between the means is $(-0.988, 2.710)$. Using the bootstrap-t method instead, it is $(-2.21, 2.24)$. ■

8.6   A Permutation Test Based on Means

The so-called *permutation test*, introduced by R. A. Fisher in the 1930s, is sometimes recommended for comparing means. The method is somewhat similar in spirit to the bootstrap, but a fundamental difference between it and the bootstrap is that the bootstrap resamples with replacement and the permutation test does not. We first outline the method in formal terms, then we illustrate the steps, and finally we indicate what this test tells us.

The permutation test based on means is applied as follows:

1. Compute the sample means for each group and label the difference $d = \bar{X}_1 - \bar{X}_2$.
2. Pool the data.
3. Randomly permute the pooled data. That is, rearrange the order of the pooled data in a random fashion.
4. Compute the sample mean for the first n_1 observations resulting from step 3, compute the sample mean of the remaining n_2 observations, and note the difference.
5. Repeat steps 3–4 B times and label the differences between the resulting sample means $d_1, \ldots, d_B$.

EXAMPLE. Imagine we have two groups with the following observations:

Group 1:	6,	19,	34,	15	
Group 2:	9,	21,	8,	53,	25

Pooling the observations yields

$$6, 19, 34, 15, 9, 21, 8, 53, 25.$$

The sample mean for group 1 is $\bar{X}_1 = 18.5$; for group 2 it is $\bar{X}_2 = 23.2$; and the difference between these means is

$$d = 18.5 - 23.2 = -4.7.$$

Next we permute the pooled data in a random fashion and for illustrative purposes we assume this yields

$$34, 21, 8, 25, 6, 19, 15, 9, 53.$$

The sample mean for the first $n_1 = 4$ observations is 22, the remaining observations have a sample mean of 20.4, and the difference between these means is $d_1 = 22 - 20.4 = 1.6$. Repeating this process of randomly permuting the pooled data B times yields B differences between the resulting sample means, which we label $d_1, \ldots, d_B$. If we want the Type I error probability to be .05, we conclude that the groups differ if the middle 95% of the values $d_1, \ldots, d_B$ do not contain d.   ■

Although the method just described is sometimes recommended for making inferences about the population means, in reality it is testing the hypothesis that the two groups being compared have identical distributions. Even under normality but unequal variances, the method fails to control the probability of a Type I error when testing the hypothesis of equal means (e.g., Boik, 1987). On the positive side, when testing the hypothesis of identical distributions, the probability of a Type I error is controlled exactly if all possible permutations of the data are used rather than just B randomly sampled permutations as is done here. An argument in favor of using the permutation test to compare means is that if distributions differ, surely the population means differ. But even if we accept this argument, the permutation test gives us little or no information about how the groups differ, let alone the magnitude of the difference between the population means, and it tells us nothing about the precision of the estimated difference between the population means based on the sample means. That is, it does not provide a confidence interval for $\mu_1 - \mu_2$. (For yet another argument in favor of the permutation test, see Ludbrook & Dudley, 1998.)

It is noted that the permutation test can be applied with any measure of location or scale, but again the method is testing the hypothesis of equal distributions. If, for example, we use variances, examples can be constructed where we are likely to reject because the distributions differ, even though the population variances are equal.

8.6.1 S-PLUS Function permg

The S-PLUS function

$$\text{permg(x, y, alpha} = 0.05, \text{est} = \text{mean, nboot} = 1000)$$

performs the permutation test just described. By default it uses means, but any measure of location or scale can be used by setting the argument est to an appropriate expression. For example, est=var would use variances rather than means.

8.7 Yuen's Method for Comparing Trimmed Means

Yuen (1974) derived a method for comparing the population γ-trimmed means of two independent groups that reduces to Welch's method for means when there is no trimming. As usual, 20% trimming ($\gamma = .2$) is a good choice for general use, but situations arise where more than 20% trimming might be beneficial (such as when the proportion of outliers in either tail of an empirical distribution exceeds 20%).

Generalizing slightly the notation in Chapter 3, let $g_j = [\gamma n_j]$, where again n_j is the sample size associated with the jth group ($j = 1, 2$) and let $h_j = n_j - 2g_j$. That is, h_j is the number of observations left in the jth group after trimming. Let

$$d_j = \frac{(n_j - 1)s_{wj}^2}{h_j(h_j - 1)},\tag{8.18}$$

where s_{wj}^2 is the γ-Winsorized variance for the jth group. Yuen's test statistic is

$$T_y = \frac{\bar{X}_{t1} - \bar{X}_{t2}}{\sqrt{d_1 + d_2}}.\tag{8.19}$$

The degrees of freedom are

$$\hat{\nu}_y = \frac{(d_1 + d_2)^2}{\frac{d_1^2}{b_1 - 1} + \frac{d_2^2}{b_2 - 1}}.$$

CONFIDENCE INTERVAL. The $1 - \alpha$ confidence interval for $\mu_{t1} - \mu_{t2}$, the difference between the population trimmed means, is

$$(\bar{X}_{t1} - \bar{X}_{t2}) \pm t\sqrt{d_1 + d_2}, \tag{8.20}$$

where t is the $1 - \alpha/2$ quantile of Student's T distribution with $\hat{\nu}_y$ degrees of freedom.

HYPOTHESIS TESTING. The hypothesis of equal trimmed means ($H_0 : \mu_{t1} = \mu_{t2}$) is rejected if

$$|T_y| \geq t.$$

As before, t is the $1 - \alpha/2$ quantile of Student's T distribution with $\hat{\nu}_y$ degrees of freedom.

The improvement in power, probability coverage, and control over Type I errors can be substantial when using Yuen's method with 20% trimming rather than Welch. For example, Wilcox (1997a, p. 111) describes a situation where when testing at the .025 level, the actual probability of rejecting with Welch's test is .092, nearly four times as large as the nominal level. Switching to Yuen's test, the actual probability of a Type I error is .042. So control over the Type I error probability is much better, but more needs to be done.

8.7.1 Comparing Medians

Although the median can be viewed as belonging to the class of trimmed means, special methods are required for comparing groups based on medians. An approach that currently seems to have practical value is as follows. Let M_j be the sample median corresponding to the jth group ($j = 1, 2$) and let S_j^2 be the McKean–Schrader estimate of the squared standard error of M_j, which is described in Section 4.11.2. Then an approximate $1 - \alpha$ confidence interval for the difference between the population medians is

$$(M_1 - M_2) \pm z_{1-\alpha/2}\sqrt{S_1^2 + S_2^2},$$

where $z_{1-\alpha/2}$ is the $1 - \alpha/2$ quantile of a standard normal distribution. Alternatively, reject the hypothesis of equal population medians if

$$\frac{|M_1 - M_2|}{\sqrt{S_1^2 + S_2^2}} \geq z_{1-\frac{\alpha}{2}}.$$

8.7.2 S-PLUS Function msmed

The S-PLUS function

$$\text{msmed}(x, y, \text{alpha} = .05)$$

has been supplied for comparing medians using the McKean–Schrader estimate of the standard error. (This function contains some additional parameters that are explained in Chapter 12.)

> **EXAMPLE.** For the data in Table 8.1, the .95 confidence interval for the difference between the medians is $(18.5, 91.5)$. ■

8.8 Bootstrap Methods for Comparing Trimmed Means

The bootstrap methods for trimmed means, described in Chapter 7, can be extended to the two-sample case. Again there are three versions of the bootstrap method that should be described and discussed.

8.8.1 The Percentile Method

Generalizing the notation in Section 8.6, generate a bootstrap sample of size n_1 from the first group, generate a bootstrap sample of size n_2 from the second group, let $\bar{X}_{t1}^*$ and $\bar{X}_{t2}^*$ be the bootstrap trimmed means corresponding to groups 1 and 2, respectively, and let

$$D^* = \bar{X}_{t1}^* - \bar{X}_{t2}^*$$

be the difference between the bootstrap trimmed means. Now suppose we repeat this process B times, yielding $D_1^*, \dots, D_B^*$. Then an approximate $1 - \alpha$ confidence interval for the difference between the population trimmed means, $\mu_{t1} - \mu_{t2}$, is

$$\left(D_{(\ell+1)}^*, D_{(u)}^* \right), \tag{8.21}$$

where, as usual, $\ell = \alpha B/2$, rounded to the nearest integer, and $u = B - \ell$. So for a .95 confidence interval, $\ell = .025B$.

HYPOTHESIS TESTING.   Reject the hypothesis of equal population trimmed means if the confidence interval given by Equation (8.21) does not contain zero.

If the amount of trimming is at least .2, the percentile bootstrap method just described is one of the most effective methods for obtaining accurate probability coverage, minimizing bias, and achieving relatively high power. But with no trimming the method performs poorly. (As noted in Chapter 7, a modification of the percentile bootstrap method performs well when working with the least squares regression estimator, even though this estimator has a finite-sample breakdown point of only $1/n$. But this modification does not perform particularly well when comparing the means of two independent groups.) The minimum amount of trimming needed to justify using a percentile bootstrap method, rather than some competing technique, has not been determined.

In subsequent chapters we will take up the problem of comparing multiple groups. To lay the foundation for one of the more effective methods, we describe the percentile bootstrap method for comparing trimmed means in another manner. Let

$$p^* = P\left(\bar{X}_{t1}^* > \bar{X}_{t2}^*\right). \tag{8.22}$$

That is, p^* is the probability that a bootstrap trimmed mean from the first group is greater than a bootstrap trimmed mean from the second. The value of p^* reflects the degree of separation between the two groups being compared, in the following sense. If the trimmed means based on the observed data are identical, meaning that $\bar{X}_{t1} = \bar{X}_{t2}$, then p^* will have a value approximately equal to .5. In fact, as the sample sizes increase, the value of p^* will converge to .5 for this special case. Moreover, if the population trimmed means are equal, then p^* will have, approximately, a uniform distribution, provided the sample sizes are not too small. That is, if $H_0 : \mu_{t1} = \mu_{t2}$ is true, p^* will have a value between 0 and 1, with all possible values between 0 and 1 equally likely if the sample sizes are not too small. (Hall, 1988a, provides relevant theoretical details and results in Hall, 1988b, are readily extended to trimmed means.) This suggests the following decision rule: Reject the hypothesis of equal trimmed means if p^* is less than or equal to $\alpha/2$ or greater than or equal to $1 - \alpha/2$. Said another way, if we let p_m^* be equal to p^* or $1 - p^*$, whichever is smaller, then reject if

$$p_m^* \leq \frac{\alpha}{2}. \tag{8.23}$$

We do not know p^*, but it can be estimated with the proportion of times a bootstrap trimmed mean from the first group is greater than a bootstrap trimmed mean from the second. That is, if A represents the number of values among $D_1^*, \ldots, D_B^*$ that are greater than zero, then we estimate p^* with

$$\hat{p}^* = \frac{A}{B}. \tag{8.24}$$

Finally, we reject the hypothesis of equal population trimmed means if $\hat{p}^*$ is less than or equal to $\alpha/2$ or greater than or equal to $1 - \alpha/2$. Or setting $\hat{p}_m^*$ to $\hat{p}^*$ or $1 - \hat{p}^*$, whichever is smaller, reject if

$$\hat{p}_m^* \leq \frac{\alpha}{2}. \tag{8.25}$$

The quantity $2\hat{p}_m^*$ is the estimated p-value.

8.8.2 Bootstrap-*t* Methods

Bootstrap-*t* methods for comparing trimmed means are preferable to the percentile bootstrap when the amount of trimming is close to zero. An educated guess is that the bootstrap-t is preferable if the amount of trimming is less than or equal to 10%, but it is stressed that this issue is in need of more research. The only certainty is that with no trimming, all indications are that the bootstrap-t outperforms the percentile bootstrap.

Bootstrap-t methods for comparing trimmed means are performed as follows:

1. Compute the sample trimmed means, $\bar{X}_{t1}$ and $\bar{X}_{t2}$, and Yuen's estimate of the squared standard errors, d_1 and d_2, given by Equation (8.18).
2. For each group, generate a bootstrap sample and compute the trimmed means, which we label $\bar{X}_{t1}^*$ and $\bar{X}_{t2}^*$. Also, compute Yuen's estimate of the squared standard error, again using Equation (8.18), which we label d_1^* and d_2^*.
3. Compute

$$T_y^* = \frac{\left(\bar{X}_{t1}^* - \bar{X}_{t2}^*\right) - \left(\bar{X}_{t1} - \bar{X}_{t2}\right)}{\sqrt{d_1^* + d_2^*}}.$$

4. Repeat steps 2 and 3 B times, yielding $T_{y1}^*, \dots, T_{yB}^*$. In terms of probability coverage, $B = 599$ appears to suffice in most situations when $\alpha = .05$.
5. Put the $T_{y1}^*, \dots, T_{yB}^*$ values in ascending order, yielding $T_{y(1)}^* \leq \cdots \leq T_{y(B)}^*$. The T_{yb}^* values ($b = 1, \dots, B$) provide an estimate of the distribution of

$$\frac{\left(\bar{X}_{t1} - \bar{X}_{t2}\right) - \left(\mu_{t1} - \mu_{t2}\right)}{\sqrt{d_1 + d_2}}.$$

6. Set $\ell = \alpha B/2$ and $u = B - \ell$, where ℓ is rounded to the nearest integer.

The equal-tailed $1 - \alpha$ confidence interval for the difference between the population trimmed means ($\mu_{t1} - \mu_{t2}$) is

$$\left(\bar{X}_{t1} - \bar{X}_{t2} - T_{y(u)}^*\sqrt{d_1 + d_2}, \bar{X}_{t1} - \bar{X}_{t2} - T_{y(\ell+1)}^*\sqrt{d_1 + d_2}\right). \qquad (8.26)$$

To get a symmetric two-sided confidence interval, replace step 3 with

$$T_y^* = \frac{\left|\left(\bar{X}_{t1}^* - \bar{X}_{t2}^*\right) - \left(\bar{X}_{t1} - \bar{X}_{t2}\right)\right|}{\sqrt{d_1^* + d_2^*}},$$

set $a = (1 - \alpha)B$, rounding to the nearest integer, in which case a $1 - \alpha$ confidence interval for $\mu_{t1} - \mu_{t2}$ is

$$\left(\bar{X}_{t1} - \bar{X}_{t2}\right) \pm T_{y(a)}^*\sqrt{d_1 + d_2}. \qquad (8.27)$$

HYPOTHESIS TESTING. As usual, reject the hypothesis of equal population trimmed means ($H_0: \mu_{t1} = \mu_{t2}$) if the $1 - \alpha$ confidence interval for the difference between the trimmed means does not contain zero. Alternatively, compute Yuen's test statistic

$$T_y = \frac{\bar{X}_{t1} - \bar{X}_{t2}}{\sqrt{d_1 + d_2}},$$

and reject if

$$T_y \leq T^*_{y(\ell+1)}$$

or if

$$T_y \geq T^*_{y(u)}.$$

When using the symmetric, two-sided confidence interval method, reject if

$$|T_y| \geq T^*_{y(a)}.$$

8.8.3 Winsorizing

Section 7.5.2 indicated that theory allows us to Winsorize the observations before taking bootstrap samples provided the amount of Winsorizing does not exceed the amount of trimming. A possible advantage of Winsorizing is shorter confidence intervals. However, we saw in Chapter 7 that if we Winsorize as much as we trim, probability coverage can be poor, at least with small to moderate sample sizes. This continues to be the case when comparing groups. But when using the percentile bootstrap method, if, for example, we trim 20% and Winsorize 10% and if the smallest sample size is at least 15, it seems that probability coverage is reasonably close to the nominal level, at least when $\alpha = .05$. Winsorizing is not recommended when using a bootstrap-t method when sample sizes are small. Perhaps this strategy provides good probability coverage with moderately large sample sizes, but this has not been determined as yet.

If a situation arises where Winsorizing makes a practical difference in terms of power and length of confidence intervals, a competing strategy is not to Winsorize but instead simply to increase the amount of trimming. The relative merits of these two strategies have not been determined.

8.8.4 S-PLUS Functions trimpb2 and yuenbt

The S-PLUS functions trimpb2 and yuenbt are supplied for applying the bootstrap methods just described. The function

trimpb2(x, y, tr = 0.2, alpha = 0.05, nboot = 2000, WIN = F, win = 0.1)

performs the percentile bootstrap method, where x is any S-PLUS variable containing the data for group 1 and y contains the data for group 2. The amount of trimming, tr, defaults to 20%, α defaults to .05, and nboot (B) defaults to 2000. The argument WIN defaults to F, for false, meaning that Winsorizing will not be done prior to generating bootstrap samples. Setting WIN equal to T, for true, Winsorizing will be done with the amount of Winsorizing determined by the argument win, which defaults to .1 (10%). This function returns the estimated significance level (or p-value), labeled sig.level, plus a $1 - \alpha$ confidence interval for the difference between the trimmed means.

The function

$$\text{yuenbt}(x, y, \text{tr} = 0.2, \text{alpha} = 0.05, \text{nboot} = 599, \text{side} = F)$$

performs the bootstrap-t method, which is based on Yuen's procedure for comparing trimmed means. The arguments are the same as before, except for the argument labeled side, which indicates whether a symmetric or equal-tailed confidence interval will be used. Side defaults to F, for false, meaning that the equal-tailed confidence interval [given by Equation (8.26)] will be computed. Setting side equal to T yields the symmetric confidence interval given by Equation (8.27).

EXAMPLE. Table 8.3 shows data from a study dealing with the effects of consuming alcohol. (The data were generously supplied by M. Earleywine.) Group 1, a control group, reflects hangover symptoms after consuming a specific amount of alcohol in a laboratory setting. Group 2 consisted of sons of alcoholic fathers. Storing the group 1 data in the S-PLUS variable A1, and the group 2 data in A2, the command trimpb2(A1,A2) returns the following output:

```
$sig.level:
[1] 0.038

$ci:
[1] 0.1666667 8.3333333
```

This says that a .95 confidence interval for the difference between the population trimmed means is (.17, 8.3). The significance level is .038, so in particular you would reject $H_0 : \mu_{t1} = \mu_{t2}$ at the .05 level. If we set the argument WIN to T, so that Winsorizing is done, then the .95 confidence interval is (0.58, 7.17), which is shorter than the confidence interval without Winsorizing. In contrast, if we use Welch's method for means, the .95 confidence interval is $(-1.6, 10.7)$. This interval contains zero, so we no longer reject, the only point being that it can make a difference which method is used. Notice that the Winsorized confidence interval using trimpb2 is substantially shorter than Welch's confidence interval, the ratio of the lengths being

$$\frac{10.7 + 1.6}{7.17 - .58} = 1.87.$$

Yet all indications are that the percentile bootstrap confidence interval generally has more accurate probability coverage.   ■

Notice that the default value for nboot (B) when using yuenbt is only 599, compared to 2000 when using trimpb2. Despite this, trimpb2 tends to have faster execution time, because it is merely computing trimmed means; yuenbt requires estimating the standard error for each bootstrap sample, which increases the execution

TABLE 8.3   Effect of Alcohol

Group 1:	0	32	9	0	2	0	41	0	0	0
	6	18	3	3	0	11	11	2	0	11
Group 2:	0	0	0	0	0	0	0	0	1	8
	0	3	0	0	32	12	2	0	0	0

TABLE 8.4   Self-Awareness Data

Group 1:	77	87	88	114	151	210	219	246	253		
	262	296	299	306	376	428	515	666	1310	2611	
Group 2:	59	106	174	207	219	237	313	365	458	497	515
	529	557	615	625	645	973	1065	3215			

time considerably. When using the bootstrap-t method (the S-PLUS function yuenbt), published papers indicate that increasing B from 599 to 999, say, does not improve probability coverage by very much, if at all, when $\alpha = .05$. This means that if we were to repeat the experiment billions of times, each time computing a .95 confidence interval, the proportion of times the resulting confidence interval contains the true difference between the population trimmed means will not be appreciably closer to .95 if we increase B from 599 to 999.

There is, however, a practical matter that should be mentioned. Consider the data in Table 8.4 and focus on group 1. Notice that the values are in ascending order. The S-PLUS function yuenbt begins by generating a bootstrap sample. The first value that it chooses might be the third value listed, which is 88. But suppose you store the data in S-PLUS in descending order instead. Then if yuenbt chooses the third value to be in the bootstrap sample, it is no longer 88 but rather 666. This means that the bootstrap sample will be altered, resulting in a different bootstrap sample trimmed mean. With B large enough, this will not change the resulting confidence interval and significance level by much. But with $B = 599$ the results might be altered enough to change your conclusion about whether to reject the null hypothesis. That is, a nearly significant result might become significant if the order of the observations is altered before invoking the bootstrap method or if we simply increase B. To reduce the likelihood of this possibility, consider using $B = 1999$ instead.

EXAMPLE. In an unpublished study by Dana (1990), the general goal was to investigate issues related to self-awareness and self-evaluation. In one portion of the study, he recorded the times individuals could keep an apparatus in contact with a specified target. The results, in hundredths of seconds, are shown in Table 8.4. Storing the data for group 1 in the S-PLUS variable G1 and storing

Continued

EXAMPLE. (*Continued*) the data for group 2 in G2, the command

$$\text{yuenbt}(G1, G2)$$

returns a .95 confidence interval of $(-312.5, 16.46)$. This interval contains zero, so we would not reject. If we increase the number of bootstrap samples (B) by setting the argument nboot to 999, now the confidence interval is $(-305.7, 10.7)$. We still do not reject, but increasing B alters the confidence interval slightly. In contrast, comparing medians via the method in Section 8.7.1, the .95 confidence interval is $(-441.4, -28.6)$, so we reject, the only point being that even among robust estimators, the choice of method can alter the conclusions reached. ■

8.8.5 Estimating Power and Judging the Sample Sizes

Imagine we compare two groups using the percentile bootstrap method with 20% trimming, as described in Section 8.8.1. If we fail to reject, this might be because there is little or no difference between the groups. Another possibility is that the population trimmed means differ by a substantively important amount but power is low. To help differentiate between these two possibilities, you can estimate how much power there was based on the data available to you.

The basic strategy is to estimate the standard errors associated with the 20% trimmed means and then to use these estimates to estimate power for a given value of the difference between the population trimmed means ($\mu_{t1} - \mu_{t2}$). The computational details, which stem from Wilcox and Keselman (in press), are shown in Box 8.1 and apply to 20% trimmed means only. Adjustments of the method when the amount of trimming is altered have not been studied. Another strategy for estimating power is to use a nested bootstrap similar to the one studied by Boos and Zhang (2000). A concern, however, is that the precision of the estimate cannot be easily assessed. That is, there is no known way of computing a reasonably accurate confidence for the actual amount of power if a nested bootstrap is used. In contrast, a confidence interval can be computed using the method in Box 8.1.

BOX 8.1 Power When Comparing 20% Trimmed Means with

a Percentile Bootstrap Method

Goal:

Given data, estimate power associated with some specified value of $\delta = \mu_{t1} - \mu_{t2}$. Alternatively, estimate the power curve based on the data at hand.

Continued

BOX 8.1 (*Continued*)

Computations:

For $i = 1, 2, \ldots, 35$, let b_i be given by

500, 540, 607, 706, 804, 981, 1176, 1402, 1681, 2008, 2353, 2769,

3191, 3646, 4124, 4617, 5101, 5630, 6117, 6602, 7058, 7459,

7812, 8150, 8479, 8743, 8984, 9168, 9332, 9490, 9607,

9700, 9782, 9839, 9868.

For example, $b_1 = 500$ and $b_4 = 706$. For any two distributions, given $\delta = \mu_{t1} - \mu_{t2}$ and an estimate of the standard error of $\bar{X}_{t1} - \bar{X}_{t2}$, namely, $S = \sqrt{d_1 + d_2}$, where d_1 and d_2 are given by Equation (8.18), power is estimated as follows. Let $v = [8\delta/S] + 1$, where $[8\delta/S]$ indicates that $8\delta/S$ is rounded down to the nearest integer, and let

$$a = 8\left(\frac{\delta}{S} - \frac{v-1}{8}\right).$$

Then power is estimated to be

$$\hat{\gamma} = \frac{b_v}{10{,}000} + a\left(\frac{b_{v+1}}{10{,}000} - \frac{b_v}{10{,}000}\right).$$

In the event $v = 36$, b_{v+1} is taken to be 10,000 in the previous equation. If $v > 36$, power is estimated to be 1.

8.8.6 S-PLUS Functions powest and pow2an

The S-PLUS function

$$\text{powest}(x, y, \text{delta})$$

estimates how much power there is when the difference between the population 20% trimmed means is delta. This is done by computing the standard errors of the sample trimmed means using the data in the S-PLUS variable x (Group 1) and the S-PLUS variable y (Group 2) and then performing the calculations in Box 8.1.

The S-PLUS function

$$\text{pow2an}(x, y, \text{ci}{=}F, \text{plotit}{=}T, \text{nboot}{=}800)$$

computes a power curve using the data in the S-PLUS variables x and y. That is, the function chooses a range of values for the difference between the population means, and for each difference it computes power using the S-PLUS function powest. By default, the power curve is plotted. To avoid the plot and get the numerical results only, set the argument plotit to F, for false. Setting the argument ci to T will result in a lower .95 confidence interval for the power curve to be computed using a bootstrap method based on nboot (*B*) bootstrap samples.

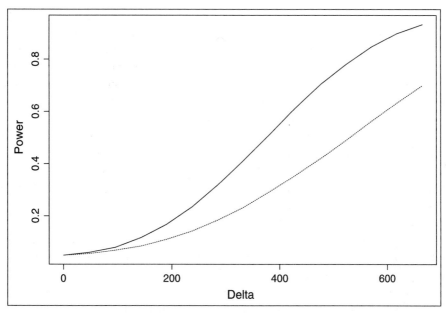

FIGURE 8.2   Estimate of the power returned by the S-PLUS function pow2an.

EXAMPLE. The S-PLUS functions powest and pow2an are illustrated with data from a reading study. (These data were generously supplied by Frank Manis.) Theoretical arguments suggest that groups should differ, but a non-significant result was obtained when comparing 20% trimmed means (or when comparing means with any of the previously described techniques). One possible explanation is that there is little or no difference between the groups, but another possibility is that power is low due to relatively large standard errors, meaning that detecting a substantively interesting difference is unlikely based on the sample sizes used. For a difference of 600 between the population 20% trimmed means, powest estimates that power is .8. Figure 8.2 shows the estimated power curve returned by pow2an. The lower, dashed line is a lower .95 confidence interval for the actual amount of power. That is, the solid line provides an approximately unbiased estimate of power, but a possibility is that power is as low as indicated by the dashed line. Based on this analysis it was concluded that power is low and that accepting the hypothesis of equal trimmed means is not warranted.   ■

8.9   Comparing MOM-Estimators, M-Estimators, and Other Measures of Location

As is probably evident, when comparing two groups, any measure of location, and indeed virtually any parameter that characterizes a distribution (such as measures of scale), can be used with a percentile bootstrap method. Basically, generate B bootstrap samples from the first group; do the same for the second, in which case the middle 95% of the differences provides an approximate .95 confidence interval. That is, proceed as was described in Section 8.8.1 when comparing trimmed means, except that the sample trimmed mean is replaced with whatever estimator you care to use. Currently, all indications are that as a general rule, if the finite-sample breakdown point of the estimator is at least .2, relatively accurate probability coverage will be obtained. So, for example, if the goal is to compare the medians of two groups, or the quartiles, the percentile bootstrap is a relatively effective method.

It is possible to use a bootstrap-t method with M-estimators. (The method requires estimating the standard error of the M-estimator, which can be done as described in Wilcox, 1997a.) However, all indications are that the resulting confidence interval is not as accurate as the confidence interval associated with the percentile bootstrap method, particularly when sampling from skewed distributions. Consequently, the rather involved computations are not described.

A negative feature of using M-estimators is that for very small sample sizes — say, less than 20 — the probability coverage may not be as accurate as the probability coverage obtained with 20% trimmed means. Also, using a bootstrap with an M-estimator is not always possible, for reasons described in Section 7.7. (Bootstrap samples can have MAD = 0.) On the positive side, the M-estimator described in Section 3.5 has a high finite-sample breakdown point, which could translate into more power, compared to using a 20% trimmed mean, when sampling is from distributions where the proportion of outliers exceeds 20%. Of course one could increase the amount of trimming to deal with this problem. And when many outliers are common, the median might be an excellent choice. An advantage of an M-estimator is that it empirically adjusts to the number of outliers, a consequence being that, compared to the median, it performs better in terms of power when sampling is from a normal distribution.

8.9.1 S-PLUS Function pb2gen

The S-PLUS function

$$pb2gen(x, y, alpha = 0.05, nboot = 2000, est = mom, \ldots)$$

computes a percentile bootstrap confidence interval for the difference between any two measures of location. As usual, x and y are any S-PLUS variables containing data and nboot is B, the number of bootstrap samples to be used. By default, $B = 2000$ is used. The argument est indicates which measure of location is to be employed. It can be any S-PLUS function that computes a measure of location and defaults to the S-PLUS function mom (written for this book), which is the MOM-estimator described in Chapter 3. The argument . . . can be used to reset certain default settings associated with the argument est. For example, if est=mean is used, means are compared. In contrast, the command

$$pb2gen(x, y, alpha = 0.05, nboot = 2000, est = mean, tr = .2)$$

would compare 20% trimmed means instead. (In this case, pb2gen and trimpb2, described in Section 8.8.4, give the same results.) The command

$$pb2gen(x, y, alpha = 0.05, nboot = 2000, est = median)$$

would compare medians.

EXAMPLE. A study was conducted comparing the EEG (electroencephalogram) readings of convicted murderers to the EEG readings of a control group with measures taken at various sites in the brain. For one of these sites the results were

Control	−0.15	−0.22	0.07	−0.07	0.02	0.24	−0.60
group:	−0.17	−0.33	0.23	−0.69	0.70	1.13	0.38
Murderers:	−0.26	0.25	0.61	0.38	0.87	−0.12	0.15
	0.93	0.26	0.83	0.35	1.33	0.89	0.58

(These data were generously supplied by A. Raine.) The sample medians are −0.025 and 0.48, respectively. Storing the data in the S-PLUS variables x1 and x2, the command

$$pb2gen(x1, x2, est = median)$$

returns a .95 confidence interval for the difference between the population medians of (−0.97, −0.085). So the hypothesis of equal population medians is rejected because this interval does not contain 0, and the data indicate that the typical measure for the control group is less than the typical measure among convicted murderers. Using the nonbootstrap method in Section 8.7.1 instead, the .95 confidence interval is (−0.89, −0.119). ■

EXAMPLE. Table 3.2 contains data on the desired number of sexual partners over the next 30 years reported by male undergraduates. The responses by 156 females are shown in Table 8.5. Does the typical response among males differ from the typical response among females? If we simply apply Student's T, we fail to reject, which is not surprising because there is an extreme outlier among the responses for males. (See the last example in Section 8.2.) But if we trim only 1% of the data, Yuen's method rejects, suggesting that the two distributions differ. However, with so little trimming, accurate confidence intervals might be difficult to obtain. Moreover, the median response among both males and females is 1, suggesting that in some sense the typical male and typical female are similar. To add perspective, we compare the .75 quantiles of the distributions, which can be estimated with the built-in S-PLUS function quantile. For example, if the responses for the males are stored in the S-PLUS variable sexm, the S-PLUS command quantile(sexm,probs=.75) estimates the .75 quantile to be 6; for females the estimate is 3. The command

$$pb2gen(sexm, sexf, est = quantile, probs = .75)$$

compares the .75 quantiles of the two groups and returns a .95 confidence interval of (1, 8). So we reject the hypothesis of equal .75 quantiles, indicating

Continued

TABLE 8.5   Desired Number of Sexual Partners for 156 Females

x:	0	1	2	3	4	5	6	7	8	10
f_x:	2	101	11	10	5	11	1	1	3	4

x:	11	12	15	20	30
f_x:	1	1	2	1	2

EXAMPLE. (*Continued*) that the groups differ among the higher responses. That is, in some sense the groups appear to be similar because they have identical medians. But if we take the .75 quantiles to be the typical response among the higher responses we might observe, the typical male appears to respond higher than the typical female.   ■

8.10   Comparing Variances or Other Measures of Scale

Although the most common approach to comparing two independent groups is to use some measure of location, situations arise where there is interest in comparing variances or some other measure of scale. For example, in agriculture, one goal when comparing two crop varieties might be to assess their relative stability. One approach is to declare the variety with the smaller variance as being more stable (e.g., Piepho, 1997). As another example, consider two methods for training raters of some human characteristic. For example, raters might judge athletic ability or they might be asked to rate aggression among children in a classroom. Then one issue is whether the variances of the ratings differ depending on how the raters were trained. Also, in some situations, two groups might differ primarily in terms of the variances rather than their means or some other measure of location. To take a simple example, consider two normal distributions both having means zero with the first having variance one and the second having variance three. Then a plot of these distributions would show that they differ substantially, yet the hypotheses of equal means, equal trimmed means, equal M-estimators, and equal medians are all true. That is, to say the first group is comparable to the second is inaccurate, and it is of interest to characterize how they differ.

There is a vast literature on comparing variances and as usual not all methods are covered here. For studies comparing various methods, the reader is referred to Conover, Johnson, and Johnson (1981), Brown and Forsythe (1974b), Wilcox (1992), plus the references they cite.

8.10.1   Comparing Variances

We begin with testing

$$H_0 : \sigma_1^2 = \sigma_2^2, \tag{8.28}$$

the hypothesis that the two groups have equal variances. Many methods have been proposed. The classic technique assumes normality and is based on the ratio of the largest sample variance to the smallest. So if $s_1^2 > s_2^2$, the test statistic is $F = s_1^2/s_2^2$; otherwise you use $F = s_2^2/s_1^2$. When the null hypothesis is true, F has a so-called F distribution, which is described in Chapter 9. But this approach has long been known to be highly unsatisfactory when distributions are nonnormal (e.g., Box, 1953), so additional details are omitted.

Currently, the most successful method in terms of maintaining control over the probability of a Type I error and achieving relatively high power is to use a slight modification of the percentile bootstrap method. In particular, set $n_m = \min(n_1, n_2)$, and, for the jth group ($j = 1, 2$), take a bootstrap sample of size n_m. Ordinarily we take a bootstrap sample of size n_j from the jth group, but when sampling from heavy-tailed distributions, and when the sample sizes are unequal, control over the probability of a Type I error can be extremely poor for the situation at hand. Next, for each group, compute the sample variance based on the bootstrap sample and set D^* equal to the difference between these two values. Repeat this $B = 599$ times, yielding 599 bootstrap values for D, which we label $D_1^*, \ldots, D_{599}^*$. As usual, when writing these values in ascending order, we denote this by $D_{(1)}^* \leq \cdots \leq D_{(B)}^*$. Then an approximate .95 confidence interval for the difference between the population variances is

$$\left(D_{(\ell)}^*, D_{(u)}^*\right), \tag{8.29}$$

where for $n_m < 40$, $\ell = 7$ and $u = 593$; for $40 \leq n_m < 80$, $\ell = 8$ and $u = 592$; for $80 \leq n_m < 180$, $\ell = 11$ and $u = 588$; for $180 \leq n_m < 250$, $\ell = 14$ and $u = 585$; and for $n_m \geq 250$, $\ell = 15$ and $u = 584$. (For results on the small-sample properties of this method, see Wilcox, in press.) Notice that these choices for ℓ and u are the same as those used in Section 7.3 when making inferences about the least squares regression slope and Pearson's correlation. The hypothesis of equal variances is rejected if the confidence interval given by Equation (8.29) does not contain zero.

Using the confidence interval given by Equation (8.29) has two practical advantages over the many alternative methods one might use to compare variances. First, compared to many methods, it provides higher power. Second, among situations where distributions differ in shape, extant simulations indicate that probability coverage remains relatively accurate, in contrast to many other methods one might use. If the standard percentile bootstrap method is used instead, then with sample sizes of 20 for both groups, the Type I error probability can exceed .1 when testing at the .05 level, and with unequal sample sizes it can exceed .15.

8.10.2 S-PLUS Function comvar2

The S-PLUS function

$$\text{comvar2}(x, y)$$

compares variances using the bootstrap method described in the previous subsection. The method can only be applied with $\alpha = .05$; modifications that allow other α values have not been derived. The arguments x and y are S-PLUS variables containing data for group 1 and group 2, respectively. The function returns a .95 confidence interval for $\sigma_1^2 - \sigma_2^2$ plus an estimate of $\sigma_1^2 - \sigma_2^2$ based on the difference between the sample variances, $s_1^2 - s_2^2$, which is labeled difsig.

8.10.3 Brown–Forsythe Method

Section 3.3.6 described a measure of scale based on the average absolute distance of observations from the median. In the notation used here, if M_1 is the median of the first group, the measure of scale for the first group is

$$\hat{\tau}_1 = \frac{1}{n_1} \sum |X_{i1} - M_1|, \tag{8.30}$$

where again $X_{11}, \ldots, X_{n_1 1}$ are the observations randomly sampled from the first group. For the second group this measure of scale is

$$\hat{\tau}_2 = \frac{1}{n_2} \sum |X_{i2} - M_2|, \tag{8.31}$$

where M_2 is the median for the second group. Notice that these measures of scale do *not* estimate the population variance (σ^2) or the population standard deviation (σ). There is a commonly recommended method for comparing groups based on these measures of scale, so it is important to comment on its relative merits.

For convenience, let

$$Y_{ij} = |X_{ij} - M_j|,$$

$i = 1, \ldots, n_j; j = 1, 2$. That is, the ith observation in the jth group (X_{ij}) is transformed to $|X_{ij} - M_j|$, its absolute distance from the median of the jth group. So the sample mean of the Y values for the jth group is

$$\bar{Y}_j = \frac{1}{n_j} \sum Y_{ij},$$

which is just the measure of scale described in the previous paragraph. Now let τ_j be the population value corresponding to $\bar{Y}_j$. That is, τ_j is the value of $\bar{Y}_j$ we would get if all individuals in the jth group could be measured. The goal is to test

$$H_0 : \tau_1 = \tau_2. \tag{8.32}$$

If we reject, we conclude that the groups differ based on this measure of dispersion.

The Brown and Forsythe (1974b) test of the hypothesis given by Equation (8.32) consists of applying Student's T to the Y_{ij} values. We have already seen, however, that when distributions differ in shape, Student's T performs rather poorly, and there are general conditions under which it does not converge to the correct answer as the sample sizes get large. We can correct this latter problem by switching to Welch's test, but problems remain when distributions differ in shape. For example, suppose we sample $n_1 = 20$ observations from a normal distribution and $n_2 = 15$ observations from the observations shown in Figure 5.6. Then when testing at the $\alpha = .05$ level, the actual probability of a Type I error is approximately .21. Like Student's T or Welch's method, the Brown–Forsythe test provides a test of the hypothesis that distributions are identical. Although it is designed to be sensitive to a reasonable measure of scale, it can be sensitive to other ways the distributions might differ. So if the goal is to compute a confidence interval for $\tau_1 - \tau_2$, the Brown–Forsythe method can be unsatisfactory if $\tau_1 \neq \tau_2$. Presumably some type of bootstrap method could improve matters, but this has not been investigated and indirect evidence suggests that practical problems will remain. Moreover, if there is explicit interest in comparing variances (σ_1^2 and σ_2^2), the Brown–Forsythe test is unsatisfactory, because $\hat{\tau}_1$ and $\hat{\tau}_2$ do not estimate the population variances, σ_1^2 and σ_2^2, respectively.

8.10.4 Comparing Robust Measures of Scale

There are at least 150 measures of scale that have been proposed. One criterion for choosing from among them is that an estimator have a relatively small standard error when sampling from any of a range of distributions. Lax (1985) compared many measures of scale in this manner where the distributions ranged between a normal and symmetric distributions with very heavy tails. In the context of hypothesis testing, having a relatively small standard error can help increase power. Two scale estimators that performed well were the percentage bend midvariance and biweight midvariance described in Section 3.3.7. These measures of scale can be compared with the S-PLUS function pb2gen, described in Section 8.9.1. (As noted in Section 3.3.8, the S-PLUS functions pbvar and bivar have been supplied for computing these two measures of scale, respectively.)

EXAMPLE. Twenty-five observations were generated from the mixed normal distribution shown in Figure 2.8, and another 25 observations were sampled from a standard normal. As explained in Section 2.7, the corresponding population variances differ considerably even though the corresponding probability curves are very similar. Storing the data in the S-PLUS variables x and y, comvar2(x, y) returned a .95 confidence interval of $(0.57, 32.2)$, so we correctly reject the hypothesis of equal population variances. The sample variances were $s_x^2 = 14.08$ and $s_y^2 = 1.23$. In contrast, the S-PLUS command

pb2gen(x,y,est=pbvar)

returns a .95 confidence interval of $(-1.59, 11.6)$ for the difference between the percentage bend midvariances. The values of the percentage bend midvariances were 2.3 and 1.7. In reality, the population values of the percentage bend midvariances differ slightly, and we failed to detect this due to low power. ■

EXAMPLE. We repeat the last example, only now we sample from the two distributions shown in Figure 8.3. (These two distributions are the normal and

Continued

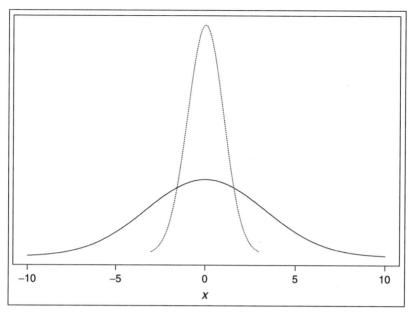

FIGURE 8.3   Two distributions that have equal variances. Testing the hypothesis of equal variances, we should not reject, even though there is a striking difference between the two distributions.

EXAMPLE. (*Continued*) mixed normal shown in Figure 2.9.) Although there is a clear and rather striking difference between these two distributions, the population means and variances are equal. So we should not reject the hypothesis of equal population variances, and with data generated by the author (with both sample sizes equal to 25) we indeed fail to reject; the .95 confidence interval returned by comvar2 is $(-13.3, 20.8)$. In contrast, comparing the percentage bend midvariances, the .95 confidence interval is $(-28.7, -1.8)$; this interval does not contain zero, so we reject. If we compare the biweight midvariances with the command

pb2gen(x,y,est=bivar),

the resulting .95 confidence interval is $(-18.8, -5.6)$ and again we reject.   ■

These two examples merely illustrate that different methods can lead to different conclusions. In the first example, it is certainly true that the two distributions being compared are very similar, in the sense described in Section 2.7. But the tails of the mixed normal differ from the tails of the normal, and in the context of measuring stability or reliability, there is a difference that might have practical importance. This difference happens to be detected in the first example of this subsection by comparing variances but not when comparing the percentage bend midvariances or the biweight midvariances. In fairness, however, for this particular situation the power associated with comparing variances is not very high. However, in the second example, clearly the distributions differ considerably in terms of scale, as indicated in Figure 8.3. Comparing the percentage bend midvariances or biweight midvariances happens to detect this, but comparing the variances does not.

8.11 Measuring Effect Size

It has long been recognized that merely rejecting the hypothesis of equal means (or any other measure of location) tells us virtually nothing about the magnitude of the difference between the two groups (e.g., Cohen, 1994). If we reject at the .001 level and the first group has a larger sample mean than the second, then we conclude that the first group has the larger population mean. But this tells us nothing about the magnitude of the difference. An article in *Nutrition Today* (19, 1984, 22–29) illustrates the importance of this issue. A study was conducted on whether a particular drug lowers the risk of heart attacks. Those in favor of using the drug pointed out that the number of heart attacks in the group receiving the drug was significantly lower than in the group receiving a placebo when testing at the $\alpha = .001$ level. However, critics of the drug argued that the difference between the number of heart attacks was trivially small. They concluded that because of the expense and side effects of using the drug, there is no compelling evidence that patients with high cholesterol levels should be put on this medication. A closer examination of the data revealed that the standard errors corresponding to the two groups were very small, so it was possible to get a statistically significant result that was clinically unimportant.

Generally, how might we measure the difference between two groups? Three approaches are considered in this section:

- Compute a confidence interval for the difference between some measure of location.
- Use a so-called standardized difference.
- Use a global comparison of the distributions.

The first approach has already been discussed, so no additional comments are given here. The second approach, which is commonly used, is typically implemented by assuming the two groups have a common variance, which we label σ^2. That is, $\sigma_1^2 = \sigma_2^2 = \sigma^2$ is assumed. Then the so-called standardized difference between the groups is

$$\Delta = \frac{\mu_1 - \mu_2}{\sigma}. \tag{8.33}$$

Assuming normality, Δ can be interpreted using results in Chapter 2. For example, if $\Delta = 2$, then the difference between the means is 2 standard deviations, and for normal distributions we have some probabilistic sense of what this means. We estimate Δ with

$$\hat{\Delta} = \frac{\bar{X}_1 - \bar{X}_2}{s_p},$$

where s_p is the pooled standard deviation given by Equation (8.2).

Unfortunately, Δ suffers from some fundamental problems. First, if groups differ, there is no reason to assume that the variances are equal. Indeed, some authorities would argue that surely they must be unequal. We could test the hypothesis of equal variances, but how much power is needed to justify the conclusion that variances are equal if we fail to reject? Another possibility is to replace σ with the standard deviation from one of the groups. That is, we might use

$$\Delta_1 = \frac{\mu_1 - \mu_2}{\sigma_1} \quad \text{or} \quad \Delta_2 = \frac{\mu_1 - \mu_2}{\sigma_2},$$

which we would estimate, respectively, with

$$\hat{\Delta}_1 = \frac{\bar{X}_1 - \bar{X}_2}{s_1} \quad \text{and} \quad \hat{\Delta}_2 = \frac{\bar{X}_1 - \bar{X}_2}{s_2}.$$

But an even more serious problem is nonnormality.

The left panel of Figure 8.1 shows two normal distributions, where the difference between the means is 1 ($\mu_1 - \mu_2 = 1$) and both standard deviations are 1. So

$$\Delta = 1.$$

Cohen (1977) defines a large effect size as one that is visible to the naked eye, and he concludes (p. 40) that for normal distributions, $\Delta = .8$ is large, $\Delta = .5$ is a medium effect size, and $\Delta = .2$ is small. Now look at the right panel of Figure 8.1. As is evident, the difference between the two distributions appears to be very similar to the difference shown in the left panel, so according to Cohen we again have a large effect size. However, in the right panel, $\Delta = .3$ because these two distributions are mixed normals with variances 10.9. This illustrates the general principle that arbitrarily small departures from normality can render the magnitude of Δ meaningless. In practical terms, if we rely exclusively on Δ to judge whether there is a substantial difference between two groups, situations will arise where we will grossly underestimate the degree to which groups differ.

Here is another concern about Δ when trying to characterize how groups differ. Look at Figure 8.3. These two distributions have equal means and equal variances, but they differ in an obvious way that might have practical importance. Although the difference between measures of location provides a useful measure of effect size, we need additional ways of gaining perspective on the extent to which groups differ.

To describe one way of measuring the degree of separation between two groups, imagine that we randomly sample an observation from one of the two distributions in Figure 8.3 and that we get the value -5. Then Figure 8.3 indicates that this observation probably came from the first group (the one with the probability density function given by the solid line) because the probability of getting a value as low as -5 from the second group is virtually zero. If we had gotten the value 0, it is more likely that the observation came from the second group because the probability of getting a value near zero is higher for the second group. More formally let $f_1(x)$ be the equation for the solid line in Figure 8.3 (which is its probability density function) and let $f_2(x)$ be the equation for the dashed line. The likelihood that the value x came from the first group is $f_1(x)$, and the likelihood that the value x came from the second group is $f_2(x)$. So if $f_1(x) > f_2(x)$, a natural rule is to decide that x came from group 1, and if $f_1(x) < f_2(x)$ to decide that x came from group 2. We do not know $f_1(x)$ and $f_2(x)$, but they can be estimated, as indicated in Section 3.7. The result is a relatively effective method for deciding whether the value x came from group 1 or 2. (e.g., Silverman, 1986). There are many other strategies one might use to decide from which group the observation came, but the method just described has been found to be relatively effective for the problem at hand.

Let Q be the probability of correctly deciding whether a randomly sampled observation came from group 1 using the strategy just outlined. Then Q provides a measure of the separation between the two groups. If the distributions are identical, then $Q = .5$. If they are completely distinct, then $Q = 1$. To add perspective, if for two normal distributions $\Delta = .8$, which is typically labeled a large effect size, then $Q = .66$. If $\Delta = .2$, then $Q = .55$. But unlike Δ, Q does not change drastically with small shifts away from a normal distribution. For example, for the left panel of Figure 8.1, $Q = .66$ and for the right panel $Q = .69$, so in both cases a large effect size is indicated. This is in contrast to Δ, which drops from 1 to .3.

It is *not* being suggested that Q be used as a measure of effect size to the exclusion of all other measures. Measuring the difference between two distributions is a complex issue that often requires several perspectives. Also, rigid adherence to the idea that $Q = .66$ is large and that $Q = .55$ is small is not being recommended. What constitutes a large difference can vary from one situation to the next. We are comparing Q to Δ merely to add perspective.

Currently, the most accurate estimate of Q is based on a rather complex bootstrap method (called the .632 estimator). A description of this estimator in a much more general context is given in Efron and Tibshirani (1993). The use of this bootstrap method when estimating Q has been investigated by Wilcox and Muska (1999). Here the computational details are relegated to Box 8.2. It is noted that the estimate of Q, $\hat{Q}_{.632}$, can be less than .5 even though we know $Q \geq .5$. So $\hat{Q}_{.632} < .5$ suggests that there is little or no difference between the groups.

BOX 8.2 A Bootstrap Estimate of Q

For each group, compute the kernel density estimator as described in Section 3.7, and label the results $\hat{f}_1(x)$ and $\hat{f}_2(x)$, respectively. Set $\hat{\eta}(X_{i1}) = 1$ if $\hat{f}_1(X_{i1}) > \hat{f}_2(X_{i1})$, otherwise $\hat{\eta}(X_{i1}) = 0$. (That is, decide X_{i1} came from group 1 if $\hat{f}_1(X_{i1}) > \hat{f}_2(X_{i1})$.) In contrast, set $\hat{\eta}(X_{i2}) = 1$ if $\hat{f}_1(X_{i2}) < \hat{f}_2(X_{i2})$, otherwise $\hat{\eta}(X_{i2}) = 0$. Generate a bootstrap sample from each group and let $\hat{\eta}^*$ be the resulting estimate of η. Repeat this process B times, yielding $\hat{\eta}_b^*$,

Continued

BOX 8.2 (*Continued*) $b = 1, \ldots, B$. Let

$$\hat{\epsilon}_1 = \frac{1}{n_1} \sum_{i=1}^{n_1} \frac{1}{B_{1i}} \sum_{b \in C_{1i}} \hat{\eta}_b^*(X_{1i}),$$

$$\hat{\epsilon}_2 = \frac{1}{n_2} \sum_{i=1}^{n_2} \frac{1}{B_{2i}} \sum_{b \in C_{2i}} \hat{\eta}_b^*(X_{2i}).$$

For the bootstrap samples obtained from the first group, C_{1i} is the set of indices of the bth bootstrap sample not containing X_{1i}, and B_{1i} is the number of such bootstrap samples. The notation $b \in C_{1i}$ means that b is an element of C_{1i}. That is, the second sum in the definition of $\hat{\epsilon}_1$ is over all bootstrap samples not containing X_{1i}. Similarly, for the bootstrap samples from the second group, C_{2i} is the set of indices of the bth bootstrap sample not containing X_{2i}, and B_{2i} is the number of such bootstrap samples. Let

$$\hat{Q}_{.632,1} = .368\hat{Q}_{ap1} + .632\hat{\epsilon}_1,$$

where

$$\hat{Q}_{ap1} = \frac{1}{n_1} \sum \hat{\eta}(X_{1i}),$$

and define $\hat{Q}_{.632,2}$ in an analogous fashion. The .632 estimator, which is computed by the S-PLUS function qhat in Section 8.11.1, is taken to be

$$\hat{Q}_{.632} = \frac{1}{n_1 + n_2} \left(n_1 \hat{Q}_{.632,1} + n_2 \hat{Q}_{.632,2} \right).$$

EXAMPLE. To illustrate a portion of the computations in Box 8.2, consider five bootstrap samples from the first group with the following observation numbers:

Bootstrap sample				
1	2	3	4	5
1	16	25	1	14
5	5	4	7	10
23	16	12	12	2
11	24	16	7	8
11	11	14	14	13
17	15	24	1	1
8	21	3	21	17

Continued

> **EXAMPLE.** (*Continued*) So the first bootstrap sample (column 1) contains observation numbers 1, 5, 23, 11, 11, 17, and 8. That is, it contains the first observation followed by the fifth observation, followed by the twenty-third observation, and so on. Note that observation 1 appears in bootstrap samples 1, 4, and 5, but not in samples 2 and 3. That is, $C_{11} = (2, 3)$, and $B_{11} = 2$, the number of elements in $C_{11} = (2, 3)$. So when $i = 1$, the second sum when computing $\hat{\epsilon}$ is over the bootstrap samples $b = 2$ and 3. That is, when $i = 1$, $\hat{\eta}_b^*(X_{1i})$ is being computed using X_{i1} values that do not appear in the bootstrap sample used to determine $\hat{\eta}_b^*$. Similarly, the second observation appears in only one bootstrap sample, the fifth. So $C_{12} = (1, 2, 3, 4)$. That is, when $i = 2$, the second sum is over $b = 1, 2, 3$, and 4. ■

8.11.1 S-PLUS Function qhat

The S-PLUS function

$$\text{qhat}(x, y)$$

estimates Q using the data stored in the S-PLUS variables x and y. (Execution time can be quite high.)

> **EXAMPLE.** For the alcohol data in Table 8.3 we rejected the hypothesis of equal 20% trimmed means. If we use a standardized difference between the two groups based on the means and the standard deviation of the first group, we get $\hat{\Delta}_1 = .4$. Using the standard deviation of the second group yields $\hat{\Delta}_2 = .6$. So taken together, and assuming normality, these results suggest a medium effect size. The S-PLUS function qhat returns
>
> `qhat.632 = .61,`
>
> supporting the view that there is a medium difference between the two groups. ■

8.11.2 The Shift Function

Again, it currently seems that no single method for characterizing how groups differ is satisfactory in all situations. A criticism of the methods covered so far is that they do not capture some of the global details of how groups differ. For example, imagine that two methods for treating depression are being compared and that the higher X happens to be the more effective the method. Further assume that the distribution associated with the first method is given by the solid line in Figure 8.3 and that the distribution for the second group is given by the dashed line. Then in terms of the population means or any other measure of location, it makes no difference which method is used. However, for the first method, there is a good chance that an individual will have a level of effectiveness less than -3; but under the second method, the probability of an effectiveness level less than −3 is virtually zero. That is, we can avoid a poor effectiveness rating by using method 2. Similarly, there is a good chance of an effectiveness rating greater than or equal to 3 using method 1, but not with method 2. That is, the relative merits of the methods change depending on where we look.

In this last example we could simply compare some measure of scale. But consider the distributions in Figure 2.10, which have equal means and variances. How might we summarize the extent to which these distributions differ? Doksum and Sievers (1976) suggest the following strategy. Rather than just compare a single measure of location, notice that we could, for example, compare the lower quartiles corresponding to these two groups. That is, we compare low values in the first group to the corresponding low values in the second. Of course, we can compare the upper quartiles as well. So, for example, if the upper quartile of the first group is 12 and the upper quartile of the second group is 8, we can say that the typical individual in the high end of the first group is better off than the comparable person in the second group. Of course, there is nothing sacred about the quartiles; we can use any quantile and get a more detailed sense about how the groups differ in contrast to using a single measure of location. If we compare the .25, .5, and .75 quantiles, we are comparing the quartiles in addition to the median. But we might also compare the .1, .2, .3, .4, .6, .7, .8, and .9 quantiles or any other set of quantiles we choose. To help convey the differences between all of the quantiles, Doksum and Sievers (1976) suggest plotting the differences between all quantiles versus the quantiles in the first group. So if x_q is the qth quantile of the first group, meaning that $P(X \leq x_q) = q$, the suggestion is to estimate all possible quantiles for each group and plot x_q versus $y_q - x_q$, where y_q is the qth quantile of the second group. If we do this for the two distributions shown in Figure 2.10, we get Figure 8.4. This says that low-scoring individuals in the second group score higher

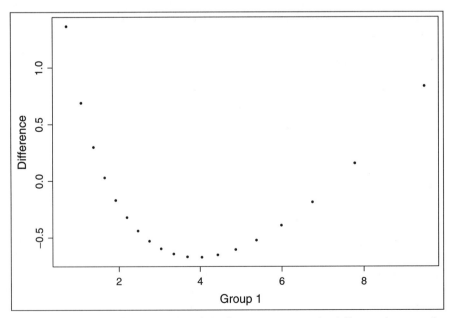

FIGURE 8.4   Plot of the quantiles of the first group versus the difference between the quantiles.

than the comparable low-scoring individuals in the first group, up to about the value 1.9. For the middle range of values in the first group, namely, between 2 and 7, the reverse is true; but for values greater than 7, again individuals in the second group score higher than comparable individuals in the first.

Doksum and Sievers also indicate how to compute a confidence band for the difference between all quantiles. That is, for each possible quantile, a confidence interval for the difference between the quantiles is computed with the property that with probability $1 - \alpha$, it will be simultaneously true that all such confidence intervals contain the true difference $(y_q - x_q)$. Said another way, their method computes a confidence interval for $y_q - x_q$ for all values of q between 0 and 1, and if we reject the hypothesis of equal quantiles when this interval does not contain zero, then the probability of *at least one* Type I error is α. The computational details of their method are not given here, but an S-PLUS function for applying the method is supplied.

8.11.3 S-PLUS Function sband

The S-PLUS function

$$sband(x,y)$$

computes an estimate of the difference between the quantiles using the data stored in the S-PLUS variables x and y and plots these differences as a function of the estimated quantiles associated with the first group, the first group being the data stored in the S-PLUS variable x. (For more details about this function plus variations of the method, see Wilcox, 1997a.)

EXAMPLE. Table 8.6 contains data from a study designed to assess the effects of ozone on weight gain in rats. (These data were taken from Doksum & Sievers, 1976.) The experimental group consisted of 22 70-day-old rats kept in an ozone environment for 7 days. A control group of 23 rats of the same age was kept in an ozone-free environment. Storing the data for the control group in the S-PLUS variable x, and storing the data for the ozone group in y, sband produces the graph shown in Figure 8.5. The + indicates the location of the median for the control group (the data stored in the first argument, x), and the lower and upper quartiles are marked with an o to the left and right of the +. For example, the median of the control group is $M = 22.7$, as indicated by the +, and the difference between the median of the ozone group (which is 11.1) and the control group is given by the solid line and is equal to -11.6. So based on the medians it is estimated that the typical rat in the ozone group gains less weight than the typical rat in the control group. However, the difference between the upper quartiles is $26.95 - 17.35 = 9.6$. Now there is more weight gain among rats in the ozone group. Looking at the graph as a whole suggests that the effect of ozone becomes more pronounced as we move along the x-axis, up to about 19, but then the trend reverses, and in fact in the upper end we see more weight

Continued

TABLE 8.6   Weight Gain (in grams) of Rats in Ozone Experiment

Control:	41.0	38.4	24.4	25.9	21.9	18.3	13.1	27.3	28.5	−16.9
Ozone:	10.1	6.1	20.4	7.3	14.3	15.5	−9.9	6.8	28.2	17.9
Control:	26.0	17.4	21.8	15.4	27.4	19.2	22.4	17.7	26.0	29.4
Ozone:	−9.0	−12.9	14.0	6.6	12.1	15.7	39.9	−15.9	54.6	−14.7
Control:	21.4	26.6	22.7							
Ozone:	44.1	−9.0								

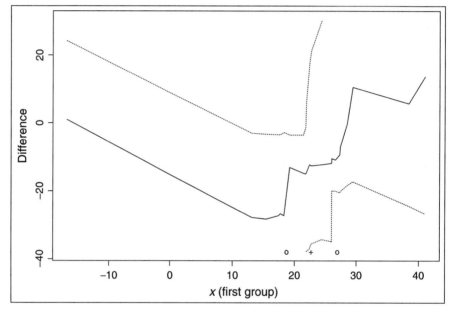

FIGURE 8.5   Example of a shift function. The dashed lines are confidence bands. The +
along the x-axis indicates the location of the median of the first group; and the o's indicate
the lower and upper quartiles.

EXAMPLE. (*Continued*) gain in the ozone group. If there are no differences
between the quantiles, the shift function should be a straight horizontal line
at 0. The dashed lines in Figure 8.5 mark the confidence band for the difference
between the quantiles. The hypothesis of equal quantiles is rejected if the lower
(upper) dashed line is above (below) zero.   ■

Notice that the left end of the lower dashed line in Figure 8.5 begins at approximately $x = 22$. This is because for $x < 22$, the lower confidence band extends
down to $-\infty$. That is, the precision of the estimated differences between the quantiles might be poor in this region based on the sample sizes used. Similarly, the
upper dashed line terminates around $x = 27$. This is because for $x > 27$, the upper
confidence band extends up to ∞.

The methods listed in this section are not exhaustive. Another approach is to examine a so-called quantile–quantile plot. If the groups have identical quantiles, a plot of the quantiles should be close to a line having slope 1 and intercept zero. Another approach to measuring effect size is the so-called *overlapping coefficient*. You estimate the distributions associated with both groups and then compute the area under the intersection of these two curves; see Clemons and Bradley (2000) for recent results on how this might be done. An area of zero corresponds to no overlap, and an area of 1 occurs when the distributions are identical and the groups do not differ in any manner whatsoever. Another useful approach is to create a boxplot for both groups on the same scale. This is easily done in S-PLUS. For example, the command boxplot(x,y) will create a boxplot for the data in both x and y. Yet another strategy is to estimate the probability that a randomly sampled observation from the first group is less than a randomly sampled observation from the second. Details about this approach will be covered in Chapter 15.

8.12 Comparing Correlations and Regression Slopes

Rather than compare measures of location or scale, situations arise where the goal is to compare correlations or regression parameters instead. That is, for every individual we have two measures and the goal is to determine whether the association for the first group differs from the association for the second. For example, in a study of schizophrenia, Dawson, Schell, Hazlett, Nuechterlein, and Filion (2000) were interested in, among other things, the association between prepulse inhibition and measures of schizophrenic symptoms. A portion of their study dealt with comparing correlations of individuals with positive symptoms to the correlation of those with negative symptoms. Also, comparing correlations or regression slopes is one strategy for determining whether a third variable modifies the association between two other variables. (See the example given at the end of this section.)

Methods for comparing correlations have been studied by Yu and Dunn (1982) and Duncan and Layard (1973), but all of these methods are known to be rather unsatisfactory, and using Fisher's r-to-z transformation is unsatisfactory, for reasons indicated in Section 6.5. Currently, the most effective procedure is to use the modified percentile bootstrap method, but adjusted to take into account the total number of observations. If we have n_1 pairs of observations for the first group, yielding a correlation of r_1, and n_2 pairs of observations for the second group, yielding a correlation of r_2, the goal is to test

$$H_0 : \rho_1 = \rho_2,$$

the hypothesis that the two groups have equal population correlation coefficients. If we reject, this indicates that the association differs for each group, but for reasons outlined in Chapter 6, how the associations differ is vague and unclear.

To apply the modified percentile bootstrap method to the present problem, let $N = n_1 + n_2$ be the total number of pairs of observations available. For the jth group, generate a bootstrap sample of n_j pairs of observations as described in Section 7.3. Let r_1^* and r_2^* represent the resulting correlation coefficients and set

$$D^* = r_1^* - r_2^*.$$

Repeat this process 599 times, yielding $D_1^*, \ldots, D_{599}^*$. Then a .95 confidence interval for the difference between the population correlation coefficients ($\rho_1 - \rho_2$) is

$$\left(D_{(\ell)}^*, D_{(u)}^*\right),$$

where $\ell = 7$ and $u = 593$ if $N < 40$; $\ell = 8$ and $u = 592$ if $40 \leq N < 80$; $\ell = 11$ and $u = 588$ if $80 \leq N < 180$; $\ell = 14$ and $u = 585$ if $180 \leq N < 250$; $\ell = 15$ and $u = 584$ if $N \geq 250$. If the resulting confidence interval does not contain zero, reject the hypothesis of equal correlations. Note that this is just a simple modification of the method used to compute a confidence interval for the slope of a regression line that was described in Chapter 7.

When using least squares regression, the slopes can be compared in a similar manner. To test

$$H_0 : \beta_1 = \beta_2,$$

where β_1 and β_2 are the slopes corresponding to the two groups, simply proceed as was done when working with Pearson's correlation, except replace r with the least squares estimate of the slope.

8.12.1 S-PLUS Functions twopcor and twolsreg

The S-PLUS function

$$\text{twopcor}(x1,y1,x2,y2)$$

computes a confidence interval for the difference between two Pearson correlations corresponding to two independent groups using the modified bootstrap method just described. The data for group 1 are stored in the S-PLUS variables x1 and y1, and the data for group 2 are stored in x2 and y2. The S-PLUS function

$$\text{twolsreg}(x1,y1,x2,y2)$$

computes a confidence interval for the difference between the slopes based on the least squares estimator described in Chapter 6.

EXAMPLE. In an unpublished study by L. Doi, there was interest in whether a measure of orthographic ability (Y) is associated with a measure of sound blending (X). Here we consider whether an auditory analysis variable (Z) *modifies* the association between X and Y. This was done by partitioning the pairs of points (X, Y) according to whether $Z \leq 14$ or $Z > 14$, and then entering the resulting pairs of points into the S-PLUS function twopcor. The .95 confidence interval for $\rho_1 - \rho_2$, the difference between the correlations, is $(-0.64, 0.14)$. This interval contains zero, so we would not reject the hypothesis of equal correlations. If we compare regression slopes instead, the .95 confidence interval is $(-0.55, 0.18)$ and again we fail to reject. It is stressed, however, that this analysis does not establish that the association does not differ for the two groups under study. A concern is that power might be low when attention is

Continued

> **EXAMPLE.** (*Continued*) restricted to Pearson's correlation or least squares regression. (Methods covered in subsequent chapters indicate that the measure of auditory analysis does modify the association between orthographic ability and sound blending.)   ■

8.13 Comparing Two Binomials

This section considers the problem of comparing the probability of success associated with two independent binomials. For example, if the probability of surviving an operation using method 1 is p_1, and if the probability of surviving using method 2 is p_2, do p_1 and p_2 differ, and if they do differ, by how much? As another example, to what degree do men and women differ in whether they believe the President of the United States is an effective leader?

Many methods have been proposed for comparing binomials, two of which are described here. These two methods were chosen based on results in Storer and Kim (1990) and Beal (1987), where comparisons of several methods were made. It is noted, however, that competing methods have been proposed that apparently have not been compared directly to the methods covered here (e.g., Berger, 1996; Coe & Tamhane, 1993). The Storer–Kim method tests $H_0 : p_1 = p_2$ using the calculations shown in Box 8.3, and Beal's method computes a $1 - \alpha$ confidence interval for $p_1 - p_2$ using the calculations in Box 8.4. The choice between these two methods is not completely clear. An appeal of Beal's method is that it provides a confidence interval and the Storer–Kim method does not. Situations arise in subsequent chapters where the Storer–Kim method has less power than Beal's method when comparing multiple groups of individuals, but when comparing two groups only, we find situations where the Storer–Kim method rejects and Beal's method does not.

BOX 8.3 Storer–Kim Methods for Comparing Two Independent Binomials

You observe r_1 successes among n_1 trials in the first group and r_2 successes among n_2 trials in the second. The goal is to test $H_0 : p_1 = p_2$. Note that the possible number of successes in the first group is any integer, x, between 0 and n_1, and for the second group it is any integer, y, between 0 and n_2. For any x between 0 and n_1 and any y between 0 and n_2, set

$$a_{xy} = 1$$

if

$$\left| \frac{x}{n_1} - \frac{y}{n_2} \right| \geq \left| \frac{r_1}{n_1} - \frac{r_2}{n_2} \right| ;$$

Continued

BOX 8.3 *(Continued)*

otherwise

$$a_{xy} = 0.$$

Let

$$\hat{p} = \frac{r_1 + r_2}{n_1 + n_2}.$$

The test statistic is

$$T = \sum_{x=0}^{n_1} \sum_{y=0}^{n_2} a_{xy} b(x, n_1, \hat{p}) b(y, n_2, \hat{p}),$$

where

$$b(x, n_1, \hat{p}) = \binom{n_1}{x} \hat{p}^x (1 - \hat{p})^{n_1 - x}$$

and $b(y, n_2, \hat{p})$ is defined in an analogous fashion. You reject if

$$T \le \alpha.$$

That is, T is the significance level.

BOX 8.4 Beal's Method for Computing a Confidence Interval for $p_1 - p_2$

Following the notation in Box 8.3, let $\hat{p}_1 = r_1/n_1$ and $\hat{p}_2 = r_2/n_2$ and let $c = z_{1-\alpha/2}^2$, where $z_{1-\alpha/2}$ is the $1 - \alpha$ quantile of a standard normal distribution. (So c is the $1 - \alpha$ quantile of a chi-squared distribution with one degree of freedom.) Compute

$$a = \hat{p}_1 + \hat{p}_2$$

$$b = \hat{p}_1 - \hat{p}_2$$

$$u = \frac{1}{4} \left(\frac{1}{n_1} + \frac{1}{n_2} \right)$$

$$v = \frac{1}{4} \left(\frac{1}{n_1} - \frac{1}{n_2} \right)$$

$$V = u\{(2 - a)a - b^2\} + 2v(1 - a)b$$

Continued

BOX 8.4 (*Continued*)

$$A = \sqrt{c\{V + cu^2(2-a)a + cv^2(1-a)^2\}}$$

$$B = \frac{b + cv(1-a)}{1+cu}.$$

The $1 - \alpha$ confidence interval for $p_1 - p_2$ is

$$B \pm \frac{A}{1+cu}.$$

8.13.1 S-PLUS Functions twobinom and twobici

The S-PLUS function

$$\text{twobinom}(r1 = \text{sum}(x), n1 = \text{length}(x), r2 = \text{sum}(y), n2 = \text{length}(y),$$

$$x = \text{NA}, y = \text{NA})$$

has been supplied to test $H_0 : p_1 = p_2$ using the Storer–Kim method in Box 8.3. The function can be used by specifying the number of successes in each group (arguments r1 and r2) and the sample sizes (arguments n1 and n2), or the data can be in the form of two vectors containing 1's and 0's, in which case you use the arguments x and y. Beal's method can be applied with the S-PLUS function

$$\text{twobici}(r1 = \text{sum}(x), n1 = \text{length}(x), r2 = \text{sum}(y), n2 = \text{length}(y), x = \text{NA},$$

$$y = \text{NA}, \text{alpha} = 0.05)$$

EXAMPLE. If for the first group we have 7 successes among 12 observations, for the second group we have 22 successes among 25 observations, the command

$$\text{twobinom}(7,12,22,25)$$

returns a significance level of .044; this is less than .05, so we would reject with $\alpha = .05$. The .95 confidence interval for $p_1 - p_2$ returned by the command

$$\text{twobici}(7,12,22,25)$$

is $(-0.61, 0.048)$; this interval contains zero, so in contrast to the Storer–Kim method we do not reject the hypothesis $H_0 : p_1 = p_2$, the only point being that different conclusions might be reached depending on which method is used. ■

EXAMPLE. In Table 8.5 we see that 101 of the 156 females responded that they want one sexual partner during the next 30 years. As for the 105 males in this study, 49 gave the response 1. Does the probability of a 1 among males differ from the probability among females? The S-PLUS function twobinom returns a significance level of .0037, indicating that the probabilities differ even with $\alpha = .0037$. The command

$$\text{twobici}(49, 105, 101, 156)$$

returns a .95 confidence interval of $(-0.33, -0.04)$, so again we reject, but there is some possibility that the difference between the two probabilities is fairly small.   ■

8.14 Exercises

1. Suppose that the sample means and variances are $\bar{X}_1 = 15$, $\bar{X}_2 = 12$, $s_1^2 = 8$, $s_2^2 = 24$ with sample sizes $n_1 = 20$ and $n_2 = 10$. Verify that $s_p^2 = 13.14$ and $T = 2.14$ and that Student's T test rejects the hypothesis of equal means with $\alpha = .05$.

2. For two independent groups of subjects, you get $\bar{X}_1 = 45$, $\bar{X}_2 = 36$, $s_1^2 = 4$, $s_2^2 = 16$ with sample sizes $n_1 = 20$ and $n_2 = 30$. Assume the population variances of the two groups are equal and verify that the estimate of this common variance is 11.25.

3. Still assuming equal variances, test the hypothesis of equal means using the data in Exercise 2 assuming random sampling from normal distributions. Use $\alpha = .05$.

4. Repeat the previous exercise, but use Welch's test for comparing means.

5. Comparing the test statistics for the preceding two exercises, what do they suggest regarding the power of Welch's test versus Student's T test for the data being examined?

6. For two independent groups of subjects, you get $\bar{X}_1 = 86$, $\bar{X}_2 = 80$, $s_1^2 = s_2^2 = 25$, with sample sizes $n_1 = n_2 = 20$. Assume the population variances of the two groups are equal and verify that Student's T rejects with $\alpha = .01$.

7. Repeat Exercise 6 using Welch's method.

8. Comparing the results of Exercises 6 and 7, what do they suggest about using Student's T versus Welch's method when the sample variances are approximately equal?

9. If for two independent groups, you get $\bar{X}_{t1} = 42$, $\bar{X}_{t2} = 36$, $s_{w1}^2 = 25$, $s_{w2}^2 = 36$, $n_1 = 24$, and $n_2 = 16$, test the hypothesis of equal trimmed means with $\alpha = .05$.

10. Referring to Exercise 9, compute a .99 confidence interval for the difference between the trimmed means.

11. For $\bar{X}_1 = 10$, $\bar{X}_2 = 5$, $s_1^2 = 21$, $s_2^2 = 29$, and $n_1 = n_2 = 16$, compute a .95 confidence interval for the difference between the means using Welch's method, and state whether you would reject the hypothesis of equal means.

12. Repeat Exercise 11, but use Student's T instead.
13. Two methods for training accountants are to be compared. Students are randomly assigned to one of the two methods. At the end of the course, each student is asked to prepare a tax return for the same individual. The returns reported by the students are

Returns

Method 1:	132	204	603	50	125	90	185	134
Method 2:	92	−42	121	63	182	101	294	36

Using Welch's test, would you conclude that the methods differ in terms of the average return? Use $\alpha = .05$.
14. Repeat Exercise 13, but compare 20% trimmed means instead.
15. You compare lawyers to professors in terms of job satisfaction and fail to reject the hypothesis of equal means or equal trimmed means. Does this mean it is safe to conclude that the typical lawyer has about the same amount of job satisfaction as the typical professor?
16. Responses to stress are governed by the hypothalamus. Imagine you have two groups of subjects. The first shows signs of heart disease and the other does not. You want to determine whether the groups differ in terms of the weight of the hypothalamus. For the first group of subjects, with no heart disease, the weights are

11.1, 12.2, 15.5, 17.6, 13.0, 7.5, 9.1, 6.6, 9.5, 18.0, 12.6.

For the other group, with heart disease, the weights are

18.2, 14.1, 13.8, 12.1, 34.1, 12.0, 14.1, 14.5, 12.6, 12.5, 19.8,

13.4, 16.8, 14.1, 12.9.

Determine whether the groups differ based on Welch's test. Use $\alpha = .05$.
17. Repeat Exercise 16, but use Yuen's test with 20% trimmed means.
18. Use Δ and Q to measure effect size using the data in the previous two exercises.
19. Published studies indicate that generalized brain dysfunction may predispose someone to violent behavior. Of interest is determining which brain areas may be dysfunctional in violent offenders. In a portion of such a study conducted by Raine, Buchsbaum, and LaCasse (1997), glucose metabolism rates of 41 murderers were compared to the rates for 41 control subjects. Results for the left hemisphere, lateral prefrontal region of the brain yielded a sample mean of 1.12 for the controls and 1.09 for the murderers. The corresponding standard deviations were 0.05 and 0.06. Verify that Student's $T = 2.45$ and that you reject with $\alpha = .05$.
20. In the previous exercise, you rejected the hypothesis of equal means. What does this imply about the accuracy of the confidence interval for the difference between the population means based on Student's T?
21. For the data in Table 8.6, if we assume that the groups have a common variance, verify that the estimate of this common variance is $s_p^2 = 236$.

22. The sample means for the data in Table 8.6 are 22.4 and 11. If we test the hypothesis of equal means using Student's T, verify that $T = 2.5$ and that you would reject with $\alpha = .05$.

23. Verify that the .95 confidence interval for the difference between the means, based on the data in Table 8.6 and Student's T, is $(2.2, 20.5)$. What are the practical problems with this confidence interval?

24. Student's T rejects the hypothesis of equal means based on the data in Table 8.6. Interpret what this means.

25. For the data in Table 8.6, the sample variances are 116.04 and 361.65, respectively. Verify that the .95 confidence interval for the difference between the means based on Welch's method is $(1.96, 20.83)$. Check this result with the S-PLUS function yuen.

26. In the previous exercise you do not reject the hypothesis of equal variances. Why is this *not* convincing evidence that the assumption of equal variances, when using Student's T, is justified?

27. The 20% Winsorized standard deviation (s_w) for the first group in Table 8.6 is 1.365 and for the second group it is 4.118. Verify that the .95 confidence interval for the difference between the 20% trimmed means, using Yuen's method, is $(5.3, 22.85)$.

28. Create a boxplot of the data in Table 8.6, and comment on why the probability coverage, based on Student's T or Welch's method, might differ from the nominal α level.

29. For the self-awareness data in Table 8.4, verify that the S-PLUS function yuenbt, with the argument tr set to 0, returns $(-571.4, 302.5)$ as a .95 confidence interval for the difference between the means.

30. For the data in Table 8.4, use the S-PLUS function comvar2 to verify that the .95 confidence interval for the difference between the variances is $(-1165766.8, 759099.7)$.

31. Describe a general situation where comparing medians will have more power than comparing means or 20% trimmed means.

32. For the data in Table 8.4, verify that the .95 confidence interval for the difference between the biweight midvariances is $(-159234, 60733)$.

33. The last example in Section 8.9 dealt with comparing males to females regarding the desired number of sexual partners over the next 30 years. Using Student's T, we fail to reject, which is not surprising because there is an extreme outlier among the responses given by males. If we simply discard this one outlier and compare groups using Student's T or Welch's method, what criticism might be made even if we could ignore problems with nonnormality?

ONE-WAY ANOVA

This chapter (and the three chapters that follow) addresses the common situation where more than two groups are to be compared based on some measure of location. We begin with a classic and commonly used method for comparing the means of independent groups under the assumption of normality and homoscedasticity. We have already seen serious practical problems with this approach when comparing two groups only. For various reasons, problems are exacerbated when comparing more than two groups, and indeed new problems are introduced. However, these problems are not completely obvious and should be described in order to motivate more modern methods. As usual, it is not assumed that the reader has any prior knowledge about the classic approach to comparing groups based on means.

To help fix ideas, we begin with a concrete example. Clinical psychologists have long tried to understand schizophrenia. One issue of interest to some researchers is whether various groups of individuals differ in terms of measures of skin resistance. In such a study, four groups of individuals were identified: (1) no schizophrenic spectrum disorder, (2) schizotypal or paranoid personality disorder, (3) schizophrenia, predominantly negative symptoms, (4) schizophrenia, predominantly positive symptoms. Table 9.1 presents the first 10 observations for each group, where the entries are measures of skin resistance (in ohms) following presentation of a generalization stimulus. (These data were supplied by S. Mednick, Dept. of Psychology, University of Southern California.) Note that the actual sample sizes in the study were larger; only the first 10 observations for each group are listed here.

You could, of course, compare all pairs of groups in terms of some measure of location. If, for example, you decide to use means, then you might simply compare the mean of the first group to the mean of the second, then compare the mean of first group to the mean of third, and so on. In symbols, you could test

$$H_0 : \mu_1 = \mu_2,$$

$$H_0 : \mu_1 = \mu_3,$$

$$H_0 : \mu_1 = \mu_4,$$

TABLE 9.1   Measures of Skin Resistance for Four Groups

No schiz.	Schizotypal	Schiz. neg.	Schiz. pos.
0.49959	0.24792	0.25089	0.37667
0.23457	0.00000	0.00000	0.43561
0.26505	0.00000	0.00000	0.72968
0.27910	0.39062	0.00000	0.26285
0.00000	0.34841	0.11459	0.22526
0.00000	0.00000	0.79480	0.34903
0.00000	0.20690	0.17655	0.24482
0.14109	0.44428	0.00000	0.41096
0.00000	0.00000	0.15860	0.08679
1.34099	0.31802	0.00000	0.87532
$\bar{X}_1 = 0.276039$	$\bar{X}_2 = 0.195615$	$\bar{X}_3 = 0.149543$	$\bar{X}_4 = 0.399699$
$s_1^2 = 0.1676608$	$s_2^2 = 0.032679$	$s_3^2 = 0.0600529$	$s_4^2 = 0.0567414$

$$n_1 = n_2 = n_3 = n_4 = 10$$

$$H_0 : \mu_2 = \mu_3,$$

$$H_0 : \mu_2 = \mu_4,$$

$$H_0 : \mu_3 = \mu_4,$$

using the methods in Chapter 8, or you might compare trimmed means or MOMs or M-estimators instead. There is, however, a technical issue that arises if you do this. Suppose there are no differences among the groups, in which case none of the six null hypotheses just listed should be rejected. To keep things simple for the moment, assume all four groups have normal distributions with equal variances, in which case Student's T test in Chapter 8 provides exact control over the probability of a Type I error when testing any single hypothesis. Further assume that each of the six hypotheses just listed are tested with $\alpha = .05$. So for *each* hypothesis, the probability of a Type I error is .05. But what is the probability of *at least one* Type I error when you perform all six tests? That is, what is the probability of making one or more mistakes and rejecting when in fact all pairs of groups being compared have equal means?

If you perform each of the tests with $\alpha = .05$, the probability of at least one Type I error will be larger than .05. The more tests you perform, the more likely you are to reject the hypothesis of equal means when in fact the groups do not differ. A common goal, then, is to test all pairs of means so that the probability of one or more Type I errors is α, where, as usual, α is some value you pick. Put another way, when comparing all pairs of groups, the goal is to have the probability of making no Type I errors equal to $1 - \alpha$.

There are two general approaches to the problem of controlling the probability of committing one or more Type I errors. The first and most popular begins by testing the hypothesis that all the groups being compared have equal means. In the illustration, the null hypothesis is written as

$$H_0 : \mu_1 = \mu_2 = \mu_3 = \mu_4.$$

More generally, when comparing J groups, the null hypothesis is

$$H_0 : \mu_1 = \mu_2 = \cdots = \mu_J. \tag{9.1}$$

If this hypothesis of equal means is rejected, you then make decisions about which pairs of groups differ (using a method described in Chapter 12.) The second general approach is to skip the methods in this chapter and use one of the appropriate techniques in Chapter 12. There are circumstances under which this latter strategy has practical advantages, but the details must be postponed for now.

9.1 Analysis of Variance (ANOVA) for Independent Groups

Assuming normality, Student's T test of the hypothesis that two independent groups have equal means can be extended to the problem of testing Equation (9.1), the hypothesis that J independent groups have equal means. Even though the goal is to test the hypothesis of equal means, the method is called *analysis of variance*, or ANOVA; it was derived by Sir Ronald Fisher. Like Student's T test, homoscedasticity is assumed, meaning that all groups have a common variance. That is, if $\sigma_1^2, \ldots, \sigma_J^2$ are the population variances of the J groups, homoscedasticity means that

$$\sigma_1^2 = \sigma_2^2 = \cdots = \sigma_J^2. \tag{9.2}$$

As was done with Student's T test, this common variance will be labeled σ_p^2. As in previous chapters, heteroscedasticity refers to a situation where not all the variances are equal.

To help convey the strategy of the traditional ANOVA method, we temporarily restrict attention to the situation where all of the sample sizes are equal. Box 9.1 summarizes the computations for this special case. (Box 9.2 covers the more general case where samples sizes might be unequal.) Here the common sample size is labeled n. In symbols, it is temporarily assumed that $n_1 = n_2 = \cdots = n_J = n$. For the schizophrenia data in Table 9.1, $n = 10$. You begin by computing the sample mean and sample variance for each group. The average of the J sample means, called the *grand mean*, is then computed and labeled $\bar{X}_G$. As indicated in Box 9.1,

$$\bar{X}_G = \frac{1}{J}(\bar{X}_1 + \cdots + \bar{X}_J).$$

Next, you compute four quantities called the *sum of squares between groups*, the *mean squares between groups*, the *sum of squares within groups*, and the *mean squares within groups*, as described in Box 9.1. For convenience, these four quantities are labeled SSBG, MSBG, SSWG, and MSWG, respectively. Finally you compute the test statistic, F, which is just MSBG divided by MSWG. If F is sufficiently large, reject the hypothesis of equal means.

BOX 9.1 Summary of How to Compute the ANOVA F-Test with
Equal Sample Sizes

Goal
Test $H_0 : \mu_1 = \cdots = \mu_J$, the hypothesis of equal means among J independent groups

Assumptions
- Random sampling
- Normality
- Equal variances

Computations
Compute the sample means, $\bar{X}_1, \ldots, \bar{X}_J$, and sample variances, $s_1^2, \ldots, s_J^2$. Then compute

$$\bar{X}_G = \frac{1}{J} \sum \bar{X}_j$$

(the grand mean),

$$N = nJ$$

(the total number of observations), and

$$SSBG = n \sum_{j=1}^{J} (\bar{X}_j - \bar{X}_G)^2,$$

$$MSBG = \frac{SSBG}{J - 1},$$

$$SSWG = (n - 1) \sum_{j=1}^{J} s_j^2,$$

$$MSWG = \frac{SSWG}{N - J} = \frac{1}{J} \sum_{j=1}^{J} s_j^2.$$

Test Statistic

$$F = \frac{MSBG}{MSWG}$$

Decision Rule
Reject H_0 if $F \geq f$, the $1 - \alpha$ quantile of an F-distribution with $\nu_1 = J - 1$ and $\nu_2 = N - J$ degrees of freedom.

When the null hypothesis is true, the exact distribution of F has been derived under the assumptions of normality and equal variances. This means that you are able to

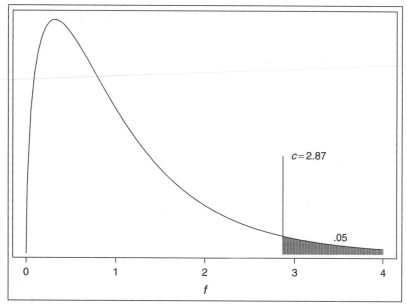

FIGURE 9.1   An F-distribution with 3 and 36 degrees of freedom. The probability that F is greater than 2.87 is .05, it is depicted by the area of the shaded region.

determine how large F must be in order to reject H_0. The distribution of F, when the null hypothesis is true, is called an *F-distribution* with degrees of freedom

$$\nu_1 = J - 1 \qquad \text{and} \qquad \nu_2 = N - J.$$

That is, the distribution depends on two quantities: the number of groups being compared, J, and the total number of observations in all of the groups, N.

For the schizophrenia illustration, there are $J = 4$ groups with a total of $N = 40$ observations, so the degrees of freedom are $\nu_1 = 4 - 1 = 3$ and $\nu_2 = 40 - 4 = 36$. Figure 9.1 shows the distribution of F with these degrees of freedom. The shaded region indicates the critical region when $\alpha = .05$; it extends from 2.87 to infinity. That is, if you want the probability of a Type I error to be .05, then reject the null hypothesis of equal means if $F \geq 2.87$, which is the .95 quantile of the F-distribution.

More generally, reject if $F \geq f$, where f is the $1 - \alpha$ quantile of an F-distribution with $\nu_1 = J - 1$ and $\nu_2 = N - J$ degrees of freedom. Tables 5–8 in Appendix B report critical values, f, for $\alpha = .1$, .05, .025, and .01 and various degrees of freedom. For example, with $\alpha = .05$, $\nu_1 = 6$, $\nu_2 = 8$, Table 6 indicates that the .95 quantile is $f = 3.58$. That is, there is a .05 probability of getting a value for F that exceeds 3.58 when in fact the population means are equal. For $\alpha = .01$, Table 8 says that the .99 quantile is 4.38. This means that if you reject when $F \geq 4.38$, the probability of a Type I error will be .01, assuming normality and that the groups have equal variances.

Now we illustrate the computations by computing F for the data in Table 9.1 and determining whether the hypothesis of equal means should be rejected with $\alpha = .05$. From Table 9.1, the sample means are 0.276039, 0.195615, 0.149543, 0.399699. The grand mean is just the average of these four sample means and is given by

$$\bar{X}_G = \frac{1}{4}(0.276039 + 0.195615 + 0.149543 + 0.399699) = 0.255224.$$

There are $n = 10$ subjects for each group, so the sum of squares between groups is

$$\text{SSBG} = n \sum_{j=1}^{J} (\bar{X}_j - \bar{X}_G)^2$$

$$= 10\{(0.276039 - 0.255224)^2 + (0.195615 - 0.255224)^2$$
$$+ (0.149543 - 0.255224)^2 + (0.399699 - 0.255224)^2\}$$

$$= 10(0.036)$$

$$= 0.36.$$

Therefore, the mean squares between groups is

$$\text{MSBG} = \frac{\text{SSBG}}{J - 1}$$

$$= \frac{0.36}{4 - 1}$$

$$= 0.12.$$

With equal sample sizes, the mean squares within groups is just the average of the sample variances:

$$\text{MSWG} = \frac{1}{J} \left(s_1^2 + \cdots + s_J^2 \right)$$

$$= \frac{1}{4}(0.1676608 + 0.032679 + 0.0600529 + 0.0567414)$$

$$= .0793.$$

Therefore,

$$F = \frac{\text{MSBG}}{\text{MSWG}}$$

$$= \frac{0.12}{0.0793}$$

$$= 1.51.$$

As already indicated, the degrees of freedom are $\nu_1 = 3$ and $\nu_2 = 36$ with a critical value $f = 2.87$. Because 1.51 is less than 2.87, you do not reject the hypothesis of equal means. This means that you do not have convincing empirical evidence that the hypothesis of equal means is unreasonable. As was the case in previous chapters, failing to reject the null hypothesis can be due to one of two reasons: The null hypothesis is true, or the null hypothesis is false but you failed to detect this because your sample size is too small to achieve reasonably high power. Methods for assessing power prior to collecting data, assuming normality, are nicely summarized by Cohen (1977). Section 9.3 describes a method for addressing power once data are available.

TABLE 9.2 ANOVA Summary Table

Source of variation	Degrees of freedom	Sum of squares	Mean square	F
Between groups	$J-1$	SSBG	MSBG	$F = \frac{\text{MSBG}}{\text{MSWG}}$
Within groups	$N-J$	SSWG	MSWG	
Totals	$N-1$	SSBG + SSWG		

TABLE 9.3 ANOVA Summary Table for the Data in Table 9.1

Source of variation	Degrees of freedom	Sum of squares	Mean square	F
Between groups	3	0.36	0.12	1.51
Within groups	36	2.8542	0.0793	
Totals	39	3.2142		

Table 9.2 outlines what is called an *analysis of variance summary table,* a common way of summarizing the computations associated with ANOVA. Table 9.3 illustrates the summary table using the data in Table 9.1.

The computations outlined in Box 9.1 are convenient for describing certain conceptual details covered in Section 9.1.1. However, an alternative method for computing *F* is a bit faster and easier and allows unequal sample sizes. The details are summarized in Box 9.2.

EXAMPLE. The computations in Box 9.2 are illustrated with the following data.

Group 1: 7, 9, 8, 12, 8, 7, 4, 10, 9, 6
Group 2: 10, 13, 9, 11, 5, 9, 8, 10, 8, 7
Group 3: 12, 11, 15, 7, 14, 10, 12, 12, 13, 14

We see that

$$A = 7^2 + 9^2 + \cdots + 14^2 = 3026,$$

$$B = 7 + 9 + \cdots + 14 = 290,$$

$$C = \frac{(7 + 9 + \cdots + 6)^2}{10} + \frac{(10 + 13 + \cdots + 7)^2}{10}$$

$$+ \frac{(13 + 11 + \cdots + 14)^2}{10} = 2890,$$

Continued

EXAMPLE. (*Continued*)

$$N = 10 + 10 + 10 = 30,$$

$$SST = 3026 - \frac{290^2}{30} = 222.67,$$

$$SSBG = 2890 - \frac{290^2}{30} = 86.67,$$

$$SSWG = 3026 - 2890 = 136,$$

$$MSBG = \frac{86.67}{3 - 1} = 43.335,$$

$$MSWG = \frac{136}{30 - 3} = 5.03.$$

So

$$F = \frac{43.335}{5.03} = 8.615.$$

■

BOX 9.2 Summary of the ANOVA *F*-Test With or Without

Equal Sample Sizes

Notation

X_{ij} refers to the *i*th observation from the *j*th group, $i = 1, \ldots, n_j$; $j = 1, \ldots, J$. (There are n_j observations randomly sampled from the *j*th group.)

Computations

$$A = \sum \sum X_{ij}^2$$

(In words, square each value, add the results, and call it *A*.)

$$B = \sum \sum X_{ij}$$

(In words, sum all the observations and call it *B*.)

$$C = \sum_{j=1}^{J} \frac{1}{n_j} \left(\sum_{i=1}^{n_j} X_{ij} \right)^2$$

Continued

BOX 9.2 (*Continued*) (Sum the observations for each group, square the result, divide by the sample size, and add the results corresponding to each group.)

$$N = \sum n_j$$

$$\text{SST} = A - \frac{B^2}{N}$$

$$\text{SSBG} = C - \frac{B^2}{N}$$

$$\text{SSWG} = \text{SST} - \text{SSBG} = A - C$$

$$\nu_1 = J - 1$$

$$\nu_2 = N - J$$

$$\text{MSBG} = \frac{\text{SSBG}}{\nu_1}$$

$$\text{MSWG} = \frac{\text{SSWG}}{\nu_2}$$

Test Statistic

$$F = \frac{\text{MSBG}}{\text{MSWG}}.$$

Decision Rule

Reject H_0 if $F \geq f$, the $1 - \alpha$ quantile of an F-distribution with $\nu_1 = J - 1$ and $\nu_2 = N - J$ degrees of freedom.

The degrees of freedom are $\nu_1 = 3 - 1 = 2$ and $\nu_2 = 30 - 3 = 27$. With $\alpha = .01$ we see, from Table 8 in Appendix B, that the critical value is $f = 4.6$. Because $22.13 > 4.6$, reject the hypothesis of equal means.

9.1.1 Some Conceptual Details

Some of the technical and conceptual details of the ANOVA F-test are important because they play a role in commonly used or recommended methods for summarizing how groups compare. It is also important to gain some sense of the strategy behind the ANOVA F-test so that its performance, relative to other methods you might use, is understood. Again, primarily for convenience, equal sample sizes are assumed. That is, $n_1 = n_2 = \cdots n_J = n$.

First look at the expression for SSBG in Box 9.1 and notice that the sum

$$\sum_{j=1}^{J} (\bar{X}_j - \bar{X}_G)^2,$$

is similar to the expression for the sample variance s^2 given in Chapter 3. Also recall that the variance of the jth sample mean is σ_j^2/n_i and because we assume all J groups have equal variances, each sample mean has variance σ_p^2/n. Moreover, if the null hypothesis of equal means is true, then

$$\frac{1}{J-1} \sum_{j=1}^{J} (\bar{X}_j - \bar{X}_G)^2,$$

is an unbiased estimate of the variance of the sample means. That is, by assumption, $\bar{X}_1, \ldots, \bar{X}_J$ each have variance σ_p^2/n and so the sample variance of these sample means estimates σ_p^2/n. In symbols,

$$E\left(\frac{1}{J-1} \sum_{j=1}^{J} (\bar{X}_j - \bar{X}_G)^2\right) = \frac{\sigma_p^2}{n}.$$

Using our rules of expected values covered in Chapter 2, it follows that

$$E(\text{MSBG}) = \sigma_p^2.$$

That is, if the null hypothesis of equal means is true, MSBG is an unbiased estimate of the assumed common variance.

Now consider the average of the sample variances:

$$\frac{1}{J} \sum_{j=1}^{J} s_j^2.$$

Assuming homoscedasticity, each of these sample variances is an unbiased estimate of σ_p^2, so in particular the average of these sample variances is an unbiased estimate of the assumed common variance. That is,

$$E(\text{MSWG}) = \sigma_p^2,$$

regardless of whether the null hypothesis of equal means is true or false. So when the null hypothesis is true, MSBG and MSWG should have similar values. In terms of the test statistic, $F = \text{MSBG}/\text{MSWG}$, if the null hypothesis is true, then F will tend to be close to 1. However, when the null hypothesis is false, MSBG will tend to be larger than MSWG. In fact, the more unequal the population means, the larger MSBG will be on average. In particular, it can be shown that

$$E(\text{MSBG}) = \sigma_p^2 + \frac{n \sum (\mu_j - \bar{\mu})^2}{J-1},$$

where $\bar{\mu}$ is the average of the population means being compared. That is,

$$\bar{\mu} = \frac{1}{J}(\mu_1 + \mu_2 + \cdots + \mu_J),$$

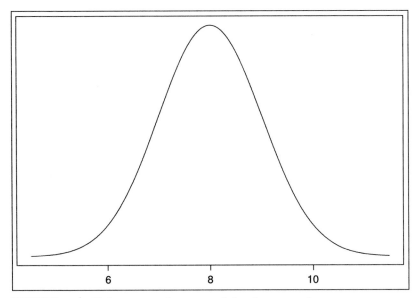

FIGURE 9.2   If three groups have normal distributions with a common variance and a common mean of 8, then the distributions will be identical, as shown. In this case, the sample means ($\bar{X}_1$, $\bar{X}_2$, and $\bar{X}_3$) will tend to be close to 8.

which is called the population *grand mean*. Consequently, if F is larger than what we would expect when the null hypothesis is true, the hypothesis of equal means should be rejected.

Now we can understand why the method in this section is called *analysis of variance* even though the goal is to test the hypothesis that the means have a common value. In essence, the ANOVA F-test compares the variation among the sample means to the variation within the groups. The variation among the sample means is measured by MSBG, and the variance for each of the groups is estimated with MSWG.

A graphical description of the strategy behind the ANOVA F-test might help. First consider the situation where the null hypothesis is true. Because normality and equal variances are assumed, all three groups will have identical distributions, as shown in Figure 9.2. In contrast, Figure 9.3 shows what the distributions might look like when the null hypothesis is false. The population means are 2, 4, and 8. The distributions are spread out, versus the situation in Figure 9.2, meaning that the sample means associated with the three groups are likely to be more spread out. In Figure 9.2, all of the sample means will tend to be close to 8, but in Figure 9.3 the sample mean for the first group will tend to be close to 2, the sample mean of the second group will tend to be close to 8, and the sample mean of the third group will tend to be close to 4. This means that if MSBG is large enough, which reflects the variation among the sample means, the hypothesis of equal means should be rejected.

Box 9.3 summarizes some properties of the ANOVA F-test that are sometimes used in applied work, particularly when measuring effect size (the extent to which the groups differ).

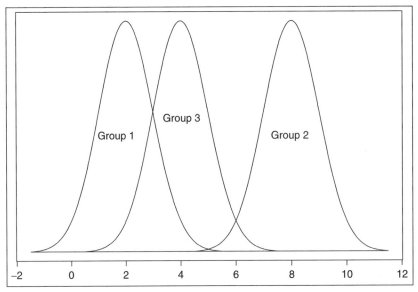

FIGURE 9.3 If the null hypothesis of equal means is false, the distributions of the three groups, still assuming normality and a common variance, might appear as shown. Compared to Figure 9.2, the sample means will tend to be more spread out.

BOX 9.3 Properties of the ANOVA Method When There Are

Equal Sample Sizes

Let $\bar{\mu}$ be the average of the population means being compared. That is,

$$\bar{\mu} = \frac{1}{J}(\mu_1 + \mu_2 + \cdots + \mu_J).$$

Then

$$E(\text{MSBG}) = \sigma_p^2 + \frac{n\sum(\mu_j - \bar{\mu})^2}{J - 1}.$$

When the null hypothesis of equal means is true,

$$\sum(\mu_j - \bar{\mu})^2 = 0,$$

so

$$E(\text{MSBG}) = \sigma_p^2.$$

In words, if the null hypothesis is true, MSBG estimates the common variance, σ_p^2; but if the null hypothesis is false, MSBG estimates a quantity that is larger than σ_p^2.

Continued

BOX 9.3 (*Continued*)
Regardless of whether the null hypothesis is true,

$$E(\text{MSWG}) = \sigma_p^2.$$

This means that if the null hypothesis is true, MSBG and MSWG estimate the same quantity and therefore will tend to have similar values. If, however, the null hypothesis is false, MSBG estimates a larger quantity than MSWG, so if $F = \text{MSBG/MSWG}$ is sufficiently large, reject the hypothesis of equal means.

EXAMPLE. *If* the null hypothesis is true, there are two ways of estimating the assumed common variance based on the results in Table 9.3. The first is with MSBG. So for the data in Table 9.1, the estimate is MSBG = 0.12. That is, if the hypothesis of equal means is true and all J groups have variances equal to σ_p^2, then an estimate of σ_p^2 is 0.12. MSWG provides a second estimate; it estimates the assumed common variance regardless of whether the means are equal, as indicated in Box 9.3. In the illustration, MSWG = 0.0793. ■

EXAMPLE. For the data in Table 9.1, is it reasonable to estimate the common variance with MSBG, assuming that the groups do indeed have equal variances? MSBG is a reasonable estimate of the assumed common variance if the null hypothesis of equal means is true. As indicated in Table 9.3, F is not large enough to reject the hypothesis of equal means, but this does not necessarily imply that it is reasonable to assume the means are equal. It might be that the means are not equal but that power was not high enough to detect this. Consequently, there is not convincing evidence that MSBG provides a reasonable estimate of the common variance. ■

We conclude this section by describing the conventional ANOVA model. Many books use the notation

$$\alpha_j = \mu_j - \bar{\mu}$$

to represent the difference between the mean of the jth group and the *grand mean*,

$$\bar{\mu} = \frac{1}{J} \sum \mu_j.$$

Another common notation is

$$\epsilon_{ij} = X_{ij} - \bar{\mu} - \alpha_j$$
$$= X_{ij} - \mu_j.$$

In words, ϵ_{ij} is the difference between the ith observation in the jth group and the corresponding mean, μ_j. That is, ϵ_{ij} is an error term: It measures the extent to which X_{ij} differs from the population mean of the jth group. Rearranging terms, this last equation becomes

$$X_{ij} = \bar{\mu} + \alpha_j + \epsilon_{ij}. \qquad (9.3)$$

The ANOVA F-test is obtained by assuming that ϵ_{ij} has a normal distribution with mean 0 and variance σ_p^2.

9.2 Dealing with Unequal Variances

Improvements on the ANOVA F-test have been phenomenal, particularly in recent years. These improvements deal with problems associated with sampling from non-normal distributions and problems due to having unequal variances. Violating the equal variance assumption associated with the F-test can result in poor power and undesirable power properties, even when sampling from a normal distribution. We saw in Chapter 8 that even if the population variances are unequal but the sample sizes are equal, Student's T controls Type I errors fairly well when sampling from normal distributions except when sample sizes are very small. In contrast, problems arise when using the ANOVA F-statistic even when the sample sizes are equal. That is, in a very real sense, as the number of groups increases, practical problems with unequal variances increase, even under normality.

Another serious problem is that even if distributions are normal but have unequal variances, the power of the F-test can be low relative to more modern methods. Moreover, it can be biased, meaning that the probability of rejecting can actually drop as the population means become unequal, and there are concerns about its ability to control the probability of a Type I error. For example, imagine you want to compare four groups, the null hypothesis of equal means is true, and you want the probability of a Type I error to be $\alpha = .05$. Situations arise where the actual probability of rejecting exceeds .27, due to comparing normal distributions that have unequal variances. When comparing six groups, the probability of a Type I error can exceed .3. That is, the actual probability of Type I error is substantially higher than the stated level of .05.

A reasonable suggestion for trying to salvage the F-test is first to test the hypothesis of equal variances and, if not significant, to assume equal variances and use F. This strategy is known to fail (Markowski & Markowski, 1990; Moser, Stevens, & Watts, 1989; Wilcox, Charlin, & Thompson, 1986). As in Chapter 8, the basic problem is that tests for equal variances do not have enough power to detect situations where violating the assumption of equal variances causes practical problems. All indications are that the F-test should be abandoned in favor of some more modern technique. The two main reasons for including the F-test here are: (1) it is commonly used because its practical problems are relatively unknown, and (2) it provides a relatively simple first step toward describing effective methods for comparing groups.

As in Chapter 8, some authorities would argue that it is virtually impossible for the null hypothesis of equal means to be true and simultaneously to have unequal variances. If we accept this argument, the probability of a Type I error is no longer an issue. But this does not salvage the F-test, because problems controlling the probability of a Type I error when variances differ reflects problems with bias.

9.2.1 Welch's Test

Many methods have been proposed for testing the equality of J means without assuming equal variances (e.g., S. Chen & Chen, 1998; Mehrotra, 1997; James, 1951; Krutchkoff, 1988; Alexander & McGovern, 1994; Fisher, 1935, 1941; Cochran & Cox, 1950; Wald, 1955; Asiribo & Gurland, 1989; Scariano & Davenport, 1986; Matuszewski & Sotres, 1986; Pagurova, 1986; Weerahandi, 1995). Unfortunately, all of these methods, plus many others, have been found to have serious practical problems (e.g., Keselman, Wilcox, Taylor, & Kowalchuk, 2000; Keselman & Wilcox, 1999). One of these problems is poor control over the probability of a Type I error and another is low power under nonnormality, a problem that cannot be escaped when using sample means. The method described here performs reasonably well under normality and heteroscedasticity and it forms the basis of a technique that deals with nonnormality. The method is due to Welch (1951) and it generally outperforms the F-test. The computational details are described in Box 9.4. (The S-PLUS function t1way, described in Section 9.4.1, contains Welch's test as a special case.)

BOX 9.4 Computations for Welch's Method

Goal
Without assuming equal variances, test $H_0 : \mu_1 = \mu_2 = \cdots = \mu_J$, the hypothesis that J independent groups have equal means.

Computations
Let

$$w_1 = \frac{n_1}{s_1^2}, \, w_2 = \frac{n_2}{s_2^2}, \ldots, w_J = \frac{n_J}{s_J^2}.$$

Next, compute

$$U = \sum w_j$$

$$\tilde{X} = \frac{1}{U} \sum w_j \bar{X}_j$$

$$A = \frac{1}{J-1} \sum w_j (\bar{X}_j - \tilde{X})^2$$

$$B = \frac{2(J-2)}{J^2 - 1} \sum \frac{(1 - \frac{w_j}{U})^2}{n_j - 1}$$

Continued

BOX 9.4 (*Continued*)

$$F_w = \frac{A}{1 + B}.$$

When the null hypothesis is true, F_w has, approximately, an F-distribution with

$$\nu_1 = J - 1$$

and

$$\nu_2 = \left[\frac{3}{J^2 - 1} \sum \frac{(1 - w_j/U)^2}{n_j - 1} \right]^{-1}$$

degrees of freedom.

Decision Rule
Reject H_0 if $F_w \geq f$, where f is the $1 - \alpha$ quantile of the F-distribution with ν_1 and ν_2 degrees of freedom.

EXAMPLE. Welch's test is illustrated with the schizophrenia data in Table 9.1, which lists the sample means and variances. Referring to Table 9.1 and Box 9.4, we see that

$$w_1 = \frac{10}{0.1676608} = 59.6, \qquad w_2 = \frac{10}{0.032679} = 306.0,$$

$$w_3 = \frac{10}{0.0600529} = 166.5, \qquad w_4 = \frac{10}{0.0567414} = 176.2.$$

Therefore,

$$U = 59.6 + 306.0 + 166.5 + 176.2 = 708.3,$$

$$\tilde{X} = \frac{1}{708.3}\{59.6(0.276039) + 306(0.195615) + 166.5(0.149543)$$

$$+ 176.2(0.399699\}$$

$$= .242,$$

$$A = \frac{1}{4 - 1}\{59.6(0.276039 - 0.242)^2 + 306(0.195615 - 0.242)^2$$

$$+ 166.5(0.149543 - 0.242)^2 + 176.2(0.399699 - 0.242)^2\}$$

$$= 2.18,$$

Continued

EXAMPLE. (*Continued*)

$$B = \frac{2(4-2)}{4^2-1}\left\{\frac{(1-59.6/708.3)^2}{9} + \frac{(1-306.0/708.3)^2}{9}\right.$$

$$\left. + \frac{(1-166.5/708.3)^2}{9} + \frac{(1-176.2/708.3)^2}{9}\right\}$$

$$= 0.0685,$$

$$F_w = \frac{2.18}{1+0.0685} = 2.04.$$

The degrees of freedom are

$$\nu_1 = 4 - 1 = 3$$

and

$$\nu_2 = \left[\frac{3}{4^2-1}(0.256)\right]^{-1}$$

$$= \frac{1}{.0512}$$

$$= 19.5$$

If you want the probability of a Type I error to be $\alpha = .05$, then, referring to Table 6 in Appendix B, the critical value is approximately 3.1. Because $F_w = 2.04$, which is less than 3.1, you fail to reject the hypothesis of equal means. ■

For the data in Table 9.1, both Welch's test and the ANOVA F-test fail to reject the hypothesis of equal means. It is stressed, however, that in applied work, situations arise where Welch's test rejects and the F-test does not. That is, in applied work, it can matter which test you use. The following example illustrates this point.

EXAMPLE. Consider the following data.

Group 1: 53 2 34 6 7 89 9 12
Group 2: 7 34 5 12 32 36 21 22
Group 3: 5 3 7 6 5 8 4 3

The ANOVA F-test yields $F = 2.7$, with a critical value of 3.24, so you do not reject. (The significance level is .09.) In contrast, Welch's test yields $W = 8$, with a critical value of 4.2, so now you reject. (The significance level is .009.) ■

9.3 Judging Sample Sizes and Controlling Power When Comparing Means

When you fail to reject the hypothesis of equal means, this might be because there are small or no differences among the groups being compared. Another possibility is that there is an important difference but that you failed to detect this due to low power. Power might be low because the sample sizes are small relative to the variances. Section 5.4.3 describes Stein's method, which might be used to help distinguish between these two possibilities. That is, given some data, you can determine how large the sample sizes must be to achieve a desired amount of power. If the sample sizes you used are small compared to what is needed to achieve high power, you have empirical evidence that the null hypothesis should not be accepted. This section describes an extension of Stein's method, called the Bishop–Dudewicz ANOVA, for judging the sample sizes when testing the hypothesis of equal means. Normality is assumed, but unlike the ANOVA F-test, homoscedasticity is not required. In fact, under normality, the method provides exact control over both Type I error probabilities and power.

Imagine that you want power to be $1 - \beta$ for some given value of

$$\delta = \sum (\mu_j - \bar{\mu})^2.$$

In case it helps, it is noted that if

$$\mu_1 = \cdots = \mu_{J-1} \qquad \text{but} \qquad \mu_J - \mu_{J-1} = a,$$

then

$$\delta = \frac{a^2(J-1)}{J}.$$

That is, if $J - 1$ of the population means are equal but the other population mean exceeds all of the other means by a, then we have a simple method for determining δ that might help when trying to specify what δ should be. For example, if for three groups you want power to be high when $\mu_1 = \mu_2 = 5$ but $\mu_3 = 6$, then $a = 1$ and $\delta = 2/3$. Given δ, α, $1 - \beta$, and n_j observations randomly sampled from the jth group, Box 9.5 shows how to determine N_j, the number of observations needed to achieve the desired amount of power.

BOX 9.5 Judging Sample Sizes

Given α, δ, and n_j observations randomly sampled from the jth group, determine N_j, the number of observations for the jth group needed to achieve power $1 - \beta$.

Continued

BOX 9.5 (*Continued*)

Let z be the $1 - \beta$ quantile of the standard normal random distribution. For the jth group, let $v_j = n_j - 1$. Compute

$$v = \frac{J}{\sum \dfrac{1}{v_j - 2}} + 2,$$

$$A = \frac{(J-1)v}{v-2}, \qquad B = \frac{v^2}{J} \times \frac{J-1}{v-2},$$

$$C = \frac{3(J-1)}{v-4}, \qquad D = \frac{J^2 - 2J + 3}{v-2},$$

$$E = B(C+D),$$

$$M = \frac{4E - 2A^2}{E - A^2 - 2A},$$

$$L = \frac{A(M-2)}{M},$$

$$c = Lf,$$

where f is the $1 - \alpha$ quantile of an F-distribution with L and M degrees of freedom,

$$b = \frac{(v-2)c}{v},$$

$$A_1 = \frac{1}{2}\left\{ \sqrt{2z} + \sqrt{2z^2 + 4(2b - J + 2)} \right\},$$

$$B_1 = A_1^2 - b,$$

$$d = \frac{v-2}{v} \times \frac{\delta}{B_1}.$$

Then

$$N_j = \max\left\{ n_j + 1, \left[\frac{s_j^2}{d}\right] + 1 \right\}. \tag{9.4}$$

For technical reasons, the number of observations needed for the jth group, N_j, cannot be smaller than $n_j + 1$. (The notation $[s_j^2/d]$ means you compute s_j^2/d and then round down to the nearest integer.)

EXAMPLE. Suppose you have three groups and you want power to be $1 - \beta = .8$ if one of the three groups has a population mean 2.74 larger than

Continued

EXAMPLE. (*Continued*) the other two. That is, $a = 2.74$, so $\delta = 5$. For illustrative purposes, assume $\alpha = .05$ and that the sample sizes are $n_1 = n_2 = n_3 = 10$, so $v_1 = v_2 = v_3 = 9$, and

$$v = \frac{3}{\frac{1}{7} + \frac{1}{7} + \frac{1}{7}} + 2 = 9.$$

The tedious calculations eventually yield a critical value of $c = 8.15$ and $d = .382$. If the sample variance for the first group is $s_1^2 = 5.43$, then N_1 is equal to either $n_1 + 1 = 11$ or $[s_1^2/d] + 1 = [5.43/.382] + 1 = 15$, whichever is larger. In this particular case $N_1 = 15$, suggesting that the original sample size of 10 is not quite satisfactory. If $N_1 = n_1 + 1$, this suggests that the available sample sizes are reasonably adequate. If the second group has a sample variance of $s_2^2 = 10$, then $N_2 = 27$, and if $s_3^2 = 20$, $N_3 = 53$. So for group 3, you have only 10 observations and about five times as many observations are required for the specified amount of power.   ■

The method just described indicates that an additional $N_j - n_j$ observations are needed for the jth group to achieve the desired amount of power. It is noted that if these additional observations can be obtained, the hypothesis of equal means can be tested without assuming homoscedasticity. More precisely, assuming normality, the probability of a Type I error will be exactly α, and power will be at least $1 - \beta$ using the Bishop–Dudewicz ANOVA method outlined in Box 9.6.

BOX 9.6 Bishop–Dudewicz ANOVA

Goal
Test $H_0 : \mu_1 = \cdots = \mu_J$ such that power is $1 - \beta$ and the probability of a Type I error is α.

Assumptions
Normality and random sampling are assumed. It is further assumed that initially you have n_j observations randomly sampled from the jth group, labeled X_{ij}, $i = 1, \ldots, n_j$, you have computed N_j as described in Box 9.5, and that you have then randomly sampled $N_j - n_j$ additional observations from the jth group, which are labeled X_{ij}, $i = n_j + 1, \ldots, N_j$. For the jth group, compute

$$T_j = \sum_{i=1}^{n_j} X_{ij},$$

$$U_j = \sum_{i=n_j+1}^{N_j} X_{ij}$$

Continued

BOX 9.6 (*Continued*)

$$b_j = \frac{1}{N_j}\left(1 + \sqrt{\frac{n_j(N_jd - s_j^2)}{(N_j - n_j)s_j^2}}\right),$$

$$\tilde{X}_j = \frac{T_j\{1 - (N_j - n_j)b_j\}}{n_j} + b_jU_j.$$

The test statistic is

$$\tilde{F} = \frac{1}{d}\sum(\tilde{X}_j - \tilde{X})^2,$$

where

$$\tilde{X} = \frac{1}{J}\sum\tilde{X}_j.$$

Decision Rule
  Reject H_0 if $\tilde{F} \geq c$, where c is the critical value given in Box 9.5.

9.3.1 S-PLUS Functions bdanova1 and bdanova2

The S-PLUS function

  bdanova1 (x, alpha = 0.05, power = 0.9, delta = NA)

performs the calculations in Box 9.5 and returns the number of observations required to achieve the specified amount of power. The argument power indicates how much power you want and defaults to .9. The argument delta corresponds to δ, and the data are stored in the S-PLUS variable x, which can be an n-by-J matrix or it can have list mode. (In the latter case it is assumed that x[[1]] contains the data for group 1, x[[2]] contains the data for group 2, and so on.)
  The S-PLUS function

  bdanova2 (x1, x2, alpha = 0.05, power = 0.9, delta = NA)

performs the second-stage analysis once the additional observations are obtained, as described in Box 9.6. Here x1 and x2 contain the first-stage and second-stage data, respectively.

9.4 Trimmed Means

Welch's heteroscedastic method for comparing means can be extended to trimmed means. That is, the goal is to test

$$H_0 : \mu_{t1} = \mu_{t2} = \cdots = \mu_{tJ},$$

the hypothesis that J independent groups have a common population trimmed mean. As in Chapter 8, trimming can greatly reduce practical problems (low power, poor control over the probability of a Type I error, and bias) associated with methods for comparing means.

Compute

$$d_j = \frac{(n_j - 1)s_{wj}^2}{h_j \times (h_j - 1)},$$

$$w_j = \frac{1}{d_j},$$

$$U = \sum w_j,$$

$$\tilde{X} = \frac{1}{U} \sum w_j \bar{X}_{tj},$$

$$A = \frac{1}{J - 1} \sum w_j (\bar{X}_{tj} - \tilde{X})^2,$$

$$B = \frac{2(J - 2)}{J^2 - 1} \sum \frac{(1 - w_j/U)^2}{h_j - 1},$$

$$F_t = \frac{A}{1 + B}. \tag{9.5}$$

When the null hypothesis is true, F_t has, approximately, an F-distribution with

$$\nu_1 = J - 1$$

$$\nu_2 = \left[\frac{3}{J^2 - 1} \sum \frac{(1 - w_j/U)^2}{h_j - 1} \right]^{-1}$$

degrees of freedom. (For $J > 2$, the expression for ν_2 reduces to $2(J - 2)/3B$.)

DECISION RULE: Reject the hypothesis of equal population trimmed means if $F_t \geq f$, the $1 - \alpha$ quantile of an F-distribution with ν_1 and ν_2 degrees of freedom.

9.4.1 S-PLUS Function t1way

The S-PLUS function

$$\text{t1way}(x, tr=.2, grp=NA)$$

tests the hypothesis of equal trimmed means using the method just described. The argument x can have list mode or it can be a matrix. In the former case, the data for group 1 are stored in the S-PLUS variable x[[1]], group 2 is stored in x[[2]], and so on. In the latter case, x is an n-by-J matrix, where column 1 contains the data for group 1, column 2 contains the data for group 2, and so forth. The argument tr indicates the amount of trimming; and when tr=0, this function performs

Welch's method for means described in Section 9.2.1. The argument grp allows you to compare a selected subset of the groups. By default all groups are used. If you set grp=c(1,3,4), then the trimmed means for groups 1, 3, and 4 will be compared, with the remaining data ignored. The function returns the value of the test statistic and the corresponding significance level (so specifying a value for α is not necessary).

If you used the S-PLUS function selby (described in Section 1.1.6) to store your data in list mode in the S-PLUS variable dat, note that you would use the S-PLUS variable data$x to analyze your data. For example, to compare the groups based on 20% trimming, use the command t1way(dat$x).

EXAMPLE. For the data in Table 9.1 and using the default amount of trimming, a portion of the output from t1way is

```
$TEST:
[1] 5.059361

$nu1:
[1] 3

$nu2:
[1] 10.82531

$siglevel:
[1] 0.01963949
```

This says that the test statistic F_t has a value of 5.06 and the significance level is .0194. So in particular you would reject the hypothesis of equal trimmed means with $\alpha = .05$ or even .02. In contrast, as previously indicated, if we compare means with Welch's method or the ANOVA F-test, we fail to reject with $\alpha = .05$, illustrating that the choice of method can alter the conclusions reached. Setting the argument tr to zero, t1way reports the results of Welch's test to be

```
$TEST:
[1] 2.038348

$nu1:
[1] 3

$nu2:
[1] 19.47356

$siglevel:
[1] 0.1417441
```

So switching from means to 20% trimmed means, the significance level drops from .14 to about .02.   ■

EXAMPLE. It is common to find situations where we get a lower significance level using trimmed means than using means. However, the reverse situation can and does occur. This point is illustrated with data taken from Le (1994), where the goal is to compare the testosterone levels of four groups of male smokers: heavy smokers (group 1), light smokers (group 2), former smokers (group 3), and nonsmokers (group 4). The data are:

G1	G2	G3	G4
.29	.82	.36	.32
.53	.37	.93	.43
.33	.77	.40	.99
.34	.42	.86	.95
.52	.74	.85	.92
.50	.44	.51	.56
.49	.48	.76	.87
.47	.51	.58	.64
.40	.61	.73	.78
.45	.60	.65	.72

The significance level using Welch's method is .0017. Using 20% trimmed means instead, the significance level is .029. One reason this is not surprising is that a boxplot for each group reveals no outliers, and it can be seen that the estimated standard errors for the means are smaller than the corresponding estimated standard errors when 20% trimming is used instead. However, a boxplot for the first group suggests that the data are skewed, which might be having an effect on the significance level of Welch's test beyond any differences among the means.   ■

9.4.2 Comparing Groups Based on Medians

The median has a relatively large standard error under normality and more generally when sampling from a light-tailed distribution, but with a sufficiently heavy-tailed distribution its standard error can be relatively small. If there is a specific interest in comparing medians, the general method for trimmed means described in this section is not recommended. That is, it is not recommended that you set tr=.5 when using the S-PLUS function t1way in Section 9.4.1 because this results in using a relatively poor estimate of the standard error of the median. A better approach is to estimate the standard error of each median with the McKean–Schrader method in Chapter 4 and then to use a slight modification of the method for trimmed means. In particular, let S_j^2 be the McKean–Schrader estimate of the squared standard error of M_j, the sample median corresponding to the jth group ($j = 1, \ldots, J$). Let

$$w_j = \frac{1}{S_j^2},$$

$$U = \sum w_j,$$

$$\tilde{M} = \frac{1}{U} \sum w_j M_j,$$

$$A = \frac{1}{J-1} \sum w_j (M_j - \tilde{M})^2,$$

$$B = \frac{2(J-2)}{J^2 - 1} \sum \frac{(1 - w_j/U)^2}{n_j - 1},$$

$$F_m = \frac{A}{1+B}. \tag{9.6}$$

DECISION RULE: Reject the hypothesis of equal population medians if $F_m \geq f$, the $1 - \alpha$ quantile of an F distribution with $\nu_1 = J - 1$ and $\nu_2 = \infty$ degrees of freedom.

9.4.3 S-PLUS Function med1way

The hypothesis of equal population medians can be tested with the S-PLUS function

med1way(x,grp),

where the argument grp can be used to analyze a subset of the groups if desired. (See Section 9.4.1.) The function returns the value of the test statistic, F_m, and the significance level.

9.5 Bootstrap Methods

This section describes how the percentile and bootstrap-t methods described in Chapter 8 can be extended to testing hypotheses based on some robust measure of location, such as MOM or trimmed means. With small sample sizes, all indications are that some type of bootstrap method has a practical advantage over the method for trimmed means in Section 9.4, assuming that we want a method that is sensitive to differences among the trimmed means only. This is particularly true for the special case where means are to be compared. With sufficiently large sample sizes, the method in Section 9.4 can be used in place of the bootstrap when comparing trimmed means, but it remains unclear how large the sample sizes must be. An argument for the method in Section 9.4 over any bootstrap techniques we might use is that if the former method rejects, we can be reasonably certain that the groups have different distributions. A possible concern, however, is that nonbootstrap methods are more sensitive to bias problems.

9.5.1 A Bootstrap-t Method

The bootstrap-t method in Chapter 8 can be extended to the problem of testing $H_0 : \mu_{t1} = \mu_{t2} = \cdots = \mu_{tJ}$, the hypothesis of equal trimmed means. The strategy is to use the available data to estimate an appropriate critical value for the test statistic,

F_t, described in Section 9.4. First, for the jth group, set

$$Y_{ij} = X_{ij} - \bar{X}_{tj}.$$

That is, for the jth group, subtract the sample trimmed from each of the observed values. Next, for the jth group, generate a bootstrap sample of size n_j from the Y_{ij} values, which we denote by Y_{ij}^*, $i = 1, \ldots, n_j$; $j = 1, \ldots, J$. So, in effect, the Y_{ij}^* values represent a random sample from distributions all of which have zero trimmed means. That is, in the bootstrap world, when working with the Y_{ij} values, the null hypothesis of equal trimmed means is true. Said another way, the observations are shifted so that each has a trimmed mean of zero, with the goal of empirically determining an appropriate critical value. The value of the test statistic F_t given by Equation (9.5) and based on the Y_{ij}^* is labeled F_t^*. The strategy is to use a collection of F_t^* values to estimate the distribution of F_t, the test statistic based on the original observations, when the null hypothesis is true. If we can do this reasonably well, then in particular we can determine an appropriate critical value.

To estimate the critical value we repeatedly generate bootstrap samples in the manner just described, each time computing the test statistic F_t^* based on the Y_{ij}^* values. Doing this B times yields $F_{t1}^*, \ldots, F_{tB}^*$. Next, put these B values in ascending order, yielding $F_{t(1)}^* \leq \cdots \leq F_{t(B)}^*$, and let u be the value of $(1 - \alpha)B$ rounded to the nearest integer. Then the hypothesis of equal trimmed means is rejected if

$$F_t \geq F_{t(u)}^*. \tag{9.7}$$

9.5.2 S-PLUS Function t1waybt

The S-PLUS function

$$(x, tr = 0.2, alpha = 0.05, grp = NA, nboot = 599)$$

performs the percentile-t bootstrap method for trimmed means that was just described. The argument x is any S-PLUS variable containing data that are stored in list mode or in a matrix. In the first case x[[1]] contains the data for group 1, x[[2]] contains the data for group 2, and so on. If x is a matrix, column 1 contains the data for group 1, column 2 the data for group 2, and so forth. The argument grp can be used to analyze a subset of the groups. For example, grp=c(2,4,5) would compare groups 2, 4, and 5 only. As usual, alpha is α and nboot is B, the number of bootstrap samples to be used.

EXAMPLE. If the data in Table 9.1 are stored in the S-PLUS variable skin, the command

$$t1waybt(skin, tr=0)$$

tests the hypothesis of equal means and returns

```
$test:
[1] 2.04
```

Continued

EXAMPLE. (*Continued*)

`$crit:`
`[1] 4.65336`

Because the test statistic $F_t = 2.04$ is less than the bootstrap estimate of the critical value, 4.65, you fail to reject. Note that in Section 9.2.1, the estimated critical value, assuming normality, was 3.1.   ■

9.5.3 Two Percentile Bootstrap Methods

There are many variations of the percentile bootstrap method that can be used to test the hypothesis that J groups have a common measure of location, but only two are described here. The first is related to a test statistic mentioned by Schrader and Hettmansperger (1980) and studied by He, Simpson, and Portnoy (1990). Let θ be any population measure of location, such as the 20% trimmed mean (μ_t) or a median, and let $\hat{\theta}_j$ be the estimate of θ based on data from the jth group ($j = 1, \ldots, J$). The test statistic is

$$H = \frac{1}{N} \sum n_j (\hat{\theta}_j - \bar{\theta})^2,$$

where $N = \sum n_j$ and

$$\bar{\theta} = \frac{1}{J} \sum \hat{\theta}_j.$$

To determine a critical value, shift the empirical distributions of each group so that the measure of location being used has a value of zero. That is, set $Y_{ij} = X_{ij} - \hat{\theta}_j$, as was done in Section 9.5.1. Then generate bootstrap samples from each group in the usual way from the Y_{ij} values and compute the test statistic based on the bootstrap samples, yielding H^*. Repeat this B times, resulting in $H_1^*, \ldots, H_B^*$, and put these B values in order, yielding $H_{(1)}^* \leq \cdots \leq H_{(B)}^*$. Then an estimate of an appropriate critical value is $H_{(u)}^*$, where $u = (1 - \alpha)B$, rounded to the nearest integer, and H_0 is rejected if $H \geq H_{(u)}^*$. (For simulation results on how this method performs when comparing M-estimators, see Wilcox, 1993b.)

The second method stems from general results derived by Liu and Singh (1997). Let

$$\delta_{jk} = \theta_j - \theta_k,$$

where for convenience it is assumed that $j < k$. That is, the δ_{jk} values represent all pairwise differences among the J groups. When working with means, for example, δ_{12} is the difference between the means of groups 1 and 2, and δ_{35} is the difference for groups 3 and 5. If all J groups have a common measure of location (i.e., $\theta_1 = \cdots = \theta_J$), then in particular

$$H_0 : \delta_{12} = \delta_{13} = \cdots = \delta_{J-1,J} = 0 \tag{9.8}$$

is true. It can be seen that the total number of δ's in Equation (9.8) is $L = (J^2 - J)/2$. For example, if $J = 3$, there are $L = 3$ values: δ_{12}, δ_{13}, and δ_{23}.

For each group, generate bootstrap samples from the *original* values and compute the measure of location of interest for each group. That is, the observations are *not* centered as was done in the previous method. Said another way, bootstrap samples are *not* generated from the Y_{ij} values but rather from the X_{ij} values. Repeat this B times. The resulting estimates of location are represented by

$$\hat{\theta}^*_{jb}(j = 1, \ldots, J; b = 1, \ldots, B)$$

and the corresponding estimates of δ are denoted by $\hat{\delta}^*_{jkb}$. (That is, $\hat{\delta}^*_{jkb} = \hat{\theta}^*_{jb} - \hat{\theta}^*_{kb}$.) The general strategy is to determine how deeply $\mathbf{0} = (0, \ldots, 0)$ is nested within the bootstrap values $\hat{\delta}^*_{jkb}$ (where $\mathbf{0}$ is a vector having length L). For the special case where only two groups are being compared, this is tantamount to determining the proportion of times $\hat{\theta}^*_{1b} > \hat{\theta}^*_{2b'}$ among all B bootstrap samples, which is how we proceeded in Section 8.8.1. But here we need special techniques for comparing more than two groups.

There remains the problem of measuring how deeply $\mathbf{0}$ is nested within the bootstrap values. Several strategies have been proposed for dealing with this problem (e.g., Liu & Singh, 1997). But in terms of Type I error probabilities and power, it remains unclear whether the choice among these methods is relevant for the problem at hand. Accordingly, only one method is described, based on a very slight modification of what is called the *Mahalanobis distance*. The details are relegated to Box 9.7, assuming familiarity with basic matrix algebra. (Appendix C summarizes the matrix algebra used in this book.)

BOX 9.7 Details About How to Test H_0 Given by Equation (9.8)

Let $\hat{\delta}_{jk} = \hat{\theta}_j - \hat{\theta}_k$ be the estimate of δ_{jk} based on the original data and let $\hat{\delta}^*_{jkb} = \hat{\theta}^*_{jb} - \hat{\theta}^*_{kb}$ based on the bth bootstrap sample ($b = 1, \ldots, B$). (It is assumed that $j < k$.) For notational convenience, we rewrite the $L = (J^2 - J)/2$ differences $\hat{\delta}_{jk}$ as $\hat{\Delta}_1, \ldots, \hat{\Delta}_L$ and the corresponding bootstrap values are denoted by $\hat{\Delta}^*_{\ell b}$ ($\ell = 1, \ldots, L$). Let

$$\bar{\Delta}^*_\ell = \frac{1}{B} \sum_{b=1}^{B} \hat{\Delta}^*_{\ell b'}$$

$$Y_{\ell b} = \hat{\Delta}^*_{\ell b} - \bar{\Delta}^*_\ell + \hat{\Delta}_\ell,$$

(so the $Y_{\ell b}$ values are the bootstrap values shifted to have mean $\hat{\Delta}_\ell$), and let

$$S_{\ell m} = \frac{1}{B - 1} \sum_{b=1}^{B} (Y_{\ell b} - \bar{Y}_\ell)(Y_{mb} - \bar{Y}_m),$$

Continued

BOX 9.7 (*Continued*)
where

$$\bar{Y}_\ell = \frac{1}{B} \sum_{b=1}^{B} Y_{\ell b}.$$

(Note that in the bootstrap world, the bootstrap population mean of $\bar{\Delta}_\ell^*$ is known and is equal to $\hat{\Delta}_\ell$.) Next, compute

$$D_b = \left(\hat{\mathbf{\Delta}}_b^* - \hat{\mathbf{\Delta}} \right) \mathrm{S}^{-1} \left(\hat{\mathbf{\Delta}}_b^* - \hat{\mathbf{\Delta}} \right)',$$

where $\hat{\mathbf{\Delta}}_b^* = (\hat{\Delta}_{1b}^*, \ldots, \hat{\Delta}_{Lb}^*)$ and $\hat{\mathbf{\Delta}} = (\hat{\Delta}_1, \ldots, \hat{\Delta}_L)$. D_b measures how closely $\hat{\mathbf{\Delta}}_b$ is located to $\hat{\mathbf{\Delta}}$. If $\mathbf{0}$ (the null vector) is relatively far from $\hat{\mathbf{\Delta}}$, reject. In particular, put the D_b values in ascending order, yielding $D_{(1)} \leq \cdots \leq D_{(B)}$, and let $u = (1 - \alpha)B$, rounded to the nearest integer.

Decision Rule
    Reject H_0 if

$$T \geq D_{(u)},$$

where

$$T = (\mathbf{0} - \hat{\mathbf{\Delta}}) \mathrm{S}^{-1} (\mathbf{0} - \hat{\mathbf{\Delta}})'.$$

Notice that with three groups ($J = 3$), $\theta_1 = \theta_2 = \theta_3$ can be true if and only if $\theta_1 = \theta_2$ and $\theta_2 = \theta_3$. So in terms of Type I errors, it suffices to test

$$H_0 : \theta_1 - \theta_2 = \theta_2 - \theta_3 = 0$$

as opposed to testing

$$H_0 : \theta_1 - \theta_2 = \theta_2 - \theta_3 = \theta_1 - \theta_3 = 0,$$

the hypothesis that all pairwise differences are zero. However, if groups differ, then rearranging the groups could alter the conclusions reached if the first of these hypotheses is tested. For example, if the groups have means 6, 4, and 2, then the difference between groups 1 and 2, as well as between 2 and 3, is 2. But the difference between groups 1 and 3 is 4, so comparing groups 1 and 3 could mean more power. That is, we might not reject when comparing group 1 to group 2 and group 2 to group 3, but we might reject if instead we compare group 1 to group 3 and group 2 to group 3. To help avoid different conclusions depending on how the groups are arranged, all pairwise differences among the groups were used in Box 9.7.

Between the two methods described in this section, it currently seems that the method in Box 9.7 is better in terms of Type I error probabilities when comparing groups based on MOM. How these two methods compare when comparing M-estimators has not been investigated as yet. The method in Box 9.7 can be used to compare trimmed means; but with a relatively small amount of trimming, it seems that the bootstrap-t method is preferable.

For the special case where the goal is to compare medians, again the method in Box 9.7 can be used. Whether it offers any practical advantages versus the method for medians in Section 9.4.2 has not been investigated.

One last comment might be helpful. Before using the percentile bootstrap methods described in this section, it is strongly recommended that the reader take into account results described in Chapter 12. There is a common convention dictating how the methods in this chapter are to be used in conjunction with those in Chapter 12, but modern insights reveal that this convention can be detrimental in some situations.

9.5.4 S-PLUS Functions b1way and pbadepth

The S-PLUS function

$$b1way(x,est=onestep,alpha=.05,nboot=599)$$

performs the first percentile bootstrap method described in the previous subsection. By default it uses an M-estimator (with Huber's Ψ). The function

$$pbadepth(x,est=mom,con=0,alpha=.05,nboot=500,op=F,allp=T,\ldots)$$

performs the other percentile bootstrap method and uses the MOM estimator by default. As usual, the argument ... can be used to reset default settings associated with the estimator being used. The argument op determines how depth is measured. By default, a Mahalanobis-type depth, outlined in Box 9.7, is used. (Setting op=T results in the minimum covariance determinant method for measuring depth, which is described in Chapter 13.) The argument allp indicates how the null hypothesis is defined. Setting allp=T, all pairwise differences are used. Setting allp=F, the function tests

$$H_0 : \theta_1 - \theta_2 = \theta_2 - \theta_3 = \cdots = \theta_{J-1} - \theta_J = 0.$$

A negative consequence of using allp=T is that in some situations, S^{-1} in Box 9.7 cannot be computed. This problem appears to be rare with $J \le 4$, but it can occur otherwise. This problem might be avoided by setting allp=F, but perhaps a better way of dealing with this problem is to use the method described in Section 12.7.3.

9.6 Random Effects Model

The ANOVA methods covered so far deal with what is called a *fixed effect* design, roughly meaning that we are interested in comparing J specific (fixed) groups. In contrast is a random effects design, where the goal is to generalize to a larger population of groups. For example, consider a study where it is suspected that the personality of the experimenter has an effect on the results. Among all the experimenters we might use, do the results vary depending on who conducts the experiment? Here, the notion of J groups corresponds to a sample of J experimenters, and for the jth experimenter we have results on n_j participants. The goal is not only to compare results among the J experimenters but to generalize to the entire population of experimenters we might use.

TABLE 9.4 Estrone Assay Measurements of a Single Blood Sample from Each of Five Postmenopausal Women

Vial	Individuals				
	P1	P2	P3	P4	P5
1	23	25	38	14	46
2	23	33	38	16	36
3	22	27	41	15	30
4	20	27	38	19	29
5	25	30	38	20	36
6	22	28	32	22	31
7	27	24	38	16	30
8	25	22	42	19	32
9	22	26	35	17	32
10	22	30	40	18	31
11	23	30	41	20	30
12	23	29	37	18	32
13	27	29	28	12	25
14	19	37	36	17	29
15	23	24	30	15	31
16	18	28	37	13	32

A study reported by Fears, Benichou, and Gail (1996) provides another illustration where 16 estrone measures (in pg/mL) from each of five postmenopausal women were taken and found to be as shown in Table 9.4. Of interest was whether the estrone levels vary among women. That is, we envision the possibility of taking many measures from each woman, but the goal is not to simply compare the five women in the study but rather to generalize to all women who might have taken part in the study.

A study by Cronbach, Gleser, Nanda, and Rajaratnam (1972, Chap. 6) provides yet another example. The Porch index of communicative ability (PICA) is a test designed for use by speech pathologists. It is intended for initial diagnosis of patients with aphasic symptoms and for measuring the change during treatment. The oral portion of the test consists of several subtests; but to keep the illustration simple, only one subtest is considered here. This is the subtest where a patient is shown an object (such as a comb) and asked how the object is used. The response by the patient is scored by a rater on a 16-point scale. A score of 6, for example, signifies a response that is "intelligible but incorrect," and a score of 11 indicates a response that is "accurate but delayed and incomplete." A concern is that one set of objects might lead to a different rating, compared to another set of objects one might use. Indeed, we can imagine a large number of potential sets of objects that might be used. To what extent do ratings differ among all of the potential sets of objects we might employ?

Let $\mu_G = E(\mu_j)$, where the expected value of μ_j is taken with respect to the process of randomly sampling a group. If all groups have a common mean, then of course no matter which J groups you happen to pick, it will be the case that

$$\mu_1 = \mu_2 = \cdots = \mu_J.$$

A more convenient way of describing the situation is to say that there is no variation among all the population means. A way of saying this in symbols is that $\sigma_\mu^2 = 0$, where

$$\sigma_\mu^2 = E(\mu_j - \mu_G)^2,$$

and again the expectation is taken with respect to the process of randomly selecting μ_j. That is, among all groups of interest, σ_μ^2 is the variance of the population means. Testing the hypothesis that all groups have the same mean is equivalent to testing

$$H_0 : \sigma_\mu^2 = 0.$$

To test H_0, the following assumptions are typically made:

1. Regardless of which group you choose, the observations within that group have a normal distribution with a common variance, σ_p^2. That is, a homogeneity of variance assumption is imposed.
2. The difference $\mu_j - \mu_G$ has a normal distribution with mean 0 and variance σ_μ^2.
3. The difference $X_{ij} - \mu_j$ is independent of the difference $\mu_j - \mu_G$.

Let MSBG and MSWG be as defined in Section 9.1, and, primarily for notational convenience, temporarily assume equal sample sizes. That is,

$$n = n_1 = \cdots = n_J.$$

Based on the assumptions just described, it can be shown that

$$E(\text{MSBG}) = n\sigma_\mu^2 + \sigma_p^2$$

and that

$$E(\text{MSWG}) = \sigma_p^2.$$

When the null hypothesis is true, $\sigma_\mu^2 = 0$ and

$$E(\text{MSBG}) = \sigma_p^2.$$

That is, when the null hypothesis is true, MSBG and MSWG estimate the same quantity, so the ratio

$$F = \frac{\text{MSBG}}{\text{MSWG}}$$

should have a value reasonably close to 1. If the null hypothesis is false, MSBG will tend to be larger than MSWG; so if F is sufficiently large, reject. It can be shown that F has an F-distribution with $J - 1$ and $N - J$ degrees of freedom when the null hypothesis is true, so reject if $F \geq f_{1-\alpha}$, where $f_{1-\alpha}$ is the $1 - \alpha$ quantile of an F-distribution with $\nu_1 = J - 1$ and $\nu_2 = N - J$ degrees of freedom. Put more simply, the computations are exactly the same as they are for the fixed effects ANOVA F-test

TABLE 9.5 Hypothetical Data Used to Illustrate a Random Effects Model

Dosage 1	Dosage 2	Dosage 3
7	3	9
0	0	2
4	7	2
4	5	7
4	5	1
7	4	8
6	5	4
2	2	4
3	1	6
7	2	1

in Section 9.1. The only difference is how the experiment is performed. Here the levels are chosen at random, whereas in Section 9.1 they are fixed.

As mentioned in Section 9.1, the fixed effects ANOVA model is often written as

$$X_{ij} = \bar{\mu} + \alpha_j + \epsilon_{ij},$$

where ϵ_{ij} has a normal distribution with mean zero and variance σ_p^2. In contrast, the random effects model is

$$X_{ij} = \mu_G + a_j + \epsilon_{ij},$$

where $a_j = \mu_j - \mu_G$. The main difference between these two models is that in the fixed effects model, α_j is an unknown *parameter*, but in the random effects model, a_j is a *random variable* that is assumed to have a normal distribution.

EXAMPLE. Suppose that for three randomly sampled dosage levels of a drug, you get the results shown in Table 9.5. To test the null hypothesis of equal means among all dosage levels you might use, compute the degrees of freedom and the F-statistic as described in Section 9.1. This yields $\nu_1 = 2$, $\nu_2 = 27$, and $F = .53$, which is not significant at the $\alpha = .05$ level. That is, among all dosage levels you might have used, you fail to detect a difference among the corresponding means. ■

9.6.1 A Measure of Effect Size

As pointed out in Chapter 8, if you test and reject the hypothesis of equal means, there remains the issue of measuring the extent to which two groups differ. As already illustrated, the significance level can be unsatisfactory. From Chapter 8, it is evident that finding an appropriate measure of effect size is a complex issue. When dealing with more than two groups, the situation is even more difficult. Measures have been proposed under the assumption of equal variances; they are far from satisfactory, but few alternative measures are available. However, measures derived under the assumption of equal variances are in common use, so it is important to discuss them here.

Suppose you randomly sample a group from among all the groups you are interested in, and then you randomly sample an individual and observe the outcome X. Let σ_X^2 be the variance of X. It can be shown that

$$\sigma_X^2 = \sigma_\mu^2 + \sigma_p^2$$

when the assumptions of the random effects model are true and where σ_p^2 is the assumed common variance among all groups we might compare. A common measure of effect size is

$$\rho_I = \frac{\sigma_\mu^2}{\sigma_\mu^2 + \sigma_p^2},$$

which is called the *intraclass correlation coefficient*. The value of ρ_I is between 0 and 1 and measures the variation among the means relative to the variation among the observations. If there is no variation among the means, in which case they have identical values, $\rho_I = 0$.

To estimate ρ_I, compute

$$n_0 = \frac{1}{J - 1}\left(N - \sum \frac{n_j^2}{N}\right),$$

where $N = \sum n_j$ is the total sample size. The usual estimate of σ_μ^2 is

$$s_u^2 = \frac{\text{MSBG} - \text{MSWG}}{n_0},$$

in which case the estimate of ρ_I is

$$r_I = \frac{s_u^2}{s_u^2 + \text{MSWG}}$$

$$= \frac{\text{MSBG} - \text{MSWG}}{\text{MSBG} + (n_0 - 1)\text{MSWG}}$$

$$= \frac{F - 1}{F + n_0 - 1}$$

For the data in Table 9.1 it was found that $F = 6.05$, $n_0 = 8$, so

$$r_I = \frac{6.05 - 1}{6.05 + 8 - 1} = .387.$$

That is, about 39% of the variation among the observations is due to the variation among the means.

Donner and Wells (1986) compared several methods for computing an approximate confidence interval for ρ_I, and their results suggest using a method derived by

Smith (1956). Smith's confidence interval is given by

$$r_I \pm z_{1-\alpha/2} V,$$

where $z_{1-\alpha/2}$ is the $1 - \alpha/2$ quantile of the standard normal distribution, read from Table 1 in Appendix B, and

$$V = \sqrt{A(B + C + D)},$$

where

$$A = \frac{2(1 - r_I)^2}{n_0^2}$$

$$B = \frac{[1 + r_I(n_0 - 1)]^2}{N - J}$$

$$C = \frac{(1 - r_I)[1 + r_I(2n_0 - 1)]}{(J - 1)}$$

$$D = \frac{r_I^2}{(J - 1)^2} \left(\sum n_j^2 - \frac{2}{N} \sum n_j^3 + \frac{1}{N^2} \left(\sum n_j^2 \right)^2 \right).$$

For equal sample sizes an exact confidence interval is available, still assuming that sampling is from normal distributions with equal variances (Searle, 1971). Let $f_{1-\alpha/2}$ be the $1 - \alpha/2$ quantile of the F-distribution with $v_1 = J - 1$ and $v_2 = N - J$ degrees of freedom. Similarly, $f_{\alpha/2}$ is the $\alpha/2$ quantile. Then an exact confidence interval for ρ_I is

$$\left(\frac{F/f_{1-\alpha/2} - 1}{n + F/f_{1-\alpha/2} - 1}, \frac{F/f_{\alpha/2} - 1}{n + F/f_{\alpha/2} - 1} \right),$$

where n is the common sample size. The tables in Appendix B give only the upper quantiles of an F-distribution, but you need the lower quantiles when computing a confidence interval for ρ_I. To determine $f_{\alpha/2, v_1, v_2}$, you reverse the degrees of freedom and look up $f_{1-\alpha/2, v_2, v_1}$, in which case

$$f_{\alpha/2, v_1, v_2} = \frac{1}{f_{1-\alpha/2, v_2, v_1}}.$$

For example, if $\alpha = .05$ and you want to determine $f_{.025}$ with $v_1 = 2$ and $v_2 = 21$ degrees of freedom, you first look up $f_{.975}$ with $v_1 = 21$ and $v_2 = 2$ degrees of freedom. The answer is 39.45. Then $f_{.025}$ with 2 and 21 degrees of freedom is the reciprocal of 39.45. That is,

$$f_{.025, 2, 21} = \frac{1}{39.45} = .025.$$

EXAMPLE. Assume that professors are rated on their level of extroversion and you want to investigate how their level of extroversion is related to student

Continued

TABLE 9.6   Students' Ratings

Group 1	Group 2	Group 3
3	4	6
5	4	7
2	3	8
4	8	6
8	7	7
4	4	9
3	2	10
9	5	9
$\bar{X}_1 = 4.75$	$\bar{X}_2 = 4.62$	$\bar{X}_3 = 7.75$

EXAMPLE. (*Continued*) evaluations of a course. Suppose you randomly sample three professors, and their student evaluations are as shown in Table 9.6. (In reality one would of course want to sample more than three professors, but the goal here is to keep the illustration simple.) To illustrate how a confidence interval for ρ_I is computed, suppose you choose $\alpha = .05$. Then $n = 8, f_{.025} = .025, f_{.975} = 4.42, F = 6.05$, and the .95 confidence interval for ρ_I is

$$\left(\frac{\frac{6.05}{4.42} - 1}{8 + \frac{6.05}{4.42} - 1}, \frac{\frac{6.05}{.025} - 1}{8 + \frac{6.05}{.025} - 1} \right) = (0.047, 0.967).$$

Hence, you can be reasonably certain that ρ_I has a value somewhere between .047 and .967. Notice that the length of the confidence interval is relatively large, since ρ_I has a value between 0 and 1. Thus, in this case, the data might be providing a relatively inaccurate estimate of the intraclass correlation. ■

In some situations you might also want a confidence interval for σ_μ^2. Methods for accomplishing this goal are available, but no details are given here. For a recent discussion of this problem, see C. Brown and Mosteller (1991).

9.6.2 A Heteroscedastic Method

One serious concern about the conventional random effects model just described is the assumption of equal variances. We have seen that violating this assumption can result in poor power and undesirable power properties in the fixed effects design, and this problem continues for the situation at hand. This section describes a method derived by Jeyaratnam and Othman (1985) for handling unequal variances. (For an alternative approach, see Westfall, 1988.) As usual, let s_j^2 be the sample variance for the jth group, let $\bar{X}_j$ be the sample mean, and let $\bar{X} = \sum \bar{X}_j/J$ be the average of the J sample means. To test $H_0 : \sigma_\mu^2 = 0$, compute

$$q_j = \frac{s_j^2}{n_j},$$

$$\text{BSS} = \frac{1}{J-1} \sum (\bar{X}_j - \bar{X})^2,$$

$$\text{WSS} = \frac{1}{J} \sum q_j,$$

in which case the test statistics is

$$F_{jo} = \frac{\text{BSS}}{\text{WSS}}$$

with

$$\nu_1 = \frac{\left(\frac{J-1}{J} \sum q_j\right)^2}{\left(\sum \frac{q_j}{J}\right)^2 + \frac{J-2}{J} \sum q_j^2}$$

and

$$\nu_2 = \frac{\left(\sum q_j\right)^2}{\sum \frac{q_j^2}{n_j - 1}}$$

degrees of freedom. In the illustration regarding students' ratings,

$$\text{BSS} = 3.13$$

$$\text{WSS} = .517$$

$$F_{jo} = 6.05.$$

(The numerical details are left as an exercise.) The degrees of freedom are $\nu_1 = 1.85$ and $\nu_2 = 18.16$, and the critical value is 3.63. Because $6.05 > 3.63$, reject and conclude there is a difference among students' ratings.

When there are unequal variances, a variety of methods have been suggested for estimating σ_μ^2, several of which were compared by P. Rao, Kaplan, and Cochran (1981). Their recommendation is that when $\sigma_\mu^2 > 0$, σ_μ^2 be estimated with

$$\hat{\sigma}_\mu^2 = \frac{1}{J} \sum \ell_j^2 (\bar{X}_j - \tilde{X})^2,$$

where

$$\ell_j = \frac{n_j}{n_j + 1}$$

$$\tilde{X} = \frac{\sum \ell_j \bar{X}_j}{\sum \ell_j}.$$

Evidently there are no results on how this estimate performs under nonnormality.

9.6.3 A Method Based on Trimmed Means

Under normality with unequal variances, the F-test can have a Type I error probability as high as .179 when testing at the $\alpha = .05$ level with equal sample sizes of 20 in each group (Wilcox, 1994a). The Jeyaratnam–Othman test statistic, F_{jo}, has a probability of a Type I error close to .05 in the same situation. However, when the normality assumption is violated, the probability of a Type I error using both F and F_{jo} can exceed .3. Another concern with both F and the Jeyaratnam–Othman method is that there are situations where power decreases even when the difference among the means increases. This last problem appears to be reduced considerably when using trimmed means. Trimmed means provide better control over the probability of a Type I error and can yield substantially higher power when there are outliers. Of course there are exceptions. Generally no method is best in all situations. But if the goal is to reduce the problems just described, an extension of the Jeyaratnam–Othman method to trimmed means has considerable practical value. (Extensions of the random effects model based on MOM have not been investigated as yet.) The computational details are relegated to Box 9.8. Readers interested in the derivation and technical details of the method are referred to Wilcox (1997a, Sec. 6.3).

BOX 9.8 Comparing Trimmed Means in a Random Effects Model

For each of the J groups, Winsorize the observations as described in Section 3.2.6 and label the results Y_{ij}. For example, if the observations for group 3 are:

$$37, 14, 26, 17, 21, 43, 25, 6, 9, 11,$$

then the 20% Winsorized values are $Y_{13} = 26$, $Y_{23} = 14$, $Y_{33} = 26$, and so forth. To test the hypothesis of no differences among the trimmed means, let h_j be the effective sample size of the jth group (the number of observations left after trimming), and compute

$$\bar{Y}_j = \frac{1}{n_j} \sum_{i=1}^{n_j} Y_{ij},$$

$$s_{wj}^2 = \frac{1}{n_j - 1} \sum (Y_{ij} - \bar{Y}_j)^2,$$

$$\bar{X}_t = \frac{1}{J} \sum \bar{X}_{tj},$$

$$\mathrm{BSST} = \frac{1}{J - 1} \sum_{j=1}^{J} (\bar{X}_{tj} - \bar{X}_t)^2,$$

$$\mathrm{WSSW} = \frac{1}{J} \sum_{j=1}^{J} \sum_{i=1}^{n_j} \frac{(Y_{ij} - \bar{Y}_j)^2}{h_j(h_j - 1)},$$

Continued

BOX 9.8 (*Continued*)

$$D = \frac{\text{BSST}}{\text{WSSW}}.$$

Let

$$q_j = \frac{(n_j - 1)s_{wj}^2}{J(b_j)(b_j - 1)}.$$

The degrees of freedom are estimated to be

$$\hat{v}_1 = \frac{\left((J-1)\sum q_j\right)^2}{\left(\sum q_j\right)^2 + (J-2)J\sum q_j^2},$$

$$\hat{v}_2 = \frac{\left(\sum q_j\right)^2}{\sum q_j^2/(b_j - 1)}.$$

Reject if $D \geq f$, the $1 - \alpha$ quantile of an F-distribution with $\hat{v}_1$ and $\hat{v}_2$ degrees of freedom.

9.6.4 S-PLUS Function rananova

The S-PLUS function

$$\text{rananova}(x, tr=.2, grp=NA)$$

performs the calculations in Box 9.8. As usual, x is any S-PLUS variable that has list mode or is a matrix (with columns corresponding to groups), tr is the amount of trimming, which defaults to .2, and grp can be used to specify some subset of the groups if desired. If grp is not specified, all groups stored in x are used. If the data are not stored in a matrix or in list mode, the function terminates and prints an error message. The function returns the value of the test statistic, D, which is stored in rananova$teststat, the significance level is stored in rananova$siglevel, and an estimate of a Winsorized intraclass correlation is returned in the S-PLUS variable rananova$rho. This last quantity is like the intraclass correlation ρ_I, but with the variance of the means and observations replaced by a Winsorized variance.

EXAMPLE. Assuming the data in Table 9.5 are stored in the S-PLUS variable data, the command rananova(data) returns.

```
$teststat:
[1] 140.0983
```

```
$df:
[1] 3.417265 36.663787
```

Continued

EXAMPLE. (*Continued*)

```
$siglevel:
[1] 0
```

```
$rho:
[1] 0.9453473
```

So we reject the hypothesis of equal trimmed means. The value for rho indicates that about 95% of the Winsorized variance among the observations is accounted for by the Winsorized variance of the *Winsorized* means. (For technical reasons, Winsorized means are used rather than trimmed means when deriving a robust analog of the intraclass correlation coefficient.) The command rananova(data,tr=0) compares means instead and returns

```
$teststat:
[1] 76.23691
```

```
$df:
[1] 3.535804 63.827284
```

```
$siglevel:
[1] 0
```

```
$rho:
[1] 0.782022
```

So again we reject. In the latter case the intraclass correlation is estimated to be .78, meaning that the variation among the means is estimated to account for 78% of the variation among all possible observations. ■

9.7 Exercises

1. For the following data, assume that the three groups have a common population variance, σ_p^2. Estimate σ_p^2.

Group 1	Group 2	Group 3
3	4	6
5	4	7
2	3	8
4	8	6
8	7	7
4	4	9
3	2	10
9	5	9
$\bar{X}_1 = 4.75$	$\bar{X}_2 = 4.62$	$\bar{X}_3 = 7.75$
$s_1^2 = 6.214$	$s_2^2 = 3.982$	$s_3^2 = 2.214$

2. For the data in the previous exercise, test the hypothesis of equal means using the ANOVA F. Use $\alpha = .05$.

3. For the data in Exercise 1, verify that Welch's test statistic is $F_w = 7.7$ with degrees of freedom $v_1 = 2$ and $v_2 = 13.4$. Then verify that you would reject the hypothesis of equal means with $\alpha = .01$. Check your results with the S-PLUS function t1way in Section 9.4.1.

4. Construct an ANOVA summary table using the following data, as described in Section 9.1, and then test the hypothesis of equal means with $\alpha = .05$.

Group 1	Group 2	Group 3	Group 4
15	9	17	13
17	12	20	12
22	15	23	17

5. In the previous exercise, what is your estimate of the assumed common variance?

6. For the data used in the preceding two exercises, verify that for Welch's test, $F_w = 3.38$ with $v_1 = 3$ and $v_2 = 4.42$.

7. Based on the results of the previous exercise, would you reject the hypothesis of equal means with $\alpha = .1$?

8. Why would you not recommend the strategy of testing for equal variances and, if not significant, using the ANOVA F-test rather than Welch's method?

9. For the data in Table 9.1, assume normality and that the groups have equal variances. As already illustrated, the hypothesis of equal means is not rejected. If the hypothesis of equal means is true, an estimate of the assumed common variance is MSBG $= .12$, as already explained. Describe a reason why you would prefer to estimate the common variance with MSWG rather than MSBG.

10. Five independent groups are compared, with $n = 15$ observations for each group. Fill in the missing values in the following summary table.

Source of variation	Degrees of freedom	Sum of squares	Mean square	F
Between groups	_____	50	_____	_____
Within groups	_____	150	_____	

11. Referring to Box 9.2, verify that for the following data, MSBG $= 14.4$ and MSWG $= 12.59$.

G1	G2	G3
9	16	7
10	8	6
15	13	9
	6	

12. Consider $J = 5$ groups with population means 3, 4, 5, 6, and 7 and a common variance $\sigma_p^2 = 2$. If $n = 10$ observations are sampled from each group, determine the value estimated by MSBG, and comment on how this differs from the value estimated by MSWG.

13. For the following data, verify that you do not reject with the ANOVA F-test with $\alpha = .05$ but that you do reject with Welch's test. What might explain the discrepancy between the two methods?

Group 1:	10	11	12	9	8	7
Group 2:	10	66	15	32	22	51
Group 3:	1	12	42	31	55	19

14. Given the following ANOVA summary table; verify that the number of groups is $J = 4$, that the total number of observations is $N = 12$, and that with $\alpha = .025$ the critical value is 5.42.

Source of variation	Degrees of freedom	Sum of squares	Mean square	F
Between groups	3	300	100	10
Within groups	8	80	10	
Total	11	428		

15. For the data in Table 9.1, the ANOVA F-test and Welch's test were not significant with $\alpha = .05$. Imagine that you want power to be .9 if the mean of one of the groups differs from the others by .2. (In the notation of Section 9.3, $a = .2$.) Verify that according to the S-PLUS function bdanova1, the required sample sizes for each group are 110, 22, 40, and 38.

16. For the data in Table 9.1, use the S-PLUS function pbadepth to compare the groups based on MOM. Verify that the significance level is .17.

17. Compare the groups in Table 9.4 with the S-PLUS function pbadepth. You should find that the function terminates with an error if allp=T is used. How might you deal with this problem given the goal of comparing groups based on MOM?

18. Store the data in Table 9.7 in an S-PLUS variable having matrix mode with two columns corresponding to the two columns shown. There are three groups, with the first column indicating to which group the value in the second column belongs. For example, the first row indicates that the value 12 belongs to group 1 and the fourth row indicates that the value 42 belongs to group 3. Use the S-PLUS function selby in Section 1.1.6 to separate the data into three groups and then store it in list mode. Then compare the three groups with Welch's test, the method in Section 9.4 with 20% trimming, and then use the

S-PLUS function pbadepth in Section 9.5.4 to compare the groups based on MOM.

19. Based on the properties summarized in Box 9.3, how would you estimate

$$\sum (\mu_j - \bar{\mu})^2?$$

20. For the data in Table 9.8, verify that the significance level, based on Welch's test, is .98. Use the S-PLUS Function t1way.

21. For the data in Table 9.8, use t1way to compare 20% trimmed means and verify that the significance level is less than .01.

TABLE 9.7   Data for Exercise 18

G	X
1	12
1	8
1	22
3	42
3	8
3	12
3	9
3	21
2	19
2	24
2	53
2	17
2	10
2	9
2	28
2	21
1	19
1	21
1	56
1	18
1	16
1	29
1	20
3	32
3	10
3	12
3	39
3	28
3	35
2	10
2	12

TABLE 9.8   Hypothetical Data for Exercises 20–22

Group 1	Group 2	Group 3	Group 4	Group 5
10.1	10.7	11.6	12.0	13.6
9.9	9.5	10.4	13.1	11.9
9.0	11.2	11.9	13.2	13.6
10.7	9.9	11.7	11.0	12.3
10.0	10.2	11.8	13.3	12.3
9.3	9.1	11.6	10.5	11.3
10.6	8.0	11.6	14.4	12.4
11.5	9.9	13.7	10.5	11.8
11.4	10.7	13.3	12.2	10.4
10.9	9.7	11.8	11.0	13.1
9.5	10.6	12.3	11.9	14.1
11.0	10.8	15.5	11.9	10.5
11.1	11.0	11.4	12.4	11.2
8.9	9.6	13.1	10.9	11.7
12.6	8.8	10.6	14.0	10.3
10.7	10.2	13.1	13.2	12.0
10.3	9.2	12.5	10.3	11.4
10.8	9.8	13.9	11.6	12.1
9.2	9.8	12.2	11.7	13.9
8.3	10.9	11.9	12.1	12.7
93.0	110.6	119.6	112.8	112.8
96.6	98.8	113.6	108.0	129.2
94.8	107.0	107.5	113.9	124.8

22. For the data in the Table 9.8, use pbadepth with the argument allp set equal to F and verify that the significance level is zero. The text mentioned a possible concern with using allp=F. Why is this not an issue here?

10

TWO-WAY ANOVA

This chapter takes up an extension of the analysis of variance method described in Chapter 9. As usual, we begin by describing basic concepts and summarizing the standard approach based on means. Then robust methods are described.

10.1 The Basics of a Two-Way ANOVA Design

The basic concepts are illustrated with a study where the goal is to understand the effect of diet on weight gains in rats. Specifically, four diets are considered that differ in: (1) amounts of protein (high and low) and (2) the source of the protein (beef versus cereal). The results for these four groups, reported in Table 10.1, are taken from Snedecor and Cochran (1967). Different rats were used in the four groups, so the groups are independent. The first column gives the weight gains of rats fed a low-protein diet, with beef the source of protein. The next column gives the weight gains for rats on a high-protein diet, again with beef the source of protein, and the next two columns report results when cereal is substituted for beef.

It is convenient to depict the population means as shown in Table 10.2. Table 10.2 indicates, for example, that μ_1 is the population mean associated with rats receiving a low-protein diet from beef. That is, μ_1 is the average weight gain if all of the millions of rats we might study are fed this diet. Similarly, μ_4 is the population mean for rats receiving a high-protein diet from cereal.

The study just described is an example of what is called a *two-way design*, meaning that there are two *independent variables*, or *factors*, being studied. Here, the first factor is *source of protein*, which has two levels: beef and cereal. The second factor is *amount of protein*, which also has two levels: low and high. A more precise description is that you have a 2-by-2 design, meaning you have two factors, both of which have two levels. If you compare three methods for increasing endurance and simultaneously take into account three different diets, you have a two-way design with both factors having three levels. More succinctly, this is called a 3-by-3 design. The first factor is method and the second is diet. As a final example, imagine you want to compare four methods for teaching statistics and simultaneously take into account

TABLE 10.1   Weight Gains (in grams) of Rats on One of Four Diets

Beef, low	Beef, high	Cereal, low	Cereal, high
90	73	107	98
76	102	95	75
90	118	97	56
64	104	80	111
86	81	98	95
51	107	74	88
72	100	74	82
90	87	67	77
95	117	89	86
78	111	58	92
$\bar{X}_1 = 79.2$	$\bar{X}_2 = 100$	$\bar{X}_3 = 83.9$	$\bar{X}_4 = 85.9$

TABLE 10.2   Depiction of the Population Means for Four Diets

	Source	
Amount	Beef	Cereal
High	μ_1	μ_2
Low	μ_3	μ_4

the amount of previous training in mathematics. If you categorize students as having poor or good training in mathematics, you have a 4-by-2 design. That is, you have a two-way design where the first factor is method of teaching and the second is previous training. If you ignore previous training and consider only the four methods, you have what is called a *one-way design* with four levels. This means you have one factor of interest (method of training) and four different methods are to be compared.

Returning to Table 10.2, you could compare these four groups by testing

$$H_0 : \mu_1 = \mu_2 = \mu_3 = \mu_4,$$

the hypothesis that all of the means are equal. However, there are other comparisons you might want to make. For example, you might want to compare the rats receiving a high-protein versus low-protein diet, *ignoring* the source of the protein. To illustrate how this might be done, imagine that the values of the population means are as shown in Table 10.3. For rats on a high-protein diet, the mean is 45 when consuming beef versus 60 when consuming cereal instead. If you want to characterize the typical weight gain for a high-protein diet while ignoring source, a natural strategy is to average the two population means, yielding $(45 + 60)/2 = 52.5$. That is, the typical rat on a protein diet gains 52.5 grams. For the more general situation depicted by Table 10.2, the typical weight gain on a high-protein diet would be $(\mu_1 + \mu_2)/2$.

TABLE 10.3 Hypothetical Population Means for Illustrating Main Effects and Interactions

	Source	
Amount	Beef	Cereal
High	$\mu_1 = 45$	$\mu_2 = 60$
Low	$\mu_3 = 80$	$\mu_4 = 90$

Similarly, the typical weight gain for a rat on a low-protein diet would be $(\mu_3 + \mu_4)/2$, which for Table 10.3 is $(80 + 90)/2 = 85$ grams. Of course, you can do the same when characterizing source of protein while ignoring amount. The typical weight gain for a rat eating beef, ignoring amount of protein, is $(45 + 80)/2 = 62.5$, and for cereal it is $(60 + 90)/2 = 75$.

As usual, you do not know the population means. What is needed is some way of testing the hypothesis that weight gain is different for a high-protein diet versus a low-protein diet, ignoring source of protein. One way of doing this is to test

$$H_0 : \frac{\mu_1 + \mu_2}{2} = \frac{\mu_3 + \mu_4}{2},$$

the hypothesis that the average of the populations means in the first row of Table 10.2 is equal to the average for the second row. If this hypothesis is rejected, then there is said to be a *main effect* for the amount of protein. More generally, a main effect for the first factor (amount) is said to exist if

$$\frac{\mu_1 + \mu_2}{2} \neq \frac{\mu_3 + \mu_4}{2}.$$

Similarly, you might want to compare source of protein while ignoring amount. One way of doing this is to test

$$H_0 : \frac{\mu_1 + \mu_3}{2} = \frac{\mu_2 + \mu_4}{2},$$

the hypothesis that the average of the means in the column for beef in Table 10.2 is equal to the average for the cereal column. If this hypothesis is rejected, then there is said to be a *main effect* for the source of protein. More generally, a main effect for the second factor is said to exist if

$$\frac{\mu_1 + \mu_3}{2} \neq \frac{\mu_2 + \mu_4}{2}.$$

Consider again the 4-by-2 design where you want to compare four methods for teaching statistics and simultaneously take into account the amount of previous training in mathematics. The means corresponding to the eight groups can be written as shown in Table 10.4. Main effects for the first factor (method) can be addressed by testing the hypothesis that the averages of the means in each row have equal values. That is, test

$$H_0 : \frac{\mu_1 + \mu_2}{2} = \frac{\mu_3 + \mu_4}{2} = \frac{\mu_5 + \mu_6}{2} = \frac{\mu_7 + \mu_8}{2}.$$

TABLE 10.4   Depiction of the Population Means for a 4-by-2 Design

	Previous Training	
Method	High	Low
1	μ_1	μ_2
2	μ_3	μ_4
3	μ_5	μ_6
4	μ_7	μ_8

Typically this is referred to as the hypothesis of *no main effects for factor A*, where *factor A* is a generic term for the first of the two factors under study. The hypothesis of no main effects for the second factor, previous training, is

$$H_0 : \frac{\mu_1 + \mu_3 + \mu_5 + \mu_7}{4} = \frac{\mu_2 + \mu_4 + \mu_6 + \mu_8}{4}.$$

That is, the average of the means in column 1 is hypothesized to be equal to the average of the means in column 2. This is called a hypothesis of no main effects for the second factor, previous training. A generic term for the second factor is *factor B*, and the hypothesis of no main effects for factor B refers to the hypothesis that the averages of the means in each column have a common value.

10.1.1 Interactions

There is one other important feature of a two-way design. Consider again a 2-by-2 design where the goal is to compare high- and low-protein diets in conjunction with two protein sources. Suppose the *population* means associated with the four groups are as depicted in Table 10.3. Now look at the first row (high amount of protein) and notice that the weight gain for a beef diet is 45 grams versus a weight gain of 60 for cereal. As is evident, there is an increase of 15 grams. In contrast, with a low-protein diet, switching from beef to cereal results in an increase of 10 grams on average. That is, in general, switching from beef to cereal results in an increase for the average amount of weight gained, but the increase differs depending on whether we look at high or low protein. This is an example of what is called an *interaction*.

In a 2-by-2 design with means as shown in Table 10.2, an *interaction* is said to exist if

$$\mu_1 - \mu_2 \neq \mu_3 - \mu_4.$$

In words, an interaction exists if for the first level of factor A the difference between the means is not equal to the difference between the means associated with the second level of factor A. *No interaction* means that

$$\mu_1 - \mu_2 = \mu_3 - \mu_4.$$

Various types of interactions arise and can be important when considering how groups differ. Notice that in the beef-versus-cereal illustration as depicted in Table 10.3, there is an increase in the average weight gain when switching from beef to cereal for both high- and low-protein diets. For high protein there is an increase in the population mean from 45 to 60, and for low protein there is an increase from 80 to 90. In both cases, though, the largest gain in weight is associated with cereal. This is an example of what is called an *ordinal interaction*. In a 2-by-2 design as depicted in Table 10.2, an *ordinal interaction* is said to exist if

$$\mu_1 > \mu_2 \quad \text{and} \quad \mu_3 > \mu_4$$

or if

$$\mu_1 < \mu_2 \quad \text{and} \quad \mu_3 < \mu_4.$$

In words, an interaction is said to be ordinal if the relative rankings remain the same; in the illustration, cereal always results in the largest weight gain, regardless of whether a low- or high-protein diet is used.

If there is a change in the relative rankings of the means, a *disordinal interaction* is said to exist. As an illustration, imagine that the population means are as follows:

Amount	Source	
	Beef	Cereal
High	80	110
Low	50	30

Observe that for the first row (a high-protein diet), the average weight gain increases from 80 to 110 as we move from beef to cereal. In contrast, for the low-protein diet, the average weight gain decreases from 50 to 30. Moreover, when comparing beef to cereal, the relative rankings change depending on whether a high- or low-protein diet is used. For a high-protein diet, cereal results in the largest gain; for a low-protein diet, beef results in a larger gain than cereal. This is an example of a disordinal interaction. In general, for the population means in Table 10.2, a disordinal interaction is said to exist if

$$\mu_1 > \mu_2 \quad \text{and} \quad \mu_3 < \mu_4$$

or if

$$\mu_1 < \mu_2 \quad \text{and} \quad \mu_3 > \mu_4.$$

Research articles often present graphical displays reflecting ordinal and disordinal interactions. Figure 10.1 is an example based on the sample means just used to illustrate a disordinal interaction. Along the x-axis we see the levels of the first factor (high and low protein). The line extending from 110 down to 30 reflects the change in means when the source of protein is beef. The other line reflects the change in means associated with cereal. As is evident, the lines cross, which reflects a disordinal interaction.

Now suppose the population means are as shown in Table 10.3. Regardless of whether rats are given a high- or low-protein diet, cereal always results in the largest average weight gain. The left panel of Figure 10.2 graphs the means. Observe that the lines are not parallel, but they do not cross, indicating an ordinal interaction.

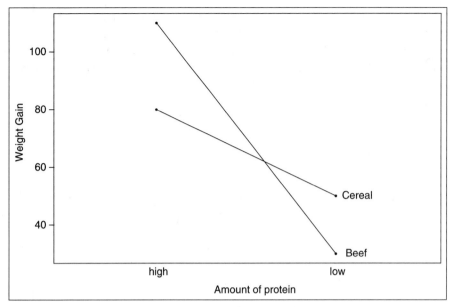

FIGURE 10.1   Example of a disordinal interaction.

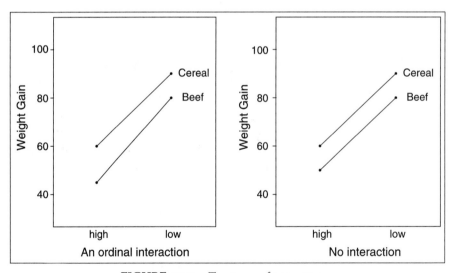

FIGURE 10.2   Two types of interactions.

Now imagine that the population means are

	Source	
Amount	Beef	Cereal
High	50	60
Low	80	90

There is no interaction because for the first row, the means increase by 10 (from 50 to 60), and the increase is again 10 for the second row. The right panel of

Figure 10.2 graphs the means. Notice that the lines are parallel. That is, when there is no interaction, a graph of the means results in parallel lines.

Notice that in the discussion of ordinal versus disordinal interactions, attention was focused on comparing means within rows. Not surprisingly, ordinal and disordinal interactions can also be defined in terms of columns. Consider again the example where the means are

	Source	
Amount	Beef	Cereal
High	80	110
Low	50	30

As previously explained, there is a disordinal interaction for rows. Now, however, look at the population means in the first column and notice that 80 is greater than 50, and for the second column 110 is greater than 30. There is an interaction because $80 - 50 \neq 110 - 30$. Moreover, the interaction is ordinal because for beef, average weight gain is largest on a high-protein diet, and the same is true for a cereal diet.

When graphing the means to illustrate a disordinal or ordinal interaction for rows, the levels of factor A (high and low amounts of protein) were indicated by the x-axis, as illustrated by Figures 10.1 and 10.2. When describing ordinal or disordinal interactions by columns, now the x-axis contains the levels of the second factor, which in this example is source of protein (beef versus cereal). Figure 10.3 illustrates what the graph looks like for the means considered here.

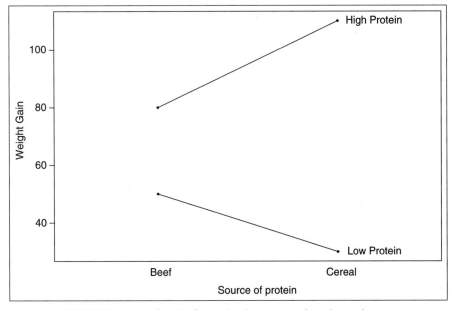

FIGURE 10.3   Graph of an ordinal interaction based on columns.

Situations also arise where there is a disordinal interaction for both rows and columns. For example, if the population means happen to be

	Source	
Amount	Beef	Cereal
High	80	110
Low	100	95

then there is a disordinal interaction for rows, because for the first row, cereal results in the largest gain, but for the second row (low amount of protein) the largest gain is for beef. Simultaneously, there is a disordinal interaction for columns, because for the first column (beef), a low-protein diet results in the largest average gain, but for the second column (cereal) a high-protein diet has the largest mean.

10.1.2 Interactions When There Are More Than Two Levels

So far, attention has been focused on a 2-by-2 design. What does no interaction mean when there are more than two levels for one or both factors? Basically, *no interaction* means that for any two levels of factor A and any two levels of factor B, there is no interaction for the corresponding cells. As an illustration, consider the population means in Table 10.5. Pick any two rows — say, the first and third. Then pick any two columns — say, the first and second. The population means for the first row and the first and second columns are 10 and 20. For the third row, the means for these two columns are 30 and 40; these four means are in boldface in Table 10.5. Notice that for these four means, there is no interaction. The reason is that for the first row, the means increase from 10 to 20, and for the third row the means increase from 30 to 40. That is, for both rows there is an increase of 10 when switching from column 1 to column 2. In a similar manner, again looking at rows 1 and 2, we see that there is an increase of 20 as we move from column 1 to column 3. That is, there is no interaction for these four means either. An interaction is said to exist among the JK means if any two rows and two columns have an interaction.

EXAMPLE. Suppose the population means for a 3-by-4 design are

Factor A	Factor B			
	Level 1	Level 2	Level 3	Level 4
Level 1	40	40	40	40
Level 2	40	40	40	40
Level 3	40	40	40	40

Is there an interaction? The answer is no, because regardless of which two rows you pick, there is an increase of 0 as you move from any one column to another.   ■

TABLE 10.5   Hypothetical Population Means Illustrating Interactions

Factor A	Factor B		
	Level 1	Level 2	Level 3
Level 1	10	20	30
Level 2	20	30	40
Level 3	30	40	50

EXAMPLE. For the population means used in this last example, is there a main effect for factor A or factor B? First consider factor A. As is evident, for any row we pick, the average of the four means is 40. That is, the average of the means is the same for all three rows; this means there is no main effect for factor A. In a similar manner, there is no main effect for factor B, because the average of the means in any column is again 40. ■

EXAMPLE. Suppose the population means for a 3-by-4 design are

Factor A	Factor B			
	Level 1	Level 2	Level 3	Level 4
Level 1	40	40	50	60
Level 2	20	20	50	80
Level 3	20	30	10	40

Is there an interaction? Looking at level 1 of factor A, we see that the means increase by 0 as we move from level 1 of factor B to level 2. The increase for level 2 of factor A is again 0, so there is no interaction for these four means. However, looking at level 1 of factor A, we see that the means increase by 10 as we move from level 1 to level 3 of factor B. In contrast, there is an increase of 30 for level 2 of factor A, which means that there is an interaction. ■

10.2 Testing Hypotheses About Main Effects and Interactions

So far, attention has been focused on explaining the meaning of main effects and interactions in terms of the population means. As usual, we do not know the population means, and this raises the issue of how we might test the hypotheses of no main effects and no interactions. As was the case in Chapter 9, the most commonly used method is based on the assumption that all groups have normal distributions with a common variance, which we again label σ_p^2.

TABLE 10.6   Sample Means for Illustrating Main Effects and Interactions

	Source	
Amount	Beef	Cereal
Low	$\bar{X}_{11} = 79.2$	$\bar{X}_{12} = 83.9$
High	$\bar{X}_{21} = 100$	$\bar{X}_{22} = 85.9$

The computations begin by computing the sample mean corresponding to the jth level of factor A and the kth level of factor B. The computations are easier to describe if we switch notation and represent this sample mean with $\bar{X}_{jk}$. That is, $\bar{X}_{jk}$ is the sample mean corresponding to the jth level of factor A and the kth level of factor B. For example, for the 2-by-2 study comparing high versus low protein and beef versus cereal, $\bar{X}_{11}$ is the sample mean for level 1 of factor A and level 1 of factor B. Referring to Table 10.1, $\bar{X}_{11}$ is the sample mean of the rats on a low-protein diet (level 1 of factor A) that consume beef (level 1 of factor B). From the first column in Table 10.1, the average weight gain for the 10 rats on this diet can be seen to be $\bar{X}_{11} = 79.2$. Level 1 of factor A and level 2 of factor B correspond to low protein from a cereal diet. The data are given in the third column of Table 10.1, and the sample mean is $\bar{X}_{12} = 83.9$. For level 2 of factor A and level 1 of factor B, $\bar{X}_{21} = 100$, the sample mean of the values in column 2 of Table 10.1. Finally, $\bar{X}_{22}$ is the sample mean for level 2 of factor A and level 2 of factor B and is 85.9. These sample means are summarized in Table 10.6.

Under normality and equal variances, a test of the hypothesis of no main effects for factor A can be performed that provides exact control over the probability of a Type I error. Box 10.1 summarizes the bulk of the calculations when all groups have a common sample size, n. (The method is not robust to violations of assumptions, particularly when the sample sizes are unequal, so the computational details for handling unequal samples are omitted.) In Box 10.1, s_{jk}^2 is the sample variance corresponding to the data used to compute $\bar{X}_{jk}$. As in the one-way design, MSWG estimates the assumed common variance, σ_p^2.

BOX 10.1 Computations for a Two-Way Design

Assumptions

1. Random sampling
2. Normality

Continued

BOX 10.1 (*Continued*)

   3. Equal variances
   4. Equal sample sizes

Notation
J is the number of levels for factor A, and K is the number of levels for factor B.

Computations

$$\bar{X}_G = \frac{1}{JK} \sum_{j=1}^{J} \sum_{k=1}^{K} \bar{X}_{jk}$$

($\bar{X}_G$ is the average of all JK sample means.)

$$\bar{X}_{j.} = \frac{1}{K} \sum_{k=1}^{K} \bar{X}_{jk}$$

($\bar{X}_{j.}$ is the average of the sample means for the jth level of factor A.)

$$\bar{X}_{.k} = \frac{1}{J} \sum_{j=1}^{J} \bar{X}_{jk}$$

($\bar{X}_{.k}$ is the average of the sample means for the kth level of factor B.)

$$A_j = \bar{X}_{j.} - \bar{X}_G, \qquad B_k = \bar{X}_{.k} - \bar{X}_G$$

$$C_{jk} = \bar{X}_{jk} - \bar{X}_{j.} - \bar{X}_{.k} + \bar{X}_G$$

$$\text{SSA} = nK \sum A_j^2, \qquad \text{SSB} = nJ \sum B_k^2$$

$$\text{SSINTER} = n \sum \sum C_{jk}^2$$

$$\text{MSWG} = \frac{1}{JK} \sum \sum s_{jk}^2, \qquad \text{SSWG} = (n-1)JK(\text{MSWG})$$

$$\text{MSA} = \frac{\text{SSA}}{J-1}, \qquad \text{MSB} = \frac{\text{SSB}}{K-1}$$

$$\text{MSINTER} = \frac{\text{SSINTER}}{(J-1)(K-1)}.$$

(Some books write MSINTER and SSINTER as MSAB and SSAB, respectively, where AB denotes the interaction of factors A and B.)

DECISION RULES: Once you complete the computations in Box 10.1, the relevant hypotheses are tested as follows:

- **Factor A**. The hypothesis of no main effects for factor A is tested with

$$F = \frac{\text{MSA}}{\text{MSWG}},$$

and you reject if $F \geq f$, the $1 - \alpha$ quantile of an F-distribution with $v_1 = J - 1$ and $v_2 = N - JK$ degrees of freedom.
- **Factor B**. The hypothesis of no main effects for factor B is tested with

$$F = \frac{\text{MSB}}{\text{MSWG}},$$

and you reject if $F \geq f$, the $1 - \alpha$ quantile of an F-distribution with $v_1 = K - 1$ and $v_2 = N - JK$ degrees of freedom.
- **Interactions**. The hypothesis of no interactions is tested with

$$F = \frac{\text{MSINTER}}{\text{MSWG}},$$

and you reject if $F \geq f$, the $1 - \alpha$ quantile of an F-distribution with $v_1 = (J - 1)(K - 1)$ and $v_2 = N - JK$ degrees of freedom.

In Box 10.1, one of the assumptions is normality. In terms of Type I errors, this assumption can be violated if all groups have identically shaped distributions and the sample sizes are reasonably large. This means, in particular, that the equal variance assumption is true. Put another way, if you reject with the ANOVA F-test, this indicates that the distributions differ, but it remains unclear how they differ and by how much. One possibility is that you reject because the populations means differ. Another possibility is that you reject primarily because the variances differ. As for power, again practical problems arise under very slight departures from normality, for reasons discussed in previous chapters. (See in particular Section 5.5.) Generally, the more groups you compare, the more likely the ANOVA F-tests described in this section will be unsatisfactory when indeed groups differ.

Table 10.7 outlines a typical ANOVA summary table for a two-way design. The notation SS in the first row stands for sum of squares, DF is degrees of freedom, and MS is mean squares.

TABLE 10.7 Typical ANOVA Summary Table for a Two-Way Design

Source	SS	DF	MS	F
A	SSA	$J - 1$	$\text{MSA} = \frac{\text{SSA}}{J-1}$	$F = \frac{\text{MSA}}{\text{MSWG}}$
B	SSB	$K - 1$	$\text{MSB} = \frac{\text{SSB}}{K-1}$	$F = \frac{\text{MSB}}{\text{MSWG}}$
INTER	SSINTER	$(J-1)(K-1)$	$\text{MSINTER} = \frac{\text{SSINTER}}{(J-1)(K-1)}$	$F = \frac{\text{MSINTER}}{\text{MSWG}}$
WITHIN	SSWG	$N - JK$	$\text{MSWG} = \frac{\text{SSWG}}{N-JK}$	

EXAMPLE. Consider the following ANOVA summary table.

Source	SS	DF	MS	F
A	200	1	200	1.94
B	300	2	150	1.46
INTER	500	2	250	2.42
WITHIN	620	6	103	

Referring to Table 10.7, we see that $J - 1$ corresponds to the value 1 in the example, indicating that $J - 1 = 1$, so the first factor has $J = 2$ levels. Similarly, the second factor has $K = 3$ levels. Table 10.7 indicates that $N - JK = 6$, but $JK = 6$, so $N - JK = N - 6 = 6$, and therefore $N = 12$. That is, the total number of observations among the 6 groups is 12. The estimate of the common variance is MSWG = 103. If we use $\alpha = .05$, then from Table 6 in Appendix B, the critical values for the three hypotheses are 5.98, 5.14, and 5.14. The F-values are less than their corresponding critical values, so you do not reject the hypotheses of no main effects for factor A, as well as for factor B and the hypothesis of no interactions. ■

10.2.1 Inferences About Disordinal Interactions

It should be mentioned that if you reject the hypothesis of no interactions, simply looking at the means is not enough to determine whether the interaction is ordinal or disordinal. Consider again the study on weight gain in rats described at the beginning of this section. Suppose that unknown to you, the population means are

Amount	Source	
	Beef	Cereal
Low	$\mu_1 = 60$	$\mu_2 = 60$
High	$\mu_3 = 50$	$\mu_4 = 70$

There is an interaction, because $60 - 60 \neq 50 - 70$, but the interaction is not disordinal. Further assume that you reject the hypothesis of no interactions based on the following sample means:

Amount	Source	
	Beef	Cereal
Low	$\bar{X}_1 = 55$	$\bar{X}_2 = 45$
High	$\bar{X}_3 = 49$	$\bar{X}_4 = 65$

That is, you have correctly concluded that there is an interaction. Notice that the sample means suggest that the interaction is disordinal, because 55 is greater than 45 but 49 is less than 65. By chance, rats on a low-protein diet with beef as the source got a smaller sample mean versus rats on a low-protein diet with cereal as the source. The important point is that to establish that a disordinal interaction exists for the rows, you must also reject the hypotheses $H_0 : \mu_1 = \mu_2$ and $H_0 : \mu_3 = \mu_4$. If, for example, $\mu_1 = \mu_2$, you do not have a disordinal interaction. Moreover, simply rejecting the hypothesis of no interaction does not tell you whether you should reject $H_0 : \mu_1 = \mu_2$ or $H_0 : \mu_3 = \mu_4$. Under normality, these hypotheses can be tested with Student's T test or Welch's test, described in Chapter 8. If both of these hypotheses are rejected and if

$$\bar{X}_1 > \bar{X}_2 \qquad \text{and} \qquad \bar{X}_3 < \bar{X}_4$$

or if

$$\bar{X}_1 < \bar{X}_2 \qquad \text{and} \qquad \bar{X}_3 > \bar{X}_4,$$

then you have empirical evidence that there is a disordinal interaction.

A similar strategy is used when checking for a disordinal interaction for columns. That is, to establish that a disordinal interaction exists, you must reject both $H_0 : \mu_1 = \mu_3$ and $H_0 : \mu_2 = \mu_4$. If both hypotheses are rejected and the sample means satisfy

$$\bar{X}_1 > \bar{X}_3 \qquad \text{and} \qquad \bar{X}_2 < \bar{X}_4$$

or if

$$\bar{X}_1 < \bar{X}_3 \qquad \text{and} \qquad \bar{X}_2 > \bar{X}_4,$$

you conclude that there is a disordinal interaction for the columns.

An illustration of how to detect a disordinal interaction is postponed until Chapter 12 because there is yet another technical issue that must be addressed: The more tests you perform, the more likely you are to reject even if none of the means differs. There is a large collection of methods for dealing with this issue, many of which will be described in Chapter 12.

10.2.2 The Two-Way ANOVA Model

There is a classic model associated with the two-way ANOVA method that generalizes the ANOVA model in Chapter 9. Although the traditional ANOVA method described in this section is relatively ineffective by modern standards, variations of the model turn out to have value (as will be seen in Chapter 15), so it is briefly described here.

The population grand mean associated with the JK groups is

$$\bar{\mu} = \frac{1}{JK} \sum_{j=1}^{J} \sum_{k=1}^{K} \mu_{jk},$$

the average of the population means. Let

$$\mu_{j.} = \frac{1}{K} \sum_{k=1}^{K} \mu_{jk},$$

be the average of the K means among the levels of factor B associated with the jth level of factor A. Similarly,

$$\mu_{.k} = \frac{1}{J} \sum_{j=1}^{J} \mu_{jk}$$

is the average of the J means among the levels of factor A associated with the kth level of factor B. The main effects associated with factor A are

$$\alpha_1 = \mu_{1.} - \bar{\mu}, \ldots, \alpha_J = \mu_{J.} - \bar{\mu}.$$

So the main effect associated with the jth level is the difference between the grand mean and the average of the means associated with the jth level, namely, $\alpha_j = \mu_{j.} - \bar{\mu}$. There are no main effects for factor A if

$$\alpha_1 = \cdots = \alpha_J = 0.$$

As for factor B, main effects are defined by

$$\beta_1 = \mu_{.1} - \bar{\mu}, \ldots, \beta_K = \mu_{.K} - \bar{\mu}.$$

The hypothesis of no main effects for factor B can be expressed as

$$H_0 : \beta_1 = \cdots = \beta_K = 0.$$

As for interactions, let

$$\gamma_{jk} = \mu_{jk} - \alpha_j - \beta_k - \bar{\mu}$$
$$= \mu_{jk} - \mu_{j.} - \mu_{.k} + \bar{\mu}.$$

Then no interactions means that

$$\gamma_{11} = \gamma_{12} = \cdots = \gamma_{JK} = 0.$$

Although the γ_{jk} terms are not very intuitive, they provide a convenient framework for deriving an appropriate test of the hypothesis that there are no interactions among any two levels of factor A and factor B.

The discrepancy between the ith observation in the jth level of factor A and kth level of factor B, versus the terms just described, is

$$\epsilon_{ijk} = X_{ijk} - \bar{\mu} - \alpha_j - \beta_k - \gamma_{jk}.$$

Rearranging terms yields

$$X_{ijk} = \bar{\mu} + \alpha_j + \beta_k + \gamma_{jk} + \epsilon_{ijk}.$$

Assuming that the error term, ϵ_{ijk}, has a normal distribution with a common variance among the groups results in the standard two-way ANOVA model, which forms the basis of the hypothesis-testing methods covered in this section.

Based on the model just described, it can be shown that

$$E(\text{MSWG}) = \sigma_p^2,$$

$$E(\text{MSA}) = \sigma_p^2 + \frac{nK}{J-1} \sum \alpha_j^2,$$

$$E(\text{MSB}) = \sigma_p^2 + \frac{nJ}{K-1} \sum \beta_k^2,$$

$$E(\text{MSINTER}) = \sigma_p^2 + \frac{n \sum \sum \gamma_{jk}^2}{(J-1)(K-1)}.$$

10.3 Heteroscedastic Methods for Trimmed Means

The tests for main effects and interactions just described assume sampling is from normal distributions with a common variance. As was the case in Chapters 8 and 9, violating the assumption of equal variances causes serious practical problems. That is, unequal variances can result in poor power properties (power can go down as the means become unequal), unsatisfactory control over the probability of a Type I error, and relatively low power versus other methods that might be used. The main reason for describing the extension of the F-test to a two-way design is that it is commonly used, but its practical problems are relatively unknown and many popular computer programs have not yet added more modern methods to their library of techniques. Another reason for describing the conventional method is to be sure readers understand the basic concepts and goals associated with two-way designs.

As in the previous two chapters, our goal is to work up to a method that gives more satisfactory results when groups differ in some manner. In previous chapters we addressed this problem by first describing a heteroscedastic method for means and then indicating how to extend the method to trimmed means. Here, for brevity, we merely describe a heteroscedastic method for trimmed means that includes the problem of comparing means as a special case. The method for dealing with main effects stems from Welch's method, described in Chapter 9. The test statistics for the main effects are computed as shown in Box 10.2. For interactions, it currently seems that an extension of a method derived by Johansen (1980), which is summarized in Box 10.3, gives more satisfactory results.

BOX 10.2 A Heteroscedastic Test for Main Effects Based on

Trimmed Means

Let n_{jk} represent the number of observations for the jth level of factor A and the kth level of B, and let h_{jk} be the number of observations left after trimming. Compute

$$R_j = \sum_{k=1}^{K} \bar{X}_{tjk}, \qquad W_k = \sum_{j=1}^{J} \bar{X}_{tjk}$$

$$d_{jk} = \frac{(n_{jk} - 1)s_{wjk}^2}{h_{jk}(h_{jk} - 1)}$$

$$\hat{v}_j = \frac{\left(\sum_k d_{jk}\right)^2}{\sum_k d_{jk}^2/(h_{jk} - 1)}, \qquad \hat{\omega}_k = \frac{\left(\sum_j d_{jk}\right)^2}{\sum_j d_{jk}^2/(h_{jk} - 1)}$$

$$r_j = \frac{1}{\sum_k d_{jk}}, \qquad w_k = \frac{1}{\sum_j d_{jk}}$$

$$\hat{R} = \frac{\sum r_j R_j}{\sum r_j}, \qquad \hat{W} = \frac{\sum w_k W_k}{\sum w_k}$$

$$B_a = \sum_{j=1}^{J} \frac{1}{\hat{v}_j}\left(1 - \frac{r_j}{\sum r_j}\right)^2, \qquad B_b = \sum_{k=1}^{K} \frac{1}{\hat{\omega}_k}\left(1 - \frac{w_k}{\sum w_k}\right)^2$$

$$V_a = \frac{1}{(J - 1)\left(1 + \frac{2(J-2)B_a}{J^2-1}\right)} \sum_{j=1}^{J} r_j(R_j - \hat{R})^2,$$

$$V_b = \frac{1}{(K - 1)\left(1 + \frac{2(K-2)B_b}{K^2-1}\right)} \sum_{k=1}^{K} w_k(W_k - \hat{W})^2.$$

Continued

BOX 10.2 (*Continued*) The degrees of freedom for factor A are

$$\nu_1 = J - 1, \qquad \nu_2 = \frac{J^2 - 1}{3B_a}.$$

The degrees of freedom for factor B are

$$\nu_1 = K - 1, \qquad \nu_2 = \frac{K^2 - 1}{3B_b}.$$

Decision Rule

Reject the hypothesis of no main effect for factor A if $V_a \geq f_{1-\alpha}$, the $1 - \alpha$ quantile of an F-distribution with the degrees of freedom for factor A. Similarly, reject for factor B if $V_b \geq f_{1-\alpha}$, where now the degrees of freedom are for factor B.

BOX 10.3 A Heteroscedastic Test of the Hypothesis of No Interactions

Based on Trimmed Means

Let

$$d_{jk} = \frac{(n_{jk} - 1)s_{wjk}^2}{b_{jk}(b_{jk} - 1)}$$

$$D_{jk} = \frac{1}{d_{jk}}$$

$$D_{.k} = \sum_{j=1}^{J} D_{jk}, \qquad D_{j.} = \sum_{k=1}^{K} D_{jk}$$

$$D_{..} = \sum D_{jk}$$

$$\tilde{X}_{tjk} = \sum_{\ell=1}^{J} \frac{D_{\ell k}\bar{X}_{t\ell k}}{D_{.k}} + \sum_{m=1}^{K} \frac{D_{jm}\bar{X}_{tjm}}{D_{j.}} - \sum_{\ell=1}^{J}\sum_{m=1}^{K} \frac{D_{\ell m}\bar{X}_{t\ell m}}{D_{..}}.$$

The test statistic is

$$V_{ab} = \sum_{j=1}^{J}\sum_{k=1}^{K} D_{jk}(\bar{X}_{tjk} - \tilde{X}_{tjk})^2.$$

Continued

> **BOX 10.3** (*Continued*) Let c be the $1 - \alpha$ quantile of a chi-squared distribution
> with $\nu = (J - 1)(K - 1)$ degrees of freedom. Reject if $V_{ab} \geq c + h(c)$, where
>
> $$h(c) = \frac{c}{2(J-1)(K-1)} \left\{ 1 + \frac{3c}{(J-1)(K-1) + 2} \right\} A,$$
>
> $$A = \sum_j \sum_k \frac{1}{f_{jk}} \left\{ 1 - D_{jk} \left(\frac{1}{D_{j.}} + \frac{1}{D_{.k}} - \frac{1}{D_{..}} \right) \right\}^2$$
>
> $$f_{jk} = h_{jk} - 3.$$
>
> (From Johansen, 1980, it might appear that this last expression should be
> $h_{jk} - 1$, but $h_{jk} - 3$ gives better control over the probability of a Type I
> error.)

10.3.1 S-PLUS Function t2way

The S-PLUS function

$$\text{t2way}(J, K, x, tr=0.2, grp=c(1:p), alpha=0.05, p=J*K)$$

tests the hypotheses of no main effects and no interaction, as described in Boxes 10.2
and 10.3. Here J and K denote the number of levels associated with factors A and B,
respectively. Like t1way, the data are assumed to be stored in x, which can be any
S-PLUS variable that is a matrix or has list mode. If stored in list mode, the first K
groups are assumed to be the data for the first level of factor A, the next K groups are
assumed to be data for the second level of factor A, and so on. In S-PLUS notation,
x[[1]] is assumed to contain the data for level 1 of factors A and B, x[[2]] is assumed
to contain the data for level 1 of factor A and level 2 of factor B, and so forth. If, for
example, a 2-by-4 design is being used, the data are stored as follows:

	Factor B			
	x[[1]]	x[[2]]	x[[3]]	x[[4]]
Factor A	x[[5]]	x[[6]]	x[[7]]	x[[8]]

For instance, x[[5]] contains the data for the second level of factor A and the first
level of factor B.

If the data are stored in a matrix, the first K columns are assumed to be the data for
the first level of factor A, the next K columns are assumed to be data for the second
level of factor A, and so on.

If the data are not stored in the assumed order, the argument grp can be
used to correct this problem. As an illustration, suppose the data are stored

as follows:

$$
\begin{array}{cc}
& \text{Factor B} \\
\text{Factor A} &
\begin{array}{cccc}
\text{x[[2]]} & \text{x[[3]]} & \text{x[[5]]} & \text{x[[8]]} \\
\text{x[[4]]} & \text{x[[1]]} & \text{x[[6]]} & \text{x[[7]]}
\end{array}
\end{array}
$$

That is, the data for level 1 of factors A and B are stored in the S-PLUS variable x[[2]], the data for level 1 of A and level 2 of B are stored in x[[3]], and so forth. To use t2way, first enter the S-PLUS command

$$grp<-c(2,3,5,8,4,1,6,7).$$

Then the command t2way(2,4,x,grp=grp) tells the function how the data are ordered. In the example, the first value stored in grp is 2, indicating that x[[2]] contains the data for level 1 of both factors A and B, the next value is 3, indicating that x[[3]] contains the data for level 1 of A and level 2 of B, and the fifth value is 4, meaning that x[[4]] contains the data for level 2 of factor A and level 1 of B. As usual, tr indicates the amount of trimming, which defaults to .2, and alpha is α, which defaults to .05. The function returns the test statistic for factor A, V_a, in the S-PLUS variable t2way$test.A, and the significance level is returned in t2way$sig.A. Similarly, the test statistics for factor B, V_b, and interaction, V_{ab}, are stored in t2way$test.B and t2way$test.AB, respectively, with the corresponding significance levels stored in t2way$sig.B and t2way$sig.AB.

As a more general example, the command

$$t2way(2,3,z,tr=.1,grp=c(1,3,4,2,5,6),alpha=.1)$$

would perform the tests for no main effects and no interactions for a 2-by-3 design for the data stored in the S-PLUS variable z, assuming the data for level 1 of factors A and B are stored in z[[1]], the data for level 1 of A and level 2 of B are stored in z[[3]], and so on. The analysis would be based on 10% trimmed means and $\alpha = .1$.

Note that t2way contains an argument p. Generally this argument can be ignored; it is used by t2way to check whether the total number of groups being passed to the function is equal to JK. If JK is not equal to the number of groups in x, the function prints a warning message. If, however, you want to perform an analysis using some subset of the groups stored in x, this can be done simply by ignoring the warning message. For example, suppose x contains data for 10 groups but that you want to use groups 3, 5, 1, and 9 in a 2-by-2 design. That is, groups 3 and 5 correspond to level 1 of the first factor and levels 1 and 2 of the second. The command

$$t2way(2,2,x,grp=c(3,5,1,9))$$

accomplishes this goal.

EXAMPLE. A total of $N = 50$ male Sprague–Dawley rats were assigned to one of six conditions, corresponding to a 2-by-3 ANOVA. (The data in this example were supplied by U. Hayes.) The two levels of the first factor have to do with whether an animal was placed on a fluid-restriction schedule one week prior to the initiation of the experiment. The other factor had to do with the injection of one of three drugs. One of the outcome measures was sucrose consumption shortly after acquisition of a LiCl-induced conditioned taste avoidance. The output from t2way appears as follows:

```
$test.A:
[1] 11.0931

$sig.A:
[1] 0.001969578

$test.B:
[1] 3.764621

$sig.B:
[1] 0.03687472

$test.AB:
[1] 2.082398

$critinter:
[1] 7.385763
```

So based on 20% trimmed means, there is a main effect for both factors A and B, but no interaction is detected.   ■

10.3.2 S-PLUS Function selby2

Chapter 1 mentioned an S-PLUS function called selby that is aimed at assisting with data manipulation when data are stored in a matrix, with some particular column indicating group membership. The function separates the data into groups and stores it in list mode, which in turn can be used by the S-PLUS functions in Chapter 9. For a two-way design, situations arise where one column of a matrix indicates the levels of factor A and another column indicates the levels of factor B. For example, suppose the data are stored in a file as follows:

A	B	X
1	3	46
1	2	23
2	1	21
1	1	35
⋮		

That is, the first column indicates the level of factor A, the second column indicates the level of factor B, and the third column is the outcome of interest. So here, the first row of data indicates that for level 1 of factor A and level 3 of factor B, the outcome is 46. Suppose these data have been stored in an S-PLUS matrix m. The problem is storing the data so that they can be fed into the function t2way. The S-PLUS function

$$\text{selby2}(m,\text{grpc},\text{coln}=NA)$$

is designed to sort the data in the matrix m into groups and store them in list mode, which in turn can be used in t2way. The argument grpc is a vector containing two values indicating which columns of the matrix m reflect the levels of factors A and B. For example,

$$\text{selby2}(m,\text{grpc}=c(1,2),\text{coln}=3)$$

indicates that there are two factors with the levels of the first factor stored in column 1, the levels of factor B stored in column 2, and the outcome variable to be analyzed is stored in column 3. So the command

$$> \text{dat}<-\text{selby2}(m,\text{grpc}=c(1,2),\text{coln}=3)$$

will determine how many levels there are for each factor and store the data in the S-PLUS variable dat$x. The variable dat$x[[1]] will contain the data for level 1 of both factors, dat$x[[2]] will contain the data for level 1 of factor A and level 2 of factor B, dat$x[[3]] will contain the data for level 1 of factor A and level 3 of factor B, and so on. That is, the data are automatically stored as described in Section 10.3.1.

10.4 Bootstrap Methods

The bootstrap methods described in previous chapters are readily extended to a two-way design. To apply the bootstrap-t method with a trimmed mean, you proceed in a manner similar to that in Chapter 9. That is, for the jth level of factor A and the kth level of factor B you subtract the trimmed mean ($\bar{X}_{tjk}$) from each of the n_{jk} observations; this is done for all JK groups. In symbols, for the jth level of factor A and the kth level of factor B, bootstrap samples are generated from $C_{1jk},\ldots,C_{n_{jk}jk}$, where $C_{ijk} = X_{ijk} - \bar{X}_{tjk}$. Said yet another way, center the data for each of the JK groups by subtracting out the corresponding trimmed mean, in which case the empirical distributions of all JK groups have a trimmed mean of zero. That is, the distributions are shifted so that the null hypothesis is true, with the goal of empirically determining an appropriate critical value. Then you generate bootstrap samples from each of these JK groups and compute the test statistics as described in Boxes 10.2 and 10.3. For the main effect associated with factor A, we label these B bootstrap test statistics as $V^*_{a1},\ldots,V^*_{aB}$; these B values provide an approximation of the distribution of V_a when the null hypothesis is true. Put these values in ascending order and label the results $V^*_{a(1)} \leq \cdots \leq V^*_{a(B)}$. If $V_a \geq V^*_{a(c)}$, where $c = (1 - \alpha)B$, rounded to the nearest integer, reject. The hypotheses of no main effect for factor B and no interaction are tested in an analogous manner.

Other robust measures of location can be compared with the percentile bootstrap method. Again there are many variations that might be used (which include important techniques covered in Chapter 12). Here, only one of these methods is described.

Let θ be any measure of location and let

$$\Upsilon_1 = \frac{1}{K}(\theta_{11} + \theta_{12} + \cdots + \theta_{1K}),$$

$$\Upsilon_2 = \frac{1}{K}(\theta_{21} + \theta_{22} + \cdots + \theta_{2K}),$$

$$\vdots$$

$$\Upsilon_J = \frac{1}{K}(\theta_{J1} + \theta_{J2} + \cdots + \theta_{JK}).$$

So Υ_j is the average of the K measures of location associated with the jth level of factor A. The hypothesis of no main effects for factor A is

$$H_0 : \Upsilon_1 = \Upsilon_2 = \cdots = \Upsilon_J,$$

and one variation of the percentile bootstrap method is to test this hypothesis using a slight modification of the method in Box 9.7. For example, one possibility is to test

$$H_0 : \Delta_1 = \cdots = \Delta_{J-1} = 0, \tag{10.1}$$

where

$$\Delta_j = \Upsilon_j - \Upsilon_{j+1},$$

$j = 1, \ldots, J - 1$. Briefly, generate bootstrap samples in the usual manner, yielding $\hat{\Delta}_j^*$, a bootstrap estimate of Δ_j. Then proceed as described in Box 9.7. That is, determine how deeply $\mathbf{0} = (0, \ldots, 0)$ is nested within the bootstrap samples. If $\mathbf{0}$ is relatively far from the center of the bootstrap samples, reject. (Chapter 12 describes another approach, where all pairwise comparisons of the rows are done instead. That is, for every j and ℓ, $j < \ell$, test $H_0 : \Upsilon_j = \Upsilon_\ell$.)

For reasons described in Section 9.5.3, the method just described is satisfactory when dealing with the probability of a Type I error; but when the groups differ, this approach might be unsatisfactory in terms of power, depending on the pattern of differences among the Υ_j values. One way of dealing with this issue is to compare all pairs of the Υ_j instead. That is, for every $j < j'$, let

$$\Delta_{jj'} = \Upsilon_j - \Upsilon_{j'},$$

and then test

$$H_0 : \Delta_{12} = \Delta_{13} = \cdots = \Delta_{J-1,J} = 0. \tag{10.2}$$

Of course, a similar method can be used when dealing with factor B.

The percentile bootstrap method just described for main effects can be extended to the problem of testing the hypothesis of no interactions. Box 10.4 outlines how to proceed.

BOX 10.4 How to Test for No Interactions Using the Percentile

Bootstrap Method

For convenience, label the JK measures of location as follows:

$$
\begin{array}{c}
\text{Factor B}\\
\begin{array}{cccc}
\theta_1 & \theta_2 & \cdots & \theta_K\\
\theta_{K+1} & \theta_{K+2} & \cdots & \theta_{2K}\\
\vdots & \vdots & \cdots & \vdots\\
\theta_{(J-1)K+1} & \theta_{(J-1)K+2} & \cdots & \theta_{JK}
\end{array}
\end{array}
$$

Factor A

If A is any r-by-s matrix and B is any t-by-u matrix, the Kronecker product of A and B, written as $A \otimes B$, is

$$
\begin{pmatrix}
a_{11}B & a_{12}B & \dots & a_{1s}B\\
& \vdots & &\\
a_{r1}B & a_{r2}B & \dots & a_{rs}B
\end{pmatrix}.
$$

Let C_J be a $(J-1)$-by-J matrix having the form

$$
\begin{pmatrix}
1 & -1 & 0 & 0 & \dots & 0\\
0 & 1 & -1 & 0 & \dots & 0\\
& & & \vdots & &\\
0 & 0 & \dots & 0 & 1 & -1
\end{pmatrix}.
$$

That is, $c_{ii} = 1$ and $c_{i,i+1} = -1$; $i = 1, \dots, J-1$, and C_K is defined in a similar fashion. A test of no interactions corresponds to testing

$$H_0 : \Psi_1 = \cdots = \Psi_{(J-1)(K-1)} = 0,$$

where

$$\Psi_L = \sum c_{L\ell}\theta_\ell,$$

$L = 1, \dots, (J-1)(K-1)$, $\ell = 1, \dots, J(K-1)$ and $c_{L\ell}$ be the entry in the Lth row and ℓth column of $C_J \otimes C_K$. So in effect we have a situation similar to that in Box 9.7. That is, generate bootstrap samples yielding $\hat{\Psi}_L^*$ values, do this B times, and then determine how deeply $\mathbf{0} = (0, \dots, 0)$ is nested within these bootstrap samples.

Continued

> **BOX 10.4** (*Continued*) A criticism of this approach is that when groups differ, not all possible tetrad differences are being tested, which might affect power. One way of dealing with this problem is, for every $j < j'$ and $k < k'$, set
>
> $$\Psi_{jj'kk'} = \theta_{jk} - \theta_{jk'} + \theta_{j'k} - \theta_{j'k'},$$
>
> and then test
>
> $$H_0 : \Psi_{1212} = \cdots = \Psi_{J-1,J,K-1,K} = 0. \tag{10.3}$$

10.4.1 S-PLUS Function pbad2way

The S-PLUS function

$$\text{pbad2way}(J, K, x, est = mom, conall = T, alpha = 0.05, nboot = 2000,$$

$$grp = NA, op = F, \ldots)$$

performs the percentile bootstrap method just described, where J and K indicate the number of levels associated with factors A and B. The argument conall=T indicates that all possible pairs are to be tested [as described by Equation (10.2)], and conall=F means that the hypotheses given by Equation (10.1) will be used instead. The remaining arguments are the same as those used in the S-PLUS function pbadepth described in Section 9.5.4.

> **EXAMPLE.** The data in Table 10.1 are used to illustrate the S-PLUS function pbad2way. Storing the data in the S-PLUS variable weight, the command
>
> $$\text{pbad2way}(2, 2, weight, est = median)$$
>
> tests all relevant hypotheses using medians. It is left as an exercise to verify that the significance levels for factors A and B are .39 and .056, respectively. The test for no interaction has a significance level of .16. ■

10.5 Testing Hypotheses Based on Medians

As was the case in Chapter 9, the heteroscedastic method for trimmed means in Boxes 10.2 and 10.3 should be modified for the special case where the goal is to compare medians. Currently, the methods described in Boxes 10.5 and 10.6 appear to perform relatively well and are based on simple modifications of the method for trimmed means plus the McKean–Schrader estimate of the standard error of the median (which was described in Section 4.11.2). Medians can also be compared using the bootstrap method described in the previous section, but the advantages of using this bootstrap method, versus the methods in Boxes 10.5 and 10.6, have not been investigated.

BOX 10.5 A Heteroscedastic Test for Main Effects Based on Medians

Let M_{jk} be the sample median for the jth level of factor A and the kth level of B, and let n_{jk} and S_{jk}^2 be the corresponding sample size and estimate of the squared standard error of M_{jk}. Here S_{jk}^2 is the McKean–Schrader estimate. Compute

$$R_j = \sum_{k=1}^{K} M_{jk}, \qquad W_k = \sum_{j=1}^{J} M_{jk},$$

$$d_{jk} = S_{jk}^2,$$

$$\hat{v}_j = \frac{(\sum_k d_{jk})^2}{\sum_k d_{jk}^2/(n_{jk}-1)}, \qquad \hat{\omega}_k = \frac{(\sum_j d_{jk})^2}{\sum_j d_{jk}^2/(n_{jk}-1)}$$

$$r_j = \frac{1}{\sum_k d_{jk}}, \qquad w_k = \frac{1}{\sum_j d_{jk}}$$

$$r_s = \sum_{j=1}^{J} r_j, \qquad w_s = \sum_{k=1}^{K} w_k,$$

$$\hat{R} = \frac{\sum_j r_j R_j}{r_s}, \qquad \hat{W} = \frac{\sum_k w_k W_k}{w_s}$$

$$B_a = \sum_{j=1}^{J} \frac{1}{\hat{v}_j}\left(1 - \frac{r_j}{\sum r_j}\right)^2, \qquad B_b = \sum_{k=1}^{K} \frac{1}{\hat{\omega}_k}\left(1 - \frac{w_k}{\sum w_k}\right)^2$$

$$V_a = \frac{\sum_j r_j (R_j - \hat{R})^2}{(J-1)\left(1 + \frac{2(J-2)B_a}{J^2-1}\right)}, \qquad V_b = \frac{\sum_k w_k (W_k - \hat{W})^2}{(K-1)\left(1 + \frac{2(K-2)B_b}{K^2-1}\right)}.$$

The degrees of freedom for factor A are $v_1 = J - 1$ and $v_2 = \infty$. For factor B the degrees of freedom are $v_1 = K - 1$ and $v_2 = \infty$.

Decision Rule

Reject the hypothesis of no main effect for factor A if $V_a \geq f_{1-\alpha}$, the $1 - \alpha$ quantile of an F-distribution with the degrees of freedom for factor A. Similarly, reject for factor B if $V_b \geq f_{1-\alpha}$, with the degrees of freedom for factor B.

BOX 10.6 A Heteroscedastic Test of the Hypothesis of No Interactions

Based On Medians

Again let $d_{jk} = S_{jk}^2$ be the McKean–Schrader estimate of the squared standard error of M_{jk}. Let

$$D_{jk} = \frac{1}{d_{jk}}$$

$$D_{.k} = \sum_{j=1}^{J} D_{jk}, \qquad D_{j.} = \sum_{k=1}^{K} D_{jk}$$

$$D_{..} = \sum_{j=1}^{J} \sum_{k=1}^{K} D_{jk}$$

$$\tilde{M}_{jk} = \sum_{\ell=1}^{J} \frac{D_{\ell k} M_{\ell k}}{D_{.k}} + \sum_{m=1}^{K} \frac{D_{jm} M_{jm}}{D_{j.}} - \sum_{\ell=1}^{J} \sum_{m=1}^{K} \frac{D_{\ell m} M_{\ell m}}{D_{..}}.$$

The test statistic is

$$V_{ab} = \sum_{j=1}^{J} \sum_{k=1}^{K} D_{jk} (\bar{X}_{jk} - \tilde{M}_{jk})^2.$$

Let c be the $1-\alpha$ quantile of a chi-squared distribution with $\nu = (J-1)(K-1)$ degrees of freedom.

Decision Rule
Reject if $V_{ab} \geq c$.

10.5.1 S-PLUS Function med2way

The computations for comparing medians, described in Boxes 10.5 and 10.6, are performed by the S-PLUS function

$$\text{med2way}(J,K,x,\text{alpha}=.05)$$

EXAMPLE. The example in Section 10.3.1 is repeated, only now medians are compared instead. The output from med2way is as follows:

```
$test.A:
[1] 8.124937
```

Continued

EXAMPLE. (*Continued*)

$sig.A:
[1] 0.004366059

$test.B:
[1] 1.805773

$sig.B:
[1] 0.1643474

$test.AB:
[1] 2.417318

$sig.AB:
[1] 0.2985974

So again there is a main effect for factor A and no interaction is found. But unlike before, the main effect for factor B is not significant at the .05 level. ■

10.6 Exercises

1. State the hypotheses of no main effects and no interactions for a 2-by-4 design with the following population means.

Factor A	Factor B			
	Level 1	Level 2	Level 3	Level 4
Level 1	μ_1	μ_2	μ_3	μ_4
Level 2	μ_5	μ_6	μ_7	μ_8

2. For the following 2-by-2 design with population means, state whether there is a main effect for factor A or for factor B and whether there is an interaction.

Factor A	Factor B	
	Level 1	Level 2
Level 1	$\mu_1 = 110$	$\mu_2 = 70$
Level 2	$\mu_3 = 80$	$\mu_4 = 40$

3. For the following 2-by-2 design with population means, determine whether there is an interaction. If there is an interaction, determine whether there is an ordinal interaction for rows.

| | Factor B | |
Factor A	Level 1	Level 2
Level 1	$\mu_1 = 10$	$\mu_2 = 20$
Level 2	$\mu_3 = 40$	$\mu_4 = 10$

4. Make up an example where the population means in a 3-by-3 design have no interaction effect but main effects for both factors exist.

5. For the following ANOVA summary table, fill in the missing values and then determine the number of levels, the total number of observations used, the estimate of the common variance, and whether the hypotheses of no main effects or no interaction should be rejected with $\alpha = .025$.

Source	SS	DF	MS	F
A	800	2	_____	_____
B	600	3	_____	_____
INTER	1200	_____	_____	_____
WITHIN	4800	36	_____	

6. For the following ANOVA summary table, fill in the missing values and then determine the number of levels, the total number of observations used, the estimate of the common variance, and whether the hypotheses of no main effects or no interaction should be rejected with $\alpha = .025$.

Source	SS	DF	MS	F
A	667	1	_____	_____
B	212.4	5	_____	_____
INTER	884	_____	_____	_____
WITHIN	3900	48	_____	

7. Imagine a study where two methods are compared for treating depression. A measure of effectiveness has been developed; the higher the measure, the more effective the treatment. Further assume there is reason to believe that males might respond differently to the methods than females. If the sample means are

| | Factor B | |
Factor A	Males	Females
Method 1	$\bar{X}_1 = 50$	$\bar{X}_2 = 70$
Method 2	$\bar{X}_3 = 80$	$\bar{X}_4 = 60$

and if the hypothesis of no main effects for factor A is rejected but the hypothesis of no interactions is not rejected, what does this suggest about which method should be used?

8. In the previous exercise, suppose the hypothesis of no interactions is rejected and in fact there is a disordinal interaction. What does this suggest about when you might use method 1 versus method 2?

9. Use your answer to Exercise 8 to make a general comment on interpreting main effects when there is a disordinal interaction.

10. Referring to Exercise 7, imagine the hypothesis of no interactions is rejected. Is it reasonable to conclude that the interaction is disordinal?

11. This exercise is based on a study, where the general goal is to study people's reactions to unprovoked verbal abuse. In the study, 40 subjects were asked to sit alone in a cubicle and answer a brief questionnaire. After the subjects had waited far longer than it took to fill out the form, a research assistant returned to collect the responses. Half the subjects received an apology for the delay and the other half were told, among other things, that they could not even fill out the form properly. Each of these 20 subjects were divided into two groups: Half got to retaliate against the research assistant by giving her a bad grade, and the other half did not get a chance to retaliate. All subjects were given a standardized test of hostility. Imagine that the sample means are

	Abuse	
Retaliation	Insult	Apology
Yes	$\bar{X}_1 = 65$	$\bar{X}_2 = 54$
No	$\bar{X}_3 = 61$	$\bar{X}_4 = 57$

Further assume that the hypotheses of no main effects are rejected and that the hypothesis of no interactions was also rejected. Interpret this result.

12. Verify that when comparing the groups in Table 10.1 based on medians and with the S-PLUS function pbad2way in Section 10.4.1, the significance levels for factors A and B are .39 and .056, respectively, and that for the test of no interaction the significance level is .16.

13. Read the data in Table 10.8 into an S-PLUS variable having matrix mode, and use the function selby2 in Section 10.3.2 to store the data in list mode. Store the output from selby2 in the S-PLUS variable dat, and then use the function t2way in Section 10.3.1 to test the hypotheses of no main effects or interactions based on means. Verify that the significance levels for factors A and B are .835 and .951, respectively.

14. For the data in the previous exercise, verify that if t2way is used to compare 20% trimmed means, the significance level when testing the hypothesis of no interaction is reported to be Inf, meaning that it is infinitely large due to division by zero. Explain why this happens based on the computations outlined in Box 10.3.

15. For the data in Exercise 13, verify that if t2way is used to compare 20% trimmed means instead, the significance levels for factors A and B are .15 and

TABLE 10.8   Data for Exercise 13

A	B	Outcome
1	1	32
1	1	21
1	3	19
1	2	21
1	3	46
1	3	33
1	3	10
1	2	11
1	2	13
1	2	12
1	2	59
1	3	28
1	1	19
1	1	72
1	1	35
2	1	33
2	1	45
2	1	31
2	2	42
2	2	67
2	2	51
2	1	19
2	2	18
2	2	21
2	3	39
2	3	63
2	3	41
2	3	10
2	3	11
2	3	34
2	3	47
2	1	21
2	1	29
2	3	26

.93, respectively. Why does the significance level for factor A drop from .835 when comparing means to .15 with 20% trimmed means instead?

16. The data used to illustrate the S-PLUS function med2way can be downloaded via anonymous ftp, as described in Section 1.2; it is stored in the file hayes.dat. Use these data to compare medians with the S-PLUS function pbad2way in Section 10.4.1.

17. The data used in the example at the end of Section 10.3.1 are stored in the
    file hayes.dat. (See Section 1.2 on how to download this file.) Compare the
    groups with med2way. Verify that you reject at the .05 level for factor A but
    not for factor B or when testing the hypothesis of no interaction. Note that in
    contrast, when comparing 20% trimmed means, you reject when dealing with
    factor B.

COMPARING DEPENDENT GROUPS

Chapters 8–10 described methods for comparing independent groups. This chapter describes methods for dealing with dependent groups or variables. For example, imagine you sample 10 married couples and want to compare husbands and wives on some measure of open-mindedness. That is, the goal is to characterize how, for a typical couple, the wife compares to her husband. If open-mindedness scores are independent among the population of married couples, we can compare husbands to their wives using the methods in Chapter 8. But there is no particular reason to assume that scores among women are independent of their spouse's score, so special methods are needed to take this into account. As another example, imagine that the endurance of athletes is measured before a particular training program is begun and again four weeks after training under some experimental method. This is an example of what is called a *repeated measures* or a *within-subjects* design, simply meaning that we repeatedly measure the same individuals over time. Of interest is whether endurance levels have changed. But because there is no particular reason to assume that endurance levels before undergoing the new training method are independent of the scores after training, the methods in Chapter 8 may not be valid. As a final example, C. R. Rao (1948) discussed a study where there was interest in the weight of cork borings from trees. Of specific interest was the difference in weight for the north, east, south, and west sides of the trees. Here we focus on the north versus east sides. Table 11.1 reports the data for 28 trees. Do the typical weights differ in some sense, and, if so, how do they differ and by how much? Because samples taken from the same tree may be dependent, again special methods are required that take this dependence into account.

When attention is restricted to comparing means, it is stressed that a plethora of methods have been proposed that are not covered in this chapter. For book-length descriptions of these techniques, which include procedures especially designed for handling longitudinal data, see Crowder and Hand (1990), Jones (1993), and Diggle, Liang, and Zeger (1994). This chapter covers the more basic methods for means that are typically used in applied research; methods that address the practical problems associated with these techniques are then described.

TABLE 11.1   Cork Boring Data

i	X_{i1} (North)	X_{i2} (East)	$D_i = X_{i1} - X_{i2}$
1	72	66	6
2	60	53	7
3	56	57	−1
4	41	29	12
5	32	32	0
6	30	35	−5
7	39	39	0
8	42	43	−1
9	37	40	−3
10	33	29	4
11	32	30	2
12	63	45	18
13	54	46	8
14	47	51	−4
15	91	79	12
16	56	68	−12
17	79	65	14
18	81	80	1
19	78	55	23
20	46	38	8
21	39	35	4
22	32	30	2
23	60	50	10
24	35	37	−2
25	39	36	3
26	50	34	16
27	43	37	6
28	48	54	−6

Before continuing, a comment about notation should be made. Notice that the second column in Table 11.1 is headed by the symbol X_{i1}. The notation X_{11} refers to the weight of the cork boring for the north side of the first tree. Similarly, X_{12} is the weight for the east side for the first tree. More generally, X_{i1} is the measurement for the north side of the ith tree, and X_{i2} is the east side of the ith tree. The sample mean for the first group (the north side) is $\bar{X}_1$, and for the second group, or east side, the average is denoted by $\bar{X}_2$. More succinctly, the mean of the jth group is

$$\bar{X}_j = \frac{1}{n} \sum_{i=1}^{n} X_{ij}.$$

The distribution associated with the jth group, ignoring all other groups, called the jth *marginal distribution*, has a population mean labeled μ_j. As was the case when working with independent groups, $\bar{X}_j$ is an unbiased estimate of μ_j under random sampling. That is, $E(\bar{X}_j) = \mu_j$.

11.1 The Paired T-Test for Means

Now we describe the most common method for comparing the means of two dependent groups. To add perspective, it is noted that the method in Chapter 8 for comparing independent groups is based on the result that the squared standard error of the difference between the sample means is just the sum of the individual squared standard errors. In symbols,

$$\text{VAR}\left(\bar{X}_1 - \bar{X}_2\right) = \frac{\sigma_1^2}{n_1} + \frac{\sigma_2^2}{n_2},$$

the point being that the difference between the sample means can be standardized, and this leads to a convenient method for comparing the means using the Laplace–Gosset strategy covered in Chapter 4. However, when the groups are dependent, $\text{VAR}(\bar{X}_1 - \bar{X}_2)$ takes on a more complicated form that is inconvenient from a technical point of view. The usual method for dealing with this technical problem is to use the differences between the pairs of observations to make inferences about the means.

Let D_i be the difference between the ith pair of observations. That is,

$$D_i = X_{i1} - X_{i2},$$

$i = 1, \ldots, n$. For example, for the first tree ($i = 1$), the weights are $X_{11} = 72$ and $X_{12} = 66$ and the difference is $D_1 = X_{11} - X_{12} = 72 - 66 = 6$. For the second tree ($i = 2$), $D_2 = X_{21} - X_{22} = 60 - 53 = 7$. It can be shown that the population mean associated with the difference scores is just the difference between the population means of the two groups or variables under study. In the illustration, if μ_1 represents the population mean for the north side of a tree and μ_2 the mean for the east side, then $\mu_1 - \mu_2$ is equal to μ_D, the population mean associated with the difference scores $D_1, \ldots, D_n$. More succinctly,

$$\mu_D = E(D) = \mu_1 - \mu_2.$$

This means that if we randomly sampled millions of trees and computed the sample mean of the difference scores, we would get the same result if instead we averaged the results for the east side and then subtracted this from the average for the north side of the trees. For the special case where the population means μ_1 and μ_2 are equal, $\mu_D = 0$. That is, testing the hypothesis $H_0 : \mu_1 - \mu_2 = 0$ is the same as testing the hypothesis $H_0 : \mu_D = 0$. Assuming normality, $H_0 : \mu_D = 0$ can be tested using Student's T-test described in Chapter 5. The calculations are summarized in Box 11.1 and this approach to comparing the means of two dependent groups is called the *paired T-test*.

EXAMPLE. For the data in Table 11.1, the sample mean and standard deviation of the D_i values are $\bar{D} = 4.36$ and $s_D = 7.93$, respectively. There are $n = 28$ pairs of observations (or D_i values), so, referring to Box 11.1, the test statistic is

$$T_D = \frac{4.36}{7.93/\sqrt{28}} = 2.9.$$

With $\nu = 28 - 1 = 27$ degrees of freedom and $\alpha = .01$, the critical value is the $1 - \alpha/2 = 1 - .01/2 = .995$ quantile of Student's T distribution, which is $c = 2.77$. Because $|T| = 2.9 \geq 2.77$, reject the hypothesis of equal means and conclude that the average weight for the north side of a tree is greater than the average weight of the east side. The .99 confidence interval for the difference between the means is

$$4.36 \pm 2.77\frac{7.93}{\sqrt{28}} = (.21, 8.5).$$

This says that, assuming normality, we can be reasonably certain that the difference between the means is at least .21 but not larger than 8.5. ■

BOX 11.1 How to Perform the Paired T-Test

Goal:
For two dependent groups, test $H_0 : \mu_1 = \mu_2$, the hypothesis that they have equal means.
  We observe n pairs observations: $(X_{11}, X_{12}), \ldots, (X_{n1}, X_{n2})$.

Assumptions:
Random sampling from normal distributions

Computations:
Begin by forming the differences between the paired observations:

$D_1 = X_{11} - X_{12}$
$D_2 = X_{21} - X_{22}$
$D_3 = X_{31} - X_{32}$
$\vdots$
$D_n = X_{n1} - X_{n2}$

Testing the hypothesis of equal means is accomplished by testing the hypothesis that the D_i values have a population mean of zero. That is, test $H_0 : \mu_D = 0$. To do this, compute the mean and variance of the

Continued

BOX 11.1 (*Continued*) D_i values:

$$\bar{D} = \frac{1}{n} \sum_{i=1}^{n} D_i$$

and

$$s_D^2 = \frac{1}{n-1} \sum_{i=1}^{n} (D_i - \bar{D})^2.$$

Next, compute

$$T_D = \frac{\bar{D}}{s_D/\sqrt{n}}.$$

The critical value is c, the $1 - \alpha/2$ quantile of Student's T distribution with $\nu = n - 1$ degrees of freedom.

Decision Rule:
The hypothesis of equal means is rejected if $|T_D| \geq c$.

Confidence Interval:
A $1 - \alpha$ confidence interval for $\mu_1 - \mu_2$, the difference between the means, is

$$\bar{D} \pm c\frac{s_D}{\sqrt{n}}.$$

As indicated in Box 11.1, the paired T-test has $n - 1$ degrees of freedom. If the groups are independent, Student's T-test in Section 8.1 has $2(n - 1)$ degrees of freedom, twice as many as the paired T-test. This means that if we compare independent groups (having equal sample sizes) using the method in Box 11.1, power will be lower than when using Student's T in Section 8.1. However, if the correlation between the observations is sufficiently high, the paired T-test will have more power instead. To provide an indication of why, it is noted that when pairs of observations are dependent, then

$$\text{VAR}\left(\bar{X}_1 - \bar{X}_2\right) = \frac{\sigma_1^2 + \sigma_2^2 - 2\rho\sigma_1\sigma_1}{n},$$

where σ_1^2 and σ_2^2 are the variances associated with groups 1 and 2, respectively. From this last equation we see that as the correlation, ρ, increases, the variance of the difference between the sample means (the squared standard error of $\bar{X}_1 - \bar{X}_2$) goes down. As noted in Chapters 5 and 8, as the standard error goes down, power goes up. In practical terms, the loss of degrees of freedom when using the paired T-test will be more than compensated for if the correlation is reasonably high. Just how high it must be in order to get more power with the paired T-test is unclear. For normal distributions, a rough guideline is that when $\rho > .25$, the paired T-test will have more power (Vonesh, 1983).

TABLE 11.2   Hypothetical Cholesterol Levels Before and After Training

X:	250,	320,	180,	240,	210,	255,	175,	280,	250,	200
Y:	230,	340,	185,	200,	190,	225,	185,	285,	210,	190

11.1.1 Assessing Power

When you fail to reject the hypothesis of equal means with the paired T-test, this might be because there is little or no difference between the groups, or perhaps there is a difference but you failed to detect it due to low power. Stein's method, described in Section 5.4.3, can be used to help distinguish between these two possibilities. You simply perform the computations in Section 5.4.3 on the D_i values. In particular, you can use the S-PLUS functions stein1 and stein2 described in Section 5.4.4.

EXAMPLE. Table 11.2 shows some hypothetical data on cholesterol levels before individuals undergo an experimental exercise program (X) and the levels after four weeks of training (Y). Applying the paired T-test, we fail to reject the hypothesis of equal means using $\alpha = .05$. But suppose we want power to be at least .8 when the difference between the means is 10. If we store the X values in the S-PLUS variable x and the Y values in the S-PLUS variable y, then the S-PLUS command stein1(x-y,10) returns the value 46. That is, a total of 46 pairs of observations are required to achieve power equal to .8 when the difference between the means is 10. Only 10 observations were used, so an additional $46 - 10 = 36$ observations are required. This analysis suggests that we should not accept the null hypothesis of equal means, because the sample size is too small to achieve a reasonable amount of power. ■

11.2 Comparing Trimmed Means

The good news about the paired T-test described in Section 11.1 is that if the observations in the first group (the X_{i1} values) have the same distribution as the observations in the second group (the X_{i2} values), so in particular they have equal means and variances, then Type I error probabilities substantially higher than the nominal level can generally be avoided. The reason is that for this special case, the difference scores (the D_i values) have a symmetric distribution, in which case methods based on means perform reasonably well in terms of avoiding Type I error probabilities substantially higher than the nominal level. However, as was the case in Chapter 8, practical problems arise when the two groups (or the two dependent variables) differ in some manner. Again, arbitrarily small departures from normality can destroy power, even when comparing groups having symmetric distributions. And if groups differ in terms of skewness, the paired T-test can be severely biased, meaning that power can actually decrease as the difference between the means gets large. Yet another problem is poor probability coverage when computing a confidence interval for the difference between the means as described in Box 11.1. If the goal is to test the hypothesis that the two variables under study have identical distributions, the paired T-test is satisfactory in terms of Type I errors. But if we reject, there is doubt as to whether this is due primarily to the difference between the means or to some other way the distributions differ.

A general strategy for addressing problems with a paired T-test is to switch to some robust measure of location, such as a trimmed mean. One possibility is simply to apply the methods for trimmed means described in Section 5.6 or the bootstrap methods in Section 7.5 to the difference scores (the D_i values). This is a reasonable approach; in some situations it offers an advantage over other strategies that might be used. But there is a technical issue that should be made clear. First it is noted that when working with means,

$$\bar{D} = \bar{X}_1 - \bar{X}_2.$$

That is, the average of the difference scores is just the difference between the averages associated with the two groups. Moreover, $\mu_D = \mu_1 - \mu_2$, which makes it possible to test $H_0 : \mu_1 = \mu_2$ simply by testing $H_0 : \mu_D = 0$. However, when using trimmed means, typically, but not always, the trimmed mean of the difference scores is not equal to the difference between the trimmed means of the marginal distributions. That is, usually, $\bar{D}_t \neq \bar{X}_{t1} - \bar{X}_{t2}$. An exception occurs when the pairs of observations are identical.

EXAMPLE. Consider the following pairs of observations:

G1:	1,	2,	3,	4,	5,	6,	7,	8,	9,	10
G2:	1,	2,	3,	4,	5,	6,	7,	8,	9,	10

Here both groups have a 20% trimmed mean of 5.5, so of course $\bar{X}_{t1} - \bar{X}_{t2} = 0$. As is evident, the difference scores (the D_i values) are all equal to zero, so the 20% trimmed mean of the difference scores is zero as well. That is, $\bar{D}_t = \bar{X}_{t1} - \bar{X}_{t2}$. However, if for the second group we rearrange the values so that now

G2:   1, 8, 2, 3, 4, 5, 6, 7, 9, 10

and the observations for G1 remain the same, then the difference scores are

0, −6, 1, 1, 1, 1, 1, 10, 0,

which have a 20% trimmed mean of 0.67. That is, $\bar{D}_t \neq \bar{X}_{t1} - \bar{X}_{t2}$. ■

More generally, if two dependent groups have identical distributions, then the population trimmed mean of the difference scores is equal to the difference between the individual trimmed means, which is zero. In symbols, $\mu_{tD} = \mu_{t1} - \mu_{t2} = 0$.

However, if the distributions differ, this equality is not necessarily true and in general it will be the case that $\mu_{tD} \neq \mu_{t1} - \mu_{t2}$. In practical terms, computing a confidence interval for μ_{tD} is not necessarily the same as computing a confidence interval for $\mu_{t1} - \mu_{t2}$. So an issue is whether one should test

$$H_0 : \mu_{tD} = 0, \tag{11.1}$$

or

$$H_0 : \mu_{t1} = \mu_{t2}. \tag{11.2}$$

The latter case will be called comparing the *marginal* trimmed means. In terms of Type I errors, currently it seems that there is no reason to prefer one approach over the other for the situation at hand. But for more than two groups, differences between these two approaches will be described later in this chapter. As for power, the optimal choice depends on how the groups differ, which of course is unknown. In some situations it will make little difference which method is used. But for the general problem of comparing groups based on some measure of location, we will see situations where the choice between these two approaches alters our conclusions and can provide different perspectives on how groups differ.

For the situation at hand, inferences about the trimmed mean of difference scores can be made with the method in Section 4.9.3, and the S-PLUS function trimci in Section 4.9.4 can be used to perform the calculations. Another option is to use the bootstrap method in Section 7.5. The remainder of this section describes how to test the hypothesis of equal trimmed means, which corresponds to Equation (11.2). In our illustration, the goal is to test the hypothesis that the north and east sides of a tree have equal population trimmed means.

To test the hypothesis of equal trimmed means using Laplace's general strategy, modified along the lines used by Gosset, we first Winsorize the observations. Here, however, it is important to keep observations paired together when Winsorizing. In symbols, fix j and let $X_{(1)j} \leq X_{(2)j} \leq \cdots \leq X_{(n)j}$ be the n values in the jth group, written in ascending order. Winsorizing the observations means computing

$$Y_{ij} = \begin{cases} X_{(g+1)j} & \text{if } X_{ij} \leq X_{(g+1)j} \\ X_{ij} & \text{if } X_{(g+1)j} < X_{ij} < X_{(n-g)j} \\ X_{(n-g)j} & \text{if } X_{ij} \geq X_{(n-g)j}, \end{cases}$$

where again g is the number of observations trimmed or Winsorized from each end of the distribution corresponding to the jth group. As usual, 20% trimming is assumed unless stated otherwise, in which case $g = [.2n]$, where $[.2n]$ means to round $.2n$ down to the nearest integer. The expression for Y_{ij} says that $Y_{ij} = X_{ij}$ if X_{ij} has a value between $X_{(g+1)j}$ and $X_{(n-g)j}$. If X_{ij} is less than or equal to $X_{(g+1)j}$, set $Y_{ij} = X_{(g+1)j}$; if X_{ij} is greater than or equal to $X_{(n-g)j}$, set $Y_{ij} = X_{(n-g)j}$.

EXAMPLE. As a simple illustration, consider these eight pairs of observations:

X_{i1}:	18	6	2	12	14	12	8	9
X_{i2}:	11	15	9	12	9	6	7	10

With 20% Winsorization, $g = 1$, and we see that $X_{(1)1} = 2$ and $X_{(2)1} = 6$. That is, the smallest value in the first row of data is 2, and Winsorizing changes its value to 6. If it had been the case that $g = 2$, then the values 6 and 2 in the first row (the two smallest values) would be increased to 8, the third smallest value in the first row. In this latter case, the first row of data would look like this:

$$Y_{i1}: \quad 18 \ 8 \ 8 \ 12 \ 14 \ 12 \ 8 \ 9.$$

Notice that the order of the values remains the same — we simply Winsorize the g smallest values. If $g = 3$, then we would have

$$Y_{i1}: \quad 18 \ 9 \ 9 \ 12 \ 14 \ 12 \ 9 \ 9.$$

Returning to $g = 1$, Winsorizing means that the largest observation is pulled down to the next largest. In our example, the 18 in the first row would become 14. Winsorizing the second row of data using this same process, 6 becomes 7 and 15 becomes 12. So after both rows are Winsorized, this yields

Y_{i1}:	14	6	6	12	14	12	8	9
Y_{i2}:	11	12	9	12	9	7	7	10

Yuen's method in Chapter 8 can be extended to the problem at hand in the following manner. Let $h = n - 2g$ be the effective sample size (the number of observations in each group after trimming). Let

$$d_j = \frac{1}{h(h-1)} \sum_{i=1}^{n} \left(Y_{ij} - \bar{Y}_j\right)^2$$

and

$$d_{12} = \frac{1}{h(h-1)} \sum_{i=1}^{n} \left(Y_{i1} - \bar{Y}_1\right)\left(Y_{i2} - \bar{Y}_2\right).$$

(The term d_{12} plays a role in adjusting for the dependence between the groups being compared.) Then the hypothesis of equal trimmed means can be tested with

$$T_y = \frac{\bar{X}_{t1} - \bar{X}_{t2}}{\sqrt{d_1 + d_2 - 2d_{12}}}, \tag{11.3}$$

which is rejected if $|T_y| \geq t$, where t is the $1 - \alpha/2$ quantile of Student's T distribution with $h - 1$ degrees of freedom. A $1 - \alpha$ confidence interval for the difference between the trimmed means $(\mu_{t1} - \mu_{t2})$ is

$$\left(\bar{X}_{t1} - \bar{X}_{t2}\right) \pm t\sqrt{d_1 + d_2 - 2d_{12}}.$$

■

11.2.1 S-PLUS Function yuend

The S-PLUS function

$$yuend(x,y,tr=.2,alpha=.05)$$

has been supplied to perform the calculations just described, where x and y are any S-PLUS variables containing the data for groups 1 and 2. As usual, the argument tr indicates the amount of trimming and defaults to .2 (20%), and alpha is α, which defaults to .05.

EXAMPLE. If the data in Table 11.1 for the north side of a tree are stored in the S-PLUS variable cork1 and the data for the east side are stored in cork2, then the S-PLUS command

$$yuend(cork1,cork2)$$

returns

```
$ci:
[1] 0.3658136 7.4119642

$siglevel:
[1] 0.03245657

$dif:
[1] 3.888889

$se:
[1] 1.66985

$teststat:
[1] 2.328885

$df:
[1] 17
```

So the .95 confidence interval for the difference between the 20% trimmed means is (0.37, 7.41); this interval does not contain zero, so you reject the hypothesis of equal trimmed means. As indicated, the significance level is .03. The estimated standard error of the difference between the sample trimmed means is 1.67. To test Equation (11.1), the hypothesis that the difference scores have a trimmed mean of zero, simply use the command

$$trimci(cork1-cork2),$$

where trimci is the S-PLUS function described in Section 4.9.4. It returns a .95 confidence interval of (0.254, 7.41), which is fairly similar to the confidence interval given by yuend.   ■

11.2.2 Comparing Medians

For the special case where the goal is to test the hypothesis that difference scores have a median of zero, simply use the method in Chapter 4. In particular, the S-PLUS function sint can be used to compute a confidence interval for the median of the difference scores. (If the data are stored in x and y, use the command sint(x-y).) As for testing $H_0 : \theta_1 = \theta_2$, the hypothesis that the population medians are equal, it currently seems that the best approach is to use a bootstrap method described in the next section of this chapter.

11.3 Bootstrap Methods

The bootstrap methods covered in previous chapters are readily extended to the problem of comparing dependent groups. However, based on what is currently known, the recommended type of bootstrap method differs in some cases. That is, the choice between percentile versus bootstrap-t is not always the same as described in Chapters 9 and 10, and there is the additional issue of whether to use difference scores. Currently, when comparing trimmed means, it seems that with 20% trimming or less, the bootstrap-t is preferable to the percentile bootstrap, particularly when comparing more than two dependent groups. (In contrast, when comparing independent groups as described in Section 9.5.3, a percentile bootstrap with 20% trimmed means appears to perform quite well in terms of controlling the probability of a Type I error.) But for 25% trimming or more, a percentile bootstrap seems better. (This is a very rough comparison and more detailed studies might alter this conclusion.) More generally, when using any location estimator with a breakdown point of at least .25 (such as MOM or the median), a percentile bootstrap method appears to perform relatively well. For the special case where the goal is to compare medians, inferences based on difference scores can be made with the method in Section 4.11. The extent to which a bootstrap method offers an advantage for this special case has not been investigated. As for comparing the medians of the marginal distributions, method RMPB2, described in the next section, can be used.

11.3.1 The Percentile Bootstrap

We begin with the percentile bootstrap. The methods in this section can be used with MOM, medians, M-estimators, and trimmed means (when the amount of trimming is sufficiently high or the amount of trimming is close to zero and n is large). As usual, we let θ be the population value for any of these measures of location, and we let $\hat{\theta}$ be the estimate of θ based on data.

But before continuing, we expand upon a point made in Section 11.2. It was noted that there are two closely related but different methods for comparing trimmed means. The first uses difference scores and the second compares marginal measures of location. This distinction carries over to the other robust measures of location considered here. For example, we could test the hypothesis that difference scores have a population value for MOM that is equal to zero, or we could test

$$H_0 : \theta_1 = \theta_2.$$

At the moment, using MOM with difference scores seems to be one of the best methods for controlling Type I error probabilities when the sample size is very small. It seems that testing $H_0 : \theta_1 = \theta_2$ also performs well in terms of Type I error probabilities, at least when the estimator being used has a reasonably high breakdown point. But choosing between these methods is not an academic matter, because they can reach different conclusions about whether groups differ. Also, using difference scores is known to yield higher power in some situations, and presumably situations arise where the reverse is true.

Method RMPB1

First consider how to compare groups based on difference scores. That is, the goal is to test

$$H_0 : \theta_D = 0,$$

the hypothesis that the typical difference score (the typical D_i value) is zero. In the context of the bootstrap, this means that we obtain a bootstrap sample by resampling with replacement n D_i values. Let $\hat{\theta}_D^*$ be the estimate of θ_D based on this bootstrap sample. Next, repeat this process B times, yielding, $\hat{\theta}_{D1}^*, \ldots, \hat{\theta}_{DB}^*$ and let $\hat{p}^*$ be the proportion of these bootstrap values that are greater than zero. So if A is the number of times $\hat{\theta}_{Db}^* > 0$, then

$$\hat{p}^* = \frac{A}{B}.$$

Then reject H_0 if

$$\hat{p}^* \leq \frac{\alpha}{2}$$

or if

$$\hat{p}^* \geq 1 - \frac{\alpha}{2}.$$

The S-PLUS function onesampb in Section 7.7.2 can be used to perform the calculations just described.

EXAMPLE. Again assume the data in Table 11.1 are stored in cork1 and cork2. Then the command

$$\text{onesampb(cork1-cork2)}$$

computes a confidence interval for the difference scores based on MOM and returns a .95 confidence interval of $(0.27, 7.82)$. So the typical difference between the north and east sides of a tree tends to be greater than zero. That is, the north side tends to have a higher cork boring weight.   ■

Method RMPB2

If instead the goal is to test

$$H_0 : \theta_1 = \theta_2,$$

bootstrap samples are generated by resampling pairs of observations. (This is in contrast to independent groups, where a bootstrap sample is obtained by resampling observations from the first group and then resampling observations from the second.) For example, suppose you observe the following five pairs of values:

6	12
19	2
23	5
32	21
15	18

Then a bootstrap sample consists of randomly resampling with replacement five pairs of values from the five pairs just listed. So the first pair, (6, 12), has probability 1/5 of being chosen each time a bootstrap pair is generated, and the same is true for the other four pairs. A bootstrap sample from these five pairs of values might look like this:

23	5
32	21
6	12
32	21
6	12

More generally, if you observe n pairs of values — say, $(X_{11}, X_{12}), \ldots, (X_{n1}, X_{n2})$ — generating a bootstrap sample means that each pair has probability $1/n$ of being selected each time a pair is chosen. That is, you randomly resample with replacement pairs of observations. The bootstrap sample of paired observations generated in this manner will be labeled $(X_{11}^*, X_{12}^*), \ldots, (X_{n1}^*, X_{n2}^*)$.

After generating a bootstrap sample of n pairs of observations, label the resulting estimates of location as $\hat{\theta}_1^*$ and $\hat{\theta}_2^*$. Repeat this B times, yielding $(\hat{\theta}_{1b}^*, \hat{\theta}_{2b}^*)$, $b = 1, \ldots, B$, and let $\hat{p}^*$ be the proportion of times the bootstrap estimate of location for the first group is greater than the estimate for the second. In symbols, $\hat{p}^*$ is an estimate of

$$P\left(\hat{\theta}_1^* > \hat{\theta}_2^*\right)$$

for a random bootstrap sample. As in Chapter 8, it is convenient to set $\hat{p}_m^*$ equal to $\hat{p}^*$ or $1 - \hat{p}^*$, whichever is smaller, in which case $2\hat{p}_m^*$ is like a significance level, and you reject the hypothesis of equal measures of location if

$$\hat{p}_m^* \leq \frac{\alpha}{2}. \tag{11.4}$$

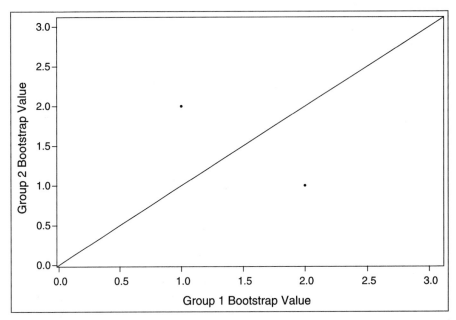

FIGURE 11.1   The lower right point corresponds to a situation where for a pair of bootstrap estimates, Group 1 has the larger value. For a point above the line, the reverse is true.

It helps to graphically illustrate the bootstrap method just described. First imagine that we generate a single bootstrap sample, compute some measure of location $\hat{\theta}$ for each group (or variable), yielding the pair $(\hat{\theta}_1^*, \hat{\theta}_2^*)$, and then plot the point corresponding to this pair of values. Then from basic principles, if $\hat{\theta}_1^* > \hat{\theta}_2^*$, $(\hat{\theta}_1^*, \hat{\theta}_2^*)$ corresponds to a point below (or to the right of) a line having a slope of 1 and an intercept of zero, as indicated in Figure 11.1. If $\hat{\theta}_1^* < \hat{\theta}_2^*$, the point $(\hat{\theta}_1^*, \hat{\theta}_2^*)$ will appear above (or to the left of) this line. And if $\hat{\theta}_1^* = \hat{\theta}_2^*$, the point $(\hat{\theta}_1^*, \hat{\theta}_2^*)$ is somewhere on the line shown in Figure 11.1.

More generally, if the null hypothesis is true and B bootstrap pairs are generated and plotted, then it should be the case that approximately half of the plotted points will be below the line having slope 1 and intercept zero, and about half should be above the line instead, as illustrated in Figure 11.2. However, if the first group has a larger population measure of location than the second ($\theta_1 > \theta_2$), then typically a majority of the bootstrap values, when plotted, will be to the right of the line having a slope of 1 and an intercept of zero, as shown in the left panel of Figure 11.3. If the reverse is true and the typical measure for group 1 is less than the typical measure for group 2, then a scatterplot of the points will appear as in the right panel of Figure 11.3. In particular, if we set $\alpha = .05$ and if 97.5% of the bootstrap values are to the right of the line in Figure 11.3, reject and conclude that the first group typically has a larger value than the second. If 97.5% of the bootstrap values are to the left of the line in Figure 11.3, reject and conclude that the first group typically has a smaller value. A $1 - \alpha$ confidence interval for $\theta_1 - \theta_2$ is

$$\left(V_{(\ell+1)}, V_{(u)}\right),$$

(11.5)

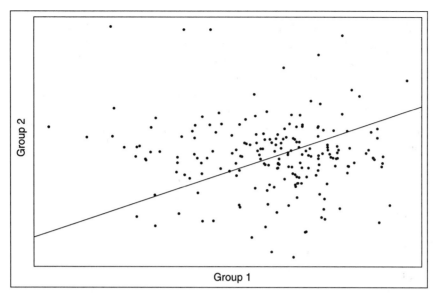

FIGURE 11.2 A plot of bootstrap values where the null hypothesis is true.

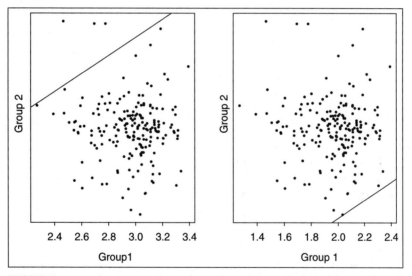

FIGURE 11.3 Bootstrap values where the null hypothesis is false. In the left panel, Group 1 has the larger measure of location. In the left panel, the reverse is true.

where $\ell = \alpha B/2$, rounded to the nearest integer, $u = B - \ell$, $V_b = \hat{\theta}_{1b}^* - \hat{\theta}_{2b}^*$, $b = 1, \ldots, B$, and $V_{(1)} \leq \ldots \leq V_{(B)}$ are the V_b values written in ascending order.

The estimated significance level when using the method just described is $2\hat{p}_m^*$, where $\hat{p}_m^*$ is given by Equation (11.4). A concern is that as the correlation between the two variables under study increases, the method can become too conservative in terms of the probability of a Type I error when using MOM or 20% trimmed means. For example, when testing at the .05 level, the actual probability of a Type I error can drop below .025 and even go as low as .01, suggesting that power might be relatively low. Currently, a method that appears to give better results in some situations is applied as follows. Set $C_{ij} = X_{ij} - \hat{\theta}_j$. That is, shift the data so that both groups have measures of location equal to zero, so in particular the null hypothesis is true. Generate bootstrap samples from the C_{ij} values and let $\hat{q}^*$ be the proportion of times, among the B bootstrap samples, that the measure of location from the first group is larger than the measure of location from the second. Because the null hypothesis is true, $\hat{q}^*$ should be approximately equal to .5. A so-called bias-adjusted significance level is

$$2\min\left(\hat{p}_a^*, 1 - \hat{p}_a^*\right),$$

where $\hat{p}_a^* = \hat{p}^* - .1(\hat{q}^* - .5)$.

11.3.2   S-PLUS Function rmmcppb

The S-PLUS function

rmmcppb(x,y=NA,alpha=.05,est=mom,dif=T,plotit=T,nboot=NA,BA=F,...)

has been supplied for comparing dependent groups based on the percentile boot-strap methods RMPB1 and RMPB2 described in the previous section. (This function contains another argument, con, which is explained in Chapter 12.) You can enter data through x, which can be a matrix or can have list mode. When x is a matrix, columns correspond to groups. (That is, column 1 is group 1, column 2 is group 2, and so forth.) Alternatively, when comparing two groups only, the data for the first group can be stored in x and the data for the second stored in y. (When comparing more than two groups, this function uses the method outlined in Section 12.8.4.) The argument est indicates which measure of location is used and defaults to MOM. The default value for the argument dif (T for true) indicates that difference scores will be used. That is, method RMPB1 in Section 11.3.1 is applied; dif=F results in using method RMPB2. When comparing two groups, this function also plots the bootstrap values if plotit=T is used. For method RMPB2, pairs of bootstrap values are plotted and the central $1 - \alpha$ percent of these values is indicated by a polygon. These centrally located values provide an approximate $1 - \alpha$ confidence region for the values of the parameters being estimated. (The following example will help clarify what this means.) Included in the plot is a line having a slope of 1 and an intercept of zero to help provide perspective on how much the groups differ. When dif=F is used, setting BA=T will cause bias-adjusted significance levels to be computed.

EXAMPLE. Figure 11.4 shows the plot created by rmmcppb, with dif=F, based on the cork data in Table 11.1. Each point represents a pair of boot-strap estimates of the measures of location, which here is taken to be MOM. The polygon contains the central $1 - \alpha$ percent of these values and provides a two-dimensional analog of a confidence interval. That is, it is estimated

Continued

EXAMPLE. (*Continued*) that (θ_1, θ_2) is somewhere inside this polygon. For example, the value $(50, 45)$ lies inside the polygon, indicating that having the typical weight for the north side of a tree equal to 50 and simultaneously having the typical weight for the east side equal to 45, is a reasonable possibility. The point $(55, 40)$ does not lie inside the polygon indicating that it is unlikely that simultaneously, $\theta_1 = 55$ and $\theta_2 = 40$. Note, however, that $\theta_1 = 55$, taken by itself, is a reasonable estimate of the typical weight for the north side of a tree. The value 55 is within the .95 confidence interval based on the S-PLUS function momci. That is, ignoring the data on the east side, the typical weight on the north side could be as high as 55. But if the typical weight on the north side is 55, Figure 11.4 suggests that it is unlikely that the typical weight on the east side is 40. In contrast, if we ignore the north side data, now 40 is a reasonable estimate for the east side. To add perspective, Figure 11.4 includes the line having slope 1 and intercept zero. If the null hypothesis is true, that is, $\theta_1 = \theta_2$, then the true values for θ_1 and θ_2 must be somewhere on this line.

Figure 11.5 shows the plot created when dif=T is used. When the typical difference score is zero, about half of the bootstrap values should be above the horizontal line having intercept zero. In this particular case, the proportion of points below this horizontal line is only .014, indicating a significance level of $2 \times .014 = .028$.

A portion of the printed output from rmmcppb (with dif=F) looks like this:

```
$output:
    psihat   sig.level   crit.sig    ci.lower   ci.upper
   5.421652     0.174        0.05    -2.178571   10.85165
```

The value under psihat, 5.42, is $\hat{\theta}_1 - \hat{\theta}_2$, the estimated difference between the typical weight for group 1 minus the typical weight for group 2. Under sig.level we see 0.174, which is the significance level. That is, the proportion of points in Figure 11.4 above the straight line is $.174/2 = .087$. If only 2.5% were above the line, you would reject at the .05 level. Under crit.sig we see the α value used and to the right is a $1 - \alpha$ confidence interval for $\theta_1 - \theta_2$: $(-2.18, 10.85)$. When using difference scores, we reject at the .05 level, but here we fail to reject, the only point being that the choice of method can make a practical difference. ■

EXAMPLE. To test the hypothesis that the two variables in Table 11.1 have equal population medians, use the command

$$\text{rmmcppb(cork,est=median,dif=F)},$$

assuming the data are stored in cork. The resulting significance level is .176. To compare 20% trimmed means, use the command

$$\text{rmmcppb(cork,est=mean,tr=.2)}.$$

■

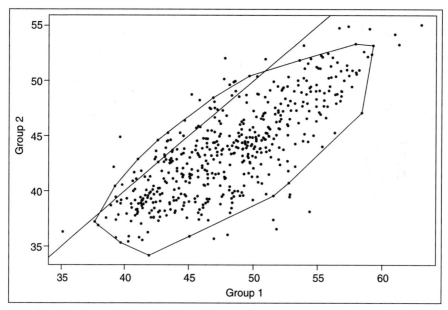

FIGURE 11.4   A plot created by the S-PLUS function rmmcppb with the argument dif=F.

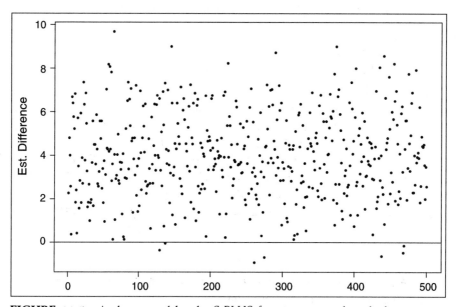

FIGURE 11.5   A plot created by the S-PLUS function rmmcppb with the argument dif=T.

11.3.3 Bootstrap-*t*

Unless the sample size is fairly large, in terms of controlling the probability of a Type I error or achieving accurate probability coverage when computing a confidence interval, the percentile bootstrap method performs poorly when working with means or when using a trimmed mean and the amount of trimming is close to zero. All indications are that when comparing dependent groups, the amount of trimming should be at least .25 when using a percentile bootstrap method to compare the marginal trimmed means. Indeed, with the amount of trimming sufficiently high, the percentile bootstrap seems preferable to a bootstrap-t in terms of Type I errors. (Some adjustments can be made to percentile bootstrap methods when the amount of trimming is between .2 and .25 that can render it relatively competitive with the bootstrap-t. The adjustments depend on the sample size, but currently, from a practical point of view, there seems to be no reason to describe them here.) Also, when comparing more than two groups, extant studies have focused on testing for equal population-trimmed means, as opposed to making inferences based on the trimmed mean of the difference scores. For this reason, the bootstrap-t for comparing dependent groups focuses on comparing trimmed means with the goal of testing $H_0 : \mu_{t1} = \mu_{t2}$. (There might be practical reasons for considering a bootstrap-t on difference scores, but this has not been established as yet.)

The goal is to estimate an appropriate critical value when the null hypothesis of equal trimmed means is true and when using the test statistic T_y given by Equation (11.3). This is done by centering the data, computing T_y based on a bootstrap sample generated from the centered data, and then repeating this process B times to get an estimate of the distribution of T_y when the null hypothesis of equal trimmed means is true. More formally, set $C_{ij} = X_{ij} - \bar{X}_{tj}$ and let C_{ij}^* be a bootstrap sample obtained by resampling with replacement n pairs of values from the C_{ij} values. Let T_y^* be the value of T_y based on the C_{ij}^* values. Repeat this process B times, yielding T_{yb}^*, $b = 1, \ldots, B$. Let $T_{y(1)}^* \leq \cdots \leq T_{y(B)}^*$ be the T_{yb}^* values written in ascending order. Set $\ell = \alpha B/2$, rounding to the nearest integer, and $u = B - \ell$. Then an estimate of the lower and upper critical values is $T_{y(\ell+1)}^*$ and $T_{y(u)}^*$, respectively. An equal-tailed $1 - \alpha$ confidence interval for $\mu_{t1} - \mu_{t2}$ is

$$\left(\bar{X}_{t1} - \bar{X}_{t2} + T_{y(\ell+1)}^* \sqrt{d_1 + d_2 - 2d_{12}}, \bar{X}_{t1} - \bar{X}_{t2} + T_{y(u)}^* \sqrt{d_1 + d_2 - 2d_{12}} \right). \quad (11.6)$$

To get a symmetric confidence interval, replace T_{yb}^* by its absolute value and set $a = (1 - \alpha)B$, rounding to the nearest integer, in which case the $(1 - \alpha)$ confidence interval for $(\mu_{t1} - \mu_{t2})$ is

$$\left(\bar{X}_{t1} - \bar{X}_{t2} \right) \pm T_{y(a)}^* \sqrt{d_1 + d_2 - 2d_{12}}.$$

(The S-PLUS function rmanovb described in Section 11.6.4 performs the calculations.)

11.4 Measuring Effect Size

Section 8.11 describes some measures of effect size and indicates why they are important. It is noted that analogs of these methods are available when comparing dependent groups. For example, a simple measure of effect size is

$$\Delta = \frac{\mu_1 - \mu_2}{\sigma_d}, \tag{11.7}$$

where σ_d is the population standard deviation of the difference scores (the D_i values). But for reasons already described, this approach can grossly underestimate the extent to which two groups differ, and it does not capture some of the global differences between the groups. Another approach is to compute a confidence interval for $\mu_1 - \mu_2$ or some robust measure of location, such as a trimmed mean or MOM.

A simple but useful way of assessing effect size is with a boxplot of the difference scores (the D_i values). When comparing dependent groups with identical distributions, the D values will tend to be symmetrically distributed about zero. So in particular, a boxplot will tend to have a median of zero. (Also see the discussion of the sign test in Chapter 15.)

Another simple measure of effect size is $\hat{p}^*$ or $2\hat{p}_m^*$, the bootstrap estimate of the significance level. These measures of effect size reflect the separation of the distributions but not the magnitude of the differences between the measures of location.

Another approach is to use an analog of Q described in Section 8.11. That is, Q represents the likelihood of correctly determining whether an observation came from group 1 or group 2. If we make an arbitrary decision, meaning that we randomly decide whether an observation came from group 1, then $Q = .5$. But suppose we decide that an observation came from group 1 using the method outlined in Box 8.2. Roughly, you first estimate the likelihood that an observation came from group 1 and then do the same for group 2. If the estimated likelihood is higher that it came from group 1, then simply decide that it indeed came from group 1. If the groups are identical, meaning they have identical (marginal) distributions, then the probability of a correct decision is $Q = .5$. The more separated the groups happen to be, the closer Q will be to 1. For dependent groups, Q can be estimated with the so-called .632 bootstrap estimator, $\hat{Q}_{.632}$, which is computed as summarized in Box 11.2. As noted in Chapter 8, a rough guideline is that $\hat{Q}_{.632} = .55$ reflects a small difference between the groups and $\hat{Q}_{.632} = .66$ is large. It is noted that $\hat{Q}_{.632} < .5$ is possible even though we know $Q \geq .5$. So $\hat{Q}_{.632} < .5$ suggests that there is little or no difference between the groups.

BOX 11.2 A Bootstrap Estimate of Q for Dependent Groups

For each group, compute the kernel density estimator as described in Section 3.7 and label the results $\hat{f}_1(x)$ and $\hat{f}_2(x)$, respectively. Set $\hat{\eta}(X_{i1}) = 1$ if $\hat{f}_1(X_{i1}) > \hat{f}_2(X_{i1})$; otherwise $\hat{\eta}(X_{i1}) = 0$. Generate a bootstrap sample by resampling n pairs of observations and let $\hat{\eta}^*$ be the bootstrap analog of $\hat{\eta}$. Repeat this process B times, yielding $\hat{\eta}_b^*$, $b = 1, \ldots, B$. Let

$$\hat{\epsilon} = \frac{1}{n} \sum_{i=1}^{n} \frac{1}{B_i} \sum_{b \in C_i} \hat{\eta}_b^*(X_{1i}).$$

Continued

> **BOX 11.2** (*Continued*)
> C_i is the set of indices of the *b*th bootstrap sample not containing the pair (X_{1i}, X_{2i}), and B_i is the number of such bootstrap samples. (The illustration in conjunction with Box 8.2 provides some details about how $\hat{\epsilon}$ is computed.) Let
>
> $$\hat{Q}_{ap} = \frac{1}{n} \sum \hat{\eta}(X_{1i}).$$
>
> The estimate of Q is
>
> $$\hat{Q}_{.632} = .368\hat{Q}_{ap} + .632\hat{\epsilon},$$

11.4.1 S-PLUS Functions qhatd

The S-PLUS function

$$qhatd(x,y,nboot{=}50)$$

computes the measure of effect size, $\hat{Q}_{.632}$, as described in Box 11.2. As usual, x and y can be any S-PLUS variables containing data.

> **EXAMPLE.** For the data in Table 11.1, qhatd returns $\hat{Q}_{.632} = .42$, suggesting that if there is a difference between the groups, it is small. However, some caution is required because the precision of the estimate of Q, based on $\hat{Q}_{.632}$, is not known. The S-PLUS command
>
> $$onesampb(x-y,est{=}median)$$
>
> returns (0, 8) as a .95 confidence interval for the median of the difference scores. This again suggests that if the groups differ, in some sense the difference might be small, because we cannot rule out the possibility that the median difference is zero. ■

11.5 Comparing Variances

As noted in Chapter 8, situations arise where the goal is to compare variances rather than some measure of location. When comparing robust measures of scale, all indications are that the percentile bootstrap method is the most effective approach. But for the special case where there is specific interest in the variances, a variation of the percentile bootstrap method is required. The method described here is based on a bootstrap analog of the so-called *Morgan–Pitman test* and currently appears to be the most satisfactory method among the many that have been proposed.

As usual, we have n pairs of randomly sampled observations: $(X_{11}, X_{12}), \ldots,$ (X_{n1}, X_{n2}). Set

$$U_i = X_{i1} - X_{i2} \qquad \text{and} \qquad V_i = X_{i1} + X_{i2},$$

$(i = 1, \ldots, n)$. That is, for the ith pair of observations, U_i is the difference and V_i is the sum. Let σ_1^2 be the variance associated with the first group (the X_{i1} values) and let σ_2^2 be the variance associated with the second. The goal is to test the hypothesis that these variances are equal. It can be shown that if

$$H_0 : \sigma_1^2 = \sigma_2^2$$

is true, then Pearson's correlation between the U and V values is zero. That is, we can test the hypothesis of equal variances by testing

$$H_0 : \rho_{uv} = 0,$$

where ρ_{uv} is Pearson's correlation between U and V. Section 6.5 describes the conventional method for testing the hypothesis of a zero correlation based on Student's T distribution. Applying this method to the U and V values yields what is known as the *Morgan–Pitman test* for equal variances. But we have already seen that the conventional test of a zero correlation has undesirable properties when variables are dependent; these problems render the Morgan–Pitman test unsatisfactory when distributions differ. To deal with this problem, we simply apply the modified percentile bootstrap as was done in Section 7.3. That is, take a bootstrap sample of n pairs of the U and V values, compute the correlation between these values, repeat this 599 times, and label the results $r_1^*, \ldots, r_{599}^*$, in which case a .95 confidence interval for ρ_{uv} is

$$\left(r_{(a)}^*, r_{(c)}^* \right),$$

where for $n < 40$, $a = 7$ and $c = 593$; for $40 \leq n < 80$, $a = 8$ and $c = 592$; for $80 \leq n < 180$, $a = 11$ and $c = 588$; for $180 \leq n < 250$, $a = 14$ and $c = 585$; while for $n \geq 250$, $a = 15$ and $c = 584$. To apply the method, you can use the S-PLUS function pcorb (described in Section 7.3.3), as illustrated in the next example.

EXAMPLE. Imagine that the data in Table 11.1 corresponding to the north side of a tree are stored in the S-PLUS variable x and that the east side data are stored in the S-PLUS variable y. Then the command pcorb(x-y,x+y) tests the hypothesis that these two dependent groups have equal variances. The function returns a .95 confidence interval of $(-0.154, 0.677)$; this interval contains zero, so you fail to reject. (The S-PLUS function comvar2d can be used as well. That is, the command comvar2d(x,y) will test the hypothesis of equal variances.) ■

11.5.1 Comparing Robust Measures of Scale

As for comparing dependent groups using a robust measure of scale, a basic percentile bootstrap method can be used. For example, if the goal is to test the hypothesis that two dependent groups have equal population percentage bend midvariances, simply proceed as described in Section 11.3, where a bootstrap sample is obtained by resampling with replacement n pairs of observations. For each group, compute the percentage bend midvariance as described by Equation (3.18) (in Box 3.1) or whichever measure of scale is of interest, and let d^* be the difference between the two estimates. Repeat this process B times, yielding $d_1^*, \ldots, d_B^*$, and then put these B values in ascending order, and label them in the usual way, namely, $d_{(1)}^* \leq \cdots \leq d_{(B)}^*$. Then a $1 - \alpha$ confidence interval for the difference between the measures of scale is $(d_{(\ell+1)}^*, d_{(u)}^*)$, where as usual $\ell = \alpha(B)/2$, rounded to the nearest integer, and $u = B - \ell$. In essence, use method RMPB2 (described in Section 11.3.1), except use a robust measure of scale rather than a robust measure of location. The calculations can be done with the S-PLUS function rmmcppb described in Section 11.3.2, with the argument dif set equal to F.

EXAMPLE. Chapter 8 describes a study where EEG measures of convicted murderers were compared to a control group. In fact measures for both groups were taken at four sites in the brain. For illustrative purposes, the first two sites for the control group are compared using the percentage bend midvariance described in Chapter 3. The data are

Site 1:	−0.15	−0.22	0.07	−0.07	0.02	0.24	−0.60	−0.17	−0.33
	0.23	−0.69	0.70	1.13	0.38				

Site 2:	−0.05	−1.68	−0.44	−1.15	−0.16	−1.29	−2.49	−1.07	−0.84
	−0.37	0.01	−1.24	−0.33	0.78				

If we store the data in the S-PLUS variables eeg1 and eeg2, the command

$$\text{rmmcppb(eeg1, eeg2, est=pbvar, dif=F)}$$

returns a significance level of .13. So we are not able to detect a difference in the variation between these two sites when testing at the .05 level. If the argument est=pbvar is replaced with est=mad so that the MAD measure of scale is used, the significance level is .25. ■

11.6 Comparing More Than Two Groups

This section describes how to compare measures of location corresponding to multiple dependent groups. One strategy is to test

$$H_0 : \theta_1 = \cdots = \theta_J. \tag{11.8}$$

Another approach is to test some appropriate hypothesis based on difference scores. A simple possibility is to test

$$H_0 : \theta_{d1} = \cdots = \theta_{d,J-1} = 0, \tag{11.9}$$

where θ is any measure of location, such as a median or MOM, and θ_{dj} is the value of θ based on the difference scores between groups j and $j + 1$ ($j = 1, \ldots, J - 1$).

That is, θ_{dj} is the population value of θ corresponding to $Y_{ij} = X_{ij} - X_{i,j+1}$. We begin with a nonbootstrap method based on trimmed means and then consider bootstrap methods that can be used with trimmed means and other robust estimators. Section 11.8 discusses the relative merits of the methods about to be described.

11.6.1 A Method for Trimmed Means

This section generalizes the methods in Sections 11.1 and 11.2 to test

$$H_0 : \mu_{t1} = \cdots = \mu_{tJ}, \tag{11.10}$$

the hypothesis that all J groups have equal trimmed means. The test statistic, labeled F, is computed as summarized in Box 11.3. You reject if $F \geq f$, where f is the $1 - \alpha$ quantile of an F-distribution with ν_1 and ν_2 degrees for freedom, where ν_1 and ν_2 are computed as described in Box 11.4.

For the special case where there is no trimming, F reduces to a standard test statistic for comparing means (which is typically referred to as an ANOVA F-test). An early approach to determining an appropriate critical value for this special case was based on assuming that the variances and correlations among the J groups follow a rather restrictive pattern called *sphericity* or *circularity* (see Huynh & Feldt, 1976). However, violating this assumption is known to cause practical problems. Methods for testing the assumption that sphericity holds have been proposed, but all indications are that such tests do not have enough power to detect situations where the assumption should be discarded (e.g., Boik, 1981; Keselman, Rogan, Mendoza, & Breen, 1980), so no details are given here. Rather, we simply rely on the method in Box 11.4 for determining a critical value that was derived without assuming sphericity. (Readers interested in more details about sphericity are referred to Kirk, 1995, as well as Rogan, Keselman, & Mendoza, 1979.) For simulation results on how the method in Box 11.3 performs with 20% trimmed means, see Wilcox, Keselman, Muska, and Cribbie, 2000.

BOX 11.3 Test Statistic for Comparing the Trimmed Means

of Dependent Groups

Winsorize the observations by computing

$$Y_{ij} = \begin{cases} X_{(g+1)j}, & X_{ij} \leq X_{(g+1)j} \\ X_{ij}, & X_{(g+1)j} < X_{ij} < X_{(n-g)j} \\ X_{(n-g)j}, & X_{ij} \geq X_{(n-g)j}. \end{cases}$$

Let $h = n - 2g$ be the effective sample size, where $g = [\gamma n]$, $[\gamma n]$ is γn rounded down to the nearest integer, and γ is the amount of trimming. Compute

$$\bar{X}_t = \frac{1}{J} \sum \bar{X}_{tj}$$

Continued

BOX 11.3 (*Continued*)

$$Q_c = (n - 2g) \sum_{j=1}^{J} \left(\bar{X}_{tj} - \bar{X}_t \right)^2$$

$$Q_e = \sum_{j=1}^{J} \sum_{i=1}^{n} \left(Y_{ij} - \bar{Y}_j - \bar{Y}_{i.} + \bar{Y}_{..} \right)^2,$$

where

$$\bar{Y}_j = \frac{1}{n} \sum_{i=1}^{n} Y_{ij}$$

$$\bar{Y}_{i.} = \frac{1}{J} \sum_{j=1}^{J} Y_{ij}$$

$$\bar{Y}_{..} = \frac{1}{nJ} \sum_{j=1}^{J} \sum_{i=1}^{n} Y_{ij}.$$

The test statistic is

$$F = \frac{R_c}{R_e},$$

where

$$R_c = \frac{Q_c}{J - 1}$$

$$R_e = \frac{Q_e}{(h - 1)(J - 1)}.$$

Decision Rule:
Reject if $F \geq f$, the $1 - \alpha$ quantile of an F-distribution with degrees of freedom computed as described in Box 11.4.

BOX 11.4 How to Compute Degrees of Freedom When

Comparing Trimmed Means

Let

$$v_{jk} = \frac{1}{n - 1} \sum_{i=1}^{n} \left(Y_{ij} - \bar{Y}_j \right) \left(Y_{ik} - \bar{Y}_{.k} \right)$$

Continued

BOX 11.4 (*Continued*)

for $j = 1, \ldots, J$ and $k = 1, \ldots, J$, where, as in Box 11.3, Y_{ij} is the Winsorized observation corresponding to X_{ij}. When $j = k$, $v_{jk} = s^2_{wj}$, the Winsorized sample variance for the jth group; and when $j \neq k$, v_{jk} is a Winsorized analog of the sample covariance.

Let

$$\bar{v}_{..} = \frac{1}{J^2} \sum_{j=1}^{J} \sum_{k=1}^{J} v_{jk}$$

$$\bar{v}_d = \frac{1}{J} \sum_{j=1}^{J} v_{jj}$$

$$\bar{v}_{j.} = \frac{1}{J} \sum_{k=1}^{J} v_{jk}$$

$$A = \frac{J^2 (\bar{v}_d - \bar{v}_{..})^2}{J - 1}$$

$$B = \sum_{j=1}^{J} \sum_{k=1}^{J} v_{jk}^2 - 2J \sum_{j=1}^{J} \bar{v}_{j.}^2 + J^2 \bar{v}_{..}^2$$

$$\hat{\epsilon} = \frac{A}{B}$$

$$\tilde{\epsilon} = \frac{n(J - 1)\hat{\epsilon} - 2}{(J - 1)\{n - 1 - (J - 1)\hat{\epsilon}\}}.$$

The degrees of freedom are

$$\nu_1 = (J - 1)\tilde{\epsilon}$$

$$\nu_2 = (J - 1)(h - 1)\tilde{\epsilon},$$

where h is the effective sample size (the number of observations left in each group after trimming).

11.6.2 S-PLUS Function rmanova

The S-PLUS function rmanova, which comes with this book, compares the trimmed means of J dependent groups using the calculations in Boxes 11.3 and 11.4. The function has the general form

$$\text{rmanova}(x, tr=.2, grp=c(1:\text{length}(x))).$$

The data are stored in x, which can be either an n-by-J matrix, with the jth column containing the data for jth group, or an S-PLUS variable having list mode. In the latter case, x[[1]] contains the data for group 1, x[[2]] contains the data for group 2, and so on. As usual, tr indicates the amount of trimming, which defaults to .2, and grp can be used to compare a subset of the groups. By default, the trimmed means of all J groups are compared. If, for example, there are five groups but the goal is to test $H_0 : \mu_{t2} = \mu_{t4} = \mu_{t5}$, then the command

$$\text{rmanova}(x, \text{grp}=c(2,4,5))$$

accomplishes this goal using 20% trimming.

EXAMPLE. The S-PLUS function rmanova is illustrated with the weight of cork borings described in the introduction to this chapter, only now all four sides of each tree are used. The data are reproduced in Table 11.3. If we store the data in the S-PLUS variable cork in a matrix having four columns, then the output from the command rmanova(cork) is

```
$test:
[1] 2.311757

$df:
[1] 2.636236  44.816010

$siglevel:
[1] 0.09624488
```

So with $\alpha = .05$, we are unable to reject the hypothesis that the typical weights of cork borings differ among the four sides of the trees. ■

11.6.3 A Bootstrap-*t* Method for Trimmed Means

When comparing trimmed means, a bootstrap-*t* method can be applied in basically the same way as described in previous chapters, only again we generate bootstrap samples in a manner consistent with dependent groups. As usual, when working with a bootstrap-*t* method, we begin by centering the data. That is, we set

$$C_{ij} = X_{ij} - \bar{X}_{tj}$$

with the goal of estimating an appropriate critical value, based on the test statistic F in Box 11.3, when the null hypothesis is true. The remaining steps are as follows:

1. Generate a bootstrap sample by randomly sampling, with replacement, n rows of data from the matrix

$$\begin{pmatrix} C_{11}, \ldots, C_{1J} \\ \vdots \\ C_{n1}, \ldots, C_{nJ} \end{pmatrix},$$

TABLE 11.3   Cork Boring Weights for the North, East, South, and West Sides of Trees

N	E	S	W
72	66	76	77
60	53	66	63
56	57	64	58
41	29	36	38
32	32	35	36
30	35	34	26
39	39	31	27
42	43	31	25
37	40	31	25
33	29	27	36
32	30	34	28
63	45	74	63
54	46	60	52
47	51	52	53
91	79	100	75
56	68	47	50
79	65	70	61
81	80	68	58
78	55	67	60
46	38	37	38
39	35	34	37
32	30	30	32
60	50	67	54
35	37	48	39
39	36	39	31
50	34	37	40
43	37	39	50
48	54	57	43

yielding

$$
\begin{pmatrix}
C_{11}^*, \dots, C_{1J}^* \\
\vdots \\
C_{n1}^*, \dots, C_{nJ}^*
\end{pmatrix}.
$$

2. Compute the test statistic F in Box 11.3 based on the C_{ij}^* values generated in step 1, and label the result F^*.
3. Repeat steps 1 and 2 B times and label the results $F_1^*, \dots, F_B^*$.
4. Put these B values in ascending order and label the results $F_{(1)}^* \leq \cdots \leq F_{(B)}^*$.

The critical value is estimated to be $F^*_{(c)}$, where $c = (1 - \alpha)B$ rounded to the nearest integer. That is, reject the hypothesis of equal trimmed means if

$$F \geq F^*_{(c)},$$

where F is the statistic given in Box 11.3 based on the X_{ij} values.

11.6.4 S-PLUS Function rmanovab

The S-PLUS function

$$\text{rmanovab}(x, tr = 0.2, alpha = 0.05, grp = 0, nboot = 599)$$

performs the bootstrap-t method just described. The arguments have their usual meaning; see, for example, Section 9.5.2.

11.7 Percentile Bootstrap Methods for Other Robust Measures of Location

When comparing independent groups, some version of the percentile bootstrap method generally performs better than the bootstrap-t, in terms of controlling the probability of a Type I error, when the amount of trimming is at least 20%. For the situation at hand, however, no percentile bootstrap method has been shown to perform better than the bootstrap-t when the amount of trimming is 20%. But in fairness, some promising percentile bootstrap methods have not been examined. When comparing 20% trimmed means, all indications are that the bootstrap-t method in Section 11.6.3 performs well, in terms of Type I errors, with sample sizes as small as 21 (Wilcox, Keselman, Muska, & Cribbie, 2000). But when comparing groups based on MOMs, medians, and M-estimators or when the amount of trimming is at least 25%, certain variations of the percentile bootstrap method provide excellent control over the probability of a Type I error.

There are in fact many variations of the percentile bootstrap method that might be used to compare dependent groups. Most are not described here, either because they are known to be relatively unsatisfactory or because little is known about how well they perform when sample sizes are small. To complicate matters, there is the issue of whether to use difference scores. So the goal in the remainder of this section is to describe the methods that currently seem to have practical value and then in Section 11.8 to comment on their relative merits.

11.7.1 Methods Based on Marginal Measures of Location

Again let θ_j be any measure of location associated with the jth group and let $\hat{\theta}_j$ be an estimate of θ_j based on the available data (the X_{ij} values). The goal is to test

$$H_0 : \theta_1 = \cdots = \theta_J,$$

the hypothesis that all J dependent groups have identical measures of location. That is, measures of location associated with marginal distributions are being compared, as opposed to using difference scores.

Method RMPB3

The first method uses the test statistic

$$Q = \sum \left(\hat{\theta}_j - \bar{\theta} \right)^2,$$

where $\bar{\theta} = \sum \hat{\theta}_j / J$ is the average of the $\hat{\theta}_j$ values. An appropriate critical value is estimated using an approach similar to the bootstrap-t technique. First, set $C_{ij} = X_{ij} - \hat{\theta}_j$. That is, shift the empirical distributions so that the null hypothesis is true. Next a bootstrap sample is obtained by resampling, with replacement, as described in step 1 of Section 11.6.3. Again we label the results

$$\begin{pmatrix} C^*_{11}, \ldots, C^*_{1J} \\ \vdots \\ C^*_{n1}, \ldots, C^*_{nJ} \end{pmatrix}.$$

For the jth column of the bootstrap data just generated, compute the measure location that is of interest and label it $\hat{\theta}^*_j$. Compute

$$Q^* = \sum \left(\hat{\theta}^*_j - \bar{\theta}^* \right)^2,$$

where $\bar{\theta}^* = \sum \hat{\theta}^*_j / J$, and repeat this process B times, yielding $Q^*_1, \ldots, Q^*_B$. Put these B values in ascending order, yielding $Q^*_{(1)} \leq \cdots \leq Q^*_{(B)}$. Then reject the hypothesis of equal measures of location if $Q > Q^*_{(c)}$, where again $c = (1 - \alpha)B$. (The S-PLUS function bd1way described in Section 11.7.2 performs these calculations.)

Method RMPB4

If the null hypothesis is true, then all J groups have a common measure of location, θ. The next method estimates this common measure of location and then checks to see how deeply it is nested within the bootstrap values obtained when resampling from the original values. That is, in contrast to method RMPB3, the data are not centered, and bootstrap samples are obtained by resampling rows of data from

$$\begin{pmatrix} X_{11}, \ldots, X_{1J} \\ \vdots \\ X_{n1}, \ldots, X_{nJ} \end{pmatrix},$$

yielding

$$\begin{pmatrix} X^*_{11}, \ldots, X^*_{1J} \\ \vdots \\ X^*_{n1}, \ldots, X^*_{nJ} \end{pmatrix}.$$

For the jth group (or column of bootstrap values) compute $\hat{\theta}_j^*$. Repeating this process B times yields $\hat{\theta}_{jb}^*$ ($j = 1, \ldots, J; b = 1, \ldots, B$). The remaining calculations are performed as outlined in Box 11.5 (and are done by the S-PLUS function ddep in Section 11.7.2). Notice that for $J = 2$, this method does not reduce to method RMPB2 described in Section 11.3.1.

BOX 11.5 Repeated-Measures ANOVA Based on the Depth

of the Grand Mean

Goal:

Test the hypothesis

$$H_0 : \theta_1 = \cdots = \theta_J.$$

1. Compute

$$S_{jk} = \frac{1}{B-1} \sum_{b=1}^{B} \left(\hat{\theta}_{jb}^* - \bar{\theta}_j^* \right) \left(\hat{\theta}_{kb}^* - \bar{\theta}_k^* \right),$$

where

$$\bar{\theta}_j^* = \frac{1}{B} \sum_{b=1}^{B} \hat{\theta}_{jb}^*.$$

(The quantity S_{jk} is the sample covariance of the bootstrap values corresponding to the jth and kth groups.)

2. Let

$$\hat{\theta}_b^* = \left(\hat{\theta}_{1b}^*, \ldots, \hat{\theta}_{Jb}^* \right)$$

and compute

$$d_b = \left(\hat{\theta}_b^* - \hat{\theta} \right) S^{-1} \left(\hat{\theta}_b^* - \hat{\theta} \right)',$$

where S is the matrix corresponding to S_{jk}, $\hat{\theta} = (\hat{\theta}_1, \ldots, \hat{\theta}_J)$, $\hat{\theta}_j$ is the estimate of θ based on the original data for the jth group (the X_{ij} values, $i = 1, \ldots, n$), and $\hat{\theta}_b = (\hat{\theta}_{1b}, \ldots, \hat{\theta}_{Jb})$. The value of d_b measures how far away the bth bootstrap vector of location estimators is from $\hat{\theta}$, which is roughly the center of all B bootstrap values.

3. Put the d_b values in ascending order: $d_{(1)} \leq \cdots \leq d_{(B)}$.

4. Let $\hat{\theta}_G = (\bar{\theta}, \ldots, \bar{\theta})$, where $\bar{\theta} = \sum \hat{\theta}_j / J$, and compute

$$D = \left(\hat{\theta}_G - \hat{\theta} \right) S^{-1} \left(\hat{\theta}_G - \hat{\theta} \right)'.$$

D measures how far away the estimated common value is from the observed measures of location (based on the original data).

5. Reject if $D \geq d_{(u)}$, where $u = (1 - \alpha)B$, rounded to the nearest integer.

11.7.2 S-PLUS Functions bd1way and ddep

The S-PLUS functions

$$\text{bd1way}(x, \text{est} = \text{onestep}, \text{nboot} = 599, \text{alpha} = 0.05)$$

and

$$\text{ddep}(x, \text{alpha} = 0.05, \text{est} = \text{mom}, \text{grp} = \text{NA}, \text{nboot} = 500, \dots)$$

perform the percentile bootstrap methods just described. The first function performs method RMPB3. By default it uses the one-step M-estimator of location (based on Huber's Ψ), but any other estimator can be used via the argument est. As usual, x is any S-PLUS variable that is a matrix or has list mode, nboot is B, the number of bootstrap samples to be used, and grp can be used to analyze a subset of the groups, with the other groups ignored. (That is, grp is used as illustrated in Section 10.3.1.) The function ddep performs method RMPB4, described in Box 11.5.

EXAMPLE. We reanalyze the data in Table 11.3 using the S-PLUS functions just described. Assuming the data are stored in the S-PLUS matrix cork, the command bd1way(cork) returns

```
$test:
17.08

$crit:
34.09
```

So comparing one-step M-estimators, we fail to reject the hypothesis that the typical weight of a cork boring is the same for all four sides of a tree. If we compare groups using MOM in conjunction with method RMPB4 in Box 11.5, the significance level is .385. ∎

11.7.3 Percentile Bootstrap Methods Based on Difference Scores

The following method, based on difference scores, has been found to have practical value, particularly in terms of controlling Type I error probabilities when sample sizes are very small. Rather than test

$$H_0 : \theta_1 = \cdots = \theta_J, \tag{11.11}$$

first consider the goal of testing the hypothesis that a measure of location associated with the difference scores $D_{ij} = X_{ij} - X_{i,j+1}$ has the value zero. That is, use the difference between the ith observation in group j and the ith observation in group $j+1, j = 1, \dots, J-1$. Let θ_j be any measure of location associated with the D_{ij} values. So, for example, θ_1 might be the population value associated with MOM based on the difference scores between groups 1 and 2, and θ_2 the population MOM value associated with difference scores between groups 2 and 3. A simple alternative to

Equation (11.11) is to test

$$H_0 : \theta_1 = \cdots = \theta_{J-1} = 0, \qquad (11.12)$$

the hypothesis that the typical difference scores do not differ and are all equal to zero. However, a criticism of this approach is that the outcome can depend on how we order the groups. That is, rather than take differences between groups 1 and 2, we could just as easily take differences between groups 1 and 3 instead, which might alter our conclusions about whether to reject. We can avoid this problem by instead taking differences among all pairs of groups. There are a total of

$$L = \frac{J^2 - J}{2}$$

such differences, which are labeled $D_{i\ell}$, $i = 1, \ldots, n$; $\ell = 1, \ldots, L$. In particular,

$$D_{i1} = X_{i1} - X_{i2},$$
$$D_{i2} = X_{i1} - X_{i3},$$
$$\vdots$$
$$D_{iL} = X_{i,J-1} - X_{iJ}.$$

EXAMPLE. For four groups ($J = 4$), there are $L = 6$ differences, given by

$$D_{i1} = X_{i1} - X_{i2},$$
$$D_{i2} = X_{i1} - X_{i3},$$
$$D_{i3} = X_{i1} - X_{i4},$$
$$D_{i4} = X_{i2} - X_{i3},$$
$$D_{i5} = X_{i2} - X_{i4},$$
$$D_{i6} = X_{i3} - X_{i4}.$$

■

The goal is to test

$$H_0 : \theta_1 = \cdots = \theta_L = 0, \qquad (11.13)$$

where θ_ℓ is the population measure of location associated with the ℓth set of difference scores, $D_{i\ell}$ ($i = 1, \ldots, n$). To test H_0 given by Equation (11.13), resample vectors of D values; but unlike the bootstrap-t, observations are not centered. That is, a bootstrap sample now consists of resampling with replacement n rows from the matrix

$$\begin{pmatrix} D_{11}, \ldots, D_{1L} \\ \vdots \\ D_{n1}, \ldots, D_{nL} \end{pmatrix},$$

yielding

$$
\begin{pmatrix}
D^*_{11}, \ldots, D^*_{1L} \\
\vdots \\
D^*_{n1}, \ldots, D^*_{nL}
\end{pmatrix}.
$$

For each of the L columns of the D^* matrix, compute MOM or whatever measure of location is of interest, and for the ℓth column label the result $\hat{\theta}^*_\ell$ ($\ell = 1, \ldots, L$). Next, repeat this B times, yielding $\hat{\theta}^*_{\ell b}$, $b = 1, \ldots, B$, and then determine how deeply the vector $\mathbf{0} = (0, \ldots, 0)$, having length L, is nested within the bootstrap values $\hat{\theta}^*_{\ell b}$. For two groups, this is tantamount to determining how many bootstrap values are greater than zero. If most are greater than (or less than) zero, we reject, as indicated in Section 11.3.1 in conjunction with method RMPB1. For more than two groups, you use a method similar to the approach in Box 11.5. The details are relegated to Box 11.6.

BOX 11.6 Repeated Measures ANOVA Based on Difference Scores

and the Depth of Zero

Goal:
Test the hypothesis, given by Equation (11.13), that all difference scores have a typical value of zero.

1. Let $\hat{\theta}_\ell$ be the estimate of θ_ℓ. Compute bootstrap estimates as described in Section 11.7.3 and label them $\hat{\theta}^*_{\ell b}$, $\ell = 1, \ldots, L; b = 1, \ldots, B$.
2. Compute the L-by-L matrix

$$
S_{\ell\ell'} = \frac{1}{B-1} \sum_{b=1}^{B} \left(\hat{\theta}^*_{\ell b} - \hat{\theta}_\ell \right) \left(\hat{\theta}^*_{\ell'b} - \hat{\theta}_{\ell'} \right).
$$

Readers familiar with multivariate statistical methods might notice that $S_{\ell\ell'}$ uses $\hat{\theta}_\ell$ (the estimate of θ_ℓ based on the original difference values) rather than the seemingly more natural $\bar{\theta}^*_\ell$, where

$$
\bar{\theta}^*_\ell = \frac{1}{B} \sum_{b=1}^{B} \hat{\theta}^*_{\ell b}.
$$

If $\bar{\theta}^*_\ell$ is used, unsatisfactory control over the probability of a Type I error can result.

Continued

BOX 11.6 (*Continued*)

3. Let $\hat{\theta} = (\hat{\theta}_1, \dots, \hat{\theta}_L)$, $\hat{\theta}_b^* = (\hat{\theta}_{1b}^*, \dots, \hat{\theta}_{Lb}^*)$, and compute

$$d_b = \left(\hat{\theta}_b^* - \hat{\theta}\right) S^{-1} \left(\hat{\theta}_b^* - \hat{\theta}\right)',$$

where S is the matrix corresponding to $S_{\ell\ell'}$.

4. Put the d_b values in ascending order: $d_{(1)} \leq \cdots \leq d_{(B)}$.

5. Let

$$0 = (0, \dots, 0),$$

having length L.

6. Compute

$$D = (0 - \hat{\theta})S^{-1}(0 - \hat{\theta})'.$$

D measures how far away the null hypothesis is from the observed measures of location (based on the original data). In effect, D measures how deeply 0 is nested within the cloud of bootstrap values.

7. Reject if $D \geq d_{(u)}$, where $u = (1 - \alpha)B$, rounded to the nearest integer.

11.7.4 S-PLUS Function rmdzero

The S-PLUS function

$$\text{rmdzero}(x, \text{est} = \text{mom}, \text{grp} = \text{NA}, \text{nboot} = \text{NA}, \dots)$$

performs the test on difference scores outlined in Box 11.6.

EXAMPLE. For the cork data in Table 11.3, rmdzero returns a significance level of .044, so in particular reject with $\alpha = .05$. That is, conclude that the typical difference score is not equal to zero for all pairs of groups. This result is in sharp contrast to comparing marginal measures of location based on MOM and the method in Box 11.5, which has a significance level of .385. Currently, it seems that the method in Box 11.5 does an excellent job of avoiding Type I error probabilities larger than the nominal level, but that in many situations it is too conservative. That is, the actual probability of a Type I error can be substantially smaller than the nominal level, suggesting that it might have relatively poor power. Switching to difference scores appears to correct this problem. ■

11.8 Comments on Which Method to Use

Several reasonable methods for comparing groups have been described, so there is the issue of which one to use. As usual, no method is perfect in all situations.

The expectation is that in many situations where groups differ, all methods based on means perform poorly relative to approaches based on some robust measure of location, such as MOM or a 20% trimmed mean. Currently, with a sample size as small as 21, the bootstrap-t method in Section 11.6.3, used in conjunction with 20% trimmed means, appears to provide excellent control over the probability of a Type I error. Its power compares reasonably well to most other methods that might be used. But as noted in previous chapters, different methods are sensitive to different features of your data, and arguments for some other measure of location, such as MOM, have been made.

The percentile bootstrap methods in Section 11.7 also do an excellent job of avoiding Type I errors greater than the nominal level, but there are indications that method RMPB3 in Section 11.7.1 can be too conservative when sample sizes are small. That is, the actual probability of a Type I error can be substantially less than α, suggesting that some other method might provide better power. Nevertheless, if there is specific interest in comparing marginal distributions with M-estimators, it is suggested that method RMPB3 in Section 11.7.1 be used and that the sample size be greater than 20. There is some indirect evidence that a larger sample size might be needed when using this method. (This is in contrast to comparing independent groups, where sample sizes greater than 20 seem to suffice.) Also, it can be used to compare groups based on MOM. But with very small sample sizes there are some indications that its power might be inadequate, at least in some situations, relative to other techniques that might be used.

Currently, among the techniques covered in this chapter, it seems that the two best methods for controlling Type I error probabilities and simultaneously providing reasonably high power are the bootstrap-t method based on 20% trimmed means and the percentile bootstrap method in Box 11.6 used in conjunction with MOM, which uses difference scores. (Other excellent options are covered in Chapter 12.) With near certainty, situations arise where some other technique is more optimal, but typically the improvement is small. However, comparing groups with MOM is not the same as comparing means, trimmed means, or M-estimators, and certainly there will be situations where some other estimator has higher power than any method based on MOM or a 20% trimmed mean. If the goal is to maximize power, several methods are contenders for routine use, but as usual, standard methods based on means are generally the least satisfactory. With sufficiently large sample sizes, trimmed means can be compared without resorting to the bootstrap-t method, but it remains unclear just how large the sample size must be. When using MOM, currently a bootstrap method is required regardless of how large the sample size might be.

As for the issue of whether to use difference scores versus robust measures of location based on the marginal distributions, each approach provides a different perspective on how groups differ, and they can give different results regarding whether groups are significantly different. There is some evidence that difference scores typically provide more power and better control over the probability of a Type I error, but more detailed study is needed to resolve this issue.

As previously mentioned, method RMPB4, outlined in Box 11.5 (and performed by the S-PLUS function ddep), is very conservative in terms of Type I errors, meaning that when testing at the .05 level, say, often the actual probability of a Type I error will be less than or equal to α and typically smaller than any other method described in this chapter. But a concern is that the actual Type I error probability can be substantially smaller than the nominal level, resulting in low power relative to many other methods you might use.

11.9 Between-by-Within, or Split-Plot, Designs

Chapter 10 covered a two-way ANOVA design involving JK independent groups, with J levels associated with the first factor and K levels with the second. A between-by-within, or split-plot, design refers to a situation where for a given level of the first factor, the measures associated with the K levels of the second factor are dependent instead. As a simple illustration, again consider the situation where endurance is measured before and after training, but now we have two training methods. Moreover, a sample of n_1 athletes undergo training method 1 and a different, independent sample of athletes undergo training method 2. So between methods, observations are independent, but they are possibly dependent between occasions. The population means are:

Method (A)	Time (B) 1	2
1	μ_{11}	μ_{12}
2	μ_{21}	μ_{22}

Moreover, the notions of main effects and interactions (described in Chapter 10) extend immediately to the situation at hand. For example, the hypothesis of no main effect for factor A (method) is

$$H_0 : \frac{\mu_{11} + \mu_{12}}{2} = \frac{\mu_{21} + \mu_{22}}{2},$$

and the hypothesis of no interaction is

$$H_0 : \mu_{11} - \mu_{12} = \mu_{21} - \mu_{22}.$$

11.9.1 Method for Trimmed Means

The computational details on how to compare trimmed means (including means as a special case) are tedious at best. Presumably most readers are more interested in applying the method versus understanding the computational details, so the bulk of the computations are relegated to Box 11.7 (assuming familiarity with basic matrix algebra). In Box 11.7, X_{ijk} represents the ith observation in level j of factor A and level k of factor B, $i = 1, \ldots, n_j; j = 1, \ldots, J$ and $k = 1, \ldots, K$. Once the quantities Q and A are computed as described in Box 11.7, let

$$c = k + 2A - \frac{6A}{k+2}.$$

When the null hypothesis is true, Q/c has, approximately, an F-distribution with $\nu_1 = k$ and $\nu_2 = k(k+2)/3A$ degrees of freedom (where k represents the number of rows corresponding to the matrix **C** in Box 11.7). For factor A, $k = J - 1$; for factor B, $k = K - 1$; and for interactions, $k = (J - 1)(K - 1)$.

DECISION RULE: Reject if $Q/c \geq f_{1-\alpha}$, the $1 - \alpha$ quantile of an F-distribution with ν_1 and ν_2 degrees of freedom. (Also see Keselman, Algina, Boik, & Wilcox, 1999).

BOX 11.7 Computations for a Split-Plot Design When Using

Trimmed Means

The hypotheses of no main effects and no interactions can be written in the form

$$H_0 : C\mu_t = 0,$$

where **C** is a k-by-JK matrix (having rank k) that reflects the null hypothesis of interest. (Here, μ_t is a column vector of population trimmed means having length JK.) Let C_J be defined as in Box 10.4 and let j'_j be a $1 \times J$ matrix of 1's. Then for factor A, $C = C_J \otimes j'_K$ and $k = J - 1$. For factor B, $C = j'_J \otimes C_K$, $k = K - 1$, and the test for no interactions uses $C = C_J \otimes C_K$, where now $k = (J - 1)(K - 1)$.

For every level of factor A, there are K dependent random variables, and each pair of these dependent random variables has a Winsorized covariance that must be estimated. For fixed j, let g_j be the number of observations trimmed from both tails. (If the amount of trimming is γ, $g_j = [\gamma n_j]$, as in Chapter 3.) The Winsorized covariance between the mth and ℓth levels of factor B is estimated with

$$s_{jm\ell} = \frac{1}{n_j - 1} \sum_{i=1}^{n_j} \left(Y_{ijm} - \bar{Y}_{.jm} \right) \left(Y_{ij\ell} - \bar{Y}_{.j\ell} \right),$$

where $\bar{Y}_{.jm} = \sum_{i=1}^{n_j} Y_{ijm} / n_j$,

$$Y_{ijk} = \begin{cases} X_{(g_j+1),jk} & \text{if } X_{ijk} \leq X_{(g_j+1),jk} \\ X_{ijk} & \text{if } X_{(g_j+1),jk} < X_{ij} < X_{(n-g_j),jk} \\ X_{(n-g_j),jk} & \text{if } X_{ijk} \geq X_{(n_j-g),jk}. \end{cases}$$

For fixed j, let $S_j = (s_{jm\ell})$, which is the matrix of Winsorized variances and covariances for level j of factor A. Let

$$V_j = \frac{(n_j - 1)S_j}{b_j(b_j - 1)}, \qquad j = 1, \ldots, J,$$

Continued

BOX 11.7 (*Continued*) and let $V = \mathrm{diag}(V_1, \ldots, V_J)$ be a block diagonal matrix. The test statistic is

$$Q = \bar{X}'C'(CVC')^{-1}C\bar{X}, \qquad (11.14)$$

where $\bar{X}' = (\bar{X}_{t11}, \ldots, \bar{X}_{tJK})$. Let $I_{K \times K}$ be a K-by-K identity matrix, let Q_j be a JK-by-JK block diagonal matrix (consisting of J blocks, each block being a K-by-K matrix), where the tth block ($t = 1, \ldots, J$) along the diagonal of Q_j is $I_{K \times K}$ if $t = j$, and all other elements are zero. (For example, if $J = 3$ and $K = 4$, then Q_1 is a 12-by-12 matrix block diagonal matrix, where the first block is a 4-by-4 identity matrix and all other elements are zero. As for Q_2, the second block is an identity matrix, and all other elements are zero.) Compute

$$A = \frac{1}{2}\sum_{j}^{J} \frac{\mathrm{tr}(\{VC'(CVC')^{-1}CQ_j\}^2) + \{\mathrm{tr}(VC'(CVC')^{-1}CQ_j)\}^2}{b_j - 1}.$$

The remaining calculations are described in the text.

11.9.2 S-PLUS Function tsplit

The S-PLUS function

$$\mathrm{tsplit}(J, K, x, tr = 0.2, grp = c(1{:}p), p = J*K)$$

performs the calculations in Box 11.4. Here, J is the number of independent groups, K is the number of dependent groups, x is any S-PLUS variable that is a matrix or has list mode, and, as usual, tr indicates the amount of trimming, which defaults to .2 if unspecified. If the data are stored in list mode, it is assumed that x[[1]] contains the data for level 1 of both factors, x[[2]] contains the data for level 1 of the first factor and level 2 of the second, and so on. If the data are not stored in the proper order, grp can be used to indicate how they are stored. For example, if a 2-by-2 design is being used, the S-PLUS command

$$\mathrm{tsplit}(2, 2, x, grp{=}c(3, 1, 2, 4))$$

indicates that the data for the first level of both factors are stored in x[[3]], the data for level 1 of factor A and level 2 of factor B are in x[[1]], and so forth. The last argument, p, can be ignored. It is needed only to satisfy certain requirements of S-PLUS. If the data are stored in a matrix, it is assumed that the first column contains the data for level 1 of both factors, the second column contains the data for level 1 of the first factor and level 2 of the second, and so forth.

11.9.3 A Bootstrap-*t* Method

To apply a bootstrap-t method when working with trimmed means, you first center the data in the usual way. In the present context, this means you compute

$$C_{ijk} = X_{ijk} - \bar{X}_{tjk},$$

$i = 1, \ldots, n_j; j = 1, \ldots, J;$ and $k = 1, \ldots, K$. That is, for the group corresponding to the jth level of factor A and the kth level of factor B, subtract the corresponding trimmed mean from each of the observations. Next, for the jth level of factor A, generate a bootstrap sample by resampling with replacement n_j vectors of observations from the data in level j of factor A. That is, for each level of factor A, you have an n_j-by-K matrix of data, and you generate a bootstrap sample from this matrix of data as described in Section 11.6.3. Label the resulting bootstrap samples C^*_{ijk}. Compute the test statistic F, based on the C^*_{ijk} values as described in Box 11.3, and label the result F^*. Repeat this B times, yielding $F^*_1, \ldots, F^*_B$, and then put these B values in ascending order, yielding $F^*_{(1)} \le \cdots \le F^*_{(B)}$. Next, compute F using the original data (the X_{ijk} values) as described in Box 11.3 and reject if $F \ge F^*_{(c)}$, where $c = (1 - \alpha)$ rounded to the nearest integer.

A crude rule that seems to apply to a wide variety of situations is: The more distributions associated with groups differ, the more beneficial it is to use some type of bootstrap method, at least when sample sizes are small. Keselman, Algina, Wilcox, and Kowalchuk (2000) compared the bootstrap-t method just described to the nonbootstrap method for a split-plot design, covered in Section 11.9.1. For the situations they examined, this rule did not apply; it was found that the bootstrap-t offered little or no advantage. Their study included situations where the correlations (or covariances) among the dependent groups differ across the independent groups being compared. However, the more complicated the design, the more difficult it becomes to consider all the factors that might influence operating characteristics of a particular method. One limitation of their study was that the differences among the covariances were taken to be relatively small. Another issue that has not been addressed is how the bootstrap-t performs when distributions differ in skewness. Having differences in skewness is known to be important when dealing with the simple problem of comparing two groups only. There is no reason to assume that this problem diminishes as the number of groups increases, and indeed there are reasons to suspect that it becomes a more serious problem. So currently, it seems that if groups do not differ in any manner or the distributions differ slightly, it makes little difference whether you use a bootstrap-t versus a nonbootstrap method for comparing trimmed means. However, if distributions differ in shape, there is indirect evidence that the bootstrap-t might offer an advantage when using a split-plot design, but the extent to which this is true is not well understood.

11.9.4 S-PLUS Function tsplitbt

The S-PLUS function

$$\text{tsplitbt}(J, K, x, tr=.2, alpha=.05, JK=J^*K, grp=c(1 : JK),}$$

$$\text{nboot=599, monitor=F)}$$

performs a bootstrap-t method for a split-plot design, as just described. The data are assumed to be arranged as indicated in conjunction with the S-PLUS function tsplit (as described in Section 11.9.2), and the arguments J, K, tr, and alpha have the same meaning as before. The argument JK can be ignored, and grp can be used to rearrange the data if they are not stored as expected by the function. (See Section 10.3.1 for an illustration on how to use grp.) The argument monitor can be used to monitor the progress of the function. If we set monitor=T, the function prints a message each time it completes a bootstrap iteration. This way you can get some sense of how long it will take before the computations are complete.

11.9.5 Using MOMs, Medians, and M-Estimators

Comparing groups based on MOMs, medians, and M-estimators in a split-plot design is possible using extensions of the bootstrap methods considered in this chapter as well as Chapters 9 and 10. Generally, bootstrap samples must be generated in a manner that reflects the dependence among the levels of factor B, and then some appropriate hypothesis is tested using some slight variation of one of the methods already described. In fact, the problem is *not* deriving a method, but rather deciding which method should be used among the many that are available. This section summarizes some approaches that are motivated by published papers, but it is noted that alternative methods might be more optimal and that for the specific situation at hand more research is needed to better understand the relative merits of different techniques. (Methods in Chapter 12 provide yet another perspective and should be considered before analyzing data.)

Again consider a two-way design where factor A consists of J independent groups and factor B corresponds to K dependent groups. First consider the dependent groups. One approach to comparing these K groups, ignoring factor A, is simply to form difference scores and then to apply the method in Box 11.6. More precisely, imagine you observe X_{ijk} ($i = 1, \ldots, n_j; j = 1, \ldots, J; k = 1, \ldots, K$). That is, X_{ijk} is the ith observation in level j of factor A and level k of factor B. Note that if we ignore the levels of factor A, we can write the data as Y_{ik}, $i = 1, \ldots, N; k = 1, \ldots, K$, where $N = \sum n_j$. Now consider levels k and k' of factor B ($k < k'$) and set

$$D_{ikk'} = Y_{ik} - Y_{ik'};$$

let $\theta_{kk'}$ be some measure of location associated with $D_{ikk'}$. Then the levels of factor B can be compared, ignoring factor A, by testing

$$\theta_{12} = \cdots = \theta_{k-1,k} = 0 \tag{11.15}$$

using the method in Section 11.7.3. In words, the null hypothesis is that the typical difference score between any two levels of factor B, ignoring factor A, is zero.

As for factor A, ignoring factor B, one approach is as follows. Momentarily focus on the first level of factor B and note that the levels of factor A can be compared using the method in Box 9.7. That is, the null hypothesis of no differences among the levels of factor A is

$$H_0 : \theta_{11} = \theta_{21} = \cdots = \theta_{J1},$$

where of course these J groups are independent, and a percentile bootstrap method can be used as described in Chapter 9. More generally, for any level of factor B — say, the kth — no main effects is

$$H_0 : \theta_{1k} = \theta_{2k} = \cdots = \theta_{Jk},$$

$(k = 1, \ldots, K)$, and the goal is to test the hypothesis that these K hypotheses are simultaneously true. Here we take this to mean that we want to test

$$H_0 : \theta_{11} - \theta_{21} = \cdots \theta_{J-1,1} - \theta_{J1} = \cdots = \theta_{J-1,K} - \theta_{JK} = 0 \qquad (11.16)$$

In this last equation, there are $C = K(J^2 - J)/2$ differences, all of which are hypothesized to be equal to zero. Generalizing the method in Box 9.7, for each level of factor A, generate bootstrap samples as is appropriate for K dependent groups (see Section 11.7.1), and then test Equation (11.16). To briefly outline the computations, label the C differences as $\delta_1, \ldots, \delta_C$ and then denote bootstrap estimates by $\hat{\delta}_c^*$ $(c = 1, \ldots C)$. For example, $\hat{\delta}_1^* = \theta_{11}^* - \theta_{21}^*$. Then we test Equation (11.16) by determining how deeply the vector $(0, \ldots, 0)$, having length C, is nested within B bootstrap values, which is done in the manner described in Box 11.6.

For factor A an alternative approach is to average the measures of location across the K levels of factor B and then to proceed in the manner described in Box 9.7. In symbols, let

$$\bar{\theta}_{j.} = \frac{1}{K} \sum_{k=1}^{K} \theta_{jk},$$

in which case the goal is to test

$$H_0 : \bar{\theta}_{1.} = \cdots = \bar{\theta}_{J.}.$$

Again for each level of factor A, you generate B samples for the K dependent groups as described in Section 11.7.1 in conjunction with method RMPB4. Let $\bar{\theta}_{j.}^*$ be the bootstrap estimate for the jth level of factor A. For levels j and j' of factor A, $j < j'$, set $\delta_{jj'}^* = \bar{\theta}_{j.}^* - \bar{\theta}_{j'.}^*$. Then you determine how deeply 0, having length $(J^2 - J)/2$, is nested within the B bootstrap values for $\delta_{jj'}^*$ using the method described in Box 11.6.

As for interactions, again there are several approaches one might adopt. Here an approach based on difference scores among the dependent groups is used. To explain, first consider a 2-by-2 design, and for the first level of factor A let $D_{i1} = X_{i11} - X_{i12}$, $i = 1, \ldots, n_1$. Similarly, for level 2 of factor A let $D_{i2} = X_{i21} - X_{i22}$, $i = 1, \ldots, n_2$, and let θ_{d1} and θ_{d2} be the population measure of location corresponding to the D_{i1} and D_{i2} values, respectively. Then the hypothesis of no interaction is taken to be

$$H_0 : \theta_{d1} = \theta_{d2},$$

which of course is the same as

$$H_0 : \theta_{d1} - \theta_{d2} = 0. \qquad (11.17)$$

Again the basic strategy for testing hypotheses is generating bootstrap estimates and determining how deeply 0 is embedded in the B values that result. For the more general case of a J-by-K design, there are a total of

$$C = \frac{J^2 - J}{2} \times \frac{K^2 - K}{2}$$

equalities, one for each pairwise difference among the levels of factor B and any two levels of factor A.

11.9.6 S-PLUS Functions sppba, sppbb, and sppbi

The S-PLUS function

$$\text{sppba}(J,K,x,\text{est}=\text{mom},\text{grp}=c(1:JK),\text{avg}=F,\text{nboot}=500,\dots)$$

tests the hypothesis of no main effects for factor A in the manner just described. Setting the argument avg to T (for true) indicates that the averages of the measures of location (the $\bar{\theta}_{j.}$ values) will be used. That is, $H_0 : \bar{\theta}_{1.} = \cdots = \bar{\theta}_{J.}$ is tested. Otherwise, the hypothesis given by Equation (11.16) is tested. The remaining arguments have their usual meaning. The S-PLUS function

$$\text{sppbb}(J,K,x,\text{est}=\text{mom},\text{grp}=c(1:JK),\text{nboot}=500,\dots)$$

tests the hypothesis of no main effects for factor B (as described in the previous section), and

$$\text{sppbi}(J,K,x,\text{est}=\text{mom},\text{grp}=c(1:JK),\text{nboot}=500,\dots)$$

tests the hypothesis of no interactions.

EXAMPLE. We examine once more the EEG measures for murderers versus a control group, only now we use the data for all four sites in the brain where measures were taken. If we label the typical measures for the control group as $\theta_{11},\dots,\theta_{14}$ and the typical measures for the murderers as $\theta_{21},\dots,\theta_{24}$, we have a 2-by-4 between-by-within design, and a possible approach to comparing the groups is testing

$$H_0 : \theta_{11} - \theta_{21} = \theta_{12} - \theta_{22} = \theta_{13} - \theta_{23} = \theta_{14} - \theta_{24} = 0.$$

This can be done with the S-PLUS function sppba with the argument avg set to F. If the data are stored in a matrix called eeg having eight columns, with the first four corresponding to the control group, then the command sppba(2,4,eeg) performs the calculations based on the MOM measure of location and returns a significance level of .098. An alternative approach is to average the value of MOM over the four brain sites for each group and then to compare these averages. That is, test $H_0 : \bar{\theta}_{1.} = \bar{\theta}_{2.}$, where $\bar{\theta}_{j.} = \sum \theta_{jk}/4$. This can be done with the command

$$\text{sppba}(2,4,\text{eeg},\text{avg}=T).$$

Now the significance level is .5, so we see that the significance level can vary tremendously depending on how we compare the groups. ■

11.9.7 The S-PLUS Function selby

Section 1.1.6 describes an S-PLUS function that is convenient when data are stored in a matrix, with one of the columns indicating the group to which an observation belongs. Basically, the function takes the data in a matrix and sorts it into groups in a manner that can be used by the S-PLUS functions written for this book. It is noted that this function can be used when there are multiple columns of data rather than just one, as illustrated in Section 1.1.6. That is, the third argument, coln, can be a vector.

EXAMPLE. Imagine that you have four dependent variables stored in columns 1–4 of the S-PLUS matrix mat and that column 5 indicates the group to which a vector of observations belongs. So, for example, the data might look like this:

v1	v2	v3	v4	G
34	42	63	19	1
26	99	45	29	1
33	42	18	32	2

$$\vdots$$

The problem is to sort the data into groups, based on the values listed in column 5 (under G), and to store the data in a manner that can be used by, for example, the functions in Section 11.9.6. The command

$$dat < -selby(mat, 5, c(1:4))$$

accomplishes this goal. (This command stores the data in dat\$x, having list mode.) So if there are five groups corresponding to the values in column 5 of mat, interactions could be investigated with the command

$$sppbi(5, 4, dat\$x).$$

▪

11.10 Exercises

1. For the data in Table 11.3, perform the paired T-test for means using the weights for the east and south sides of the trees. Verify that the significance level is .09.
2. Repeat the previous exercise, but use 20% trimmed means instead, using the difference scores in conjunction with the S-PLUS function trimci in Section 4.9.4. Note that the significance level is .049.
3. If in Exercise 1 you compare the marginal 20% trimmed means with the S-PLUS function yuend in Section 11.2.1, verify that now the significance level is .121.
4. Generally, why is it possible to get a different significance level comparing the marginal trimmed means than when making inferences about the trimmed mean of the difference scores?
5. Based on what is currently known, would you expect more power when comparing the marginal trimmed means or when making inferences about the trimmed mean of the difference scores?

6. Repeat Exercise 1, but now use a percentile bootstrap method based on MOM and the difference scores. That is, use with the S-PLUS function rmmcppb. Verify that the significance level is .092.

7. Repeat the last exercise, except now compare the marginal measures of location based on MOM. So in the S-PLUS function rmmcppb, set dif=F. Verify that the significance level is .578.

8. Based on the S-PLUS output from rmmcppb, if you were to repeat the last exercise, except adding 6.75 to every observation in the second group, would you reject with $\alpha = .05$? Verify your answer by adding 6.75 to every observation in the second group and invoking rmmcpp.

9. For the data in Exercise 1, verify that the .95 confidence interval for the median of the difference scores, based on the method in Chapter 4, is $(-7, .55)$.

10. Compare the marginal medians of the data in Exercise 1 using the S-PLUS function rmmcppb in Section 11.3.2. Verify that the .95 confidence interval for the difference between the population medians is $(-14, 4.5)$.

11. In this chapter and in Chapter 8 reference was made to a study dealing with EEG measures for murderers versus a control group. In another portion of this study, the measures for four sites in the brain were found to be as reported in Table 11.4. (These observations differ from those used in previous illustrations.) Using difference scores, compare site 1 and site 3 for murderers using both means and 20% trimmed means. (Use the S-PLUS function trimci.) Verify that the significance levels are .61 and .27, respectively.

12. For the data in Table 11.4, compare all four sites for murderers with the S-PLUS function rmanova in Section 11.6.2. Verify that you reject with both means and 20% trimmed means with $\alpha = .01$.

TABLE 11.4 EEG Measures

	Murderers				Controls		
Site 1	Site 2	Site 3	Site 4	Site 1	Site 2	Site 3	Site 4
−0.19	−1.05	−0.04	−1.48	0.45	−1.57	0.96	−0.76
0.39	−1.39	1.08	−0.95	0.24	−1.12	1.25	−0.31
0.09	−0.49	0.19	−2.14	0.33	−1.53	−0.64	−1.57
0.54	−0.76	1.06	−0.33	0.81	−0.28	0.29	−1.25
0.78	−0.36	−0.36	−1.09	1.30	−0.53	−0.05	−0.98
0.59	−1.17	0.80	−1.06	0.46	−1.09	0.09	−2.35
0.04	−1.75	0.11	−1.41	−0.01	−1.98	1.07	−0.94
0.38	−0.83	1.05	−0.29	1.11	−0.84	0.88	−1.62
0.25	−0.40	−0.07	−1.90	0.16	−1.25	0.28	−0.55
0.01	−1.06	0.50	−0.07	1.02	−1.07	0.00	−1.31
0.40	−1.36	0.54	−0.63	0.67	−0.92	−0.08	−2.18
0.52	−1.30	1.69	−0.22	1.37	−0.69	0.62	−0.86
1.35	−0.45	0.01	−1.22	0.59	−0.64	−0.02	−0.06
0.02	−0.86	−0.07	−1.65	0.66	−1.43	−0.48	−0.93

13. For the data in Table 11.4, compare all groups using the S-PLUS function tsplit in Section 11.9.2. Verify that when comparing murderers to the control group, the main effect has a significance level of .9955.

14. For the data in Table 11.4, compare murderers versus controls using MOM and difference scores as outlined in Section 11.9.5. That is, use the S-PLUS function sppba in Section 11.9.6. Verify that the significance level is .138, and compare this result with the previous exercise.

15. In the previous exercise, you could have used averages rather than difference scores when comparing the murderers to the controls using the approach described in Section 11.9.6. Using the S-PLUS function sppba with the argument avg set equal to T, verify that the significance level is .89.

16. Compare the results from the previous three exercises and comment on finding the optimal method for detecting a true difference among groups.

17. For the data in Table 11.4, use the S-PLUS function sppbi to test the hypothesis of no interactions based on MOM. Verify that the significance level is .1.

12

MULTIPLE COMPARISONS

Chapters 9–11 describe how to test the hypothesis that two or more groups have a common measure of location. When working with means, for example, one goal was to test

$$H_0 : \mu_1 = \cdots = \mu_J, \tag{12.1}$$

the hypothesis that J groups have equal population means. It is common, however, to want to know more about how the groups compare: Which groups differ? How do they differ, and by how much? When addressing these questions, what role should the methods in Chapters 9–11 play? A very common strategy is first to test Equation (12.1) and, if a nonsignificant result is obtained, to stop and fail to declare any of the groups to be different. One goal in this chapter is to cover modern insights into the relative merits of this approach.

The other general goal is to describe methods for controlling what is called the *familywise error rate* (FWE) (sometimes called the *experimentwise error rate*) that is, the probability of making at least one Type I error when performing multiple tests. To elaborate, imagine you have four independent groups and for the moment assume normality and homoscedasticity. Suppose that for each pair of means we test the hypothesis of equal means using Student's T. That is, the goal is to test

$$H_0 : \mu_1 = \mu_2,$$
$$H_0 : \mu_1 = \mu_3,$$
$$H_0 : \mu_1 = \mu_4,$$
$$H_0 : \mu_2 = \mu_3,$$
$$H_0 : \mu_2 = \mu_4,$$
$$H_0 : \mu_3 = \mu_4.$$

If we test each of these hypotheses at the .05 level ($\alpha = .05$), then of course the probability of a Type I error will be .05 for each test. But what is the probability of *at least one* Type I error among the six hypotheses of interest? That is, what is the probability of erroneously concluding that one or more pairs of means differ when in fact none differ at all? Determining this probability is complicated by the fact that the individual tests are not all independent. For example, when testing the first two hypotheses, the corresponding Student's T-tests will be dependent because both have the first sample mean in the numerator of the test statistic. That is, $\bar{X}_1$ is used both times. If all six tests were independent, then we could use the binomial probability function to determine the probability of at least one Type I error. But because there is dependence, special methods are required. Bernhardson (1975) describes a situation where the probability of at least one Type I error (FWE) can be as high as .29 for the situation at hand. When comparing all pairs of 10 groups, this probability can be as high as .59.

12.1   Homoscedastic Methods for the Means of Independent Groups

We begin by describing classic methods for comparing means that assume normality and equal variances. Problems with these classic methods have long been established (e.g., Wilcox, 1996c), but they are commonly used; their pitfalls remain relatively unknown, so they are included here merely for completeness and future reference.

12.1.1   Fisher's Least Significant Difference (LSD) Method

One of the earliest strategies for comparing multiple groups is the so-called *least significant difference* (LSD) method due to Sir Ronald Fisher. Assuming normality and homoscedasticity, first perform the ANOVA F-test in Section 9.1. If a significant result is obtained, apply Student's T to all pairs of means, but unlike the approach in Chapter 8, typically the assumption of equal variances is taken advantage of by using the estimate of the assumed common variance when performing Student's T. Under normality and homoscedasticity, this has the advantage of increasing the degrees of freedom, which in turn can mean more power.

To be more concrete, suppose the ANOVA F-test in Section 9.1 is significant for some specified value of α and let MSWG (described in Box 9.1) be the estimate of the assumed common variance. To test

$$H_0 : \mu_j = \mu_k, \tag{12.2}$$

the hypothesis that the mean of the jth group is equal to the mean of the kth group, compute

$$T = \frac{\bar{X}_j - \bar{X}_k}{\sqrt{\text{MSWG}\left(\frac{1}{n_j} + \frac{1}{n_k}\right)}}. \tag{12.3}$$

When the assumptions of normality and homoscedasticity are met, T has a Student's T-distribution with $v = N - J$ degrees of freedom, where J is the number of groups being compared and $N = \sum n_j$ is the total number of observations in all J groups. So when comparing the jth group to the kth group, you reject the hypothesis

TABLE 12.1   Hypothetical Data for Three Groups

G1	G2	G3
3	4	6
5	4	7
2	3	8
4	8	6
8	7	7
4	4	9
3	2	10
9	5	9

of equal means if

$$|T| \geq t_{1-\alpha/2},$$

where $t_{1-\alpha/2}$ is the $1 - \alpha/2$ quantile of Student's T-distribution with $N - J$ degrees of freedom.

EXAMPLE. For the data in Table 12.1, it can be shown that MSWG $= 4.14$, the sample means are $\bar{X}_1 = 4.75$, $\bar{X}_2 = 4.62$, and $\bar{X}_3 = 7.75$, and the F-test is significant, so according to Fisher's LSD procedure, you would proceed by comparing each pair of groups with Student's T-test. For the first and second groups,

$$T = \frac{4.75 - 4.62}{\sqrt{4.14(\frac{1}{8} + \frac{1}{8})}} = .128.$$

The degrees of freedom are $\nu = 21$, and with $\alpha = .05$, Table 4 in Appendix B says that the critical value is 2.08. Therefore, you fail to reject. That is, the F-test indicates that there is a difference among the three groups, but Student's T suggests that the difference does not correspond to groups 1 and 2. For groups 1 and 3,

$$T = \frac{4.75 - 7.75}{\sqrt{4.14(\frac{1}{8} + \frac{1}{8})}} = -2.94,$$

and because 2.94 is greater than the critical value, 2.08, reject. That is, conclude that groups 1 and 3 differ. In a similar manner, you conclude that groups 2 and 3 differ as well, because $T = 3.08$.   ■

When the assumptions of normality and homoscedasticity are true, Fisher's method controls FWE when $J = 3$. That is, the probability of at least one Type I error will be less than or equal to α. However, when there are more than three groups ($J > 3$), this is no longer true (Hayter, 1986). To gain some intuition as to why, suppose four groups are to be compared, the first three have equal means, but the mean of the fourth group is so much larger than the other three means that power is close to 1. That is, with near certainty, you will reject with the ANOVA F-test and proceed to compare all pairs of means with Student's T at the α level. So in particular you will test

$$H_0 : \mu_1 = \mu_2,$$
$$H_0 : \mu_1 = \mu_3,$$
$$H_0 : \mu_2 = \mu_3,$$

each at the α level, and the probability of at least one Type I error among these three tests will be greater than α.

12.1.2 The Tukey–Kramer Method

Tukey was the first to propose a method that controls FWE. He assumed normality and homoscedasticity and obtained an exact solution when all J groups have equal sample sizes. Kramer (1956) proposed a generalization that provides an approximate solution when the sample sizes are unequal, and Hayter (1984) showed that when there is homoscedasticity and sampling is from normal distributions, Kramer's method is conservative. That is, it guarantees that FWE will be less than or equal to α.

When comparing the jth group to the kth group, the Tukey–Kramer $1 - \alpha$ confidence interval for $\mu_j - \mu_k$ is

$$(\bar{X}_j - \bar{X}_k) \pm q\sqrt{\frac{\mathrm{MSWG}}{2}\left(\frac{1}{n_j} + \frac{1}{n_k}\right)}, \tag{12.4}$$

where n_j is the sample size of the jth group, MSWG is the mean square within groups, which estimates the assumed common variance (see Box 9.1), and q is a constant read from Table 9 in Appendix B, which depends on the values of α, J (the number of groups being compared), and the degrees of freedom,

$$\nu = N - J,$$

where again N is the total number of observations in all J groups. Under normality, equal variances and equal sample sizes, the *simultaneous probability coverage* is exactly $1 - \alpha$. That is, with probability $1 - \alpha$, it will be simultaneously true that the confidence interval for $\mu_1 - \mu_2$ will indeed contain $\mu_1 - \mu_2$, the confidence interval for $\mu_1 - \mu_3$

TABLE 12.2   Ratings of Three Types of Cookies

Method 1	Method 2	Method 3
5	6	8
4	6	7
3	7	6
3	8	8
4	4	7
5	5	
3	8	
4	5	
8		
2		

will indeed contain $\mu_1 - \mu_3$, and so on. You reject $H_0 : \mu_j = \mu_k$ if

$$\frac{|\bar{X}_j - \bar{X}_k|}{\sqrt{\frac{\text{MSWG}}{2}\left(\frac{1}{n_j} + \frac{1}{n_k}\right)}} \geq q.$$

EXAMPLE. Table 12.2 shows some hypothetical data on the ratings of three brands of cookies. Each brand is rated by a different sample of individuals. There are a total of $N = 23$ observations, so the degrees of freedom are $\nu = 23 - 3 = 20$, the sample means are $\bar{X}_1 = 4.1$, $\bar{X}_2 = 6.125$, and $\bar{X}_3 = 7.2$, the estimate of the common variance is $\text{MSWG} = 2.13$, and with $\alpha = .05$, Table 9 in Appendix B indicates that $q = 3.58$. The confidence interval for $\mu_1 - \mu_3$ is

$$(4.1 - 7.2) \pm 3.58\sqrt{\frac{2.13}{2}\left(\frac{1}{10} + \frac{1}{5}\right)} = (-5.12, -1.1).$$

This interval does not contain zero, so you reject the hypothesis that typical ratings of brands 1 and 3 are the same. You can compare brand 1 to brand 2 and brand 2 to brand 3 in a similar manner, but the details are left as an exercise. ■

12.1.3 A Step-Down Method

All-pairs power refers to the probability of detecting all true differences among all pairwise differences among the means. For example, suppose you want to compare four groups where $\mu_1 = \mu_2 = \mu_3 = 10$ but $\mu_4 = 15$. In this case, all-pairs power refers to the probability of rejecting $H_0 : \mu_1 = \mu_4$ and $H_0 : \mu_2 = \mu_4$ and $H_0 : \mu_3 = \mu_4$. Still assuming normality and homoscedasticity, it is possible to achieve higher all-pairs power than with the Tukey–Kramer method in Section 12.1.2 using what is called a *step-down* technique. One price for this increased power is that you can no longer compute confidence intervals for the differences among the pairs of means. Box 12.1 summarizes the details.

BOX 12.1 Summary of the Step-Down Procedure

The goal is to perform all pairwise comparisons of the means of J independent groups such that the familywise Type I error probability is α.

1. Test $H_0 : \mu_1 = \cdots = \mu_J$ at the $\alpha_J = \alpha$ level of significance. Assuming normality and homoscedasticity, this is done with the ANOVA F-test in Chapter 9. If you fail to reject, stop; otherwise continue to the next step.

2. For each subset of $J - 1$ means, test the hypothesis that these means are equal at the $\alpha_{J-1} = \alpha$ level of significance. If all such tests are nonsignificant, stop. Otherwise continue to the next step.

3. For each subset of $J - 2$ means, test the hypothesis that they are equal at the $\alpha_{J-2} = 1 - (1 - \alpha)^{(J-2)/J}$ level of significance. If all of these tests are nonsignificant, stop; otherwise continue to the next step.

4. In general, test the hypothesis of equal means, for all subsets of p means, at the $\alpha_p = 1 - (1 - \alpha)^{p/J}$ level of significance, when $p \leq J - 2$. If all of these tests are nonsignificant, stop and fail to detect any differences among the means; otherwise continue to the next step.

5. The final step consists of testing all pairwise comparisons of the means at the $\alpha_2 = 1 - (1 - \alpha)^{2/J}$ level of significance. In this final step, when comparing the jth group to the kth group, you either fail to reject, you fail to reject by implication from one of the previous steps, or you reject.

EXAMPLE. Consider $J = 5$ methods designed to increase the value of a client's stock portfolio, which we label methods A, B, C, D, and E. Further assume that when comparing these five methods, you are willing to sacrifice confidence intervals to enhance your all-pairs power. Assume that you want the familywise error rate to be $\alpha = .05$. The first step is to test

$$H_0 : \mu_1 = \mu_2 = \mu_3 = \mu_4 = \mu_5$$

at the $\alpha_5 = \alpha = .05$ level of significance, where the subscript 5 on α_5 indicates that in the first step, all $J = 5$ means are being compared. If you fail to reject H_0, stop and decide that there are no pairwise differences among the five methods. If you reject, proceed to the next step, which consists of testing the equality of the means for all subsets of four groups. In the illustration, suppose the F-test for equal means is applied as described in Chapter 9, yielding

Continued

EXAMPLE. (*Continued*) $F = 10.5$. Assuming the critical value is 2.6, you would reject and proceed to the next step. That is, you test

$$H_0 : \mu_1 = \mu_2 = \mu_3 = \mu_4$$

$$H_0 : \mu_1 = \mu_2 = \mu_3 = \mu_5$$

$$H_0 : \mu_1 = \mu_2 = \mu_4 = \mu_5$$

$$H_0 : \mu_1 = \mu_3 = \mu_4 = \mu_5$$

$$H_0 : \mu_2 = \mu_3 = \mu_4 = \mu_5.$$

In this step you test each of these hypotheses at the $\alpha_4 = \alpha = .05$ level of significance, where the subscript 4 indicates that each test is comparing the means of four groups. Note that both the first and second steps use the same significance level, α. If in the second step, all five tests are nonsignificant, you stop and fail to detect any pairwise differences among the five methods; otherwise you proceed to the next step. In the illustration, suppose the values of your test statistic, F, are 9.7, 10.2, 10.8, 11.6, and 9.8, with a critical value of 2.8. So you reject in every case, but even if you reject in only one case, you proceed to the next step.

The third step consists of testing all subsets of exactly three groups, but this time you test at the

$$\alpha_3 = 1 - (1 - \alpha)^{3/5}$$

level of significance, where the subscript 3 is used to indicate that subsets of three groups are being compared. In the illustration, this means you test

$$H_0 : \mu_1 = \mu_2 = \mu_3$$

$$H_0 : \mu_1 = \mu_2 = \mu_4$$

$$H_0 : \mu_1 = \mu_3 = \mu_4$$

$$H_0 : \mu_1 = \mu_2 = \mu_5$$

$$H_0 : \mu_1 = \mu_3 = \mu_5$$

$$H_0 : \mu_1 = \mu_4 = \mu_5$$

$$H_0 : \mu_2 = \mu_3 = \mu_4$$

$$H_0 : \mu_2 = \mu_3 = \mu_5$$

$$H_0 : \mu_2 = \mu_4 = \mu_5$$

$$H_0 : \mu_3 = \mu_4 = \mu_5$$

Continued

EXAMPLE. (*Continued*) using $\alpha_3 = 1 - (1 - .05)^{3/5} = .030307$. If none of these hypotheses is rejected, you stop and fail to detect any pairwise difference among all pairs of methods; otherwise you continue to the next step.

The final step is to compare the jth group to the kth group by testing

$$H_0 : \mu_j = \mu_k.$$

This time you test at the

$$\alpha_2 = 1 - (1 - \alpha)^{2/5}$$

level. In the illustration, $\alpha_2 = .020308$. In this final stage, you make one of three decisions: You fail to reject H_0, you fail to reject H_0 due to the results from a previous step, or you reject. To clarify the second decision, suppose you fail to reject $H_0 : \mu_1 = \mu_3 = \mu_4$. Then by implication, you would conclude that $\mu_1 = \mu_3$, $\mu_1 = \mu_4$, and $\mu_3 = \mu_4$, *regardless* of what you got in the final step. That is, $H_0 : \mu_1 = \mu_3$, $H_0 : \mu_1 = \mu_4$, and $H_0 : \mu_3 = \mu_4$ would be declared not significant by implication, even if they were rejected in the final step. This might seem counterintuitive, but it is necessary if you want to control the familywise Type I error probability. Table 12.3 summarizes the results.   ■

As stressed in Chapter 9, the ANOVA F-test performs rather poorly when the normality or homoscedasticity assumption is violated. One particular problem is low power under arbitrarily small departures from normality. When comparing means with a step-down procedure, there is a sense in which this problem is exacerbated. To illustrate why, imagine you are comparing four groups; all of the groups have unequal means, the first three groups have normal distributions, but the fourth has the mixed normal described in Section 2.7. Then the ANOVA F-test, applied to all four groups, can have low power, which in turn can mean that the step-down method described here has low power as well. In fact, even a single outlier in one group can mask substantial differences among the other groups being compared. (Dunnett & Tamhane, 1992, describe results on a step-up method that has practical advantages under normality and homoscedasticity, but it suffers from the same problem just described.)

12.2 ANOVA F Versus Multiple Comparisons

It is common practice to use Tukey's method only if the ANOVA F-test is significant. More generally, when using any multiple comparison procedure to compare groups based on some measure of location θ, it is common first to test

$$H_0 : \theta_1 = \cdots = \theta_J \tag{12.5}$$

and, if a nonsignificant result is obtained, to fail to detect any differences among the groups. Testing this omnibus test obviously plays a central role in a step-down method, but there is an important negative consequence of this strategy: Many modern multiple comparison procedures are designed to control FWE without first testing Equation (12.5). One example is Tukey's procedure. Under normality, homoscedasticity, and

TABLE 12.3   Illustration of the Step-Down Procedure

Groups	F	α	ν_1	Critical value	Decision
ABCDE	11.5	$\alpha_5 = .05$	4	2.61	Significant
ABCD	9.7	$\alpha_4 = .05$	3	2.84	Significant
ABCE	10.2				Significant
ABDE	10.8				Significant
ACDE	11.6				Significant
BCDE	9.8				Significant
ABC	2.5	$\alpha_3 = .0303$	2	3.69	Not significant
ABD	7.4				Significant
ACD	8.1				Significant
ABE	8.3				Significant
ACE	12.3				Significant
ADE	18.2				Significant
BCD	2.5				Not significant
BCE	9.2				Significant
BDE	8.1				Significant
CDE	12.4				Significant
AB	5.1	$\alpha_2 = .0203$	1	5.85	Not significant by implication
AC	6.0				Not significant by implication
AD	19.2				Significant
AE	21.3				Significant
BC	1.4				Not significant by implication
BD	6.0				Not significant by implication
BE	15.8				Significant
CD	4.9				Not significant by implication
CE	13.2				Significant
DE	3.1				Not significant

equal sample sizes, it guarantees that FWE will be exactly equal to α. But if Tukey's method is used contingent on rejecting with the ANOVA *F*, this is no longer true — FWE will be less than α, indicating that power will be reduced (Bernhardson, 1975). Generally, most modern multiple comparison procedures are designed so that FWE will be approximately equal to α. That is, they do not require that you first apply one of the methods in Chapters 9–11 and reject the hypothesis that groups have a common measure of location. Indeed, if you use modern methods only when you first reject, you run the risk of lowering the actual Type I error probability by an unknown amount, which in turn might mask true differences among the groups being compared. Robust and heteroscedastic ANOVA methods remain relevant, however, because they might be useful in some type of step-down or step-up technique.

12.3 Heteroscedastic Methods for the Means of Independent Groups

Consistent with earlier chapters, when groups have identical distributions, the methods in the previous section appear to provide reasonably good control over the probability of a Type I error. But when distributions differ in some manner, they suffer from the same problems associated with the homoscedastic methods described in previous chapters: Poor power, undesirable power properties (bias, meaning that power can decrease as the difference among the means increases), and poor probability coverage. This section describes heteroscedastic methods that reduce these problems, but serious practical concerns remain, some of which are not addressed even with very large sample sizes. However, the methods in this section set the stage for effective techniques.

12.3.1 Dunnett's T3

For multiple comparison procedures based on means, Dunnett (1980a, 1980b) documented practical problems with methods that assume homoscedasticity and then compared several heteroscedastic methods, two of which stood out when sampling from normal distributions. Although nonnormality can ruin these methods, they are important because they provide a basis for deriving substantially improved techniques.

Dunnett's so-called T3 procedure is just Welch's method described in Section 8.3, but with the critical value adjusted so that FWE is approximately equal to α when sampling from normal distributions. Let s_j^2 be the sample variance for the jth group, again let n_j be the sample size and set

$$q_j = \frac{s_j^2}{n_j}, j = 1, \ldots, J.$$

When comparing group j to group k, the degrees of freedom are

$$\hat{v}_{jk} = \frac{(q_j + q_k)^2}{\frac{q_j^2}{n_j - 1} + \frac{q_k^2}{n_k - 1}}.$$

The test statistic is

$$W = \frac{\bar{X}_j - \bar{X}_k}{\sqrt{q_j + q_k}},$$

and you reject $H_0 : \mu_j = \mu_k$ if $|W| \geq c$, where the critical value, c, is read from Table 10 in Appendix B. (This table provides the .05 and .01 quantiles of what is called the *Studentized maximum modulus distribution*.) When using Table 10, you need to know the total number of comparisons you plan to perform. When performing all pairwise comparisons, the total number of comparisons is

$$C = \frac{J^2 - J}{2}.$$

In the illustration, there are $J = 3$ groups, so the total number of comparisons is

$$C = \frac{3^2 - 3}{2} = 3.$$

If you have $J = 4$ groups *and* you plan to perform all pairwise comparisons, $C = (4^2 - 4)/2 = 6$.

EXAMPLE. Suppose the goal is to compare five groups to a control group and that only these five comparisons are to be done. That is, the goal is to test $H_0 : \mu_j = \mu_6$, $j = 1, \ldots, 5$, so $C = 5$. If $\alpha = .05$ and the degrees of freedom are 30, the critical value is $c = 2.73$. If you have five groups and plan to do all pairwise comparisons, $C = 10$; and with $\alpha = .01$ and $\nu = 20$, the critical value is 3.83. ■

A confidence interval for $\mu_j - \mu_k$, the difference between the means of groups j and k, is given by

$$(\bar{X}_j - \bar{X}_k) \pm c\sqrt{\frac{s_j^2}{n_j} + \frac{s_k^2}{n_k}}.$$

By design, the simultaneous probability coverage will be approximately $1 - \alpha$, under normality, when computing C confidence intervals and c is read from Table 10 in Appendix B.

EXAMPLE. Table 9.1 reports skin resistance for four groups of individuals. If the goal is to compare all pairs of groups with $\alpha = .05$, then $C = 6$, and the confidence interval for the difference between the means of the first two groups is $(-0.35, 0.52)$; this interval contains zero, so you fail to detect a difference. The degrees of freedom are 12.3, the critical value is $c = 3.07$, the test statistic is $W = 0.56$, and again you fail to reject, because $|W| < 3.07$. It is left as an exercise to show that for the remaining pairs of means, you again fail to reject. ■

12.3.2 Games–Howell Method

An alternative to Dunnett's T3 is the Games and Howell (1976) method. When comparing the jth group to the kth group, you compute the degrees of freedom, $\hat{\nu}_{jk}$, exactly as in Dunnett's T3 procedure, and then you read the critical value, q, from Table 9 in Appendix B. (Table 9 reports some quantiles of what is called the Studentized range distribution.) The $1 - \alpha$ confidence interval for $\mu_j - \mu_k$ is

$$(\bar{X}_j - \bar{X}_k) \pm q \sqrt{\frac{1}{2}\left(\frac{s_j^2}{n_j} + \frac{s_k^2}{n_k}\right)}.$$

You reject $H_0 : \mu_j = \mu_k$ if this interval does not contain zero, which is the same as rejecting if

$$\frac{|\bar{X}_j - \bar{X}_k|}{\sqrt{\frac{1}{2}\left(\frac{s_j^2}{n_j} + \frac{s_k^2}{n_k}\right)}} \geq q.$$

Under normality, the Games–Howell method appears to provide more accurate probability coverage than Dunnett's T3 method when all groups have a sample size of at least 50. A close competitor under normality is Dunnett's (1980b) C method, but no details are given here.

EXAMPLE. Imagine you have three groups, with $\bar{X}_1 = 10.4$, $\bar{X}_2 = 10.75$,

$$\frac{s_1^2}{n_1} = .11556,$$

$$\frac{s_2^2}{n_2} = .156.$$

Then $\hat{\nu} = 19$ and with $\alpha = .05$, $q = 3.59$, so the confidence interval for $\mu_1 - \mu_2$ is

$$(10.4 - 10.75) \pm 3.59 \sqrt{\frac{1}{2}(.11556 + .156)} = (-.167, 0.97).$$

This interval contains 0, so you do not reject the hypothesis of equal means. ■

12.3.3 Alternative Methods Based on Adjustments of α

Dunnett's T3 and the Games–Howell method use adjusted critical values to attempt to control FWE. There is a collection of methods for controlling FWE that adjust the p-values in a more direct fashion that should be mentioned. The easiest to use is based on the Bonferroni inequality, which, if C hypotheses are to be tested, test each hypothesis at the α/C level. Provided the probability of a Type I error can be controlled for each of the individual tests, FWE will be at most α.

Other approaches are based on what are called *sequentially rejective* methods. For example, Hochberg's (1988) method is applied as follows. Let $P_1, \ldots, P_C$ be the p-values associated with the C tests, put these p-values in descending order, and label the results $P_{[1]} \geq P_{[2]} \geq \cdots \geq P_{[C]}$. The method is applied in steps, where the total number of steps is at most C. Beginning with $k = 1$ (step 1), reject all hypotheses if

$$P_{[k]} \leq \frac{\alpha}{k}.$$

That is, reject all hypotheses if the largest p-value is less than or equal to α. If $P_{[1]} > \alpha$, proceed as follows:

1. Increment k by 1. If

$$P_{[k]} \leq \frac{\alpha}{k},$$

   stop and reject all hypotheses having a p-value less than or equal to $P_{[k]}$.
2. If $P_{[k]} > \alpha/k$, repeat step 1.
3. Repeat steps 1 and 2 until you reject or all C hypotheses have been tested.

Benjamini and Hochberg (1995) proposed a similar method, only in step 1 of Hochberg's method, $P_{[k]} \leq \alpha/k$ is replaced by

$$P_{[k]} \leq \frac{(C - k + 1)\alpha}{C}.$$

Results in Williams, Jones, and Tukey (1999) support the use of the Benjamini–Hochberg method over Hochberg. Both of these procedures have power greater than or equal to the Bonferroni method. However, the Bonferroni method can be used to compute confidence intervals, and currently there is uncertainty about how one should compute confidence intervals when using either of the two sequentially rejective methods just described.

EXAMPLE. Suppose six hypotheses are tested with the Benjamini–Hochberg method based on the following results:

Number	Test	p-value	
1	$H_0 : \mu_1 = \mu_2$	$P_1 = .010$	$P_{[5]}$
2	$H_0 : \mu_1 = \mu_3$	$P_2 = .015$	$P_{[3]}$
3	$H_0 : \mu_1 = \mu_4$	$P_3 = .005$	$P_{[6]}$
4	$H_0 : \mu_2 = \mu_3$	$P_4 = .620$	$P_{[1]}$
5	$H_0 : \mu_2 = \mu_4$	$P_5 = .130$	$P_{[2]}$
6	$H_0 : \mu_3 = \mu_4$	$P_6 = .014$	$P_{[4]}$

Because $P_{[1]} > .05$, fail to reject the fourth hypothesis. Had it been the case that $P_{[1]} \leq .05$, you would stop and reject all six hypotheses. Because you did not reject, set $k = 2$; and because $C = 6$, we see that

$$\frac{(C - k + 1)\alpha}{C} = \frac{5(.05)}{6} = .0417.$$

Continued

EXAMPLE. (*Continued*) Because $P_{[2]} = .130 > .0417$, fail to reject the fifth hypothesis and proceed to the next step. Incrementing k to 3,

$$\frac{(C - k + 1)\alpha}{C} = \frac{4(.05)}{6} = .0333,$$

and because $P_{[3]} = .015 \leq .0333$, reject this hypothesis and the remaining hypotheses having p-values less than or equal to .0333. That is, reject hypotheses 1, 2, 3, and 6 and fail to reject hypotheses 4 and 5. If the Bonferroni inequality had been used instead, we see that $.05/6 = .00833$, so only hypothesis 3 would be rejected.   ■

A criticism of the Benjamini–Hochberg method is that situations can be found where some hypotheses are true, some are false, and the probability of at least one Type I error will exceed α among the hypotheses that are true (Hommel, 1988; cf. Keselman, Cribbie, & Holland, 1999). In contrast, Hochberg's method does not suffer from this problem. However, the Benjamini–Hochberg method does have the following property. When C hypotheses are tested, let Q be the proportion of hypotheses that are true and rejected. That is, Q is the proportion of Type I errors among the null hypotheses that are correct. If all hypotheses are false, then of course $Q = 0$, but otherwise Q can vary from one experiment to the next. That is, if we repeat a study many times, the proportion of erroneous rejections will vary. The *false-discovery rate* is the expected value of Q. That is, if a study is repeated infinitely many times, the false-discovery rate is the average proportion of Type I errors among the hypotheses that are true. Benjamini and Hochberg (1995) show that their method ensures that the false-discovery rate is less than or equal to α. The Benjamini–Hochberg method can be improved if the number of true hypotheses is known. Of course it is not known how many null hypotheses are in fact correct, but Benjamini and Hochberg (2000) suggest how this number might be estimated, which can result in higher power. (For related results, see Finner and Roters, 2002; Sarkar, 2002.)

12.4 Linear Contrasts

When dealing with a two-way design, as described in Chapter 10, it is convenient to describe relevant multiple comparison procedures in the context of what are called *linear contrasts*. For a J-by-K design, let $L = JK$ represent the total number of groups. (So in a one-way design, $L = J$.) By definition, a *linear contrast* is any linear combination of the means among L groups having the form

$$\Psi = \sum_{\ell=1}^{L} c_\ell \mu_\ell, \tag{12.6}$$

where $c_1, \ldots, c_L$, called *contrast coefficients*, are constants that sum to 0. In symbols, Ψ is a linear contrast if

$$\sum c_\ell = 0.$$

EXAMPLE. Section 10.1 describes a two-by-two design dealing with weight gain in rats. The two factors are source (beef versus cereal) and amounts of protein (low versus high), and here the population means are represented as follows:

	Source	
Amount	Beef	Cereal
Low	μ_1	μ_2
High	μ_3	μ_4

Consider the hypothesis of no main effect for the first factor, amount of protein. As explained in Chapter 10, the null hypothesis is

$$H_0 : \frac{\mu_1 + \mu_2}{2} = \frac{\mu_3 + \mu_4}{2}.$$

Rearranging terms in this last equation, the null hypothesis can be written as a linear contrast, namely,

$$H_0 : \mu_1 + \mu_2 - \mu_3 - \mu_4 = 0.$$

That is $\Psi = \mu_1 + \mu_2 - \mu_3 - \mu_4$, the contrast coefficients are $c_1 = c_2 = 1$, $c_3 = c_4 = -1$, and the null hypothesis is $H_0 : \Psi = 0$.   ■

EXAMPLE. Now consider a three-by-two design:

	Factor B	
Factor A	1	2
1	μ_1	μ_2
2	μ_3	μ_4
3	μ_5	μ_6

An issue that is often of interest is not just whether there is a main effect for factor A, but which levels of factor A differ and by how much. That is, the goal is to compare level 1 of factor A to level 2, level 1 to level 3, and level 2 to level 3. In symbols, the three hypotheses of interest are

$$H_0 : \frac{\mu_1 + \mu_2}{2} = \frac{\mu_3 + \mu_4}{2},$$

Continued

EXAMPLE. (*Continued*)

$$H_0 : \frac{\mu_1 + \mu_2}{2} = \frac{\mu_5 + \mu_6}{2},$$

$$H_0 : \frac{\mu_3 + \mu_4}{2} = \frac{\mu_5 + \mu_6}{2}.$$

In terms of linear contrasts, the goal is to test

$$H_0 : \Psi_1 = 0,$$
$$H_0 : \Psi_2 = 0,$$
$$H_0 : \Psi_3 = 0,$$

where

$$\Psi_1 = \mu_1 + \mu_2 - \mu_3 - \mu_4,$$

$$\Psi_2 = \mu_1 + \mu_2 - \mu_5 - \mu_6,$$

$$\Psi_3 = \mu_3 + \mu_4 - \mu_5 - \mu_6.$$

The interactions can be written as linear contrasts as well. For example, the hypothesis of no interaction for the first two rows corresponds to

$$H_0 : \mu_1 - \mu_2 = \mu_3 - \mu_4,$$

which is the same as testing

$$H_0 : \Psi_4 = 0,$$

where

$$\Psi_4 = \mu_1 - \mu_2 - \mu_3 + \mu_4.$$

Similarly, for rows 1 and 3, the hypothesis of no interaction is

$$H_0 : \Psi_5 = 0,$$

where

$$\Psi_5 = \mu_1 - \mu_2 - \mu_5 + \mu_6.$$

In general, there are a collection of C linear contrasts that one might want to test; the goal is to devise a method for performing these C tests in a manner that controls FWE.   ■

12.4.1 Scheffé's Homoscedastic Method

Assuming normality and homoscedasticity, Scheffé's classic method can be used to test C hypotheses about C linear contrasts such that FWE is less than or equal to

α — regardless of how large C might be! Let Ψ be any specific hypothesis and let

$$\hat{\Psi} = \sum_{\ell=1}^{L} c_\ell \bar{X}_\ell. \tag{12.7}$$

That is, estimate the mean, μ_ℓ, of the ℓth group with $\bar{X}_\ell$, the sample mean of the ℓth group, and then plug this estimate into Equation (12.6) to get an estimate of Ψ. Then the confidence interval for Ψ is

$$(\hat{\Psi} - S, \hat{\Psi} + S), \tag{12.8}$$

where

$$S = \sqrt{(L-1)f_{1-\alpha}(\text{MSWG}) \sum \frac{c_\ell^2}{n_\ell}},$$

MSWG is the mean square within groups (described in Chapters 9 and 10) that estimates the assumed common variance, and $f_{1-\alpha}$ is the $1-\alpha$ quantile of an F-distribution with $\nu_1 = L - 1$ and $\nu_2 = N - L$ degrees of freedom, where $N = \sum n_j$ is the total number of observations in all L groups. (For a one-way design with J levels, $L = J$; for a J-by-K design, $L = JK$.) In particular, $H_0 : \Psi = 0$ is rejected if the confidence interval given by Equation (12.8) does not contain zero.

EXAMPLE. For the special case where all pairwise comparisons of J independent groups are to be performed, Scheffé's confidence interval for the difference between the means of the jth and kth groups, $\mu_j - \mu_k$, is

$$(\bar{X}_j - \bar{X}_k) \pm S,$$

where

$$S = \sqrt{(J-1)f_{1-\alpha}(\text{MSWG}) \left(\frac{1}{n_j} + \frac{1}{n_k} \right)},$$

and $f_{1-\alpha}$ is the ANOVA F critical value based on $\nu_1 = J - 1$ and $\nu_2 = N - J$ degrees of freedom. ■

Scheffé's method remains one of the more popular multiple comparison procedures in applied work, but it suffers from the same problems associated with other homoscedastic methods already covered. Even under normality but heteroscedasticity, problems arise (Kaiser & Bowden, 1983). Also, under normality and homoscedasticity, the Tukey–Kramer method should give shorter confidence intervals than the Scheffé method, but for certain collections of linear contrasts the reverse is true (Scheffé, 1959, p. 76).

12.4.2 Heteroscedastic Methods

Two heteroscedastic methods for linear contrasts are covered in this section.

Welch–Šidák Method

The first contains Dunnett's T3 method as a special case and is called the Welch–Šidák method. (The computations are performed by the S-PLUS function lincon described in Section 12.6.1.) Again let L represent the total number of groups being compared, let Ψ be any linear contrast of interest, and let C represent the total number of contrasts to be tested. An expression for the squared standard error of $\hat{\Psi}$ is

$$\sigma_{\hat{\Psi}}^2 = \sum \frac{c_\ell^2 \sigma_\ell^2}{n_\ell},$$

where σ_ℓ^2 and n_ℓ are the variance and sample size of the ℓth group, respectively. An estimate of this quantity is obtained simply by replacing σ_ℓ^2 with s_ℓ^2, the sample variance associated with the ℓth group. That is, estimate $\sigma_{\hat{\Psi}}^2$ with

$$\hat{\sigma}_{\hat{\Psi}}^2 = \sum \frac{c_\ell^2 s_\ell^2}{n_\ell}.$$

Let

$$q_\ell = \frac{c_\ell^2 s_\ell^2}{n_\ell}.$$

The degrees of freedom are estimated to be

$$\hat{\nu} = \frac{\left(\sum q_\ell\right)^2}{\sum \frac{q_\ell^2}{n_\ell - 1}}.$$

The test statistic is

$$T = \frac{\hat{\Psi}}{\hat{\sigma}_{\hat{\Psi}}}.$$

The critical value, c, is a function of $\hat{\nu}$ and C (the total number of hypotheses you plan to perform) and is read from Table 10 in Appendix B. Reject if $|T| \geq c$, and a confidence interval for Ψ is

$$\hat{\Psi} \pm c\hat{\sigma}_{\hat{\Psi}}. \tag{12.9}$$

Kaiser–Bowden Method

A heteroscedastic analog of Scheffé's method was derived by Kaiser and Bowden (1983). The computations are exactly the same as in the Welch–Šidák method, except the squared critical value is now

$$A = (L - 1)\left(1 + \frac{L - 2}{\hat{v}_2}\right)f,$$

where f is the $1 - \alpha$ quantile of an F-distribution with $v_1 = L - 1$ and

$$\hat{v}_2 = \frac{(\sum q_\ell)^2}{\sum \frac{q_\ell^2}{n_\ell - 1}}.$$

That is, $\hat{v}_2 = \hat{v}$, the estimated degrees of freedom used by the Welch–Šidák method. The confidence interval for Ψ is

$$\hat{\Psi} \pm \sqrt{A \sum q_\ell},$$

where again $q_\ell = c_\ell^2 s_\ell^2 / n_\ell$. Generally, the Welch–Šidák method will provide shorter confidence intervals and more power than the Kaiser–Bowden method. But the critical values in Table 10 in Appendix B for the Welch–Šidák method are limited to $\alpha = .05$ and .01 and at most 28 hypotheses.

EXAMPLE. For the special case where all pairwise comparisons among J independent groups are to be performed, the Kaiser–Bowden method for computing a confidence interval for $\mu_j - \mu_k$ is

$$(\bar{X}_j - \bar{X}_k) \pm \sqrt{A\left(\frac{s_j^2}{n_j} + \frac{s_k^2}{n_k}\right)},$$

where now

$$A = (J - 1)\left(1 + \frac{J - 2}{\hat{v}_{jk}}\right)f,$$

$\hat{v}_{jk}$ is as given in Section 12.3.1 (in conjunction with Dunnett's T3 procedure), and f is the $1 - \alpha$ quantile of an F-distribution with $v_1 = J - 1$ and $v_2 = \hat{v}_{jk}$ degrees of freedom. ■

EXAMPLE. Consider a 2-by-2 design and suppose the sample means are $\bar{X}_1 = 10, \bar{X}_2 = 14, \bar{X}_3 = 18, \bar{X}_4 = 12$; the sample variances are $s_1^2 = 20, s_2^2 = 8, s_3^2 = 12, s_4^2 = 4$; and the sample sizes are $n_1 = n_2 = n_3 = n_4 = 4$. Further assume you want to test three hypotheses with the Welch–Šidák method: no

Continued

EXAMPLE. (*Continued*) main effects for factor A, no main effects for factor B, and no interaction. In terms of linear contrasts, the goal is to test

$$H_0 : \Psi_1 = 0,$$
$$H_0 : \Psi_2 = 0,$$
$$H_0 : \Psi_3 = 0,$$

where

$$\Psi_1 = \mu_1 + \mu_2 - \mu_3 - \mu_4,$$
$$\Psi_2 = \mu_1 + \mu_3 - \mu_2 - \mu_4,$$
$$\Psi_3 = \mu_1 - \mu_2 - \mu_3 + \mu_4.$$

Moreover, the probability of at least one Type I error is to be .05 among these three tests. For the first hypothesis, the estimate of Ψ_1 is

$$\hat{\Psi}_1 = 10 + 14 - 18 - 12 = -6.$$

The estimate of the squared standard error is

$$\hat{\sigma}^2_{\hat{\Psi}_1} = \frac{1^2(20)}{4} + \frac{1^2(8)}{4} + \frac{(-1)^2(12)}{4} + \frac{(-1)^2(4)}{4}$$
$$= 11,$$

the degrees of freedom for the Welch–Šidák method can be shown to be 9.3; and with $C = 3$, the critical value is approximately $c = 2.87$. The test statistic is

$$T = \frac{-6}{\sqrt{11}} = -1.8,$$

and because $|T| < 2.87$, fail to reject. If instead the Kaiser–Bowden method is used, because there is a total of $L = 4$ groups, the critical value is

$$c = \sqrt{(4-1)\left(1 + \frac{4-2}{9.3}\right)3.81} = 3.7$$

which is considerably larger than the critical value based on the Welch–Šidák method. The other two hypotheses can be tested in a similar manner, but the details are left as an exercise. ■

EXAMPLE. In Chapter 10 it was noted that when investigating the possibility of a disordinal interaction, it can become necessary to test a collection of relevant hypotheses. In the previous example, note that $\bar{X}_1 < \bar{X}_2$ but that $\bar{X}_3 > \bar{X}_4$. To establish that there is a disordinal interaction, you must be able to reject both $H_0 : \mu_1 = \mu_2$ and $H_0 : \mu_3 = \mu_4$. One way to control FWE among these two hypotheses is with the Welch–Šidák method. ■

12.5 **Judging Sample Sizes**

As noted in Chapter 8, failing to reject some hypothesis might be because there is little or no difference between the groups, or there might be an important difference that was missed due to low power. So issues of general importance are determining whether the sample size used was sufficiently large to ensure adequate power and how large the sample size should have been if power is judged to be inadequate; there is the related problem of controlling the length of confidence intervals. This section describes two methods for accomplishing the last of these goals when working with linear contrasts.

12.5.1 Tamhane's Procedure

The first of the two methods is due to Tamhane (1977). A general form of his results can be used to deal with linear contrasts, but here attention is restricted to all pairwise comparisons among J independent groups. The goal is achieve confidence intervals having some specified length $2m$, given s_j^2, the sample variance from the jth group based on n_j observations ($j = 1, \ldots, J$). The computational details are summarized in Box 12.2.

BOX 12.2 Summary of Tamhane's Method

Goal

Compute confidence intervals for all pairwise differences among J independent groups such that the simultaneous probability coverage is equal to $1 - \alpha$ and the length of each confidence interval is $2m$. Normality is assumed but unequal variances are allowed.

Compute

$$A = \sum \frac{1}{n_j - 1}$$

$$\nu = \left[\frac{J}{A}\right],$$

where the notation $[J/A]$ means you compute J/A and round down to the nearest integer. Next, determine b from Table 11 in Appendix B with ν degrees of freedom. (Table 11 gives the quantiles of the range of independent Student T variates.) Note that Table 11 assumes $\nu \leq 59$. For larger degrees of freedom, use Table 9 in Appendix B instead. Let

$$d = \left(\frac{m}{b}\right)^2.$$

Continued

BOX 12.2 (*Continued*) Letting s_j^2 be the sample variance for the *j*th group, the total number of observations required from the *j*th group is

$$N_j = \max\left\{ n_j + 1, \left[\frac{s_j^2}{d}\right] + 1 \right\}.$$

Once the additional observations are available, compute the generalized sample mean, $\tilde{X}_j$, for the *j*th group as described and illustrated in Box 9.6 in connection with the Bishop–Dudewicz ANOVA. The confidence interval for $\mu_j - \mu_k$ is

$$\left(\tilde{X}_j - \tilde{X}_k - m, \; \tilde{X}_j - \tilde{X}_k + m \right).$$

EXAMPLE. Assume that for three groups, $n_1 = 11, n_2 = 21, n_3 = 41$, and the goal is to compute confidence intervals having length 4 and having simultaneous probability coverage $1 - \alpha = .95$. So

$$A = \frac{1}{10} + \frac{1}{20} + \frac{1}{40} = .175,$$

$$v = \left[\frac{3}{.175}\right] = [17.14] = 17,$$

and $b = 3.6$. Confidence intervals with length 4 means that $m = 2$, so

$$d = \left(\frac{2}{3.6}\right)^2 = .3086.$$

If the sample variance for the first group is $s_1^2 = 2$, then

$$N_1 = \max\left\{ 11 + 1, \left[\frac{2}{.3086}\right] + 1 \right\}$$

$$= \max(12, [6.4] + 1)$$

$$= 12.$$

You already have $n_1 = 11$ observations, so you need $12 - 11 = 1$ more. If you get $\tilde{X}_1 = 14$ and $\tilde{X}_2 = 17$, the confidence interval for the difference between the means is

$$(14 - 17 - 2, 14 - 17 + 2) = (-5, -1).$$

■

12.5.2 S-PLUS Function tamhane

Tamhane's two-stage multiple comparison procedure can be applied with the S-PLUS function

$$\text{tamhane}(x, x2 = NA, cil = NA, crit = NA).$$

The first-stage data are stored in x (in a matrix or in list mode) and the second-stage data in x2. If x2 contains no data, the function prints the degrees of freedom, which can be used to determine h as described in Box 12.2. Once h has been determined, store it in the argument crit. If no critical value is specified, the function terminates with an error message; otherwise it prints the total number of observations needed to achieve confidence intervals having length cil (given by the third argument). (So in the notation of Section 12.5.1, the value of cil divided by 2 corresponds to m.) If it finds data in the argument x2, confidence intervals are computed as described in Box 12.2 and returned by the function tamhane in the S-PLUS variable ci.mat; otherwise, the function returns ci.mat with the value NA.

EXAMPLE. For the data in Table 9.1, if the goal is to perform all pairwise comparisons such that the confidence intervals have length 1 (so $m = .5$) and FWE is to be .05, then the degrees of freedom are 9 and, from Table 11 in Appendix B, $h = 4.3$. If the data are stored in the S-PLUS variable skin, the command

$$\text{tamhane}(\text{skin}, \text{cil}{=}1, \text{crit}{=}4.3)$$

returns

```
$n.vec:
[1]  13  11  11  11
```

indicating that the required sample sizes are 13, 11, 11, and 11, respectively. ■

12.5.3 Hochberg's Procedure

This section describes another two-stage procedure derived by Hochberg (1975). In contrast to Tamhane's procedure, Hochberg's method allows the possibility of no additional observations being required in the second stage, uses the usual sample mean (rather than the generalized sample mean), and ensures that the lengths of the confidence intervals are at most $2m$. When using Tamhane's approach, the confidence intervals have length exactly equal to $2m$. (As before, m is some constant chosen by the investigator.) Box 12.3 summarizes the computations for the more general case, where C linear contrasts are to be tested, n_j represents the sample size for the jth group in the first stage, and s_j^2 is the sample variance for the jth group based on these n_j observations. (Hochberg's original derivation assumed equal sample sizes in the first stage, but the adjustment for unequal sample sizes in Box 12.3 appears to perform well.)

BOX 12.3 Summary of Hochberg's Method

As in Tamhane's procedure, read the critical value h from Table 11 in Appendix B, with the degrees of freedom, ν, computed as in Box 12.2. If $\nu > 59$, use Table 9 instead. Compute

$$d = \left(\frac{m}{h}\right)^2.$$

The total number of observations you need to sample from the jth group is

$$N_j = \max\left(n_j, \left[\frac{s_j^2}{d}\right] + 1\right). \tag{12.10}$$

Sample an additional $N_j - n_j$ observations from the jth group, and compute the sample mean, $\bar{X}_j$, based on all N_j values. For all pairwise comparisons, the confidence interval for $\mu_j - \mu_k$ is

$$(\bar{X}_j - \bar{X}_k) \pm hb,$$

where

$$b = \max\left(\frac{s_j}{\sqrt{N_j}}, \frac{s_k}{\sqrt{N_k}}\right).$$

Be sure to notice that the sample variances used to compute b are not recomputed once the additional observations are available. For technical reasons, you use the sample variances based on the initial n observations.

As for the linear contrast $\Psi = \sum c_j \mu_j$, sum the positive c_j values and label the result a. Again N_j is given by Equation (12.10), except

$$d = \left(\frac{m}{ha}\right)^2.$$

Let

$$b_j = \frac{c_j s_j}{\sqrt{N_j}}.$$

Let A be the sum of the positive b_j values and C be the sum of the negative b_j values. Compute

$$D = \max(A, -C),$$

$$\hat{\Psi} = \sum c_j \bar{X}_j,$$

in which case the confidence interval for Ψ is

$$\hat{\Psi} \pm hD.$$

EXAMPLE. To illustrate Hochberg's method, imagine that with four groups, each having a sample size of $n = 25$, one of your goals is to compute a confidence interval for $H_0 : \Psi = 0$ with $\alpha = .05$, where

$$\Psi = \mu_1 + \mu_2 - \mu_3 - \mu_4,$$

and the length of the confidence interval is to be at most $2m = 8$. So $\nu = 24$ and, from Table 11 in Appendix B, $b = 3.85$. We see that the sum of the positive contrast coefficients is $a = 2$, so

$$d = \left(\frac{4}{3.85(2)} \right)^2 = .2699.$$

If $s_1^2 = 4$, $s_2^2 = 12$, $s_3^2 = 16$, and $s_4^2 = 20$, then

$$N_1 = \max \left(25, \left[\frac{4}{.2699} \right] + 1 \right) = 25.$$

Hence, no additional observations are required. The sample sizes for the other three groups are $N_2 = 45$, $N_3 = 60$, and $N_4 = 75$.

Once the additional observations are available, compute

$$b_1 = \frac{1 \times 2}{\sqrt{25}} = .4.$$

Similarly, $b_2 = .5164$, $b_3 = -.5164$, and $b_4 = -.5164$. The sum of the positive b_j values is $A = b_1 + b_2 = .4 + .5164 = 0.9164$. The sum of the negative b_j values is $C = -1.0328$, so $-C = 1.0328$,

$$D = \max(.9164, 1.0328) = 1.0328,$$

$$bD = 3.85 \times 1.0328 = 3.976.$$

If, after the additional observations are sampled, the sample means are $\bar{X}_1 = 10$, $\bar{X}_2 = 12$, $\bar{X}_3 = 8$, and $\bar{X}_4 = 18$, then

$$\hat{\Psi} = 10 + 12 - 8 - 18 = -4,$$

and the confidence interval is

$$-4 \pm 3.976 = (-7.976, -0.024).$$

Thus, you would reject H_0. ■

12.5.4 S-PLUS Function hochberg

The S-PLUS function

$$\text{hochberg}(x, x2 = NA, cil = NA, crit = NA)$$

performs Hochberg's two-stage procedure. The arguments are used in the same manner as in Section 12.5.2.

12.6 Methods for Trimmed Means

The serious practical problems with methods based on means, described in previous chapters, extend to the situation at hand. One way of addressing these problems is to switch to generalizations of the Welch–Šidák and Kaiser–Bowden methods to trimmed means. Letting L represent the total number of groups, details are summarized in Box 12.4. (For a J-by-K design, $L = JK$.)

The method in Box 12.4 can be used to compute confidence intervals for all pairwise differences of the trimmed means for the special case of J independent groups. That is, the goal is to compute a confidence interval for $\mu_{tj} - \mu_{tk}$, for all $j < k$, such that the simultaneous probability coverage is approximately $1 - \alpha$. The confidence interval for $\mu_{tj} - \mu_{tk}$ is

$$(\bar{X}_{tj} - \bar{X}_{tk}) \pm c\sqrt{d_j + d_k},$$

where

$$d_j = \frac{(n_j - 1)s_{wj}^2}{h_j(h_j - 1)},$$

h_j is the number of observations left in group j after trimming, and c is read from Table 10 in Appendix B with

$$\hat{v} = \frac{(d_j + d_k)^2}{\frac{d_j}{h_j - 1} + \frac{d_k}{h_k - 1}}$$

degrees of freedom.

BOX 12.4 Heteroscedastic Tests of Linear Contrasts Based

on Trimmed Means

Let $\mu_{t1}, \ldots, \mu_{tL}$ be the trimmed means corresponding to L independent groups, and let

$$\Psi = \sum_{\ell=1}^{L} c_\ell \mu_{t\ell}$$

be some linear contrast of interest. It is assumed that there is a total of C such linear contrasts; for each the goal is to test $H_0 : \Psi = 0$ with FWE equal to α.

Continued

BOX 12.4 (*Continued*) Compute

$$\hat{\Psi} = \sum_{\ell=1}^{L} c_\ell \bar{X}_{t\ell},$$

$$d_\ell = \frac{c_\ell^2 (n_\ell - 1) s_{w\ell}^2}{h_\ell (h_\ell - 1)},$$

where $s_{w\ell}^2$ and h_ℓ are the Winsorized variance and effective sample size (the number of observations left after trimming) of the ℓth group, respectively. An estimate of the squared standard error of $\hat{\Psi}$ is

$$S_e = \sum d_\ell.$$

Letting

$$G = \sum \frac{d_\ell^2}{h_\ell - 1},$$

the estimated degrees of freedom are

$$\hat{\nu} = \frac{S_e^2}{G}.$$

Then a confidence interval for Ψ, based on a trimmed analog of the Welch–Šidák method, is

$$\hat{\Psi} \pm c\sqrt{S_e},$$

where c is read from Table 10 in Appendix B (and is again a function of C and $\hat{\nu}$).

As for a trimmed analog of the Kaiser–Bowden method, the confidence interval is

$$\hat{\Psi} \pm \sqrt{AS_e},$$

where

$$A = (L - 1)\left(1 + \frac{L - 2}{\hat{\nu}}\right) f,$$

and f is the $1 - \alpha$ quantile of an F-distribution with $\nu_1 = L - 1$ and $\nu_2 = \hat{\nu}$ degrees of freedom.

12.6.1 S-PLUS Function lincon

The S-PLUS function

$$lincon(x, con=0, tr=.2, alpha=.05, KB=F)$$

tests hypotheses for a collection of linear contrasts based on trimmed means. The default value for con is zero, indicating that all pairwise comparisons are to be done.

Other linear contrasts can be specified by storing a matrix of linear contrast coefficients in con. By assumption, the rows of con correspond to groups, and the columns correspond to the linear contrasts. That is, column 1 contains the contrast coefficients for the first linear contrast of interest, column 2 has the contrast coefficients for the second linear contrast, and so forth. So in general, con must be a matrix having J rows and C columns, where C is the number of linear contrasts to be tested. If $\alpha = .05$ or $.01$, and simultaneously the number of contrasts is less than or equal to 28, lincon uses the generalization of the Welch–Šidák method to trimmed means as described in Box 12.4; otherwise it uses the extension of the Kaiser–Bowden method. Setting the argument KB to T forces the function to use Kaiser–Bowden.

EXAMPLE. If the data in Table 9.1 are stored in the S-PLUS variable skin, the command lincon(skin) returns:

Group	Group	test	crit	se	df
1	2	0.2909503	3.217890	0.11539200	9.627086
1	3	0.8991534	3.306229	0.08713196	8.568838
1	4	2.2214991	3.307845	0.08703807	8.550559
2	3	1.1074715	3.406642	0.10105753	7.609956
2	4	1.5823636	3.408688	0.10097658	7.593710
3	4	4.0624997	3.190010	0.06688001	9.999866

Group	Group	psihat	ci.lower	ci.upper
1	2	-0.03357333	-0.4048921	0.33774542
1	3	0.07834500	-0.2097332	0.36642319
1	4	-0.19335500	-0.4812635	0.09455346
2	3	0.11191833	-0.2323484	0.45618510
2	4	-0.15978167	-0.5039794	0.18441602
3	4	-0.27170000	-0.4850479	-0.05835214

So, for example, the test statistic for comparing group 1 to group 2 is 0.29, the critical value is 3.2, the estimate of $\Psi_1 = \mu_{t1} - \mu_{t2}$ is -0.03, and the confidence interval for this difference is $(-0.40, 0.34)$. ■

EXAMPLE. A classic problem is comparing treatment groups to a control group. That is, rather than perform all pairwise comparisons, the goal is to compare group 1 to group J, group 2 to group J, and so on. In symbols, the goal is to test

$$H_0 : \mu_{t1} = \mu_{tJ},$$
$$H_0 : \mu_{t2} = \mu_{tJ},$$
$$\vdots$$
$$H_0 : \mu_{t,J-1} = \mu_{tJ}$$

Continued

EXAMPLE. (*Continued*) with FWE equal to α. For illustrative purposes, assume $J = 4$. Then to compare the first three groups to the control (group 4) with the S-PLUS function lincon, store the matrix

$$\begin{pmatrix} 1 & 0 & 0 \\ 0 & 1 & 0 \\ 0 & 0 & 1 \\ -1 & -1 & -1 \end{pmatrix}$$

in some S-PLUS variable — say, mmat. This can be done with the S-PLUS command

$$\text{mmat} < -\text{matrix}(c(1,0,0,-1,0,1,0,-1,0,0,1,-1), \text{ncol}=3).$$

Then the command

$$\text{lincon}(x, \text{con}=\text{mmat})$$

will perform the relevant comparisons. For the schizophrenia data in Table 9.1, the results are

```
$test:
      con.num      test      crit            se           df
            1 -2.221499  2.919435  0.08703807  8.550559
            2 -1.582364  2.997452  0.10097658  7.593710
            3 -4.062500  2.830007  0.06688001  9.999866
$psihat:
      con.num     psihat    ci.lower      ci.upper
            1 -0.1933550  -0.4474570    0.0607470
            2 -0.1597817  -0.4624542    0.1428908
            3 -0.2717000  -0.4609709   -0.0824291
```

So again the conclusion is that groups 3 and 4 differ.   ■

EXAMPLE. Consider a 3-by-3 ANOVA design with population trimmed means labeled in the usual way:

	Factor B		
Factor A	1	2	3
1	μ_{t1}	μ_{t2}	μ_{t3}
2	μ_{t4}	μ_{t5}	μ_{t6}
3	μ_{t7}	μ_{t8}	μ_{t9}

All pairwise comparisons for the main effects of factor A are given by the three

Continued

EXAMPLE. (*Continued*) linear contrasts

$$\Psi_1 = \mu_{t1} + \mu_{t2} + \mu_{t3} - \mu_{t4} - \mu_{t5} - \mu_{t6},$$

$$\Psi_2 = \mu_{t1} + \mu_{t2} + \mu_{t3} - \mu_{t7} - \mu_{t8} - \mu_{t9},$$

$$\Psi_3 = \mu_{t4} + \mu_{t5} + \mu_{t6} - \mu_{t7} - \mu_{t8} - \mu_{t9}.$$

So the matrix of contrast coefficients is

$$
\begin{pmatrix}
1 & 1 & 0 \\
1 & 1 & 0 \\
1 & 1 & 0 \\
-1 & 0 & 1 \\
-1 & 0 & 1 \\
-1 & 0 & 1 \\
0 & -1 & -1 \\
0 & -1 & -1 \\
0 & -1 & -1
\end{pmatrix}.
$$

■

12.6.2 S-PLUS Function mcp2atm for Two-Way Designs

As indicated in the previous section, multiple comparisons for a two-way ANOVA design can be performed by specifying an appropriate set of linear contrasts. For convenience the S-PLUS function

mcp2atm(J,K,x,tr = 0.2,con = 0, alpha = 0.05, grp = NA, op = F)

is provided for performing all pairwise comparisons for the main effects as well as for all interactions. This function creates the appropriate linear contrast coefficients for you and calls the function lincon. By default, FWE is set at α when performing the $M_a = (J^2 - J)/2$ pairwise comparisons for factor A; the same is done for the $M_b = (K^2 - K)/2$ pairwise comparisons for factor B. As for interactions, FWE is based on all $M_i = M_a M_b$ interactions. If op=T is used, FWE is controlled for all $M = M_a + M_b + M_i$ tests to be performed.

12.6.3 Linear Contrasts Based on Medians

The method for testing linear contrasts based on trimmed means should not be used for the special case where the goal is to compare medians, but a modification of the method for trimmed means appears to perform relatively well when the goal is to compare medians. Again let L indicate the number of groups to be compared and let M_ℓ represent the sample median of the ℓth group. Let $\theta_1, \ldots, \theta_L$ be the population medians, and now let

$$\Psi = \sum_{\ell=1}^{L} c_\ell \theta_\ell$$

be some linear contrast of interest. Compute

$$\hat{\Psi} = \sum_{\ell=1}^{L} c_\ell M_\ell,$$

and let

$$S_e^2 = \sum c_\ell^2 S_\ell^2,$$

where S_ℓ^2 is the McKean–Schrader estimate of the squared standard error of M_ℓ. Then an approximate $1 - \alpha$ confidence interval for Ψ is

$$\hat{\Psi} \pm cS_e,$$

where c is read from Table 10 in Appendix B with degrees of freedom $\nu = \infty$ and where C is again the number of hypotheses to be tested. As usual, this method is designed so that FWE will be approximately $1 - \alpha$.

12.6.4 S-PLUS Functions msmed and mcp2med

The S-PLUS function

$$\text{msmed}(x, y = NA, con = 0, alpha = 0.05)$$

can be used just like the function lincon, ignoring the second argument y. So if data are stored in a matrix called mat having six columns corresponding to six groups, msmed(mat) will perform all pairwise comparisons using medians. (If only two groups are being compared, the second argument, y, can be used as indicated in Section 8.7.2.) The S-PLUS function

$$\text{mcp2med}(J, K, x, con = 0, alpha = 0.05, grp = NA, op = F)$$

performs multiple comparisons among main effects and interactions and is used exactly like the function mcp2atm in Section 12.6.2; the only difference is that now medians are compared rather than trimmed means.

12.7 Bootstrap Methods

As was the case in Chapter 8, when comparing means, all indications are that the bootstrap-*t* method performs better than the percentile bootstrap. However, with at least 20% trimming or when using some other robust measure of location, extant results support the use of some type of percentile bootstrap method instead. When comparing groups with trimmed means, the minimum amount of trimming required to justify switching from the bootstrap-*t* to the percentile bootstrap is unknown. That is, with 15% or perhaps 10% trimming it might be preferable to use some variation of the percentile bootstrap, but this has not been established. But with 20% trimming, all indications are that a percentile bootstrap has practical value. When comparing groups using MOMs or M-estimators, currently some type of percentile bootstrap method is recommended.

12.7.1 Bootstrap-*t*

As usual, let $\Psi_1, \ldots, \Psi_C$ indicate C linear contrasts of interest, which include all pairwise comparisons of the groups as a special case. A bootstrap-*t* method for testing the hypothesis that each of these C linear contrasts is zero is outlined in Box 12.5.

BOX 12.5 Bootstrap-*t* Method for Trimmed Means

Goal

Test $H_0 : \Psi = 0$, for each of C linear contrasts such that FWE is α.

1. For each of the L groups, generate a bootstrap sample, $X_{i\ell}^*$, $i = 1, \ldots, n_{\ell}$; $\ell = 1, \ldots, L$. For each of the L bootstrap samples, compute the trimmed mean, $\bar{X}_{t\ell}^*$,

$$\hat{\Psi}^* = \sum c_\ell \bar{X}_{t\ell}^*,$$

$$d_\ell^* = \frac{c_\ell^2 (n_\ell - 1)(s_{w\ell}^*)^2}{h_\ell (h_\ell - 1)},$$

where $(s_{w\ell}^*)^2$ is the Winsorized variance based on the bootstrap sample taken from the ℓth group and h_ℓ is the effective sample size of the ℓth group (the number of observations left after trimming).

2. Compute

$$T^* = \frac{|\hat{\Psi}^* - \hat{\Psi}|}{\sqrt{A^*}},$$

where $\hat{\Psi} = \sum c_\ell \bar{X}_{t\ell}$ and $A^* = \sum d_\ell^*$. The results for each of the C linear contrasts are labeled $T_1^*, \ldots, T_C^*$.

3. Let

$$T_q^* = \max \{T_1^*, \ldots, T_C^*\}.$$

In words, T_q^* is the maximum of the C values $T_1^*, \ldots, T_C^*$.

4. Repeat steps 1–3 B times, yielding T_{qb}^*, $b = 1, \ldots, B$.

Let $T_{q(1)}^* \leq \cdots \leq T_{q(B)}^*$ be the T_{qb}^* values written in ascending order, and let $u = (1 - \alpha)B$, rounded to the nearest integer. Then the confidence interval for Ψ is

$$\hat{\Psi} \pm T_{q(u)}^* \sqrt{A},$$

where $A = \sum d_\ell$, $d_\ell = \frac{c_\ell^2 (n_\ell - 1)s_{w\ell}^2}{h_\ell (h_\ell - 1)}$, and the simultaneous probability coverage is approximately $1 - \alpha$.

12.7.2 S-PLUS Function linconb

The S-PLUS function

$$\text{linconb}(x, \text{con} = 0, \text{tr} = 0.2, \text{alpha} = 0.05, \text{nboot} = 599)$$

has been supplied to perform the bootstrap-t method for trimmed means just described. As usual, the argument con can be used to specify the contrast coefficients. If con is not passed to the function, all pairwise comparisons are performed.

12.7.3 Sequentially Rejective Methods for MOMs and M-Estimators

When comparing groups using MOMs or M-estimators, the bootstrap method in Section 8.8.1 can be generalized to control FWE in a variety of ways. Currently, a slight modification of a sequentially rejective method derived by Rom (1990) appears to perform well when sample sizes are less than or equal to 100. With larger sample sizes, perhaps a direct application of Rom's method is preferable, but this issue needs further research.

Consider the problem of all pairwise comparisons among J groups based on any robust measure of location. As usual the measure of location associated with the jth group is labeled θ_j; here the goal is to test for every $j < k$ the hypothesis

$$H_0 : \theta_j = \theta_k \tag{12.11}$$

with FWE equal to α. The total number of hypotheses to be tested is

$$C = \frac{J^2 - J}{2}.$$

Compute the test statistic $\hat{p}^*$ described in Section 8.8.1 for comparing the jth group to the kth group. That is, among the B bootstrap samples, $\hat{p}^*$ is the proportion of times the bootstrap estimate for the jth group is greater than the bootstrap estimate of the kth. There are a total of C $\hat{p}^*$ values and for convenience they are labeled $\hat{p}_c^*$, ($c = 1, \ldots, C$). So $\hat{p}_1^*$, for example, is the proportion of times among B bootstrap samples that a bootstrap estimate of θ_1 is larger than a bootstrap estimate of θ_2, $\hat{p}_2^*$ is the estimate when comparing group 1 to group 3, and $\hat{p}_C^*$ is the estimate when comparing group $J - 1$ to group J. Let

$$\hat{p}_{mc}^* = \min\left(\hat{p}_c^*, 1 - \hat{p}_c^*\right).$$

The value $2\hat{p}_{mc}^*$ represents an estimated p-value for the corresponding hypothesis, as was explained in Chapter 8.

The modification of Rom's method is applied as follows. Put the $\hat{p}_{mc}^*$ values in *descending* order, yielding $\hat{p}_{m[1]}^* \geq \hat{p}_{m[2]}^* \geq \cdots \geq \hat{p}_{m[C]}^*$. So, for example, $\hat{p}_{m[1]}^*$ is the largest of the C values just computed and $\hat{p}_{m[C]}^*$ is the smallest. Decisions about the individual hypotheses are made as follows. If $\hat{p}_{m[1]}^* \leq \alpha_1$, where α_1 is read from Table 12.4, reject all C of the hypotheses. Put another way, if the largest

TABLE 12.4   Values of α_c for $\alpha = .05$ and $.01$

c	$\alpha = .05$	$\alpha = .01$
1	.02500	.00500
2	.02500	.00500
3	.01690	.00334
4	.01270	.00251
5	.01020	.00201
6	.00851	.00167
7	.00730	.00143
8	.00639	.00126
9	.00568	.00112
10	.00511	.00101

estimated p-value, $2\hat{p}^*_{m[1]}$, is less than or equal to α, reject all C hypotheses. If $\hat{p}^*_{m[1]} > \alpha_1$ but $\hat{p}^*_{m[2]} \leq \alpha_2$, fail to reject the hypothesis associated with $\hat{p}^*_{m[1]}$, but the remaining hypotheses are rejected. If $\hat{p}^*_{m[1]} > \alpha_1$ and $\hat{p}^*_{m[2]} > \alpha_2$ but $\hat{p}^*_{m[3]} \leq \alpha_3$, fail to reject the hypotheses associated with $\hat{p}^*_{m[1]}$ and $\hat{p}^*_{m[2]}$, but reject the remaining hypotheses. In general, if $\hat{p}^*_{m[c]} \leq \alpha_c$, reject the corresponding hypothesis and all other hypotheses having smaller $\hat{p}^*_m$ values. For other values of α (assuming $c > 1$) or for $c > 10$, use

$$\alpha_c = \frac{\alpha}{c}$$

(which corresponds to a slight modification of a sequentially rejective method derived by Hochberg, 1988.) This will be called *method SR.*

Method SR is unusual, in the sense that familiarity with many multiple comparison procedures suggests a slightly different approach. In particular, a natural guess at how to proceed is to compute the estimated significance level for the cth hypothesis, $2\hat{p}^*_{mc}$, and then to use Rom's method outlined in Section 12.8.2. But a practical concern is that with small to moderate sample sizes, now FWE will be substantially smaller than intended. In fact, if the goal is to have FWE equal to α, typically the actual FWE will be less than $\alpha/2$, which can affect power. A better approach would be to use the Benjamini–Hochberg method in Section 12.3.3; but again, with small sample sizes the actual FWE level can be too small compared to the nominal level. For large sample sizes, the Benjamini–Hochberg method might be preferable, but this issue has received little attention.

Simulation studies indicate that method SR performs well in terms of controlling FWE when sample sizes are less than 100. However, a criticism of method SR is that it is unknown what happens to FWE as all sample sizes get large. If all pairwise comparisons among four groups are to be performed and all groups have equal sample sizes of 100, then FWE is approximately .06. With sample sizes of 200, FWE is approximately .074, so for large sample sizes, perhaps Rom's method is preferable.

EXAMPLE. To illustrate method SR, imagine you test five hypotheses corresponding to the linear contrasts $\Psi_1, \ldots, \Psi_5$ and get $\hat{p}^*_{m1} = .02$, $\hat{p}^*_{m2} = .005$, $\hat{p}^*_{m3} = .23$, $\hat{p}^*_{m4} = .002$, $\hat{p}^*_{m5} = .013$, respectively. Further assume that you want FWE to be .05. The largest of these five values is .23, which corresponds to $H_0 : \Psi_3 = 0$. That is, $\hat{p}^*_{m[1]} = .23$; with $\alpha = .05$, $\alpha_1 = .025$; this is less than .23, so you fail to reject $H_0 : \Psi_3 = 0$. Had it been the case that $\hat{p}^*_{m[1]}$ was less than or equal to .025, you would stop and reject all five hypotheses. The next largest $\hat{p}^*$ value is .02, which corresponds to $H_0 : \Psi_1 = 0$; this is less than $\alpha_2 = .025$, so $H_0 : \Psi_1 = 0$ is rejected. Moreover, the remaining hypotheses are rejected regardless of what their $\hat{p}^*$ value happens to be. ■

12.7.4 S-PLUS Functions pbmcp and mcp2a

The S-PLUS function

$$\text{pbmcp}(x, \text{alpha} = 0.05, \text{nboot} = \text{NA}, \text{grp} = \text{NA}, \text{est} = \text{mom},$$
$$\text{con} = 0, \text{bhop} = F, \ldots)$$

performs multiple comparisons using method SR described in the previous section. By default, all pairwise comparisons are performed, but a collection of linear contrasts can be specified instead via the argument con, which is used as illustrated in Section 12.6.1. The function computes $2\hat{p}^*_m$, the estimated significance level for each hypothesis, and lists it in the column headed sig.test. The appropriate critical value based on method SR, which is taken from Table 12.4, is listed under sig.crit. (The value listed is $2\alpha_c$, which is compared to $2\hat{p}^*_m$ as previously described.) At the end of the output is a value for sig.num, the number of significant results. When all groups have sample sizes of at least 100, it might be preferable to set the argument bhop to T. This causes the significant critical levels to be computed via the Benjamini–Hochberg method in Section 12.3.3.

EXAMPLE. For the data in Table 9.1, the S-PLUS function pbmcp returns

```
$output:
     con.num      psihat sig.test sig.crit  ci.lower    ci.upper
[1,]       1 -0.03790389   0.6495  0.05000 -0.359385  0.43023143
[2,]       2  0.10149361   0.4945  0.05000 -0.184108  0.55663000
[3,]       3 -0.14128764   0.0770  0.02040 -0.549206  0.27054725
[4,]       4  0.13939750   0.3520  0.03380 -0.167386  0.37823232
[5,]       5 -0.10338375   0.2390  0.02540 -0.532423  0.12117667
[6,]       6 -0.24278125   0.0130  0.01702 -0.587199 -0.03574371
```

Continued

EXAMPLE. (*Continued*)

`$con:`

	[,1]	[,2]	[,3]	[,4]	[,5]	[,6]
[1,]	1	1	1	0	0	0
[2,]	-1	0	0	1	1	0
[3,]	0	-1	0	-1	0	1
[4,]	0	0	-1	0	-1	-1

`$num.sig:`
`[1] 1`

So one significant result was obtained and corresponds to the sixth hypothesis where group 3 is compared to group 4. That is, among individuals with schizophrenia, the typical measure of skin resistance among those with predominantly negative symptoms is lower than among those with predominantly positive symptoms instead. In this particular case, similar results are obtained when comparing 20% trimmed means with the function linconb (described in Section 12.7.2), but no significant results are obtained when comparing medians with the S-PLUS msmed in Section 12.6.4.

For convenience when dealing with a two-way ANOVA design, the S-PLUS function

mcp2a(J,K,x,est=mom,con=0,alpha=.05,nboot=NA,grp=NA,...)

performs all pairwise multiple comparisons among the rows and the columns and then does all tetrad differences relevant to interactions.   ■

12.7.5 A Percentile Bootstrap Method for 20% Trimmed Means

The 20% trimmed mean has received considerable attention in recent years, because it performs nearly as well as the mean when distributions are normal and it can handle a fair degree of heavy-tailedness (situations where outliers are likely to appear). In particular, in contrast to medians, power remains relatively high under normality. Although arguments for preferring MOM can be made (e.g., MOM can handle more outliers), the 20% trimmed mean remains one of the better measures of location. When performing all pairwise comparisons based on 20% trimmed means, a special variation of the percentile bootstrap method currently seems best. The method is basically the same as the percentile bootstrap method in Section 12.7.3, except that a single critical value is used for testing all C hypotheses; this critical value is designed specifically for comparing 20% trimmed means. An approximation of the critical value for $\alpha = .05$ is

$$p_{\text{crit}} = \frac{0.0268660714}{C} - 0.0003321429,$$

which is based on results in Wilcox (2001d). Now you reject $H_0 : \Psi_c = 0$ if $2\hat{p}_{mc}^* \leq 2p_{\text{crit}}$.

It is possible that, in terms of Type I errors, the method in Section 12.7.3 is as good as and perhaps slightly better than the approach mentioned here when comparing 20% trimmed means. That is, you might be able to use the S-PLUS function pbmcp with est = tmean (a 20% trimmed mean), but this issue needs further investigation before any recommendations can be made.

12.7.6 S-PLUS Function mcppb20

The S-PLUS function

$$\text{mcppb20(x, crit} = \text{NA, con} = 0, \text{tr} = 0.2, \text{alpha} = 0.05, \text{nboot} = 2000,$$

$$\text{grp} = \text{NA)}$$

performs the multiple comparison procedure just described. If no value for crit, the critical value, is specified, the function chooses a critical value based on results in Wilcox (2001d). For situations where a critical value has not been determined, the function approximates the critical value with p_{crit} given above if $\alpha = .05$. Otherwise it uses α/C as the critical value, a strategy that stems from the Bonferroni method covered in Section 12.8.1.

12.8 Methods for Dependent Groups

Multiple comparison methods for independent groups typically take advantage of the independence among the groups in some manner. When comparing dependent groups instead, generally some modification of the methods for independent groups must be made.

12.8.1 Bonferroni Method

One of the simplest and earliest methods for comparing dependent groups is based on what is known as the *Bonferroni inequality*. The strategy is simple: If you plan to test C hypotheses and want FWE to be α, test each of the individual hypotheses at the α/C level of significance. So, for example, when comparing groups having normal distributions, if you plan to test five hypotheses and want FWE to be at most .05, perform the five paired T-tests at the .01 level. If each of the resulting confidence intervals has probability coverage .99, then the simultaneous probability coverage will be greater than .95. That is, with probability at least .95, all of the confidence intervals will contain the true value of the parameter being estimated. Moreover, if the individual tests are able to control the probability of a Type I error, then the Bonferroni method controls FWE.

> **EXAMPLE.** Consider J dependent groups and imagine that all pairwise comparisons are to be performed. Then the number of hypotheses to be tested
>
> *Continued*

EXAMPLE. (*Continued*) is $C = (J^2 - J)/2$. So if, for each pair of groups, the hypothesis of equal (population) trimmed means is tested as described in Section 11.2, and if the goal is to have FWE equal to .05, perform each test at the .05/C level. ■

12.8.2 Rom's Method

Several improvements on the Bonferroni method have been published, and one that stands out is a so-called sequentially rejective method derived by Rom (1990), which has been found to have good power relative to several competing methods (e.g., Olejnik, Li, Supattathum, & Huberty, 1997). To apply it, compute significance levels for each of the C tests to be performed and label them $P_1, \ldots, P_C$. Next, put the significance levels in descending order, which are now labeled $P_{[1]} \geq P_{[2]} \geq \cdots \geq P_{[C]}$. Proceed as follows:

1. Set $k = 1$.
2. If $P_{[k]} \leq d_k$, where d_k is read from Table 12.5, stop and reject all C hypotheses; otherwise, go to step 3.
3. Increment k by 1. If $P_{[k]} \leq d_k$, stop and reject all hypotheses having a significance level less than or equal to d_k.
4. If $P_{[k]} > d_k$, repeat step 3.
5. Continue until you reject or all C hypotheses have been tested.

An advantage of Rom's method is that its power is greater than or equal to that of the Bonferroni approach. In fact, Rom's method always rejects as many or more hypotheses. A negative feature is that confidence intervals are not readily computed.

A closely related method was derived by Hochberg (1988) where, rather than use Table 12.5, use $d_k = \alpha/(C - k + 1)$. For $k = 1$ and 2, d_k is the same as in

TABLE 12.5   Critical Values, d_k, for Rom's Method

k	$\alpha = .05$	$\alpha = .01$
1	.05000	.01000
2	.02500	.00500
3	.01690	.00334
4	.01270	.00251
5	.01020	.00201
6	.00851	.00167
7	.00730	.00143
8	.00639	.00126
9	.00568	.00112
10	.00511	.00101

TABLE 12.6 Illustration of Rom's method

Number	Test	Significance level	
1	$H_0 : \mu_1 = \mu_2$	$P_1 = .010$	$P_{[5]}$
2	$H_0 : \mu_1 = \mu_3$	$P_2 = .015$	$P_{[3]}$
3	$H_0 : \mu_1 = \mu_4$	$P_3 = .005$	$P_{[6]}$
4	$H_0 : \mu_2 = \mu_3$	$P_4 = .620$	$P_{[1]}$
5	$H_0 : \mu_2 = \mu_4$	$P_5 = .130$	$P_{[2]}$
6	$H_0 : \mu_3 = \mu_4$	$P_6 = .014$	$P_{[4]}$

Rom's method. An advantage of Hochberg's method is that it does not require special tables and can be used with $k > 10$.

EXAMPLE. Imagine you want to perform all pairwise comparisons among four dependent groups and you apply some method for means and get the significance levels shown in Table 12.6. Further assume that you want FWE to be .05. The largest significance level is .62; this is greater than .05, so you fail to reject the corresponding hypothesis, $H_0 : \mu_2 = \mu_3$. The next largest significance level is .130; this is greater than $d_2 = .025$, so you fail to reject $H_0 : \mu_2 = \mu_4$. The next largest significance level is .015; this is less than $d_3 = .0167$, so you stop and reject the corresponding hypothesis as well as those having smaller significance levels. ■

12.8.3 Linear Contrasts Based on Trimmed Means

More generally, a collection of linear contrasts can be tested when working with trimmed means corresponding to dependent groups. First consider a single linear contrast based on the marginal trimmed means of L groups:

$$\Psi = \sum_{\ell=1}^{L} c_\ell \mu_\ell.$$

Then $H_0 : \Psi = 0$ can be tested as outlined in Box 12.6. Alternatively, a generalization of difference scores can be used instead. That is, set

$$D_i = \sum_{\ell=1}^{L} c_\ell X_{i\ell}$$

and test the hypothesis that the population trimmed mean of the D_i values is zero. When testing C such hypotheses, FWE can be controlled with Rom's method.

BOX 12.6 How to Test a Linear Contrast Based on the

Marginal Trimmed Means of Dependent Groups

Goal

Test $H_0 : \Psi = 0$ for each of C linear contrasts such that FWE is α. (There are L groups.)

Let Y_{ij} ($i = 1, \ldots, n$; $j = 1, \ldots, L$) be the Winsorized values, which are computed as described in Box 11.3. Let

$$A = \sum_{j=1}^{L} \sum_{\ell=1}^{L} c_j c_\ell d_{j\ell},$$

where

$$d_{j\ell} = \frac{1}{h(h-1)} \sum_{i=1}^{n} (Y_{ij} - \bar{Y}_j)(Y_{i\ell} - \bar{Y}_\ell)$$

and h is the number of observations left in each group after trimming. Let

$$\hat{\Psi} = \sum_{\ell=1}^{L} c_\ell \bar{X}_{t\ell}.$$

Test statistic:

$$T = \frac{\hat{\Psi}}{\sqrt{A}}.$$

Decision Rule

Reject if $|T| \geq t$, where t is the $1 - \alpha/2$ quantile of a Student's T-distribution with $\nu = h - 1$ degrees of freedom. When testing more than one linear contrast, FWE can be controlled with Rom's method.

12.8.4 S-PLUS Function rmmcp

The S-PLUS function

$$\text{rmmcp}(x, \text{con} = 0, \text{tr} = 0.2, \text{alpha} = 0.05, \text{dif} = T)$$

performs multiple comparisons among dependent groups using trimmed means and Rom's method for controlling FWE. By default, difference scores are used. Setting dif=F results in comparing marginal trimmed means. (When α differs from both .05 and .01, FWE is controlled with Hochberg's method as described in Section 12.8.2.)

EXAMPLE. Imagine a two-way ANOVA design where husbands and wives are measured at two different times. Then $L = 4$, all four groups are dependent, and this is an example of a two-way ANOVA with a within-subjects design on both factors. That is, the levels of factor A are possibly dependent, as are the levels of factor B. Assume that for wives, the trimmed means are μ_{t1} and μ_{t2} at times 1 and 2, respectively and that the trimmed means at times 1 and 2 for the husbands are μ_{t3} and μ_{t4}, respectively. If the goal is to detect an interaction by testing

$$H_0 : \mu_{t1} - \mu_{t2} = \mu_{t3} - \mu_{t4},$$

the linear contrast is $\Psi = \mu_{t1} - \mu_{t2} - \mu_{t3} + \mu_{t4}$. To use rmmcp, first store the contrast coefficients in some S-PLUS variable. For example, use the command

$$\text{mat} < -\text{matrix}(c(1,-1,-1,1)).$$

Then if the data are stored in the S-PLUS matrix m1, the command

$$\text{rmmcp(m1,con=mat,dif=F)}$$

will perform the computations.   ■

12.8.5 Percentile Bootstrap Methods

A percentile bootstrap method for multiple comparisons among dependent groups can be performed using a simple combination of techniques already described. First consider all pairwise comparisons where the goal is to compare group j to k by testing

$$H_0 : \theta_j = \theta_k \tag{12.12}$$

for all $j < k$. As usual, the goal is to have FWE equal to α. Generate bootstrap samples as described in Section 11.7.1 in conjunction with method RMPB4. For any two specific groups, compute a significance level (or p-value) as indicated in Section 11.3.1 in connection with method RMPB2. That is, compute the proportion of bootstrap values from the first group that are greater than the bootstrap values from second group, label the result $\hat{p}^*$, set

$$\hat{p}_m^* = \min(\hat{p}^*, 1 - \hat{p}^*),$$

in which case the estimated significance level is $2\hat{p}_m^*$. Then FWE is controlled using a modification of Rom's method. In particular, let $\hat{p}_{mc}^*$ be the value of $\hat{p}_m^*$ when performing the cth comparison, $c = 1, \dots, C$. Again, if all pairwise comparisons are being performed, $C = (J^2 - J)/2$. Now proceed as in Section 12.7.3. That is, put the $\hat{p}_{mc}^*$ in descending order, yielding $\hat{p}_{m[1]}^* \geq \cdots \geq \hat{p}_{m[C]}^*$. Then use Table 12.4 as illustrated in Section 12.7.3.

Section 11.3.1 mentioned an adjustment of $\hat{p}^*$ used in conjunction with method RMPB2. It was called a bias-adjusted critical value and labeled $\hat{p}_a^*$. Here, when testing the cth hypothesis, $\hat{p}_a^*$ is labeled $\hat{p}_{ca}^*$. If this adjustment is used, the estimated significance level is $2\min(\hat{p}_{ca}^*, 1 - \hat{p}_{ca}^*)$. Now it seems that a direct application of Rom's method can be used to control FWE. When using the method in the previous paragraph, there are some indications that it might not be quite satisfactory with large sample sizes, but switching to $\hat{p}_{ca}^*$ and using Rom's method seems to correct this.

12.8.6 Using Difference Scores

Another approach is to use difference scores. That is, when comparing group j to group k, set

$$D_{ijk} = X_{ij} - X_{ik},$$

the difference between the ith pair of observations. (Now bootstrap samples, D_{ijk}^*, are obtained by resampling n vectors of observations from the n-by-C matrix of difference scores, where $C = (J^2 - J)/2$.) Then, when testing the hypothesis that the typical difference is zero, you control FWE using either the Bonferroni method or Rom's technique.

To elaborate, it is again convenient to relabel the D_{ijk}^* values as D_{ic}^*, $c = 1, \ldots, C$. So here, $c = 1$ corresponds to comparing group 1 to group 2, $c = 2$ is comparing group 1 to group 3, and so on. For the cth comparison, let $\hat{p}_c^*$ be the proportion of times among B bootstrap resamples that $D_{ic}^* > 0$. As usual, let

$$\hat{p}_{mc}^* = \min(\hat{p}_c^*, 1 - \hat{p}_c^*),$$

in which case $2\hat{p}_{mc}^*$ is the estimated significance level for the cth comparison. Then put the p-values in descending order and make decisions about which hypotheses are to be rejected using method SR outlined in Section 12.7.3. That is, once the $\hat{p}_{mc}^*$ is computed, reject the hypothesis corresponding to $\hat{p}_{mc}^*$ if $\hat{p}_{mc}^* \leq \alpha_c$, where α_c is read from Table 12.4. Alternatively, reject if the estimated significance level $2\hat{p}_{mc}^* \leq 2\alpha_c$.

As for linear contrasts, consider any specific linear contrast with contrast coefficients $c_1, \ldots, c_J$, set

$$D_i = \sum c_j X_{ij},$$

and let θ_d be the typical (population) value of this sum. Then $H_0 : \theta_d = 0$ can be tested by generating a bootstrap sample from the D_i values, repeating this B times, computing $\hat{p}^*$, the proportion of bootstrap estimates that are greater than zero, in which case $2\min(\hat{p}^*, 1 - \hat{p}^*)$ is the estimated significance level. Then FWE can be controlled in the manner just outlined.

When comparing groups using MOM, at the moment it seems that the method based on difference scores often provides the best power versus testing Equation (12.12). Both approaches do an excellent job of avoiding Type I error probabilities greater than the nominal α level, still using MOM. But when testing Equation (12.12), the actual Type I error probability can drop well below the nominal level in situations where the method based on difference scores avoids this problem. This suggests that the method based on difference scores will have more power; and indeed, there are situations where this is the case even when the two methods have comparable Type I error probabilities. It is stressed, however, that a comparison of these methods, in terms of power, needs further study, and perhaps the bias-adjusted critical value mentioned in Section 12.8.5 helps increase power when testing Equation (12.12), but this issue has not yet been investigated.

12.8.7 S-PLUS Function rmmcppb

The S-PLUS function

$$\text{rmmcppb}(x, y = \text{NA, alpha} = 0.05, \text{con} = 0, \text{est} = \text{mom, plotit} = T,$$

$$\text{dif} = T, \text{grp} = \text{NA, nboot} = \text{NA, BA} = F, \ldots)$$

performs multiple comparisons among dependent groups using the percentile bootstrap methods just described. The argument dif defaults to T (for true), indicating that difference scores will be used. If dif=F, difference scores are not used. For example, when comparing all pairs of groups, hypotheses given by Equation (12.12) will be tested instead. If dif=F and BA=T, the significance levels are computed as described in Section 11.3.1 and Rom's method is used to control FWE. (With BA=F, a slight modification of Rom's method is used instead.) If no value for con is specified, then all pairwise differences will be tested as given by Equation (12.12). As usual, if the goal is test hypotheses other than all pairwise comparisons, con can be used to specify the linear contrast coefficients. (See Section 12.6.1 for an illustration of how to use con.)

12.9 Analyzing Between-by-Within Designs

There are various ways of performing multiple comparisons when dealing with a between-by-within (or split-plot) design using a combination of methods already described. A few specific possibilities are summarized here in the hope that one of them will match the needs of the reader. We begin with nonbootstrap methods for trimmed means and then consider bootstrap methods for other measures of location.

As in Chapter 11, it is assumed that levels of factor B correspond to dependent groups and that the levels of factor A are independent. First consider factor A. A simple approach is simply to average over the levels of factor B. So X_{ijk} becomes $Y_{ij} = \sum_k X_{ijk}/K$. Then all pairs of levels of factor A can be compared as based on the Y_{ij} values described in Section 12.6, and the computations can be performed with the S-PLUS function lincon in Section 12.6.1.

An alternative approach, one that provides more detail about how groups differ, is to perform all pairwise comparisons among the levels of factor A for each level of factor B. So, for any k, the goal is to test

$$H_0 : \mu_{tjk} = \mu_{tj'k},$$

for all $j < j'$ and $k = 1, \ldots, K$. The total number of hypotheses is $K(J^2 - J)/2$, and it is desired to control FWE among all of these tests. For fixed k, because independent groups are being compared, one approach is simply to create the appropriate linear contrasts and use the S-PLUS function lincon in Section 12.6.1. (For convenience, the S-PLUS function bwamcp, described in the next section, creates these linear contrasts for you.)

As for factor B, one approach is to ignore the levels of factor A and to test hypotheses based on the trimmed means corresponding to the difference scores associated with any two levels of factor B. So now there are $(K^2 - K)/2$ hypotheses to be tested, where each hypothesis is that the difference scores have a trimmed mean of zero. One way of controlling FWE is with the method in Section 12.8.2. From Chapter 11, another approach is not to take difference scores, but rather to use the marginal trimmed means.

A more detailed approach is as follows. Consider the jth level of factor A. Then there are $(K^2 - K)/2$ pairs of groups that can be compared. If, for each of the J levels of factor A, all pairwise comparisons are performed, the total number of comparisons is $J(K^2 - K)/2$. In symbols, the goal is to test

$$H_0 : \mu_{tjk} = \mu_{tjk'},$$

for all $k < k'$ and $j = 1, \ldots, J$. And of course an alternative approach is to use difference scores instead.

As for interactions, take any two levels of factor A — say, j and j' — and do the same for factor B — say, levels k and k'. Form the difference scores

$$D_{ij} = X_{ijk} - X_{ijk'}, \quad \text{and} \quad D_{ij'} = X_{ij'k} - X_{ij'k'},$$

and let μ_{tj} and $\mu_{tj'}$ be the population trimmed means associated with these difference scores. Then one way of stating the hypothesis of no interaction for these specific levels of factors A and B is with

$$H_0 : \mu_{tj} = \mu_{tj'},$$

and of course this can be done for any two levels of factors A and B. The goal is to test this hypothesis for all $j < j'$ and $k < k'$ in a manner that controls FWE, and this might be done as described in Section 12.8.2.

12.9.1 S-PLUS Functions bwamcp, bwbmcp, and bwimcp

Three S-PLUS functions are supplied for applying the methods just described. The first is

$$\text{bwamcp}(J, K, x, tr = 0.2, JK = J * K, grp = c(1 : JK),$$
$$\text{alpha} = 0.05, KB = F, op = T).$$

The default value for the argument op is T, meaning that the hypotheses $H_0 : \mu_{tjk} = \mu_{tj'k}$ for all $j < j'$ and $k = 1, \ldots, K$ are tested. In essence, the function creates the appropriate set of linear contrasts and calls the S-PLUS function lincon. (The argument KB is used as described in Section 12.6.1.) Setting op=F results in averaging the data over the levels of factor B and performing all pairwise comparisons corresponding to the levels of factor A.

The S-PLUS function

$$\text{bwbmcp}(J, K, x, \text{tr} = 0.2, JK = J * K, \text{grp} = c(1:JK), \text{con} = 0,$$
$$\text{alpha} = 0.05, \text{dif} = T, \text{pool} = F)$$

compares the levels of factor B instead. If pool=T is used, it simply pools the data for you and then calls the function rmmcp. If dif=F is used, the marginal trimmed means are compared instead. By default, pool=F, meaning that $H_0 : \mu_{tjk} = \mu_{tjk'}$ is tested for all $k < k'$ and $j = 1, \ldots, J$. For each level of factor A, the function simply selects data associated with the levels of factor B and sends them to the S-PLUS function rmmcp described in Section 12.8.3.

As for interactions, the S-PLUS function

$$\text{bwimcp}(J, K, x, \text{tr} = 0.2, JK = J * K, \text{grp} = c(1:JK), \text{alpha} = 0.05)$$

can be used.

12.9.2 Bootstrap Methods

This section describes methods aimed at testing a collection of linear contrasts based on some bootstrap method. One advantage is that it is easy to test some variations of the methods described at the beginning of this section. Consider, for example, factor A and let $\bar{\theta}_{j.} = \sum \theta_{jk}/K$ be the average measure of location for the jth level. That is, for a fixed level of factor A, $\bar{\theta}_{j.}$ is the average of the measures of location across the levels of factor B. Then an approach to comparing the levels of factor A is to test

$$H_0 : \bar{\theta}_{j.} = \bar{\theta}_{j'.} \tag{12.13}$$

for every $j < j'$. That is, for all pairs of rows, compare the average measure of location among the dependent groups. There are $C = (J^2 - J)/2$ such comparisons, the individual tests can be performed as described in Chapter 11, and the resulting significance levels can be used in conjunction with the modified Rom's method to control FWE. (That is, use method SR in Section 12.7.3.) Of course, another possibility is to focus on the kth level of factor B and test

$$H_0 : \theta_{jk} = \theta_{j'k} \tag{12.14}$$

for all $j < j'$ and then to do this for all K levels for factor B (i.e., $k = 1, \ldots, K$). So now there is a total of $C = K(J^2 - J)/2$ tests to be performed.

As for factor B, again a simple approach is to ignore the levels of factor A, simply view the data as coming from K dependent groups, and use the methods in Section 12.8.1 or 12.8.2. As before, difference scores can be used or marginal measures of location can be compared. In the latter case, again ignore factor A and let θ_k be the population measure of location associated with the kth level of factor B. Then, for every $k < k'$, test

$$H_0 : \theta_k = \theta_{k'}$$

and control FWE using method SR in Section 12.7.3.

As for Interactions, first consider a two-by-two design — say, two independent groups measured at two different times. For the first group, let θ_{d1} be the population value for some measure of location based on the time 1 measure minus the time 2 measure. For example, θ_{d1} might be the population M-estimators corresponding to the difference scores between times 1 and 2. Similarly, for group 2, let θ_{d2} be the difference for group 2. Then an approach to interactions is to test

$$H_0 : \theta_{d1} = \theta_{d2}.$$

This can be done by first taking a bootstrap sample based on the difference scores associated with group 1, repeating this for group 2, and labeling the difference between these two bootstrap estimates $\hat{\delta}^*$. Then repeat this B times and let $\hat{p}^*$ be the proportion of times $\hat{\delta}^*$ is greater than zero. Then $2\hat{p}^*$ is the estimated significance level. For the general case of J independent groups and K dependent groups, this process can be repeated for any two levels of factor A and any two levels of factor B, and FWE can again be controlled by adopting method SR in Section 12.7.3.

12.9.3 S-PLUS Functions spmcpa, spmcpb, and spmcpi

The S-PLUS function

$$\text{spmcpa}(J, K, x, est = mom, JK = J * K, grp = c(1:JK),$$
$$avg = F, nboot = NA, \dots)$$

performs pairwise comparisons for factor A of a split-plot design as described in the previous section. Setting est=tmean results in using 20% trimmed means. The argument avg indicates whether measures of location are to be averaged. That is, the goal is to test the hypotheses given by Equation (12.13) if avg=T is used. If avg is not specified (in which case it defaults to F, for false), the hypotheses given by Equation (12.14) will be tested. The function determines B if the argument nboot is omitted. Otherwise, the arguments have their usual meaning.

The S-PLUS function

$$\text{spmcpb}(J, K, x, est = mom, JK = J * K, grp = c(1:JK),$$
$$dif=T, nboot = NA, \dots)$$

performs pairwise comparisons among the levels of factor B. Setting est=tmean results in using 20% trimmed means. The argument dif=T indicates that difference scores will be used. Setting dif=F results in marginal measures of location being compared. The S-PLUS function

$$\text{spmcpi}(J,K,x,est=mom,JK=J*K,grp=c(1:JK),nboot=NA,\dots)$$

tests hypotheses related to no interactions.

When using difference scores, the following might help when reading the output. For every two levels of factor B, the function creates difference scores. In effect the number of levels for factor B becomes $(K^2 - K)/2$ and the contrast coefficients reported correspond to the total number of parameters, which is $J(K^2 - K)/2$.

EXAMPLE. Imagine that for each of two independent groups, the goal is to compare K measures in a manner that controls FWE. This problem is a special case of a split-plot or a between-by-within subjects design. For example, one group might receive an experimental drug at four different times, a control group receives a placebo, and the goal is to compare the groups at time 1, at time 2, at time 3, and at time 4. That is, four hypotheses are to be performed and the goal is to have FWE equal to α. If the data are stored in the S-PLUS variable x, the command bwamcp(2,4,x) will perform the analysis. ■

EXAMPLE. Another example stems from an illustration in Chapter 11 where EEG for murderers was measured at four sites in the brain and the same was done for a control group, in which case the goal might be, among other things, to determine which sites differ between the two groups. When working with means under the assumption of normality, the problem is simple: Compare each site with Welch's method and control FWE with Rom's procedure. To compare groups with outliers removed (using MOM), use the S-PLUS function spmcpa described earlier. If the EEG data are stored in the S-PLUS function eeg, the command spmcpa(2, 4, eeg) returns:

```
$output:
      con.num        psihat sig.test crit.sig   ci.lower    ci.upper
[1,]     1       -0.54895604    0.020   0.0254 -0.9635714 -0.03111111
[2,]     2       -0.02000000    0.802   0.0500 -0.7363889  0.60989011
[3,]     3        0.16000000    0.550   0.0338 -0.5131319  0.91923077
[4,]     4       -0.01275641    0.902   0.0500 -0.7785714  0.65279221

$con:
       [,1]  [,2]  [,3]  [,4]
[1,]     1     0     0     0
[2,]     0     1     0     0
[3,]     0     0     1     0
[4,]     0     0     0     1
[5,]    -1     0     0     0
[6,]     0    -1     0     0
[7,]     0     0    -1     0
[8,]     0     0     0    -1

$num.sig:
[1] 1
```

The contrast coefficients are reported in $con. For example, among the eight groups, the first column indicates that measures of location corresponding to the first and fifth groups are compared. (That is, for level 1 of factor B, compare

Continued

> **EXAMPLE.** (*Continued*) levels 1 and 2 of factor A.) The value of num.sig is 1, meaning that one difference was found based on whether the value listed under sig.test is less than or equal to the corresponding value listed under crit.sig. The data indicate that the typical EEG for murderers differs from the control group at the first site where measures were taken.   ■

12.10 Exercises

1. Assuming normality and homoscedasticity, what problem occurs when comparing multiple groups with Student's T-test?

2. For five independent groups, assume that you plan to do all pairwise comparisons of the means and you want FWE to be .05. Further assume that $n_1 = n_2 = n_3 = n_4 = n_5 = 20$, $\bar{X}_1 = 15$, $\bar{X}_2 = 10$, $s_1^2 = 4$, $s_2^2 = 9$, $s_3^2 = s_4^2 = s_5^2 = 15$; test $H_0 : \mu_1 = \mu_2$ using (a) Fisher's method (assuming the ANOVA F-test rejects), (b) Tukey–Kramer, (c) Dunnett's T3, (d) Games–Howell, (e) Scheffé's method, (f) Kaiser–Bowden.

3. Repeat the previous exercise, but now with $n_1 = n_2 = n_3 = n_4 = n_5 = 10$, $\bar{X}_1 = 20$, $\bar{X}_2 = 12$, $s_1^2 = 5$, $s_2^2 = 6$, $s_3^2 = 4$, $s_4^2 = 10$, and $s_5^2 = 15$.

4. You perform six tests and get significance levels .07, .01, .40, .001, .1, and .15. Based on the Bonferroni inequality, which would be rejected with FWE equal to .05?

5. For the previous exercise, if you use Rom's method, which tests would be rejected?

6. You perform five tests and get significance levels .049, .048, .045, .047, and .042. Based on the Bonferroni inequality, which would be rejected with FWE equal to .05?

7. Referring to the previous exercise, which would be rejected with Rom's procedure?

8. Imagine you compare four groups with Fisher's method and you reject the hypothesis of equal means for the first two groups. If the largest observation in the fourth group is increased, what happens to MSWG? What does this suggest about power when comparing groups 1 and 2 with Fisher's method?

9. Repeat the previous exercise, but with the Tukey–Kramer and Scheffé's methods instead.

10. For the data in Table 9.1, each group has 10 observations, so when using Tamhane's method to compute confidence intervals for all pairwise differences, the degrees of freedom are 9 and the value of h in Table 11 of Appendix B is 4.3. Verify that if the goal is to have confidence intervals with FWE equal to .05 and lengths .5, the required sample sizes for each group are 50, 11, 18, and 17.

11. Repeat the previous exercise using the S-PLUS function hochberg. Verify that the required sample sizes are 50, 10, 18, and 17.

12. Use the S-PLUS function lincon to verify the results in the first example of Section 12.6.1.

13. Use the S-PLUS function msmed to compare all pairs of groups based on medians and the data in Table 9.1. Verify that the output from msmed is

$test:

	Group	Group	test	crit	se
[1,]	1	2	0.1443384	2.63	0.2742167
[2,]	1	3	0.4313943	2.63	0.3025886
[3,]	1	4	0.5795942	2.63	0.3019699
[4,]	2	3	0.9624726	2.63	0.1767479
[5,]	2	4	0.7709186	2.63	0.1756865
[6,]	3	4	1.4059718	2.63	0.2173265

$psihat:

	Group	Group	psihat	ci.lower	ci.upper
[1,]	1	2	-0.039580	-0.7607699	0.6816099
[2,]	1	3	0.130535	-0.6652731	0.9263431
[3,]	1	4	-0.175020	-0.9692008	0.6191608
[4,]	2	3	0.170115	-0.2947319	0.6349619
[5,]	2	4	-0.135440	-0.5974955	0.3266155
[6,]	3	4	-0.305555	-0.8771238	0.2660138

14. Use the S-PLUS function msmedse to compute the standard errors for the median corresponding to the four groups in Table 9.1. Compare the results to the estimated standard error for the 20% trimmed means returned by trimse. Now note that when comparing groups 3 and 4, you reject with 20% trimmed means, as was illustrated in Section 12.6.1, but not with medians, as illustrated in Exercise 13.

15. Perform all pairwise comparisons of the groups in Table 11.3 using the S-PLUS function rmmcppb in Section 12.8.4. Use the MOM estimate of location and use difference scores. Verify that a difference between groups 2 and 3 is found with FWE set at .05.

16. A. Thompson and Randall–Maciver (1905) report four measurements of male Egyptian skulls from five different time periods. The first was maximal breadth of skull, and the five time periods were 4000 bc, 3300 bc, 1850 bc, 200 bc, and 150 ad. A portion of the output from lincon, when comparing means (with the argument tr set equal to zero), is

$psihat:

	Group	Group	psihat	ci.lower	ci.upper
[1,]	1	2	-1.0000000	-4.728058	2.72805781
[2,]	1	3	-3.1000000	-6.402069	0.20206887
[3,]	1	4	-4.1333333	-7.563683	-0.70298384
[4,]	1	5	-4.8000000	-8.729236	-0.87076449
[5,]	2	3	-2.1000000	-5.258496	1.05849612
[6,]	2	4	-3.1333333	-6.427215	0.16054796

	Group	Group	psihat	ci.lower	ci.upper
[7,]	2	5	-3.8000000	-7.615383	0.01538268
[8,]	3	4	-1.0333333	-3.813622	1.74695545
[9,]	3	5	-1.7000000	-5.103517	1.70351741
[10,]	4	5	-0.6666667	-4.193848	2.86051459

So when comparing means, significant results are obtained when comparing group 1 to group 4 and group 1 to group 5. The output from pbmcp (when comparing groups based on MOM) is

```
$output:
```

	con.num	psihat	sig.test	sig.crit	ci.lower	ci.upper
[1,]	1	-0.65134100	0.5910	0.05000	-4.358466	2.75714286
[2,]	2	-3.62997347	0.0110	0.01460	-7.172174	-0.05952381
[3,]	3	-3.70689655	0.0050	0.01278	-7.266667	-0.42142857
[4,]	4	-4.37356322	0.0005	0.01022	-8.733333	-1.23333333
[5,]	5	-2.97863248	0.0145	0.01702	-6.040134	0.19333333
[6,]	6	-3.05555556	0.0215	0.02040	-6.347619	0.42450142
[7,]	7	-3.72222222	0.0030	0.01136	-7.540404	-0.56410256
[8,]	8	-0.07692308	0.8545	0.05000	-3.433333	3.26638177
[9,]	9	-0.74358974	0.3580	0.02540	-4.817949	2.26819923
[10,]	10	-0.66666667	0.4475	0.03380	-4.869565	2.50000000

```
$con:
```

	[,1]	[,2]	[,3]	[,4]	[,5]	[,6]	[,7]	[,8]	[,9]	[,10]
[1,]	1	1	1	1	0	0	0	0	0	0
[2,]	-1	0	0	0	1	1	1	0	0	0
[3,]	0	-1	0	0	-1	0	0	1	1	0
[4,]	0	0	-1	0	0	-1	0	-1	0	1
[5,]	0	0	0	-1	0	0	-1	0	-1	-1

```
$num.sig:
[1] 5
```

Interpret the results and contrast them with the analysis based on means.

17. Among the five groups in the previous exercise, only one group was found to have an outlier based on any of the boxplot rules in Section 3.4. What might explain why more significant differences are found when comparing groups based on MOM versus the mean?

13

ROBUST AND EXPLORATORY REGRESSION

Chapter 7 indicates how to make inferences about regression parameters when using the least squares estimator if standard assumptions (normality and homoscedasticity) are violated. A comparable method for making inferences about Pearson's correlation is also described there, but there are other fundamental problems with least squares regression and Pearson's correlation that need to be addressed. One basic concern is that outliers can greatly distort both of these methods. A second concern is that heteroscedasticity can grossly inflate the standard error of the ordinary least squares estimator, relative to other estimators one might use, even under normality. This means that in terms of power (the probability of detecting an association), using least squares regression can be relatively ineffective. A related problem is getting an accurate and relatively short confidence interval for the slope and intercept. This chapter describes some of the tools one might use to address these problems. Some issues related to multiple predictors are discussed, but the emphasis in this chapter is on simple regression, meaning that there is only one predictor. (Chapter 14 expands upon strategies for dealing with multiple predictors.)

13.1 Detecting Outliers in Multivariate Data

First consider the problem of detecting outliers in bivariate data. As in Chapter 6, imagine we have n pairs of observations, which we label $(X_1, Y_1), \ldots, (X_n, Y_n)$. At first glance the problem might appear to be trivial: Simply apply one of the outlier detection methods in Chapter 3 to the X values and do the same to the Y values. Comrey (1985) gives a rough indication of why this approach can be unsatisfactory with the following example. It is not unusual for an individual to be young or for an individual to have hardening of the arteries. But it is unusual for someone to be both young and have hardening of the arteries. More formally, this approach suffers from a fundamental problem: If the points are rotated, values that were declared outliers might no longer be declared outliers, and points that were not declared outliers might

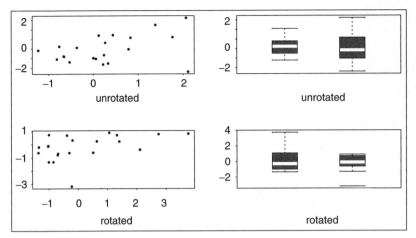

FIGURE 13.1   For multivariate data, simply checking for outliers among the marginal distributions can lead to different conclusions depending on how the points are rotated.

become outliers. That is, if you look at a scatterplot of points on a sheet of paper, declaring a point an outlier should not depend on whether you hold your head up straight versus tilting your head slightly to the right.

To illustrate this important point, look at the upper left panel of Figure 13.1. The points in this plot were created by generating 20 values from a standard normal curve for both X and the error term (ϵ) and setting

$$Y_i = X_i + \epsilon_i.$$

That is, points were generated according to the standard regression model discussed in Chapter 6, where the regression line has a slope of 1 and an intercept of zero. Then an additional point was added at $X = 2.1$ and $Y = -2.4$; it appears in the lower right corner of the scatterplot. To the right of the scatterplot are boxplots for the X and Y values. As is evident, no outliers are found, and none are found using any of the other outlier detection methods in Chapter 3. Note, however, that the point $(X, Y) = (2.1, -2.4)$ is unusual by construction. The reason is that when $X = 2.1$, Y should have a value that is reasonably close to 2.1 as well (because the regression line is simply $Y = X$). But given that $X = 2.1$, $Y = -2.4$ is located 4.5 standard deviations away from the regression line, which makes it unusual. Indeed, a casual glance at the scatterplot suggests that it is somehow removed from the bulk of the points.

The bottom left portion of Figure 13.1 shows the same points rotated by 45 degrees. So in effect the points are rotated so that now they are centered around a regression line having a slope of zero. The lower right panel of Figure 13.1 shows the resulting boxplots of the rotated points. Now an outlier is found among the Y values, which corresponds to the unusual point in the scatterplot of the unrotated points. What we need is an outlier detection method that takes into account the overall structure of the scatterplot. In particular, outliers should remain outliers under any rotation of the points we might make.

In the context of regression, the problem just illustrated is important, because when fitting a straight line to data, even when no outliers are found among the X values and none are found among the Y values, it is possible for a few points to be separated from the bulk of the observations in a way that has an inordinate effect on the least squares regression line. That is, in a very real way, points can be outliers in a scatterplot even though they are not deemed outliers when attention is restricted to the X values or the Y values. As an illustration, again consider the points in the scatterplot in the upper left panel of Figure 13.1, only now we add two points at $(X, Y) = (2.1, -2.4)$. For the original 20 points, the least squares slope of this line is $b_1 = 1.063$. So in this particular case, the least squares regression line provides a fairly accurate estimate of the true slope, which is 1. What is particularly important is that the two points added at $(X, Y) = (2.1, -2.4)$ have a tremendous influence on the least squares regression line — the estimate of the slope drops from 1.063 to 0.316.

A criticism of the illustration just given might be that the two points added to the scatterplot should have some influence on the estimated slope. However, another point of view is that a few unusual values should not mask a true association. If we test the hypothesis that the slope is zero using the conventional method covered in Section 6.3.1, then, based on the original 20 values, we reject with $\alpha = .001$. But when the two unusual values are added to the data, the significance level increases to .343. In this particular case, simply restricting the range of X to values less than 2.1 corrects this problem. But the simple strategy of restricting the range of X is not always effective when trying to detect associations that might be masked by outliers. Indeed, conventional wisdom is that restricting the range of X can actually mask an association, and this is in fact a realistic concern, as will be illustrated in Section 13.4.

13.1.1 A Relplot

An outlier detection method that satisfies our goal of dealing with the rotation of points and taking into account the overall structure of the data is the so-called *relplot* proposed by Goldberg and Iglewicz (1992); it is a bivariate analog of the boxplot. The somewhat involved computations are not particularly important for present purposes and therefore not given. (Computational details can be found in Goldberg & Iglewicz, 1992; and Wilcox, 1997a, Section 7.6.) However, familiarity with a relplot helps convey other concepts and strategies used in this chapter. The basic idea is first to compute a measure of location that has a reasonably high breakdown point. That is, any reasonable measure of location should be embedded in the central portion of a scatterplot of the data, and we want to avoid having a measure of location that is not near the center due to a few points that are unusually separated from the bulk of the observations. (Recall from Chapter 3 that we need estimators with a reasonably high breakdown point when searching for outliers, which continues to be the case here.) The particular measure of location used by Goldberg and Iglewicz is related to the M-estimator described in Chapter 3. They also compute a measure of covariance that is based primarily on the centrally located points. Based on these measures of location and covariance, a relplot creates two ellipses. The inner ellipse contains the central half of the points; points outside the outer ellipse are declared outliers.

13.1.2 S-PLUS Function relplot

The S-PLUS function relplot, written for this book, has the form

$$relplot(x, y, plotit = T)$$

and computes a relplot. As usual, x and y are any S-PLUS variables containing data. The function also returns a correlation coefficient based on the centrally located data (labeled mrho), but it plays no important role in this book and in fact suffers from a practical problem illustrated momentarily.

EXAMPLE. To illustrate how a relplot can alter our perceptions about the association between two variables, first look at Figure 13.2, which shows a scatterplot of data from a study where the goal is to predict reading ability. (These data were generously supplied by L. Doi. The X values are stored in column 4 of the file read.dat, and the Y values are in column 8.) Also shown is the least squares regression line, which has an estimated slope of -0.032. Testing the hypothesis of a zero slope using the conventional method in Section 6.3.1, the significance level is .764 (which is the same significance level obtained when testing $H_0 : \rho = 0$ with Student's T) and the .95 confidence interval for the slope is $(-0.074, 0.138)$. Using the modified bootstrap method in Section 7.3, the .95 confidence interval is $(-0.27, 0.11)$. So again no association is detected between the two variables under study. ■

Figure 13.3 shows a relplot of the same data. The inner ellipse contains the central half of the data; points outside the outer ellipse are declared outliers. As is evident,

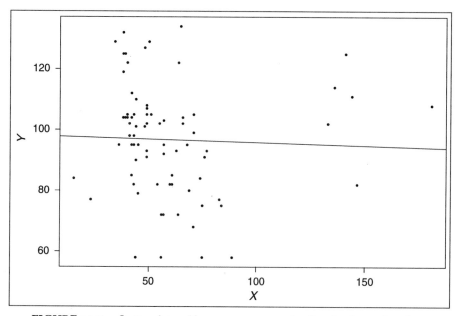

FIGURE 13.2   Scatterplot and least squares regression line for the reading data.

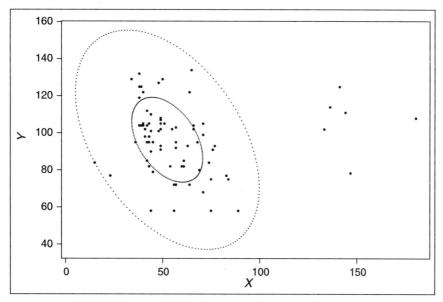

FIGURE 13.3   Relplot for the reading data in Figure 13.2.

we get a rather different impression about the association between these two variables versus the scatterplot in Figure 13.2. Figure 13.3 suggests that for the bulk of the points there might be a negative association that is masked when using least squares due to the outliers in the right portion of the scatterplot. (Methods covered later in this chapter support this conclusion.)

EXAMPLE. A criticism of the relplot is that outliers might affect our overall sense of how points are associated, depending on where the outliers happen to be located. Consider again the data in the upper portion of Figure 13.1 — only momentarily we ignore the unusual point that was added at $(X, Y) = (2.1, -2.4)$. Figure 13.4 shows a relplot of the data (and it reports a correlation of .66, which is reasonably close to Pearson's correlation, $r = .68$). ■

Now we add two points at $(X, Y) = (2.1, -2.4)$. Figure 13.5 shows the relplot. It correctly identifies the two outliers, but the association among the points not declared outliers is less pronounced. In particular, the relplot correlation drops from .66 to .32 and Pearson's correlation becomes $r = .21$.

13.1.3 MVE and MCD Estimators

The relplot is certainly an improvement on the simple strategy of checking for outliers among the X values only (ignoring the Y values) and then doing the same for the Y values. However, in addition to the concern just illustrated, a limitation is that it has not been extended to situations where we have more than two measures for each individual. There is an interesting alternative called a *bagplot* (Rousseeuw, Ruts,

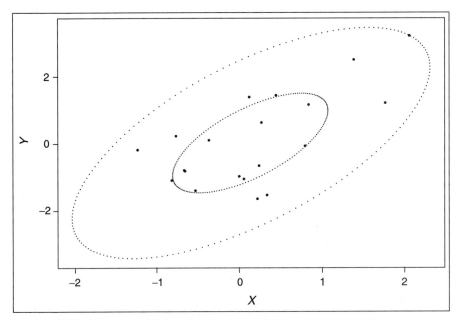

FIGURE 13.4 Relplot where both X and Y have normal distributions.

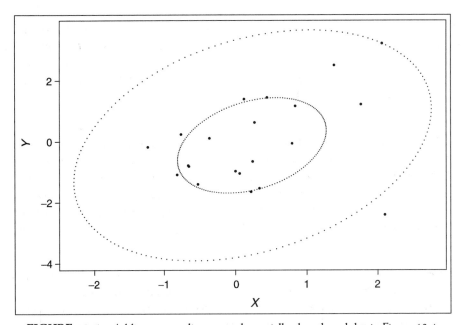

FIGURE 13.5 Adding two outliers can substantially alter the relplot in Figure 13.4.

& Tukey, 1999), but it too is restricted to the bivariate case. This is not to say that the bagplot has no practical value, but to conserve space attention is now restricted to methods that have been generalized to more than two variables. (The bagplot might soon meet this requirement if certain technical difficulties can be addressed.) But before describing multivariate alternatives to the relplot, we first need to consider robust analogs of the covariance between X and Y.

Recall from Chapter 6 that the sample covariance between X and Y is

$$s_{xy} = \frac{1}{n-1} \sum_{i=1}^{n} (X_i - \bar{X})(Y_i - \bar{Y}).$$

This quantity is the numerator of Pearson's correlation, and it has a finite-sample breakdown point of only $1/n$. That is, a single unusual point can have an inordinate influence on its value. One general strategy for dealing with outliers in bivariate data hinges on analogs of s_{xy} that can handle a large number of outliers. One such strategy arises as follows. Consider all ellipses that contain half of the data points. The inner ellipse in Figure 13.3 is one example for the reading data. The strategy behind the *minimum-volume ellipsoid* estimator, typically called the *MVE* estimator, is to search among all of the ellipses containing half of the data and identify the one having the smallest area. Once this ellipse is identified, a robust analog of the usual covariance (s_{xy}) is the covariance based only on the points inside this particular ellipse, and the mean of these points provides a measure of location with a high breakdown point. (The finite-sample breakdown point is .5.) Although the details as to how to find the ellipse with the smallest area are not straightforward, S-PLUS has a built-in function that performs the calculations for you. (It has the form cov.mve(m), where m is a matrix having n rows.) Moreover, the estimate is automatically rescaled so that when sampling from normal distributions, it estimates the variance and Pearson's correlation. (A comparable function, called MVE, can be found in SAS.)

Given the covariance and measure of location just described, one can measure the distance of each point from the center of the scatterplot using an analog of what is called the *Mahalanobis distance*. This distance can be used to judge how far away a point happens to be from the centrally located portion of a scatterplot. Rousseeuw and van Zomeren (1990) proposed a rule for deciding if a point is sufficiently far from the center to be declared an outlier. For readers familiar with matrix algebra, the details are relegated to Box 13.1 for the more general case where we have p measures for each individual. (Basic matrix algebra is summarized in Appendix C.)

BOX 13.1 How to Detect Outliers Using the MVE or MCD

Measures of Covariance

We have a sample of p measures for each of n individuals, which is denoted by $X_i' = (X_{i1}, \ldots, X_{ip})$, $i = 1, \ldots, n$. Let C and M be the center and covariance matrix, respectively, of the data determined by the MVE estimator (which is computed by the built-in S-PLUS function cov.mve) or by the MCD estimator (which is computed by the built-in S-PLUS function cov.mcd). Let

$$D_i = \sqrt{(X_i - C)'M^{-1}(X_i - C)}.$$

Continued

BOX 13.1 (*Continued*)

The value D_i measures how far the point X_i' is from the center of the data and is a generalization of what is called the *Mahalanobis distance*. (See for example Mardia, Kent, & Bibby, 1979, for details about the Mahalanobis distance.) The point X_i' is declared an outlier if $D_i > \sqrt{\chi_{.975,p}^2}$, the square root of the .975 quantile of a chi-squared distribution with p degrees of freedom.

An alternative to the MVE estimator is the so-called *minimum-covariance determinant* (MCD) estimator. To convey the basic strategy, we first must describe the notion of a generalized variance (introduced by S. Wilks in 1934), which is intended to measure the extent to which a scatterplot of points is tightly clustered together. For bivariate data, this measure of dispersion is given by

$$s_g^2 = s_x^2 s_y^2 (1 - r^2), \tag{13.1}$$

where s_x^2 and s_y^2 are the sample variances associated with the X and Y values, respectively, and r is Pearson's correlation between X and Y. (In the multivariate case, s_g^2 is the determinant of the covariance matrix.) Recall from Chapter 3 that the smaller the variance of the X values, the more tightly clustered together are the X values, and of course a similar result applies to the Y values. We have seen that a single outlier can inflate the sample variance tremendously, but what is more important here is that the sample variance can be small only if the X values are tightly clustered together with no outliers. And in Chapter 6 we saw that the correlation is sensitive to how far points happen to be from the regression line around which they are centered. When the sample variances are small and Pearson's correlation is large, the generalized variance will be small.

To provide some perspective, the left panel of Figure 13.6 shows a scatterplot of 100 points for which $s_g^2 = 0.82$. The right panel shows a scatterplot of another 100 points, except that they are more tightly clustered around the line $Y = X$, and the generalized variance has decreased to $s_g^2 = 0.23$. The left panel of Figure 13.7 shows another 100 points that were generated in the same manner as those shown in the left panel of Figure 13.6, except that the variance of the X values was reduced from 1 to 0.5. Now $s_g^2 = 0.20$. In the right panel of Figure 13.7, the points are more tightly clustered together, and the generalized variance has decreased to $s_g^2 = 0.06$.

Now consider any subset of the data containing half of the points. The strategy behind the MCD estimator is to search among all such subsets and identify the one with the smallest generalized variance. Then the MCD measure of location and covariance is just the mean and covariance of these points. (For results supporting the use of the MCD estimator over the MVE estimator, see Woodruff & Rocke, 1994.) As with the MVE estimator, computing the MCD measure of location and covariance is a nontrivial, computer-intensive problem, but S-PLUS has a built-in function that performs the calculations for you (called mcd.cov); SAS has an analog of this function called MCD. (For a description of the algorithm used, see Rousseeuw & van Driesen, 1999.) Once these measures of location and scale are available, you can measure the relative distance of a point from the center using the method in Box 13.1, and

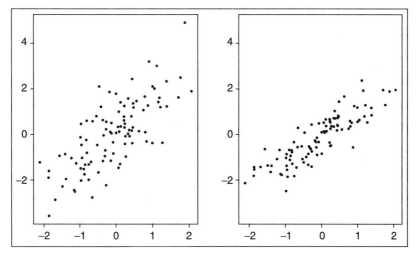

FIGURE 13.6   The more tightly points are clustered around a line, the smaller the generalized variance. The right panel has a smaller generalized variance than the left panel.

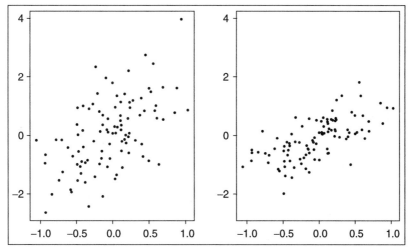

FIGURE 13.7   The generalized variance is also related to the variance of X and Y. In the left panel, $s_g^2 = .2$; in the right panel, $s_g^2 = .06$.

these distances can be used to detect outliers. (For an extension of this method, see Rocke & Woodruff, 1996.)

13.1.4 S-PLUS Function out

The S-PLUS function

$$\text{out(m,mcd=F,plotit=T)}$$

detects outliers using the MVE method when mcd=F is used; otherwise, the MCD method is used. It is common for the MCD method to find more outliers than the method based on MVE.

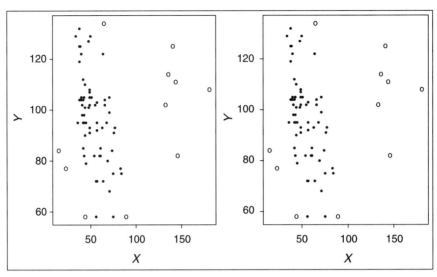

FIGURE 13.8   Output from the S-PLUS function out. The left panel is with mcd=F, meaning that the MVE method is used; the right panel is with mcd=T, meaning that the MCD method is used. Both methods return the same results in this particular case.

EXAMPLE. If the reading data used to create Figure 13.2 are stored in the S-PLUS variable blob, the command out(blob) creates the plot shown in the left panel of Figure 13.8. Points marked with a circle are declared outliers. The right panel shows the plot created by the function out with MCD=T. So in this particular instance, both methods flag the same points as outliers, and more points are declared outliers than with the relplot.   ■

13.1.5 The Minimum Generalized Variance Method

It might seem that we could simply discard any outliers detected by the MVE or MCD methods and estimate the regression line with the data that remain. That is, use a method similar in spirit to the MOM estimator in Section 3.5.2. However, many variations of this approach are known to be unsatisfactory — they can mask the overall association (cf. Fung, 1993). This is somewhat expected based on properties of Pearson's correlation coefficient, r, covered in Chapter 6. In particular, we saw that restricting the range of X or Y can greatly influence r, so it is not too surprising that if we focus on the middle 50% of the data only, we might be misled regarding the association between X and Y.

EXAMPLE. As an illustration, the MVE and MCD methods are applied to the original 20 points in Figure 13.1. Both X and Y were generated from normal distributions, with the regression line between X and Y having a slope

Continued

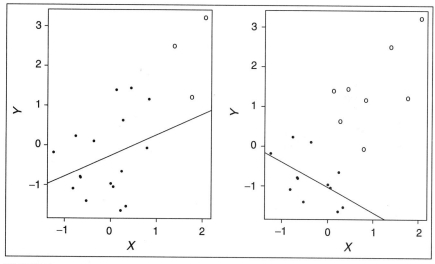

FIGURE 13.9   Eliminating outliers using the MVE or MCD method and fitting a least squares regression line to the data that remain can result in a poor estimate of the true slope, which is 1 in this particular case.

EXAMPLE. (*Continued*) equal to 1. The left panel of Figure 13.9 shows the plot created by out plus the least squares regression line based on the points not flagged as outliers. This line has a slope of .55, so it poorly estimates the true slope. The right panel shows the results when the MCD method is applied instead. As is evident, MCD finds more outliers. But what is perhaps more important, if the outliers found by MCD are discarded, the true association between X and Y is completely lost, as indicated by the least squares regression line based on the points not declared outliers.　　　　■

EXAMPLE. To add perspective, the process used to generate Figure 13.9 was repeated 500 times, and each time the least squares estimate of the slope was computed using the points not flagged as outliers. So again the true slope is 1. The first boxplot in the left panel of Figure 13.10 shows the estimated slopes when using MVE to detect outliers; the second boxplot is based on the least squares estimate of the slope using all of the data instead. (That is, outliers are not discarded.) The second panel shows a boxplot of the least squares estimate when outliers detected by the MCD are removed; again, the other boxplot is based on the estimated slopes when outliers are not removed. As is evident, discarding outliers and applying least squares is a relatively unsatisfactory strategy in this particular case, because the least squares estimate based on all of the data tends to be closer to the true slope being estimated. More generally, where

Continued

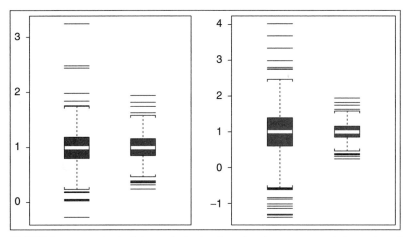

FIGURE 13.10   In the left panel, the first boxplot shows estimated slopes, based on least squares, with outliers detected by the MVE method removed. The other boxplot shows the estimated slopes when retaining all points. The right panel is the same as the left, but with the MVE method replaced by MCD.

EXAMPLE. (*Continued*) observations do not have a normal distribution, discarding outliers and fitting a least squares regression line to the remaining data can be highly unsatisfactory.   ■

There are many strategies for discarding outliers and fitting a line to the data that remain. Currently, most seem to offer little or no improvement over other regression estimators described in this chapter. However, a variation of this strategy, where outliers are discarded and then a line is fit to the remaining points, does have practical value, at least in some situations, and is based on a different approach to detecting outliers. When searching for outliers, of particular importance is detecting so-called *bad leverage points*. A *leverage point* is an outlier among the X values. A *regression outlier* is a point with a relatively large residual. A *bad leverage point* is a leverage point that is also a regression outlier. A *good leverage point* is a leverage point that is not a regression outlier. That is, a good leverage point is a point that is reasonably close to the regression line, as illustrated in Figure 13.11. Good leverage points lower the standard error of the least squares estimate of the slope without giving a distorted indication of the association among the bulk of the observations. Bad leverage points can result in a poor fit to the majority of the data, even when using various robust estimators. So the hope is to be able to eliminate the effects of bad leverage points yet achieve a relatively accurate estimate of the slope and intercept, even under normality and homoscedasticity. In some situations this means that we want to avoid identifying and discarding so many points that the true association is lost. There are direct methods for detecting regression outliers (Rousseeuw & van Zomeren, 1990), but currently it seems that an indirect method performs best, based on the criterion of achieving a relatively low standard error, when estimating the slope.

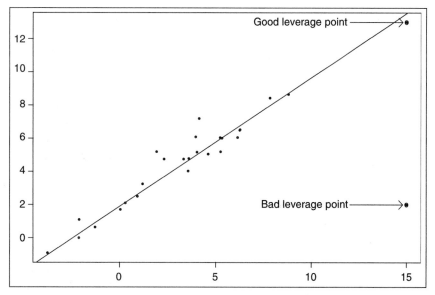

FIGURE 13.11   Illustration of a good versus bad leverage point.

For p-variate data, an outlier detection method that has been found to have practical value, in the context just described, begins by determining the p centrally located points and then determining how close each point is to the center using the notion of a generalized variance. There are many ways of finding the center of a cloud of points, some of which use some multivariate analog of the median (e.g., Small, 1990), and perhaps for the situation at hand the method used here can be improved upon; this issue is in need of further investigation.

The steps for applying this alternative outlier detection method, called the *MGV method*, are as follows.

1. Initially, all n points are described as belonging to set A.
2. Find the p points that are most centrally located. In the bivariate case, for the ith pair of points, compute

$$d_i = \sum_{j=1}^{n} \sqrt{\frac{(X_j - X_i)^2}{\text{MAD}_x^2} + \frac{(Y_j - Y_i)^2}{\text{MAD}_y^2}}, \tag{13.2}$$

where MAD_x and MAD_y are the values of MAD for the X and Y values, respectively. The two most centrally located points are taken to be the two points having the smallest d_i values. For the more general case where we have p measures for each individual, $(X_{i1}, \ldots, X_{ip})$, $i = 1, \ldots, n$,

$$d_i = \sum_{j=1}^{n} \sqrt{\sum_{\ell=1}^{p} \frac{(X_{j\ell} - X_{i\ell})^2}{\text{MAD}_\ell^2}}, \tag{13.3}$$

where MAD_ℓ is the value of MAD based on $X_{1\ell}, \ldots, X_{n\ell}$.

3. Remove the centrally located points from set A and put them into set B. At this step, the generalized variance of the points in set B is zero. (When dealing with p measures, any p distinct points will have a generalized variance of zero.)

4. If the ith point in set A is put in set B, the generalized variance of the points in set B will be changed to some value that is labeled s_{gi}^2. That is, associated with every point in A is the value s_{gi}^2, which is the resulting generalized variance when it, and it only, is placed in set B. Compute s_{gi}^2 for every point in A.

5. Among the s_{gi}^2 values computed in the previous step, permanently remove the point associated with the smallest s_{gi}^2 value from set A and put it in set B. That is, find the point in set A that is most tightly clustered together with the points in set B. Once this point is identified, permanently remove it from A and leave it in B henceforth.

6. Repeat steps 4 and 5 until all points are now in set B.

The first p points removed from set A have a generalized variance of zero, which is labeled $s_{g(1)}^2 = \cdots = s_{g(p)}^2 = 0$. When the next point is removed from A and put into B (using steps 4 and 5), the resulting generalized variance of set B is labeled $s_{g(p+1)}^2$; continuing this process, each point has associated with it some generalized variance when it is put into set B. Note that by construction, $s_{g(1)}^2 \leq s_{g(2)}^2 \leq \cdots \leq s_{g(n)}^2$.

Based on the process just described, the ith point has associated with it one of the ordered generalized variances just computed. For example, in the bivariate case, associated with the ith point (X_i, Y_i) is some value $s_{g(j)}^2$ indicating that the ith point was removed in the jth step of the process used to compute the values $s_{g(1)}^2 \leq s_{g(2)}^2 \leq \cdots \leq s_{g(n)}^2$. For convenience, the generalized variance associated with the ith point, $s_{g(j)}^2$, is labeled D_i. The p deepest points have D values of zero. Points located at the edges of a scatterplot have the highest D values, meaning that they are relatively far from the center of the cloud of points. Moreover, we can detect outliers simply by applying one of the outlier detection rules in Chapter 3 to the D_i values. Note, however, that we would not declare a point an outlier if D_i is small, only if D_i is large. If we use the rule based on the median and MAD, for example, then according to Equation (3.22), the point (X_i, Y_i) is declared an outlier if

$$\frac{|D_i - M_D|}{\text{MAD}_D/.6745} > 2.24, \tag{13.4}$$

where M_D and MAD_D are the median and the value of MAD, respectively, based on the D values.

Of course, an alternative to Equation (13.4) is some type of boxplot rule. Currently, in the context of regression, the boxplot rule described in Section 3.4.4 has received the most attention and will be used henceforth. (When trying to estimate regression parameters, the effect of using Equation (13.4) has not been studied.) In particular, declare the ith point an outlier if

$$D_i > q_2 - 1.5(\text{IQR}), \tag{13.5}$$

where IQR $= q_2 - q_1$ and where q_1 and q_2 are the ideal fourths based on the D_i values. For the more general case where there are p variables, replace Equation (13.5) with

$$D_i > M_D + \sqrt{\chi^2_{.975,p}}(\text{IQR}),$$

where $\sqrt{\chi^2_{.975,p}}$ is the square root of the .975 quantile of a chi-squared distribution with p degrees of freedom.

13.1.6 S-PLUS Function outmgv

The S-PLUS function

$$\text{outmgv}(x,y = NA, \text{plotit} = T, \text{outfun} = \text{outbox}, \ldots)$$

applies the MGV outlier detection method just described. If the second argument is not specified, it is assumed that x is a matrix with p columns corresponding to the p variables under study. So, for example, in the bivariate case, x could be a matrix having n rows and two columns. If the second argument, y, is specified, the function combines the data in x with the data in y and checks for outliers among these $p + 1$ variables. In particular, the data do not have to be stored in a matrix; they can be stored in two vectors (x and y) and the function combines them into a single matrix for you. If plotit=T is used and bivariate data are being studied, a plot of the data will be produced, with outliers marked by a circle. The argument outfun can be used to change the outlier detection rule applied to the depths of the points (the D_i values in the previous section). By default, the boxplot rule based on Equation (13.5) is used.

EXAMPLE. Consider the following five pairs of points:

X:	6	22	19	29	33
Y:	11	7	42	22	26

To find the two centrally located points, first note that for the X values MAD is 7 and for the Y values MAD is 11. For the first pair of points, $(X, Y) = (6, 11)$,

$$d_1 = \sqrt{\frac{(6-6)^2}{7^2} + \frac{(11-11)^2}{11^2} + \cdots + \frac{(6-33)^2}{7^2} + \frac{(11-26)^2}{11^2}} = 6.73.$$

In a similar manner, $d_2 = 4.896$, $d_3 = 5.76$, $d_4 = 4.523$, and $d_5 = 5.36$. The two smallest d_i values are 4.896 and 4.523, which correspond to the points (29, 22) and (22, 7). ■

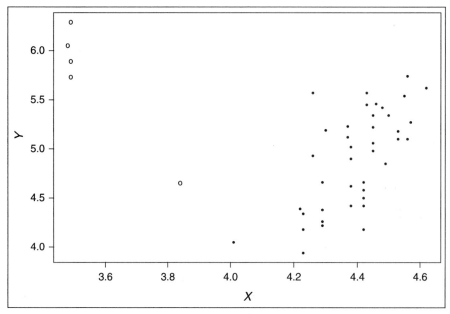

FIGURE 13.12   For the star data, the MGV outlier detection method flags the points indicated by a circle as outliers.

> **EXAMPLE.** If the star data in Figure 6.3 are stored in the S-PLUS variables starx and stary, the command
>
> outmgv(starx,stary)
>
> produces the plot shown in Figure 13.12. If the data are stored in the n-by-2 matrix mstar, the command outmgv(mstar) will again produce Figure 13.12. ■

13.1.7 A Variation of the MGV Method for Large Sample Sizes

The MGV outlier detection method has practical value when fitting a straight line to data. However, a criticism is that as n gets large, execution time increases substantially using the software provided with this book. (Much faster software could be written, but this has not been done as yet.) When fitting a line to data, a regression method based in part on a variation of the MGV method performs well and has faster execution time. This alternative method begins by determining for each point the generalized variance when it is removed. That is, we remove the ith point, (X_i, Y_i), and compute the generalized variance for the remaining $n - 1$ points, which we label as D_i. So, for example, D_1 is the generalized variance if (X_1, Y_1) is eliminated from the data, and D_2 is the generalized variance when (X_1, Y_1) is put back and (X_2, Y_2) is removed instead. Extreme points will tend to have smaller D values versus points near the center of the data. So points with unusually small D values are declared outliers, and this will be called method $MGVF$ or the *inward depth* method. Here, the boxplot rule for detecting

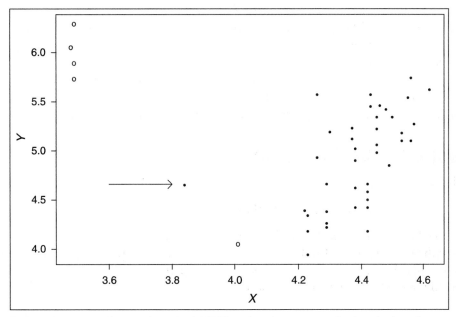

FIGURE 13.13 Output from the MGVF outlier detection method applied to the star data. The MGVF method misses what appears to be an obvious outlier, but despite this it has practical value when estimating the slope.

outliers (described in Section 3.4.4) will be applied to the D values unless stated otherwise.

It is stressed that if the main goal is to detect outliers, the method just described can be unsatisfactory. For example, if we apply the MGVF method to the data in Figure 13.12, we get the plot shown in Figure 13.13. Notice the point indicated by the arrow. It is declared an outlier in Figure 13.12 and it certainly seems to be relatively far from the majority of points. However, it is not declared an outlier by method MGVF. But despite this shortcoming, method MGVF will be seen to have practical value when we take up regression.

For completeness, there are several other approaches to measuring how deeply a point is embedded in a scatterplot (see, for example, Liu & Singh, 1997). Perhaps some of these techniques will be found to have practical value versus the measures of depth described here, but this remains to be seen. Also, there are other methods for detecting outliers that might have more practical value when fitting a straight line to data. For example, Fung (1993) begins by declaring points an outlier with the MVE method in Section 13.1.3. Then an iterative method is used to determine whether any of these outliers should be put back into the set of points not declared an outlier. (Some preliminary checks do not support the use of this method, but a more detailed study is needed.) Another possibility is to modify the MVE or MCD methods so that rather than use the central 50% of the data to determine the center and correlations, some higher proportion is used instead. This lowers the breakdown point, but in terms of estimating the association between two variables, perhaps this approach has practical value. So far, variations of this approach have been found to be relatively unsatisfactory. Another possibility is to use method MGVF but rather than use the boxplot rule on the D values, use instead the method in Section 3.4.2 which is based on the median and MAD. This strategy now flags the point marked by the arrow in Figure 13.12 as an outlier, but an additional point is marked an outlier beyond the five outliers found in Figure 13.11. Other approaches to detecting outliers in multivariate data were recently proposed by Rocke and Woodruff (1996); Poon, Lew, and Poon (2000); and Peña and Prieto (2001). Perhaps they can be used effectively in regression, relative to the other methods considered here, but this remains to be determined.

13.1.8 S-PLUS Function outmgvf

The S-PLUS function

$$\text{outmgvf}(x,y = NA, \text{plotit} = T, \text{outfun} = \text{outbox}, \dots)$$

checks for outliers using the MGVF method. The arguments x and y are used as described in Section 13.1.6. By default a boxplot rule is applied to the measures of depth, but this can be altered with the argument outfun. For example, outfun=out would use MAD and the median instead.

13.1.9 A Projection Method for Detecting Outliers

If the main goal is to detect outliers, all of the methods described so far are open to criticism. One issue has to do with the so-called *outside rate per observation*. This is just the expected or average proportion of outliers among n randomly sampled vectors. When working with a single random variable and when sampling from a normal distribution, a goal has been to achieve an outside rate per observation roughly equal to .05. The basic boxplot rule in Section 3.4.4 achieves this goal reasonably well (Hoaglin, Iglewicz, & Tukey, 1986), but the outside rate per observation is a bit unstable as the sample size increases from small to moderately large values. The method in Section 3.4.5 was developed to help correct this problem.

It seems that for most outlier detection methods aimed at multivariate data, little or nothing is known about their outside rate per observation. Checks on this rate when sampling from bivariate normal data indicate that the rate can be well above .05 when using the MVE method, and it is even higher when using MCD when variables are correlated (cf. Fung, 1993). These methods are well known and now easy to apply, but for some purposes an alternative strategy might be in order. This section outlines one approach for which the outside rate per observation is roughly equal to .05. The method reflects a blend of techniques that have been proposed, and it appears to have practical value for a wide range of situations.

The method begins by computing the Donoho and Gasko (1992) estimate of the median of the data (which can be done with the S-PLUS function dmean). The computational details are too involved to give here, but an outline of the strategy might help. For simplicity, attention is restricted to the bivariate case, but the method can be extended to more than two variables. The Donoho–Gasko median is based on something called *halfspace depth*, which is a method for measuring how deeply a point is nested within a scatterplot of all the data. For any point in a scatterplot, consider any line going through this point. As is evident, a certain proportion of the points in a scatterplot will be on or above this line, and a certain proportion will be on or below it. For convenience, the smaller of these two proportions is labeled P_m. Now, among the infinitely many lines going through some specific point there will be a minimum P_m value; this is called the *halfspace depth* of the point. (For bivariate data, halfspace depth can be computed exactly using the algorithm in Rousseeuw & Ruts, 1996. For more than two variables, an approximation has been derived by Rousseeuw & Struyf, 1998.) A high halfspace depth indicates that a point is deeply nested within the cloud of data. The Donoho–Gasko median is the average of all points having the largest depth. (For a single variable it reduces to the usual sample median.) An alternative and much simpler approach when trying to find the center of a scatterplot is to compute the median for each of the variables under study, but there are theoretical concerns about this strategy that go well beyond the scope of this book.

Before continuing, the notion of a projection of a point onto a line is needed. Consider any line through a scatterplot of data. For convenience, call this line $\mathcal{L}$. Now consider any point in the scatterplot, say, (X, Y). If we draw a line through this point that is perpendicular to the line $\mathcal{L}$, it will intersect with $\mathcal{L}$ at some point, say, (X_p, Y_p). The point (X_p, Y_p) is the (orthogonal) projection of (X, Y) onto the line $\mathcal{L}$.

Consider any point among the scatterplot of the data, and form the line connecting this point with the Donoho–Gasko median. Then project all points onto this line. (For computational details, see for example, Graybill, 1983, Section 4.4.) The distance between the projected points can be used to check for outliers. One possibility is simply to apply a boxplot rule, but a slight modification is needed to achieve an outside rate per observation reasonably close to .05. In particular, use Equation (3.25) on the distances of the projected points, but with $k = 2.4$ when dealing with bivariate data. (For the general case of p-variate data, use $k = \sqrt{\chi^2_{.95,p}}$.) Equation (3.26) is not used, because only points with large distances are declared outliers. As in Section 3.4.5, use the interquartile range based on the ideal fourths. The process just described is repeated for every point in the scatterplot. That is, n projections are considered. Any point is declared an outlier if it is found to be an outlier for any of the projections. (Complete computational details can be found in Wilcox, 2002.)

13.1.10 S-PLUS Function outpro

The S-PLUS function

$$\text{outpro}(m, \text{gval} = NA, \text{plotit} = T, \text{op}=T)$$

checks for outliers using the projection method just described. Here m is assumed to be a matrix having two or more columns. The argument gval is k and defaults to $\sqrt{\chi^2_{.95,p}}$ if not specified. If op=T is used with bivariate data and plotit=T, the function creates a scatterplot of the data and draws a polygon containing the centrally located points and ignoring the outliers. (This polygon is the .5 depth contour as described by Liu, Parelius, & Singh, 1999. It encompasses approximately half of the data, corresponding to the points having the highest halfspace depths.) Setting op=F, the function does not ignore the outliers when drawing the .5 depth contour. (An illustration will be given in Figure 13.21.)

13.2 Some Robust Regression Methods

This section summarizes some robust regression estimators that appear to have considerable practical value. All of the regression methods in this chapter can be used when there are multiple (p) predictors. However, explaining the basic strategy behind some of these methods is easier if we first focus on the single predictor case. As in Chapter 6, if there is a single predictor, it is assumed that

$$Y_i = \beta_0 + \beta_1 X_i + \epsilon \tag{13.6}$$

and X and ϵ are independent and that for p predictors the standard regression model is that

$$Y_i = \beta_0 + \beta_p X_{ip} + \cdots + \beta_1 X_{i1} + \epsilon,$$

$i = 1, \ldots, n$. Typically it is assumed that $E(\epsilon) = 0$, but here it is assumed that ϵ has a median of zero instead. The goal is to estimate the unknown slope and intercept (β_1 and β_0). With p predictors there are $p + 1$ parameters to be estimated: $\beta_0, \ldots, \beta_p$.

Although it is clear that the blind use of least squares regression is highly unsatisfactory, all indications are that no single regression method is always optimal among the many situations encountered in practice. That is, regression method A might have substantial advantages relative to method B in some situations, but situations arise where the reverse can happen as well. To complicate matters, several criteria are used to compare different regression methods, making it difficult and seemingly impossible to single out one method for routine use — several methods need to be considered. So for the moment we merely describe some regression methods and then try to convey their relative merits. At a minimum it is suggested that the methods in this section be given serious consideration, and it is recommended that several of the estimators in Section 13.3 be considered as well.

Another important point is that not all regression estimators are covered in this chapter. The omission of some methods was not arbitrary, but some experts might argue passionately that additional methods should have been included in this chapter. There might be merit to these arguments. Moreover, views about regression estimators continue to evolve. Simultaneously, some would argue that too many estimators are covered in this chapter and that providing such a seemingly bewildering array of methods will only confuse individuals learning about modern techniques. Currently, it seems that familiarity with multiple methods is a must. As will be illustrated, even among the better robust estimators, different results can be obtained with different methods, and choosing a method that provides relatively short confidence intervals is a nontrivial task, as will become evident. During the exploratory phases of an investigation, it seems that several estimators should be considered. An educated guess is that some of the estimators in this section can be ignored in most applications, but there is no compelling evidence that this can be done safely, so the goal is to cover a reasonable number of estimators in the hope that at least one of them will be valuable to the reader. Some authorities would argue that simply looking at a scatterplot enables us to tell which points are influential and that this should tell us which estimator should be used. But some of the illustrations later in this chapter suggest that dealing with regression is not always that simple.

In case it helps, a brief list of some of the estimators omitted from this chapter is given here. Maronna and Morgenthaler (1986) discuss how one might approach regression via robust covariances. A variation and extension of this method is discussed by Wilcox (1997a). This chapter describes two M-estimators, but certain variations of this approach are not covered. In particular, GM-based estimators (with Mallows weights) are not discussed. Recent results on this estimator are reported by Bianco, Boente, and Rienzo (2000), but work reviewed by Wilcox (1997a) suggests that it does not compete well with other estimators when there is heteroscedasticity. Despite this, perhaps arguments can be made for using this estimator in applied work, but this remains to be established. For a survey of results related to M-estimators, see Maronna, Yohai, and Zamar (1993). Some methods approach heteroscedasticity assuming that it can be modeled using some *known* function (e.g., Carroll & Ruppert, 1982; Giltinan, Carroll, & Ruppert, 1986). Such situations are not discussed here. For methods that deal with heteroscedasticity by attempting to estimate the optimal weights in weighted least squares, see Cohen, Dalal, and Tukey (1993) as well as Wilcox (1996a). (Rank-based approaches are briefly discussed in Chapter 15.)

13.2.1 The Theil–Sen (TS) Estimator

As is evident, any two distinct points determine a line. A method proposed by Theil (1950) and Sen (1968) estimates the slope of a regression line by computing the slope for all pairs of points having distinct X values and then computing the median of these slopes; the result will be labeled b_{1ts}. More formally, let X_i and $X_{i'}$ be any two X values such that $X_i > X_{i'}$. The slope corresponding to the two points (X_i, Y_i) and $(X_{i'}, Y_{i'})$ is

$$b_{1ii'} = \frac{Y_i - Y_{i'}}{X_i - X_{i'}}. \tag{13.7}$$

Computing the slope for all pairs of points having $X_i > X_{i'}$, the median of these slopes is the Theil–Sen estimate of β_1 and is labeled b_{1ts}. The intercept is estimated with

$$b_{0ts} = M_y - b_{1ts}M_x,$$

where M_y and M_x are the sample medians corresponding to the Y and X values, respectively.

When there is one predictor, the finite-sample breakdown point of the Theil–Sen estimator is approximately .29 (Dietz, 1989), meaning that about 29% of the data must be altered to make the resulting estimate of the slope and intercept arbitrarily large or small. A negative feature is that the finite-sample breakdown point has not been established when there are two or more predictors, but it appears to decrease as p gets large. An advantage is that its standard error can be tens, even hundreds, of times

TABLE 13.1   Boscovich's Data on Meridian Arcs

X:	0.0000	0.2987	0.4648	0.5762	0.8386
Y:	56,751	57,037	56,979	57,074	57,422

smaller than the ordinary least squares estimator when the error term is hetero-scedastic. Even when the error term has a normal distribution but is heteroscedastic, the Theil–Sen estimator can be substantially more accurate.

EXAMPLE. The computation of the Theil–Sen estimator is illustrated with the data in Table 13.1, which were collected about 200 years ago and analyzed by Roger Boscovich. The data deal with determining whether the earth bulges at the center, as predicted by Newton, or whether it bulges at the poles as suggested by Cassini. Here X is a transformed measure of latitude and Y is a measure of arc length. (Newton's prediction implied that $\beta_1/(3\beta_0) \approx 1/230$.) For the first two pairs of points, the estimated slope is

$$\frac{57,037 - 56,751}{0.2987 - 0} = 1560.1.$$

Computing the slope for the remaining nine pairs of points and taking the median yields 756.6. It is left as an exercise to verify that the intercept is estimated to be 56,685. Interestingly, $b_{1ts}/(3b_{0ts}) = 0.0044$, which is fairly close to Newton's prediction: $1/230 = 0.0043$. (Least squares gives a very similar result.)   ■

13.2.2 S-PLUS Function tsreg

The S-PLUS function

$$\text{tsreg}(x,y)$$

(written for this book) computes the Theil–Sen estimate of the slope and intercept. Here, x can be any n-by-p matrix of predictors and y is any S-PLUS variable containing the Y values. For the data in Table 13.1 this function returns

```
$coef:
[1] 56685.3235   756.6191

$residuals:
[1]   65.67654   125.67443   -58.00000   -47.28736   102.17580
```

where $residuals are the residuals given by $r_i = Y_i - b_{1ts}X_i - b_{0ts}$.

It is noted that generalizations of the Theil–Sen estimator to multiple predictors are available. One approach, which is used by the S-PLUS function tsreg, is based on the so-called *Gauss–Seidel algorithm*, which is described in a general context by Hastie and Tibshirani (1990), but no details are given here. Other extensions have been considered (e.g., Hussain & Sprent, 1983; Wilcox, 1998), but currently there is no known reason for preferring them to the method used here.

13.2.3 M-Estimators

Section 3.2.8 introduced M-estimators of location, which stem from the realization that different ways of measuring closeness lead to different measures of location. To quickly review, if the goal is to choose a constant c that is close to all of the values $X_1, \ldots, X_n$, and if we measure the overall closeness of c to these n values with

$$\sum (X_i - c)^2,$$

the sum of all squared differences, then the value of c that minimizes this sum is $c = \bar{X}$, the sample mean. If we use

$$\sum |X_i - c|$$

to measure closeness, this leads to taking c to be the median. M-estimators of location consider a broad class of measures that might be used to measure closeness, and members of this class have been identified that have desirable properties. In particular, if we use Huber's Ψ (described in Section 3.2.8), we get an estimator that has a finite-sample breakdown point of .5, and the accuracy of the estimator (its standard error) compares well to the sample mean in the event sampling is from a normal distribution. But unlike the sample mean, this particular M-estimator can maintain a relatively small standard error when sampling is from a heavy-tailed distribution instead, where outliers are fairly common.

The basic idea behind M-estimators of location is readily extended to regression. Rather than measure the fit of a regression line with the sum of squared residuals, M-estimators are based on a broad class of functions that includes least squares as a special case. This approach leads to choosing the slope and intercept to be the values satisfying

$$\sum \Psi \left(\frac{r_i}{\hat{\tau}} \right) = 0, \tag{13.8}$$

where Ψ is some function to be determined, $r_1, \ldots, r_n$ are the residuals corresponding to some choice for the slope and intercept, and $\hat{\tau}$ is some measure of scale based on the residuals. [Equation (13.8) is a generalization of Equation (3.11).]

In the context of regression, many M-estimators (choices for Ψ) have been proposed and studied. Several have excellent theoretical properties, but no attempt is made to list all of these methods here. (Readers interested in more details are referred to Coakley & Hettmansperger, 1993; Staudte & Sheather, 1990; Rousseeuw & Leroy, 1987; Wilcox, 1997a.) Rather, two methods are described in Box 13.2. The first is based on Huber's Ψ in conjunction with what are called *Schweppe weights*. Roughly, the strategy is to look for outliers among the X values (using the h_i values given in Box 13.2) and to make adjustments if any outliers are found. The method also makes an adjustment if any unusually large residuals are found. The second method described in Box 13.2 was derived by Coakley and Hettmansperger (1993).

BOX 13.2 Two Regression M-Estimators

The first method begins by setting $k = 0$ and

$$b_i = \frac{1}{n} + \frac{(X_i - \bar{X})^2}{\sum (X_i - \bar{X})^2}.$$

Compute the least squares estimate of the intercept and slope as described in Chapter 6. Denote them by b_{0k} and b_{1k}, respectively. Proceed as follows:

1. Compute the residuals, $r_{i,k} = y_i - b_{0k} - b_{1k}X_{i1}$, let M_k be equal to the median of the largest $n - 1$ of the $|r_{i,k}|$, $\hat{\tau}_k = 1.48M_k$, and let $e_{i,k} = r_{i,k}/\hat{\tau}_k$.

2. Form weights,

$$w_{ik} = \frac{\sqrt{1 - b_i}}{e_{ik}} \Psi \left(\frac{e_{ik}}{\sqrt{1 - b_i}} \right),$$

   where

$$\Psi(x) = \max[-K, \min(K, x)]$$

   is Huber's Ψ with $K = 2\sqrt{2/n}$.

3. Use the w_{ik} values to obtain a weighted least squares estimate of the slope and intercept. That is, find the values $b_{0,k+1}$ and $b_{1,k+1}$ that minimize

$$\sum w_{ik}r_i^2.$$

   (S-PLUS and other software have built-in functions for computing weighted least squares estimates of the slope and intercept.) Increase k by 1.

4. Repeat steps 1–3 until convergence. That is, iterate until the change in the estimates of the slope and intercept are small.

The second estimator, called the *Coakley–Hettmansperger estimator*, is described for the general case of p predictors. Begin by computing the LTS estimator in Section 13.3.7, a vector having length $p + 1$. Compute the residuals, r_i $(i = 1, \ldots, n)$, and let

$$\hat{\tau} = 1.4826 \left(1 + \frac{5}{n - p} \right) \times \text{med}\{|r_i|\}$$

and

$$w_i = \min \left\{ 1, \left[\frac{b}{(\mathbf{x}_i - \mathbf{m}_x)'\mathbf{C}^{-1}(\mathbf{x}_i - \mathbf{m}_x)} \right]^{a/2} \right\},$$

> **BOX 13.2** (*Continued*)
> where the quantities $\mathbf{m_x}$ and $\mathbf{C}$ are the minimum volume ellipsoid (MVE) estimators of location and covariance associated with the predictors. (The notation $\mathrm{med}\{|r_i|\}$ refers to the median of the values $|r_1|, \ldots, |r_n|$.) Let $\Psi'(r_i/\hat{\tau}w_i) = 1$ if $|r_i/\hat{\tau}w_i| \leq K$; otherwise $\Psi'(r_i/\hat{\tau}w_i) = 0$. Coakley and Hettmansperger suggest using $K = 1.345$. Let $\mathbf{W} = \mathrm{diag}(w_i)$ and $\mathbf{B} = \mathrm{diag}(\Psi'(r_i/\hat{\tau}w_i))$. The Coakley–Hettmansperger estimator is
>
> $$b_{\mathrm{ch}} = b_{\mathrm{lts}} + (\mathbf{X'BX})^{-1}\mathbf{X'W}\Psi\left(\frac{r_i}{w_i\hat{\tau}}\right)\hat{\tau}.$$

A serious criticism of the first M-estimator in Box 13.2 is that the finite-sample breakdown point is only $2/n$. That is, it can handle one outlier, but two might cause practical problems. One way of addressing this concern is to switch to the Coakley–Hettmansperger estimator. It begins by computing the LTS estimate of the slope and intercept as described in Section 13.3.7. Then it forms weights based on how deeply each X_i is nested within all of the X values. It then uses these weights, in conjunction with Huber's Ψ, to adjust the initial estimate of the slope and intercept. The result is an estimator with the highest possible breakdown point of .5. It can be substantially more accurate than the first method outlined in Box 13.2, and it enjoys excellent theoretical properties. For the special case of a single predictor, often it seems that some other estimator is a bit more satisfactory, but it remains one of the many estimators that should be given serious consideration. A possible argument for the Coakley–Hettmansperger estimator is that software is available that has relatively fast execution time, even when the sample size is fairly large and there are multiple predictors.

13.2.4 S-PLUS Functions bmrg and chreg

The S-PLUS functions

$$\mathrm{bmreg}(x,y) \quad \text{and} \quad \mathrm{chreg}(x,y)$$

compute the two M-estimators just described. The function bmreg performs the calculations for the first method in Box 13.2; chreg computes the Coakley–Hettmansperger estimator.

13.2.5 MGV and MGVF Regression

The so-called *MGV regression estimator* first checks for outliers using Equation (13.5). If any outliers are found, they are discarded, and the Theil–Sen estimator is applied to the data that remain. (So this approach is similar in spirit to the class of skipped estimators mentioned in Section 3.5.2.) All indications are that this does a relatively good job of eliminating any points that cause contamination bias when using Theil–Sen, where, roughly, *contamination bias* refers to the ability of a very small number of unusual values to result in a poor fit to the bulk of the observations. (More details are given in Section 13.4.) If instead you apply the least squares estimator after eliminating outliers, the accuracy of your estimate can be relatively poor, particularly when there is heteroscedasticity. That is, it might seem that least squares might be salvaged by first eliminating outliers; but based on the standard error of this method, it cannot be recommended (Wilcox, 2001b). Of course, some alternative to both Theil–Sen and least squares might be used after outliers are removed, but so far Theil–Sen seems to be a very effective choice.

As will be illustrated, the MGV estimator appears to have practical value when dealing with contamination bias, and it seems to compete reasonably well in terms of achieving a relatively small standard error. But a practical concern is that as the sample size increases, execution time can become unacceptably high when using the software written for this book, particularly when computing confidence intervals using the bootstrap method in Section 13.5. An alternative approach is to replace the MGV outlier detection method with the inward method described in Section 13.1.7. That is, discard any outliers found by the method in Section 13.1.7 and again apply Theil–Sen to the data that remain. This will be called the *MGVF regression estimator*.

13.2.6 S-PLUS Functions mgvreg and mgvfreg

The S-PLUS function

$$mgvreg(x,y,regfun=tsreg,outfun=outbox)$$

computes the MGV estimate of the slope and intercept as just described. The argument regfun indicates which estimator is applied after outliers are removed and defaults to Theil–Sen. The argument outfun indicates which boxplot rule will be applied to the depths of the points (the D values described in Section 13.1.5). By default, a boxplot rule is used. The function

$$mgvfreg(x,y,regfun=tsreg,outfun=outbox)$$

is like mgvreg, except that the inward outlier detection method (described in Section 13.1.7) is used instead.

13.3 More Regression Estimators

When there is a single predictor, it seems that often the Theil–Sen estimator is a good choice, but there are indications that with multiple predictors it might not compete well with other methods. (This issue is in need of more study.) Generally, the methods in the previous section seem to provide reasonably good alternatives to least squares, but currently the only certainty is that exceptions can occur. Accordingly, some additional estimators are described here that might prove to be useful.

13.3.1 S-Estimators and a Modification of Theil–Sen

S-estimators of regression parameters search for the slope and intercept values that minimize some measure of scale associated with the residuals. Least squares, for example, minimizes the variance of the residuals and is a special case of S-estimators. The hope is that by replacing the variance with some measure of scale that is relatively insensitive to outliers, we will obtain estimates of the slope and intercept that are relatively insensitive to outliers as well. As noted in Chapter 3, there are many measures of scale. The main point is that if, for example, we use the percentage bend midvariance (described in Section 3.3.7), situations arise where the resulting estimate of the slope and intercept has advantages over other regression estimators we might use. This is not to say that other measures of scale never provide a more satisfactory estimate of the regression parameters. But for general use, it currently seems that the percentage bend midvariance is a good choice.

Here a simple approximation of the S-estimator is used. (There are other ways of computing S-estimators, e.g., Croux, Rousseeuw, & Hössjer, 1994; Ferretti et al., 1999. Perhaps they have practical advantages, but it seems that this possibility has not been explored.) As with the Theil–Sen estimator, consider any two points such that $X_i < X_{i'}$ and again let $b_{1ii'}$ be the corresponding slope given by Equation (13.7). For convenience we let K represent the total number of slopes that can be computed in this manner. In the event all n of the X values are distinct, there are a total of $K = n(n-1)/2$ such slopes. Next, let

$$v_j = Y_j - b_{1ii'}X_j,$$

$j = 1, \ldots, n$. That is, for each of the n points and the slope corresponding to X_i and $X_{i'}$ ($b_{1ii'}$), compute the difference between Y_j and $b_{1ii'}X_j$, ($j = 1, \ldots, n$) and label the results $v_1, \ldots, v_n$. Let $S_{ii'}$ be some measure of scale based on the v values just computed, repeat this process for all K slopes, and let $S_{\min}$ be the smallest of the corresponding $S_{ii'}$ values. The final estimate of the slope is the value of $b_{1ii'}$ corresponding to $S_{\min}$, which we label b_1. The intercept is taken to be

$$b_0 = M_y - b_1 M_x,$$

where M_x and M_y are the medians of the X and Y values, respectively. This will be called *method STS*. (As with the Theil–Sen estimator, the Gauss–Seidel method is used to handle multiple predictors.)

13.3.2 S-PLUS Function stsreg

The S-PLUS function

$$\text{stsreg(x,y,sc=pbvar,...)}$$

computes the STS estimator (the S-type modification of the Theil–Sen estimator) just described. The arguments x and y are used as described in Section 13.2.2. The argument sc indicates which measure of scale will be applied to the residuals and defaults to the percentage bend midvariance. The argument ... can be replaced by arguments related to the chosen measure of scale, sc.

> **EXAMPLE.** For the data in Table 13.1, the S-PLUS function stsreg returns $b_1 = 490.5$ and $b_0 = 56,809$. In contrast, the Theil–Sen estimate of the slope is 756.6, the only point being that these two methods can give substantially different results.   ■

13.3.3 An Extension of S-Type Estimators

An appeal of the STS estimator in Section 13.2.3 is that it can provide a reasonable fit to the majority of the points in situations where many other estimators provide a poor fit instead. There are exceptions, but there are situations where it has practical value. However, even when this estimator gives a reasonably good fit to the majority of points, it can provide a relatively poor estimate of the true slope when the error term is heteroscedastic. That is, its standard error can be relatively high compared to other estimators that might be used.

One approach toward this problem is to use the STS estimator in Section 13.2.3 as a preliminary fit to data, with the goal of detecting points with unusually large residuals. Such points can result in a poor fit to the bulk of the observations and a highly inaccurate estimate of the true slope (the slope we would get if all individuals could be measured). So one strategy is first to check for points that have unusually large residuals, called *regression outliers*, to remove them, and then to fit a line to the data that remain. (For an extensive comparison of methods for detecting regression outliers when there is homoscedasticity, see Wisnowski, Montgomery, & Simpson, 2001.) One specific strategy for fitting a line to data after outliers are removed is simply to use the least squares estimator covered in Chapter 6. But this approach can perform poorly when the error term is heteroscedastic and should be used with caution. The Theil–Sen estimator performs relatively well when there is heteroscedasticity, little accuracy is lost, versus least squares, in the event that the error term is homoscedastic, so here the Theil–Sen estimator is used to fit a line to data once regression outliers are removed.

The specific method used to detect regression outliers is applied as follows. First, fit a line to the data using the S-type modification of the Theil–Sen estimator (the STS estimator) described in Section 13.3.1. Let b_1 and b_0 be the resulting estimates of the slope and intercept and let

$$r_i = Y_i - b_1 X_1 - b_0,$$

$(i = 1, \ldots, n)$ be the usual residuals. Let M_r be the median of the residuals and let MAD_r be the median of the values $|r_1 - M_r|, \ldots, |r_n - M_r|$. Then the ith point (X_i, Y_i) is declared a regression outlier if

$$|r_i - M_r| > \frac{2(MAD_r)}{.6745}. \tag{13.9}$$

The final estimate of the slope and intercept is obtained by applying the Theil–Sen estimator to those points not declared regression outliers. When there are

p predictors, again compute the residuals based on STS and use Equation (13.9) to eliminate any points with large residuals. This will be called *method TSTS*.

13.3.4 S-PLUS Function tstsreg

The estimator just described is computed by the S-PLUS function

$$tstsreg(x, y, sc = pbvar, \ldots),$$

which was written for this book. The argument sc indicates which measure of scale will be used when method STS is employed to detect regression outliers, and the default measure of scale is the percentage bend midvariance.

13.3.5 Least Median of Squares (LMS) Estimator

For the general case of p predictors, the *least median of squares* (LMS) estimator simply chooses the values $b_0, b_1, \ldots, b_p$ so as to minimize the median of the squared residuals. That is, choose $b_0, b_1, \ldots, b_p$ so as to minimize

$$\text{MED}\left(r_1^2, \ldots, r_n^2\right),$$

the median of the values $r_1^2, \ldots, r_n^2$, where

$$r_i = Y_i - b_0 - b_p X_{ip} \ldots, b_1 X_{i1},$$

$(i = 1, \ldots, n)$ are the residuals.

The LMS estimator has the highest possible breakdown point, .5. Its main use has been as an exploratory tool, such as when trying to detect regression outliers. Its standard error generally compares poorly to many other estimators, and it can give a poor fit to the majority of points, so this method should be used cautiously. However, despite its many shortcomings, situations do arise where in the preliminary stages of analysis it gives a reasonable fit to data when many other methods do not.

13.3.6 S-PLUS Function lmsreg

The LMS (least median of squares) estimator can be computed with the built-in S-PLUS function

$$lmsreg(x, y).$$

13.3.7 Least Trimmed Squares (LTS) Estimator

Rather than minimize the sum of the squared residuals or the median of the squared residuals, another approach is to minimize the sum of the trimmed squared residuals instead. That is, now the slope and intercept are taken to be the values that minimize

$$\sum_{i=1}^{h} r_{(i)}^2, \tag{13.10}$$

where $r_{(1)}^2 \leq r_{(2)}^2 \leq \ldots \leq r_{(n)}^2$ are the squared residuals written in ascending order. This is called the *least trimmed squares* (LTS) estimator. Typically, $b = [n/2] + 1$ is used to achieve the highest possible breakdown point, which is approximately .5. (The notation $[n/2]$ means that $n/2$ is rounded down to the nearest integer.)

Consideration has been given to increasing the efficiency of the LTS estimator by lowering its breakdown point. That is, investigations have been conducted to see whether increasing b in the previous paragraph results in situations where the standard error of the LTS estimator competes more favorably with other estimators that might be used. In particular, consideration has been given to using $b = [(1 - \gamma)n]$ with $\gamma = .2$ and .25 (Wilcox, 2001b). Generally, LTS has a smaller standard error than least squares when the error term is sufficiently heteroscedastic and $b = [n/2] + 1$ is used. Switching to $\gamma = .2$ or .25 does not appear to improve its performance appreciably, so this variation of the LTS estimator is not considered henceforth.

13.3.8 S-PLUS Function ltsreg

The built-in S-PLUS function

$$\text{ltsreg}(x,y)$$

computes the LTS (least trimmed squares) estimator just described with $b = [n/2] + 1$, the highest amount of trimming.

13.3.9 Least Trimmed Absolute (LTA) Value Estimator

A close variation of the LTS estimator is the least trimmed absolute (LTA) value estimator. Rather than choose the intercept and slope so as to minimize Equation (13.10), the goal is to minimize

$$\sum_{i=1}^{b} |r|_{(i)}, \tag{13.11}$$

where $|r|_{(i)}$ is the ith smallest absolute residual and b is as defined as in Section 13.3.7. (For recent results on the LTA estimator, see Hawkins & Olive, 1999.) Like LTS, the LTA estimator can have a much smaller standard error than the least squares estimator, but its improvement over the LTS estimator seems to be marginal at best, at least based on what is currently known. (For some comparisons of LTA and LTS, see Wilcox, 2001b.)

13.3.10 S-PLUS Function ltareg

The S-PLUS function

$$\text{ltareg}(x,y)$$

computes the LTA estimate of the slope and intercept.

13.3.11 Deepest Regression Line

The outlier detection methods covered in Section 13.1 are based in part on measuring how deeply a point is embedded in a scatterplot. Today there are comparable (numerical) methods for measuring how deeply a line is embedded in a scatterplot. This leads to yet another method of fitting a line to data: Search for the line that is most deeply nested within a cloud of points. Rousseeuw and Hubert (1999) have examined how this might be done and provided an algorithm for implementing the technique. Box 13.3 provides a brief outline of how this approach is applied, but complete details go well beyond the scope of this book. A negative feature is that the software provided with this book allows one predictor only.

BOX 13.3 A Brief Outline of the Deepest Regression Line Estimator

Let b_1 and b_0 be any choice for the slope and intercept, respectively, and let r_i $(i = 1, \ldots, n)$ be the corresponding residuals. Roughly, this candidate fit is called a nonfit if you can find some partition or splitting of the X values such that all of the residuals for the lower X values are negative (positive) but that for all of the higher X values the residuals are positive (negative). So, for example, if all of the points lie above a particular straight line, in which case all of the residuals are positive, this line is called a nonfit. More formally, a candidate fit is called a nonfit if and only if a value for v can be found such that

$$r_i < 0 \qquad \text{for all } X_i < v$$

and

$$r_i > 0 \qquad \text{for all } X_i > v$$

or

$$r_i > 0 \qquad \text{for all } X_i < v$$

and

$$r_i < 0 \qquad \text{for all } X_i > v.$$

The regression depth of a fit (b_1, b_0) relative to $(X_1, Y_1), \ldots, (X_n, Y_n)$ is the smallest number of observations that need to be removed to make (b_1, b_0) a nonfit. The deepest regression estimator corresponds to the values of b_1 and b_0 that maximize regression depth.

13.3.12 S-PLUS Function depreg

The S-PLUS function

$$depreg(x,y)$$

computes the deepest regression line. A negative feature of this function is that execution time can be relatively high compared to many other estimators you might use.

13.4 Comments on Choosing a Regression Estimator

Again it is stressed that no single regression estimator is optimal in all situations. However, this does not mean that choosing an estimator is an academic matter. The standard least squares estimator covered in Chapter 6 is satisfactory when variables are independent. But when there is an association, different methods can lead to drastically different conclusions, and the routine use of least squares, to the exclusion of all other estimators that might be used, can be disastrous. Although authorities are not in agreement as to which method is best, familiarity with their relative merits might help the reader choose a method. For some purposes, particularly in the exploratory phases of an investigation, multiple methods might be considered. However, when testing hypotheses, there is the problem of controlling the probability of a Type I error. If multiple tests are performed using different regression estimators, the familywise error rate can be controlled along the lines covered in Chapter 12. A simple strategy would be to use the Bonferroni method in Section 12.8.1; or a sequentially rejective method might be used as described in Section 12.8.2. However, a consequence of such adjustments is lower power versus testing a single hypothesis.

Given the goal of finding a straight line that gives a good fit to the bulk of the points, a minimum requirement is that an estimator have a reasonably high finite-sample breakdown point. This eliminates least squares plus the first of the two M-estimators covered in Section 13.2.3. However, having a high finite-sample breakdown point is not sufficient to guarantee a good fit to the majority of points. That is, some estimators can be highly influenced by a few unusual points even though they have a high breakdown point. Said yet another way, some estimators guard against complete disaster, meaning that a few unusual values cannot result in estimates that are arbitrarily large (or small). But some of these estimators might poorly reflect the association among the vast majority of the data, depending on where the outliers are located. So as a general rule, regardless of which estimator is used, it seems prudent always to graphically check how well a line fits the data.

EXAMPLE. For the data in Figure 13.2, the least squares estimate of the slope is -0.02. In contrast, both TS and LTA estimate the slope to be -0.28, the MGV estimate is -0.69, and the STS estimate is -0.8, the only point being that the choice of an estimator can make a practical difference. In this particular case, LTS and the Coakley–Hettmansperger estimate give results similar to least squares. As was indicated in Section 13.1.2 (see Figure 13.3), the rightmost six points are outliers. If we eliminate these outliers by restricting the range of X to values less than 100, the least squares estimate is now -0.48, with a significance level less than .001, and the Coakley–Hettmansperger estimate is -0.64.

Continued

> **EXAMPLE.** (*Continued*) However, simply restricting the range of the X values does not necessarily deal with problems due to heteroscedasticity and outliers among the Y values. (Restricting the range of the Y values and applying standard hypothesis-testing methods results in using the wrong standard error.) Even without restricting the range, the MGV and STS estimates suggest that the association might be more negative for the bulk of the observations than indicated by least squares when the range of X is restricted. ■

Next it is illustrated that although robust estimators can be substantially more accurate than least squares, situations do arise where some robust estimators are greatly influenced by a few outliers, despite having a high breakdown point. That is, it can make a difference which robust method is used, and switching to some robust method does not necessarily eliminate the need to check the fit to data graphically.

> **EXAMPLE.** Twenty points were generated as in Figure 13.4, so the true slope is 1. Then two aberrant points were added to the data at $(X, Y) = (2.1, -2.4)$. Figure 13.14 shows a scatterplot of the points plus the LTS regression line, which has an estimated slope of -0.94. So in this case, LTS is a complete disaster in terms of detecting how the majority of the points were generated. The least squares estimate is -0.63. The Coakley–Hettmansperger estimator relies on LTS as an initial estimate of the slope, and despite its high breakdown point the estimate of the slope is -0.65. In contrast, the MGV estimate of the slope is $.97$, nearly equal to the true slope, 1. The deepest regression line estimate is $.66$, and the STS estimate performs poorly in this particular case, the estimate being -0.98. The LMS estimate of the slope is 1.7, so it performs poorly as well in this particular instance. Again, this is not to suggest that LMS, LTS, STS, and the Coakley–Hettmansperger estimators be excluded from consideration. Rather, the point is that despite the robust properties they enjoy, they can perform poorly in some situations where other methods do well. Also, although MGV does very well here, this is not to suggest that it be used to the exclusion of all other methods. ■

In Figure 13.14, it is evident that the rightmost point is an outlier, and eliminating it by restricting the range of the X values improves matters considerably for the estimators that performed poorly in this particular case. But it cannot be stressed too strongly that there is more to robust regression than simply restricting the range of the X values, as will be illustrated.

To add perspective, the process used to generate the data in Figure 13.14 was repeated 500 times, and estimates of the slope were computed using least squares, the M-estimator with Schweppe weights (estimator bmreg), the Coakley–Hettmansperger estimator (chreg), the Theil–Sen estimator (tsreg), and the deepest regression line. Boxplots of the results are shown in Figure 13.15. Notice that the

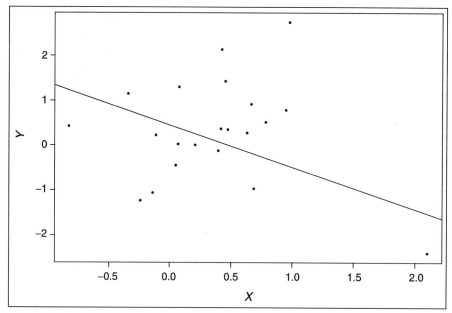

FIGURE 13.14   Twenty points where the true slope is 1, plus two outliers. The solid line is the LTS regression line, which performs poorly in this particular case.

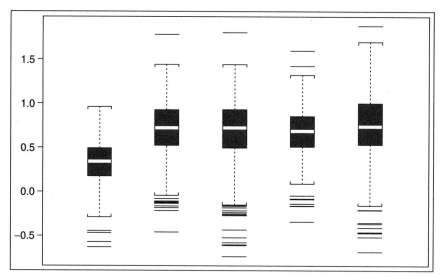

FIGURE 13.15   Boxplots of the estimated slope using (from left to right) least squares, M-estimator (with Schweppe weights), the Coakley–Hettmansperger estimate, Theil–Sen, and the deepest regression line. The true slope is 1, but all five estimators are influenced by two outliers.

medians of all these estimators differ from 1, the value being estimated. This illustrates that these estimators can be sensitive to a type of *contamination bias*. That is, despite having a reasonably high finite-sample breakdown point, it is possible for a few unusual points to result in a poor fit to the bulk of the observations. So these

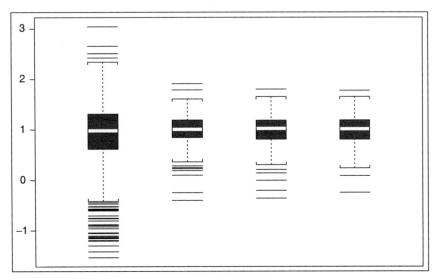

FIGURE 13.16 As in Figure 13.15, the only estimators used here are LTS with a breakdown point of .5, LTS with a breakdown point of .2, TSTS, and MGV. All four estimators have a median value approximately equal to the true slope, despite the outliers.

estimators can provide substantial advantages versus least squares, but they do not eliminate all practical concerns.

Figure 13.16 shows the results when using LTS with a breakdown point of .5, LTS with a breakdown point of .2, TSTS, and the MGV estimator in Section 13.2.5. (The LTA estimator gives results similar to LTS.) In contrast to the estimators in Figure 13.15, the median of all the estimators is approximately 1. So in this particular situation, these estimators do a better job of avoiding contamination bias. Note that there is considerably more variation among the LTS estimates based on a breakdown point of .5. This illustrates that under normality, this estimator is substantially less accurate than the others, at least on average. Again, this is not to suggest that LTS has no practical value.

There is some evidence that the STS estimator generally gives a better fit to the majority of points versus LTS and LMS. In particular, it seems common to encounter situations where STS is less affected by a few aberrant points. However, exceptions occur, as is illustrated next, so again it seems that multiple methods should be considered.

EXAMPLE. Figure 13.17 shows 20 points that were generated in the same manner as in Figure 13.14. So the two aberrant points located at $(X, Y) = (2.1, -2.4)$ are positioned relatively far from the true regression line, which again has a slope of 1 and an intercept of zero. Also shown in Figure 13.17 are the STS, LMS, and MGV estimates of the regression line. In this particular case, STS performs poorly; the estimated slope is $-.23$. The estimated slope

Continued

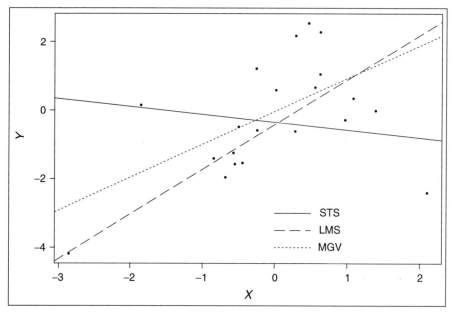

FIGURE 13.17   Illustration that some robust estimators can give substantially different results from others, depending on where the outliers happen to be.

EXAMPLE. (*Continued*) and intercept based on the MGV estimator are 0.96 and −0.03, respectively, which are closer to the true values than are the other estimates considered here. The estimated slope based on the TSTS estimator is .58, and least squares returns an estimate of .4. So once again we see that the choice of which robust estimator to use can make a substantial difference in how the association between X and Y is summarized.   ■

To provide more perspective about the relative merits of various robust estimators, some comparisons are made of their small-sample accuracy versus the least squares estimator, including situations where there is heteroscedasticity. This is done in terms of the standard error of the least squares estimate of the slope divided by the standard error of some competing estimator. So if, for example, this ratio is less than 1, least squares tends to be more accurate. A ratio of .5, for instance, means that least squares has a standard error that is only half as large, and a ratio of 3 indicates that it is three times as large instead. Here it is assumed that $Y = X + \lambda(X)\epsilon$, where the function $\lambda(X)$ reflects heteroscedasticity. Setting $\lambda(X) = 1$ corresponds to the homoscedastic case. Table 13.2 shows estimates of these ratios for the estimators TS, MGVF, MGV, the deepest regression line estimator (T^*), and TSTS, where VP 1 corresponds to $\lambda(X) = 1$, VP 2 is where $\lambda(X) = X^2$, and VP 3 is

$$\lambda(X) = \left(1 + \frac{2}{(|X| + 1)}\right)\epsilon. \ [1]$$

---

[1] The estimated ratios of the standard errors in Table 13.2 are based on simulations with 5000 replications.

TABLE 13.2 Estimated Ratios of Standard Errors, X Distribution Symmetric, $n = 20$

X	ϵ	VP	TS	MGVF	MGV	T^*	TSTS
N	N	1	0.91	0.91	0.91	0.76	0.88
		2	2.64	2.64	2.62	3.11	2.36
		3	202.22	201.18	196.31	187.89	135.70
N	SH	1	4.28	4.28	4.27	4.42	3.51
		2	10.67	10.66	10.94	11.03	8.66
		3	220.81	220.29	214.31	228.59	121.35
N	AL	1	1.13	1.13	1.13	0.92	1.05
		2	3.21	3.21	3.21	3.69	2.84
		3	183.74	182.69	177.53	146.89	106.70
N	AH	1	8.89	8.84	8.85	16.41	7.05
		2	26.66	26.59	27.07	25.81	20.89
		3	210.37	209.85	204.25	182.20	103.04
SH	N	1	0.81	0.80	0.72	0.61	0.76
		2	40.57	40.55	42.30	55.47	27.91
		3	41.70	40.54	34.44	40.08	22.57
SH	SH	1	3.09	3.08	2.78	2.88	2.41
		2	78.43	78.41	83.56	90.84	47.64
		3	38.70	38.03	31.93	45.29	17.80
SH	AL	1	0.99	0.98	0.87	0.73	0.90
		2	46.77	46.74	49.18	63.60	31.46
		3	39.32	38.68	32.70	31.70	19.76
SH	AH	1	6.34	6.31	5.64	6.75	4.62
		2	138.53	138.49	146.76	108.86	78.35
		3	43.63	43.22	37.34	39.38	18.40

N = normal; SH = symmetric, heavy-tailed; AL = asymmetric, light-tailed; AH = asymmetric, heavy-tailed.

So for VP 2, the error term has more variance, corresponding to extreme X values, and VP 3 is a situation where the error term has more variance when the value of X is near its median. The results are limited to situations where the distribution for X is symmetric; but very similar results are obtained when X has an asymmetric distribution instead. In Table 13.2, the distributions for ϵ are taken to be normal (N), symmetric and heavy-tailed (SH), asymmetric and relatively light-tailed (AL), and asymmetric and relatively heavy-tailed (AH). (For precise information about these four distributions, see Wilcox, 2001b.)

The results in Table 13.2 can be roughly summarized as follows. If the error term is homoscedastic and simultaneously has a light-tailed distribution, the ordinary least squares estimator covered in Chapter 6 competes well against the alternative estimators considered here. However, as the distribution of the error term becomes more heavy-tailed (meaning outliers become more common), the least squares estimator becomes unsatisfactory. Moreover, if the error term is sufficiently heavy-tailed, the least squares estimator is disastrous, even when there is homoscedasticity. In fact, even when the error term is normal but heteroscedastic, the least squares estimator can be highly unsatisfactory. Note that for VP 3, situations arise where the least squares estimator has a standard error more than 100 times larger than competing estimators.

Said another way, if two variables are independent, then power is not an issue when testing the hypothesis that the slope is zero (because the null hypothesis is true). However, if the variables are dependent, then the results in Table 13.2 suggest that the choice of estimator can make a substantial difference in terms of the likelihood of detecting a true association.

For the situations in Table 13.2, the MGV and MGVF estimators do not seem to have a striking advantage over the other robust estimators considered. It is stressed, however, that while several estimators compete well with least squares, it seems to be easy to find fault with any estimator that has been proposed. For example, the execution time required for MGV and MGVF increases substantially as the sample gets large, which makes them impractical for many situations based on the software and computer hardware currently available. Also, MGVF tends to have faster execution time than MGV, so the results in Table 13.2 might seem to suggest using MGVF over MGV. However, situations can be constructed where MGV gives a good fit to the bulk of the points when all of the other estimators give a unsatisfactory fit instead. Again, it seems that no single estimator is always ideal.

It is noted that the M-estimators in Section 13.2.3 also offer an advantage over least squares when there is heteroscedasticity. If there is reasonable certainty that the Coakley–Hettmansperger estimator is not being affected by contamination bias, it is a possible option and is fairly fast in terms of execution time.

So which estimator should be used? A rough strategy might be as follows. First plot the points and then plot the estimated regression line using several of the estimators considered in this book. (The S-PLUS command

$$\text{abline}(\text{mgvreg}(x,y)\$\text{coef}),$$

for example, will plot the line corresponding to the MGV estimator.) The goal at this point is merely to make sure that a reasonable fit to the data is obtained. (At this stage, it is strongly recommended that a smooth of the data, described in Section 14.1, be checked as well.) In some cases, simply restricting the range of the X values might improve the fit and provide relatively short confidence intervals for the slope and intercept. If several estimators appear to give an adequate fit, then one strategy is to choose the estimator that will have relatively high power and relatively short confidence intervals, assuming of course that probability coverage will be adequate. The results in Table 13.2 illustrate that the choice of estimator might make a practical difference, and it will be illustrated that in applied work this is indeed the case. Although no single estimator is optimal in all situations, there seems to be general agreement on the worst possible strategy: Apply standard least squares regression and assume all is well. At a minimum, use an estimator with a reasonably high breakdown point. Also, it is strongly recommended that an estimator perform reasonably well (in terms of achieving a relatively low standard error) when there is heteroscedasticity.

13.5 Hypothesis Testing and Confidence Intervals

Confidence intervals for the slope and intercept can be computed with a percentile bootstrap method with bootstrap samples generated as described in Section 7.3. When there is one predictor, if we observe $(X_1, Y_1), \ldots, (X_n, Y_n)$, a bootstrap sample is obtained by resampling with replacement n pairs of these points, with each pair of points having probability $1/n$ of being resampled. The resulting bootstrap sample will be labeled $(X_1^*, Y_1^*), \ldots, (X_n^*, Y_n^*)$. When there are p predictors, resample n vectors of observations from

$$(X_{11}, \ldots, X_{1p}, Y_1), \ldots, (X_{n1}, \ldots, X_{np}, Y_n).$$

Next, compute the slope and intercept based on one of the robust estimators in this chapter, applied to the bootstrap sample just obtained, and repeat this process B times. If we label the resulting estimates of the slope as $b_{11}^*, \ldots, b_{1B}^*$, then an approximate $1 - \alpha$ confidence interval for the slope is given by the middle 95% of these bootstrap estimates:

$$\left(b_{1(\ell+1)}^*, b_{1(u)}^* \right),$$

where $\ell = \alpha B/2$, rounded to the nearest integer, and $u = B - \ell$. If $\hat{p}^*$ is the proportion of bootstrap estimates greater than zero, the significance level (when testing $H_0 : \beta_1 = 0$) is

$$2 \min \left(\hat{p}^*, 1 - \hat{p}^* \right).$$

A confidence interval and significance level for the intercept, and the other slope coefficients when $p > 1$, can be computed in the same manner.

13.5.1 S-PLUS Function regci

The S-PLUS function

$$\text{regci(x,y,regfun=tsreg,nboot=599,alpha=.05)}$$

computes percentile bootstrap confidence intervals using the method just described. As indicated, the default regression estimator is tsreg (the Theil–Sen estimator). This default estimator was chosen based on its ability to improve substantially upon the least squares estimator and because its execution time is fairly fast. But it is stressed that situations arise where some other estimator might offer a better fit to the majority of the data and yield substantially shorter confidence intervals.

> **EXAMPLE.** The diabetes data in Exercise 15 of Chapter 6 serve to illustrate the output from regci when the least trimmed squares estimator is used. Assuming the data are stored in the S-PLUS variables x and y, the command regci(x,y,regfun=ltsreg) returns
>
> *Continued*

EXAMPLE. (*Continued*)

```
$regci:
              [,1]         [,2]
[1,]    1.22786964    1.64758295
[2,]   -0.00727429    0.03702609

$sig.level:
[1]    0.0000000    0.1866667

$se:
[1]    0.11215894    0.01120903
```

So the .95 confidence interval for the intercept is (1.23, 1.65) with a significance level of 0.0, and the confidence interval for the slope is (−0.007, .04) with a significance level of 0.187. The estimated standard errors for the intercept and slope are 0.11 and 0.011, respectively. ■

EXAMPLE. Sometimes restricting the range of X values eliminates outliers that mask an association or create contamination bias, but exceptions are encountered. Consider again the diabetes data. One goal was to understand how age is related to C-peptide concentrations, and here we focus on children under the age of seven (for reasons motivated by results covered in Chapter 14). The MGV and MVE outlier detection methods in Sections 13.1.5 and 13.1.3 find one outlier. However, to eliminate it by restricting the range of the X values would entail eliminating points that are not flagged as outliers. (The MCD outlier detection method finds an additional outlier, but to eliminate both outliers by restricting the range of X values would again entail eliminating points that are not flagged as outliers.) If we simply apply the S-PLUS function regci using the default Theil–Sen estimator, the .95 confidence interval is (0.0, .11), with a significance level of .023. So with $\alpha = .05$ we reject the hypothesis of a zero slope. Using the Coakley–Hettmansperger estimator, the significance level is .063, so now we fail to reject. As for the S-estimator in Section 13.3.1, the significance level is .07. Using least squares in conjunction with the method in Chapter 7, the .95 confidence interval is (−0.001, .106), so we fail to reject, although we come close when testing at the .05 level. With the MGV estimator, the .95 confidence interval is (.004, .14) with a significance level of .013, about half the significance level based on Theil–Sen, which merely illustrates that the choice of method can make a practical difference. ■

EXAMPLE. Figure 6.3 shows some star data where the least squares regression line provides a poor fit to the bulk of the observations. It was noted that if we

Continued

EXAMPLE. (*Continued*) eliminate the five points having the smallest X values (values less than 4), which are flagged as outliers among the X values using the boxplot methods in Chapter 3, a much better fit to the data is obtained. If we compute a .95 confidence interval for the slope using least squares and the percentile bootstrap method in Section 7.3, we get (1.85, 3.95). Using the Theil–Sen estimator (tsreg) or the MGVFREG estimator, the .95 confidence interval based on the S-PLUS function regci is (2.11, 4.00). So in this particular case we get a slightly longer confidence interval using least squares. The least squares estimate of the slope is 2.93. The Theil–Sen estimate, as well as the estimate returned by MGVFREG, is 3.07, and MGVREG estimates the slope to be 3.22. However, if we use the outlier detection rule in Section 3.4.2 (based on MAD and M), an additional outlier among the X values is found: 4.01. If we restrict X to those values greater than 4.01, the least squares estimate of the slope increases slightly to 2.98 and Theil–Sen increases to 3.22, the same value returned by MGVREG. ■

EXAMPLE. Here is another illustration that restricting the range of the X values can decrease the strength of an association. Fifty values for the error term (ϵ) were generated on a computer from a standard normal distribution, the same was done for X, and $Y = .35X + \epsilon$ was computed, yielding 50 values for Y. Pearson's correlation between X and Y was found to be .26 with a significance level of .026 using Student's T-test of $H_0 : \rho = 0$ described in Chapter 6. Restricting the range of X to values less than 1.3 results in eliminating five values, and the significance level increases to .094. Restricting the range of X so that $|X| \leq 1.5$ results in the elimination of six points with the significance level now equal to .14. ■

EXAMPLE. Table 6.3 reports data on aggression in the home (X) versus a measure of cognitive functioning among children (Y). In Figure 6.9, it was illustrated that if we split the data into two groups according to whether X has a value greater than or equal to 25, we get two strikingly different impressions about the association between X and Y. For $X \geq 25$ the least squares estimate of the slope is close to zero, but for $X < 25$ the estimated slope is -0.19. In Section 7.3 we saw that the least squares slope is significantly different from zero (based on the modified percentile bootstrap method) when using all of the data. So a case might be made that there is a negative association between X and Y up to about $X = 25$ and that for $X \geq 25$ there appears to be little or no association at all. But is it adequate to describe the association as linear and negative for $X < 25$, or is it more accurate to say that the association is primarily due to heteroscedasticity? Figure 13.18 shows the plot created by the

Continued

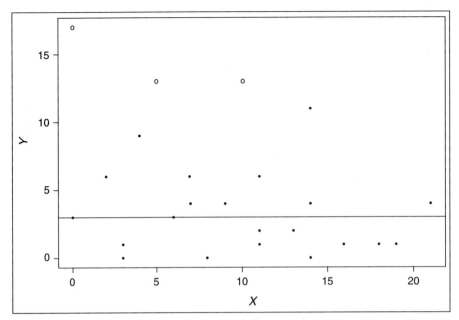

FIGURE 13.18   For the aggression data, with $X < 25$, the horizontal line is the MGV regression line, which suggests that the typical value of Y changes very little with X and that the apparent dependence between X and Y is perhaps better explained by other perspectives (such as heteroscedasticity).

EXAMPLE. (*Continued*) S-PLUS function outmgv when attention is restricted to $X < 25$. The horizontal line is the regression line based on the MGV estimator. (Theil–Sen, the Coakley–Hettmansperger estimator, and MGVF return estimated slopes of -0.11, -0.13, and -0.11, respectively.) If we use the method in Section 8.10.1 to compare the variance of the Y values corresponding to $X < 25$ versus $X \geq 25$, the .95 confidence interval for the difference between the variances is $(2.49, 32.34)$, so we reject at the .05 level of significance, indicating heteroscedasticity. In this particular case, our ability to reject hinges on using a measure of scale that is sensitive to outliers. If, for example, we use the percentage bend midvariance instead (a robust measure of scale described in Chapter 3), now we fail to reject. Also, with $X < 25$, the least squares regression line is no longer significantly different from zero; the .95 confidence interval (based on the modified bootstrap method in Chapter 7) is $(-0.51, 0.17)$. Using Theil–Sen, in conjunction with the S-PLUS function regci in Section 13.5.1, we get a .95 confidence interval of $(-0.43, 0.11)$ and again we fail to reject. Of course, failing to reject the hypothesis of a zero slope might be due to low power resulting from a reduction of the sample size.   ■

In summary, there is evidence that recall scores are associated with aggression in the home, but it seems that some care must be taken when describing what this association might be. In particular, there is evidence of decreasing variability among the Y values as X gets large, in the sense that relatively large Y values become less likely as X increases. In terms of typical Y values as X increases, again there are some indications of a negative association. But it seems that if the apparent negative association is real, it is weaker in some sense compared to the decrease in variation and number of outliers among the Y values as X gets large. For example, the estimated slope based on the MGV estimator is exactly equal to zero for the data in Figure 13.18. That is, *typical* recall scores are estimated to have no association with aggression in the home; but despite this, relatively high recall test scores appear to become less likely as X gets large.

EXAMPLE. Figure 13.19 shows a scatterplot of the same points shown in Figures 13.2 and 13.3. As previously noted, the least squares estimate of the slope is highly nonsignificant, but it is fairly evident the six rightmost points are outliers. If we ignore these outliers by restricting attention to points having $X < 100$, the least squares regression line is the dashed line in Figure 13.19, which has an estimated slope of -0.48, and the .95 confidence interval for the slope (using the modified percentile bootstrap method in Section 7.3) is $(-0.87, -0.17)$. The Theil–Sen estimate of the slope is -0.6, and the bootstrap confidence interval for the slope (using the S-PLUS function regci) is $(-0.90, -0.27)$. In this particular case, both methods yield confidence intervals having approximately the same length. ■

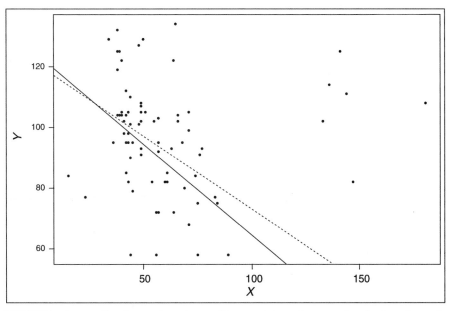

FIGURE 13.19   For the reading data in Figure 13.2, outliers can be eliminated simply by restricting the range of X values. In this particular case, various methods give similar confidence intervals for the slope based on the remaining data.

We have seen that in some situations, restricting the range of the X values can do a reasonable job of salvaging least squares. But it is important to realize that simply restricting the range of the X values does not always salvage least squares, especially when computing a confidence interval or testing hypotheses. In particular, situations arise where least squares yields an estimated slope similar to other robust estimators, but the length of the confidence interval can be substantially larger.

EXAMPLE. Consider the data in Figure 13.20, which is from the reading study mentioned in connection with Figure 13.2. The solid, nearly horizontal line is the least squares regression line, and the dashed line is based on the TSTS estimator in Section 13.3.3. It is evident that the upper six points are outliers, but if we simply eliminate them and apply the conventional methods in Chapter 6 to test the hypothesis of a zero slope, we are using the wrong standard error. Still using least squares, if a .95 bootstrap confidence interval for the slope is computed as described in Chapter 7, we get $(-0.417, 0.414)$. As is evident, this confidence interval is not remotely close to rejecting the hypothesis of a zero slope. In contrast, the .95 confidence interval based on the Theil–Sen estimator (using the S-PLUS function regci) is $(-0.428, 0)$ — about half the length of the confidence interval based on least squares — with a significance level of .056; the estimated slope is -0.23. The MGV estimator estimates the slope to be -0.35 and the TSTS estimate is -0.426. The .95 confidence interval based on

Continued

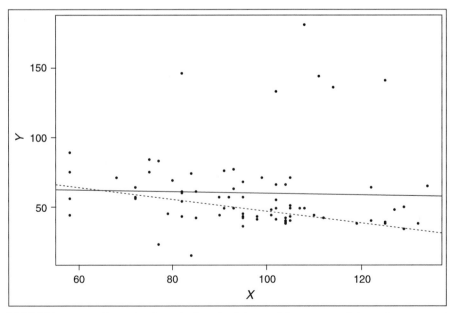

FIGURE 13.20   Example where the length of the confidence interval depends in a crucial way on the regression estimator used.

> **EXAMPLE.** (*Continued*) this last estimator is $(-0.6, -0.125)$ with a significance level of .003. The .95 confidence interval based on the MGV estimator is $(-0.56, -0.125)$ with a significance level less than .001. ■

13.6 Robust Measures of Correlation

This section summarizes some robust analogs of Pearson's correlation, r. We begin with two relatively simple measures of association and then consider how they might be improved.

13.6.1 Winsorized Correlation

The Winsorized correlation coefficient is obtained by Winsorizing the n pairs of observations as described in Section 11.2 and in Box 11.3. (For technical reasons, trimming is a less satisfactory approach to defining a robust correlation.) The Winsorized correlation between X and Y is just Pearson's correlation applied to the Winsorized values. The resulting correlation coefficient will be labeled r_w.

Letting ρ_w be the population analog of r_w, it can be shown that when X and Y are independent, $\rho_w = 0$. If we assume independence — implying homoscedasticity — a simple test of

$$H_0 : \rho_w = 0$$

is to compute

$$T_w = r_w \sqrt{\frac{n-2}{1-r_w^2}}. \tag{13.12}$$

Let

$$\nu = n - 2g - 2,$$

where, as in Chapter 3, $g = [\gamma n]$ and γ is the amount of Winsorizing. (Remember that $[\gamma n]$ means to compute γn and then to round down to the nearest integer.) Then reject if $|T_w| \geq t_{1-\alpha/2}$, the $1-\alpha/2$ quantile of Student's T-distribution with ν degrees of freedom. Setting the amount of Winsorizing to zero (i.e., using $\gamma = 0$) reduces T_w to the test statistic covered in Section 6.5.

13.6.2 Percentage Bend Correlation

A criticism of the Winsorized correlation is that the amount of Winsorizing is fixed in advance rather than determined by your data. One solution is to use what is called the *percentage bend correlation*, r_{pb}, which is computed as described in Box 13.4. Under independence, the population value of r_{pb}, ρ_{pb}, is zero. To test $H_0 : \rho_{pb} = 0$, compute

$$T_{pb} = r_{pb}\sqrt{\frac{n-2}{1-r_{pb}^2}} \tag{13.13}$$

and reject if $|T_{pb}| \geq t_{1-\alpha/2}$, where $t_{1-\alpha/2}$ is the $1 - \alpha/2$ quantile of Student's T-distribution with $n - 2$ degrees of freedom.

BOX 13.4 How to Compute the Percentage Bend Correlation

For the observations $X_1, \ldots, X_n$, let $\hat{\theta}$ be the sample median. Choose a value for β between 0 and 1 and compute

$$W_i = |X_i - \hat{\theta}|,$$
$$m = [(1 - \beta)n],$$

where the notation $[(1 - \beta)n]$ is $(1 - \beta)n$ rounded down to the nearest integer. Using $\beta = .2$ appears to be a good choice in most situations. (The value of β determines the finite-sample breakdown point of a measure of scale used to detect outliers.) Let $W_{(1)} \leq \cdots \leq W_{(n)}$ be the W_i values written in ascending order and let

$$\hat{\omega}_x = W_{(m)}.$$

Let i_1 be the number of X_i values such that $(X_i - \hat{\theta})/\hat{\omega}_x < -1$ and let i_2 be the number of X_i values such that $(X_i - \hat{\theta})/\hat{\omega}_x > 1$. Compute

$$S_x = \sum_{i=i_1+1}^{n-i_2} X_{(i)}$$

$$\hat{\phi}_x = \frac{\hat{\omega}_x(i_2 - i_1) + S_x}{n - i_1 - i_2}.$$

Set $U_i = (X_i - \hat{\phi}_x)/\hat{\omega}_x$. Repeat these computations for the Y_i values, yielding $V_i = (Y_i - \hat{\phi}_y)/\hat{\omega}_y$. Let

$$\Psi(x) = \max[-1, \min(1, x)].$$

Set $A_i = \Psi(U_i)$ and $B_i = \Psi(V_i)$. The percentage bend correlation is estimated to be

$$r_{pb} = \frac{\sum A_i B_i}{\sqrt{\left(\sum A_i^2\right)\left(\sum B_i^2\right)}}.$$

13.6.3 S-PLUS Functions wincor and pbcor

The S-PLUS functions

$$\text{wincor}(x,y,tr=.2) \quad \text{and} \quad \text{pbcor}(x,y,beta=.2)$$

compute the Winsorized and percentage bend correlations. The argument tr determines how much Winsorizing is done and defaults to .2. This function also tests $H_0 : \rho_w = 0$ using T_w given by Equation (13.12). Setting tr=0 results in the conventional test based on Pearson's correlation, which is described in Section 6.5. The function pbcor tests the hypothesis of independence using Equation (13.13).

The practical advantages of the Winsorized and percentage bend correlation are that they limit the influence of a few unusual X values, and they do the same for the Y values. However, a concern is that neither one takes into account the overall structure of the data, as is illustrated next.

EXAMPLE. Look again at the scatterplot in Figure 13.4 and recall that Pearson's correlation is $r = .68$. Figure 13.5 shows the same data but with two unusual values added at $(X, Y) = (2.1, -2.4)$, which causes r to drop to .21. For the data in Figure 13.4, the Winsorized correlation is $r_w = .59$ with a significance level of .011 when testing $\rho_w = 0$ with Equation (13.12). But in Figure 13.5, the Winsorized correlation is only $r_w = .33$ with a significance level of .14. As for the percentage bend correlation, in Figure 13.4, $r_{pb} = .57$ with a significance level of .008; but in Figure 13.5, $r_{pb} = .25$ with a significance level of .26. Of course, one might argue that the association between X and Y is weaker in Figure 13.5. Certainly this is true in some sense, but simultaneously it is erroneous to conclude that there is no association. ■

13.6.4 Heteroscedastic Bootstrap Confidence Intervals for Robust Correlations

Chapter 6 noted that heteroscedasticity can result in undesirable properties when using the conventional method for testing the hypothesis that Pearson's correlation is zero. These same concerns apply when using the methods in Sections 13.6.1 and 13.6.2. When these methods reject, it is reasonable to conclude that the variables under study are dependent, but the reason for rejecting remains a bit unclear. One strategy is to replace the methods in Sections 13.6.1 and 13.6.2 with techniques that allow heteroscedasticity. This can be done by using a bootstrap to compute a confidence interval for some robust correlation coefficient. Essentially, proceed as described in Section 13.5 but with the estimated slope replaced by the correlation of interest.

13.6.5 S-PLUS Function corb

The S-PLUS function

$$\text{corb}(x,y,corfun=pbcor,nboot=599)$$

tests the hypothesis of a zero correlation using a heteroscedastic percentile bootstrap method. By default, the percentage bend correlation is used.

EXAMPLE. For the reading data in Figure 13.2, the test of a zero percentage bend correlation, based on Equation (13.12), has a significance level of .014. Is heteroscedasticity playing a role here? Using a percentile bootstrap method instead, a .95 confidence interval for the percentage bend correlation is $(-0.63, -0.19)$; this interval does not contain zero, so again we reject.   ■

13.6.6 Correlation with Outliers Removed

We saw in Section 13.1.5 that simply discarding outliers using the MVE or MCD methods and then applying Pearson's correlation to the remaining data can mask an association. We have also seen that the MGVF outlier detection method can miss outliers. However, it generally detects any blatant outliers, so another strategy is first to check for outliers using MGVF, to remove any that are found, and then to apply the percentage bend correlation to the data that remain. By using the percentage bend correlation, you guard against any outliers that were not found by MGVF but that might have an inordinate influence on the correlation. However, a criticism of this approach is that there are no published papers on how one might test the hypothesis of independence based on the resulting correlation coefficient. A bootstrap method might give good control over the probability of a Type I error, but more research is needed before this approach can be recommended.

Yet another strategy is to remove outliers with the projection method described in Section 13.1.9 and to compute Pearson's correlation on the data that remain. One appeal of this method is that under normality, the resulting estimate of the population correlation coefficient, ρ, is relatively accurate. That is, if our criterion is to get a good estimate of ρ in the event data follow a bivariate normal distribution, this is one of the few methods that performs reasonably well. Also, there is a simple test of a zero correlation that performs well under homoscedasticity.

Let r_p be Pearson's correlation based on the data not flagged as outliers using the projection method. Let

$$T_p = r_p \sqrt{\frac{n-2}{1-r_p^2}}.$$

Then, when testing at the .05 level, reject the hypothesis of zero correlation when $|T_p| \geq c$, where

$$c = \frac{6.947}{n} + 2.3197.$$

The value of r_p can be computed quickly with $n = 100$, but as the sample size increases, execution time will get high with the software currently available. There are types of heteroscedasticity where the method does not control the probability of a Type I error. However, if the only goal is to control the probability of a Type I error when there is independence, all indications are that T_p is satisfactory (Wilcox, 2002).

13.6.7 S-PLUS Functions ocor and scor

The S-PLUS function

$$\text{ocor}(x, y, \text{corfun} = \text{pbcor}, \text{outfun} = \text{outmgvf}, \text{pcor} = F, \text{plotit} = F)$$

computes a correlation coefficient by removing outliers detected by the function outfun, which defaults to the MGVF method, and then applying the function indicated by the argument corfun. By default, corfun is the percentage bend correlation. To use the Winsorized correlation, set corfun=wincor. To apply Pearson's correlation instead, set the argument pcor to T, or you can use corfun=pcor. To plot the data, set plotit to T.

The S-PLUS function

$$\text{scor}(x, y = \text{NA}, \text{plotit} = T)$$

computes Pearson's correlation after outliers, identified with the S-PLUS function outpro, are removed. If no values for y are specified, it is assumed that x is a matrix with p columns of data. The function tests the hypothesis of a zero correlation at the .05 level. If plotit=T is used, the same plot created by outpro is produced. (Methods for controlling FWE, the familywise error rate, have not been investigated as yet.)

EXAMPLE. If the star data in Figure 13.12 are stored in the S-PLUS variables starx and stary, the command

$$\text{ocor}(\text{starx}, \text{stary})$$

returns a correlation of .63. This is in contrast to Pearson's correlation, $r = -.21$. If we eliminate points where starx is less than 4.1, now Pearson's correlation is $r = .65$. Often, restricting the range of X values gives similar results to robust methods, but this is not always the case. Figure 13.21 shows the plot created by scor (which is the same plot created by outpro). Here, $r_p = .68$, $|T_p| = 6.26$, and the .05 critical value is 2.47, so reject and conclude there is dependence. The command

$$\text{ocor}(\text{starx}, \text{stary}, \text{outfun} = \text{out})$$

removes outliers with the MVE method and returns a value of .71, but now no simple test of a zero correlation has been derived. (A bootstrap method might be used, but this has not been investigated as yet.) So in some instances using MVE (or MCD) can lower the correlation, but situations arise where it increases the correlation as well. ■

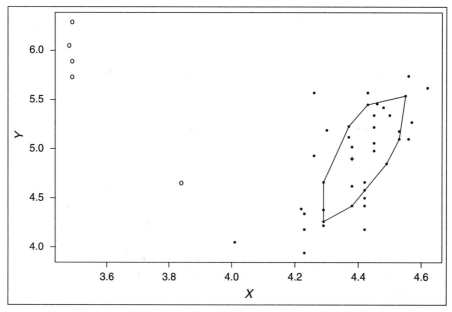

FIGURE 13.21   Plot created by the S-PLUS function scor based on the star data.

13.6.8 Explanatory Power

This section describes a general approach to measuring association based on what is called *explanatory power*. Roughly, it measures the variation among the predicted Y values (based on X) relative to the variation of Y when the predictor X is ignored. Let $S(Y)$ indicate some (population) measure of variability. For example, $S(Y)$ might be the variance of Y (σ_y^2), or it could be the Winsorized variance, the percentage bend midvariance, or even MAD. Let $\hat{Y}$ be any method for predicting Y given some value for X. For example, $\hat{Y}$ might be the least squares estimate of Y given X, or it could be any of the robust regression estimators described in this chapter. In its most general form *explanatory power* is defined to be

$$\eta^2 = \frac{S(\hat{Y})}{S(Y)}. \tag{13.14}$$

If we use the sample variance to measure variation and least squares regression to predict Y given X, then η^2 is just ρ^2, the square of Pearson's correlation, which is the population value of the coefficient of determination covered in Chapter 6. Note that η^2 can be turned into a correlation coefficient by taking the square root and multiplying by the sign of the estimated slope. That is, if the regression line has a negative slope, make η negative as well; otherwise, η is taken to be positive.

To provide some graphical intuition about explanatory power, look at the left panel of Figure 13.22, which shows 10 points all falling on the straight line having slope 1 and intercept zero. As is evident, $\hat{Y} = Y = X$, and the variance of both the Y and the $\hat{Y}$ values is 115.2. That is, the explanatory power is 1. In the right panel of

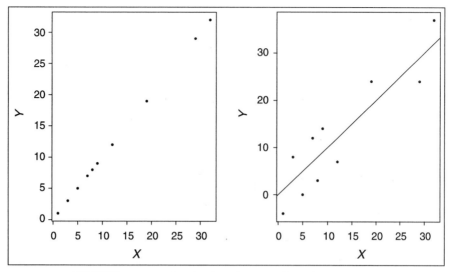

FIGURE 13.22   Graphical explanation of explanatory power.

Figure 13.21, 10 points are centered around the straight line having slope 1 and intercept zero, but they are not on the line. The predicted Y values (the $\hat{Y}_i$ values) fall on the straight line and again have variance 115.2. In contrast, the observed Y values have variance 159.6, so the explanatory power is $115.2/159.6 = .72$.

A natural estimate of η^2, based on the least squares regression line and the usual sample variance, is

$$\hat{\eta}_1^2 = \frac{\sum (\hat{Y}_i - \tilde{Y})^2}{\sum (Y_i - \bar{Y})^2}, \tag{13.15}$$

where $\tilde{Y} = \sum \hat{Y}_i/n$, the average of the predicted Y values. A little algebra shows that this estimate is just r^2, the coefficient of determination. However, results in Doksum and Samarov (1995) as well as Wilcox (2000a) suggest estimating explanatory power by computing the squared correlation between the values $(Y_i, \hat{Y}_i)$. That is, estimate η^2 with

$$\hat{\eta}_2^2 = (\mathrm{COR}(\hat{Y}, Y))^2. \tag{13.16}$$

If the goal is to use the variance as a measure of explanatory power, then use Pearson's correlation. If instead you want to use a Winsorized variance, use the Winsorized correlation, and use the percentage bend correlation for the percentage bend midvariance.

Before ending this section, it is remarked that yet another approach to measuring association stems from the following connection between Pearson's correlation and the least squares regression line:

$$r = b_1 \frac{s_x}{s_y}. \tag{13.17}$$

So a reasonable generalization is

$$r_g = b_g \frac{\hat{\tau}_x}{\hat{\tau}_y},\tag{13.18}$$

where now b_g is any estimate of the slope and $\hat{\tau}_x^2$ is an estimate of some measure of variation among the X values, such as the Winsorized variance or the percentage bend midvariance described in Chapter 3.

13.6.9   S-PLUS Functions epow and ecor

The S-PLUS function

$$\text{epow}(x,y, \text{pcor} = F, \text{regfun} = \text{tsreg}, \text{corfun} = \text{pbcor}, \text{outkeep} = F,$$

$$\text{outfun} = \text{outmgvf}, \text{varfun} = \text{pbvar}, \text{op} = T)$$

estimates explanatory power (η^2) as just described. (When there are p predictors, x is assumed to be an n-by-p matrix.) If you want to use Pearson's correlation between $\hat{Y}$ and Y to estimate explanatory power, set the argument pcor to T; otherwise the function uses the measure of correlation specified by corfun, which defaults to the percentage bend correlation. The argument regfun indicates which regression estimator will be used; the default method is the Theil–Sen regression line. The measure of variation (varfun) defaults to the percentage bend midvariance. The argument op controls how η^2 is estimated. By default (op=T), the squared correlation between the $\hat{Y}$ and Y values is used. Should a situation arise where there is some reason to use Equation (13.15), use op=F. (The argument varfun is ignored if op=T.) The argument outkeep indicates whether outliers are to be retained when estimating explanatory power. If this argument is F, the data are checked for outliers using the function outfun (which defaults to the method in Section 13.1.7) and outliers are removed.

The S-PLUS function

$$\text{ecor}(x,y,\text{pcor} = F, \text{regfun} = \text{tsreg}, \text{corfun} = \text{pbcor}, \text{outkeep} = F,$$

$$\text{outfun} = \text{outmgvf})$$

computes the explanatory power correlation coefficient. The arguments are the same as those used by epow. The function ecor assumes that x is a vector or a matrix having a single column of numbers. This is in contrast to epow, which allows x to be a matrix with multiple predictors.

EXAMPLE. For the data in Figure 13.19, first consider the situation where outliers are kept. The command

$$\text{epow}(x,y,\text{outkeep}=T)$$

Continued

EXAMPLE. (*Continued*)
returns a value of only .078. However, restricting the range of X to values less than 100 increases the estimate substantially to .23. If you use instead the command epow(x,y) so that again the range of X values is not restricted, outliers are removed and now the estimate is .28. That is, by removing outliers, in effect the range of X values is restricted, except that now we get a slightly larger measure of explanatory power because even when attention is restricted to $X < 100$, a few outliers are detected. So roughly, there is a fair amount of explanatory power among the points clustered together, but there are outliers that, when included, reduce the explanatory power considerably. Because the slope of the regression line is negative, if we were to convert the estimate .078 into a correlation, we would get $-1\sqrt{.078} = -.28$, which is the value returned by the S-PLUS function ecor. In this particular case, the percentage bend correlation is also equal to $-.28$. ■

EXAMPLE. Again the aggression data in Table 6.3 are analyzed, except that now all of the data are used and consideration is given to fitting the data with $Y = \beta_0 + \beta_1 X + \beta_2 X^2$. Figure 13.23 shows a scatterplot of the data plus the least squares regression line (the dashed line in the figure) and the regression line based on the Coakley–Hettmansperger estimator. For both methods,

Continued

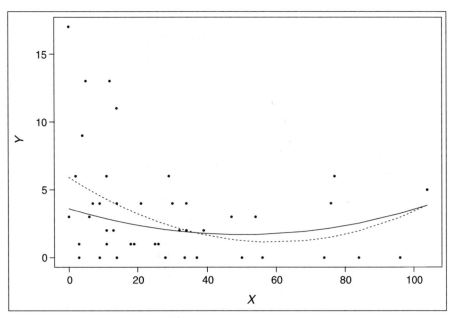

FIGURE 13.23 Quadratic regression line fit to the aggression data.

EXAMPLE. (*Continued*) the hypothesis $H_0 : \beta_2 = 0$ is not rejected. The explanatory power based on least squares is .098; for the Coakley–Hettmansperger estimator, in conjunction with the percentage bend variance, it is .051. If the quadratic term is ignored, the estimates are now .088 and .040, respectively, suggesting a modest improvement in the fit when the quadratic term is included.   ■

13.6.10 Detecting Associations in Contingency Tables

Section 2.3 introduced the notion of a contingency table. Special methods for analyzing contingency tables have been developed, and there are many books devoted exclusively to this topic. These books are typically classified under the analysis of categorical data, recent examples of which are Agresti (1990, 1996), Andersen (1997), and Powers and Xie (1999). Although it is impossible to cover the many important issues here, a few comments about detecting associations might help.

Consider a two-by-two contingency table. To be concrete, look at the hypothetical data in Table 13.3, which deals with blood pressure and personality types. For example, there were eight individuals who were classified as having a type A personality and simultaneously having high blood pressure. A fundamental issue is determining whether personality type and blood pressure are dependent. An approach to this problem is to test the assumption that they are independent. If you are able to reject, you conclude that they are dependent, and this means that knowing whether a person has a type A personality provides you with some information about that person's probability of having high blood pressure.

For the four cells in Table 13.3, let n_{11} be the number of individuals falling in the first row and first column. In Table 13.3, $n_{11} = 8$. More generally, n_{ij} represents the entry in the ith row and jth column. So, for example, $n_{21} = 5$. Let

$$n_{i+} = n_{i1} + n_{i2} \qquad \text{and} \qquad n_{+j} = n_{1j} + n_{2j}.$$

In the example, $n_{1+} = 8 + 67 = 75$ and $n_{+2} = 87$. The total number of observations is represented by n, which is 100 in Table 13.3.

A classic test of the hypothesis of independence is based on

$$X^2 = \frac{n(n_{11}n_{22} - n_{12}n_{21})^2}{n_{1+}n_{2+}n_{+1}n_{+2}}.$$

TABLE 13.3   Hypothetical Results on Personality versus Blood Pressure

Personality	Blood pressure		Total
	High	Not high	
A	8	67	75
B	5	20	25
Total	13	87	100

When the null hypothesis of independence is true, X^2 has, approximately, a chi-squared distribution with 1 degree of freedom. For the more general case where there are R rows and C columns,

$$X^2 = \sum_{i=1}^{R} \sum_{j=1}^{C} \frac{n \left(n_{ij} - \frac{n_{i+} n_{+j}}{n} \right)^2}{n_{i+} n_{+j}}$$

and the degrees of freedom are

$$\nu = (R - 1)(C - 1).$$

Now

$$n_{i+} = \sum_j n_{ij} \quad \text{and} \quad n_{+j} = \sum_i n_{ij}.$$

If X^2 equals or exceeds the $1 - \alpha$ quantile of a chi-squared distribution with ν degrees of freedom, which is read from Table 3 in Appendix B, you reject. In the illustration

$$X^2 = \frac{100[8(20) - 67(5)]^2}{75(25)(13)(87)} = 1.4.$$

With $\nu = 1$ degree of freedom and $\alpha = .05$, the critical value is 3.84; because $1.4 <$ 3.84, you fail to reject. This means that you are unable to detect any dependence between personality type and blood pressure. Generally, the chi-squared test of independence performs reasonably well in terms of Type I errors (e.g., Hosmane, 1986), but difficulties can arise, particularly when the number of observations in any of the cells is relatively small. For instance, if any of the n_{ij} values is less than or equal to 5, problems might occur in terms of Type I errors. There are a variety of methods for improving upon the chi-squared test, but details are not given here. Interested readers can refer to Agresti (1990).

It is noted that measuring the strength of the association in a contingency table is a nontrivial matter. An early approach was to use the so-called *phi coefficient*:

$$\phi = \frac{X}{\sqrt{n}}.$$

But this measure, plus all other functions of X^2, have been found to have little value as measures of association (e.g., Fleiss, 1981). Effective methods have been devised, but the details go beyond the scope of this book.

13.7 Exercises

1. The following data were collected from 29 lakes in Florida by the U.S. Environmental Protection Agency (and are taken from Stromberg, 1993).

NIN:	5.548	4.896	1.964	3.586	3.824	3.111	3.607	3.557	2.989	18.053
	3.773	1.253	2.094	2.726	1.758	5.011	2.455	0.913	0.890	2.468	4.168
	4.810	34.319	1.531	1.481	2.239	4.204	3.463	1.727

TW:	0.137	2.499	0.419	1.699	0.605	0.677	0.159	1.699	0.340	2.899
	0.082	0.425	0.444	0.225	0.241	0.099	0.644	0.266	0.351	0.027	0.030
	3.400	1.499	0.351	0.082	0.518	0.471	0.036	0.721

NIN is the average influent nitrogen concentration and TW is the water retention time. Plot the points, and verify that six points are declared outliers by the MVE and MCD methods but that the MGV outlier detection method in Section 13.1.5 detects only two outliers, namely, points 10 and 23. (In this particular case, the MGV method will flag the same points as the MVE and MCD methods if, in the the argument list of outmgv, the option se=T is used. This option standardizes the data.)

2. In the study mentioned in the previous exercise, there was a third variable of interest, Y: the mean annual nitrogen concentration. One particular goal was to understand how Y is related to NIN and TW. The Y values are

    2.590 3.770 1.270 1.445 3.290 0.930 1.600 1.250 3.450 1.096 1.745
    1.060 0.890 2.755 1.515 4.770 2.220 0.590 0.530 1.910 4.010
    1.745 1.965 2.550 0.770 0.720 1.730 2.860 0.760

    Create a scatterplot of Y versus NIN. Verify that the least squares regression line has a slope of .012, and plot this line with the S-PLUS command

    abline(lsfit(NIN,Y)$coef),

    assuming that the data are stored in the S-PLUS variables NIN and Y. As will be evident, the two largest NIN values are outliers. Verify that if they are eliminated, then now the least squares regression line has a slope of .56. Plot this line.

3. Repeat the previous exercise, except use TW to predict Y and verify that the six rightmost points are outliers.

4. The results in the previous exercise suggest restricting the range of TW values to those less than 1. Verify that if we eliminate these values and the corresponding Y values, and if the S-PLUS function out is used to check for outliers among the scatterplot of the remaining Y and TW values, there are outliers among the Y values. Also verify that the least squares estimate of the slope is -1.975 versus -1.88 using Theil–Sen.

5. In the previous exercise, why would it be improper to eliminate the outliers among the Y values by restricting the range of Y values and computing a confidence interval for the slope using the conventional method in Section 6.3.1?

6. In the previous two exercises, why might some robust estimator provide a substantially different confidence interval for the slope than the conventional method in Section 6.3.1? What must be done to determine whether a robust

method does indeed give a shorter confidence interval? Apply this method using Theil–Sen and the MGV estimators.

7. The star data in Figures 6.3 and 13.12 are examined again, only now we consider the problem of predicting the surface temperature given the light intensity. The data are

Y:	4.37 4.56 4.26 4.56 4.30 4.46 3.84 4.57 4.26 4.37 3.49 4.43 4.48 4.01 4.29 4.42 4.23 4.42 4.23 3.49 4.29 4.29 4.42 4.49 4.38 4.42 4.29 4.38 4.22 3.48 4.38 4.56 4.45 3.49 4.23 4.62 4.53 4.45 4.53 4.43 4.38 4.45 4.50 4.45 4.55 4.45 4.42
X:	5.23 5.74 4.93 5.74 5.19 5.46 4.65 5.27 5.57 5.12 5.73 5.45 5.42 4.05 4.26 4.58 3.94 4.18 4.18 5.89 4.38 4.22 4.42 4.85 5.02 4.66 4.66 4.90 4.39 6.05 4.42 5.10 5.22 6.29 4.34 5.62 5.10 5.22 5.18 5.57 4.62 5.06 5.34 5.34 5.54 4.98 4.50

Plot the data, and note that not all outliers can be eliminated simply by restricting the range of the X values. Using all of the data, verify that the .95 confidence interval for the slope, using least squares and the modified percentile bootstrap method in Chapter 7, is $(-0.33, 0.14)$ and so you fail to reject. Then verify that when using the Theil–Sen estimate and the S-PLUS function regci, the .95 confidence interval is $(0.0, 0.18)$ with a significance level of .017. Finally, verify that when using the MGV estimator, the .95 confidence interval is $(0.05, 0.22)$ with a significance level of .013.

8. For the data in the previous exercise, the least trimmed squares estimate of the slope is .169, the STS estimator in Section 13.2.3 estimates the slope to be .149, and the MGV estimate is .161. It might be argued that these estimates are similar enough that it makes no practical difference which estimator is used. What is wrong with this argument?

9. Verify the results in Section 13.2.2.

10. The MGV outlier detection method can fail to declare points outliers that are clearly unusual and flagged as outliers by other methods. Why is this desirable?

11. For the data in Exercise 7, the MGVF estimate of the slope is only .107 and the Coakley–Hettmansperger estimate is .13. The latter estimator has the highest possible breakdown point, .5. Both of these estimates are smaller than the least trimmed squares and the MGV estimates reported in Exercise 8. What might explain this?

12. For the data in Table 13.1, the percentage bend correlation is $r_{pb} = .9367$. Verify that you reject $H_0 : \rho_{pb} = 0$ with $\alpha = .05$.

13. The earthquake data in Table 13.4 were taken from a brochure published by the Southern California Earthquake Center. For the first two variables (magnitude and fault length) check for outliers among the data using the MVE, MCD, projection, and MGV methods. (Use the S-PLUS functions out, outpro, and outmgv, and be sure to examine the plots created by these functions.) Verify that the MCD method identifies five points as outliers, the MVE method identifies four, and the MGV and projection methods identify

TABLE 13.4　Earthquake Data

Magnitude	Fault length (kilometers)	Duration (seconds)
7.8	360	130
7.7	400	110
7.5	75	27
7.3	70	24
7.0	40	7
6.9	50	15
6.7	16	8
6.7	14	7
6.6	23	15
6.5	25	6
6.4	30	13
6.4	15	5
6.1	15	5
5.9	20	4
5.9	6	3
5.8	5	2

only two. Imagine we want to predict fault length based on magnitude. If we used the strategy of removing outliers and fitting a line to the remaining data, why might the MGV method be preferable when fitting a regression line to the bulk of the observations?

14. Repeat the previous exercise, but now use magnitude versus duration.

15. For the data in Table 13.4, remove the two points with the largest magnitude and check for outliers when examining magnitude and fault length with duration ignored. Why might it be a poor idea to remove outliers and use the MGV regression estimator on the data that remain?

16. For the data in Table 13.4, based on a plot of magnitude versus fault length, would you expect Pearson's correlation to be larger or smaller than the correlation returned by the S-PLUS function ocor in Section 13.6.7? Check your response by actually computing the correlations.

17. The percentage bend correlation between magnitude and fault length is .897. Verify that you reject the hypothesis of a zero correlation with $\alpha = .05$.

18. The average LSAT scores (X) for the 1973 entering classes of 15 American law schools, and the corresponding grade point averages (Y), are as follows.

X: 576 635 558 578 666 580 555 661 651 605 653 575 545 572 594

Y: 3.39 3.30 2.81 3.03 3.44 3.07 3.00 3.43 3.36 3.13 3.12 2.74 2.76 2.88 2.96

Create a boxplot for the X values and note that no outliers are found. Do the same for the Y values. Now verify that the function relplot also finds no outliers but that outliers are detected when using MGVF, MVE, and MCD.

19. For the data in the previous exercise, verify that the .95 confidence interval for the slope, using least squares and the bootstrap method in Section 7.3.1, is $(0.0022, 0.0062)$ and that the .95 confidence interval for the slope using the S-PLUS function regci in Section 13.5.1 (with the Theil–Sen estimator) is $(0.0031, 0.0061)$. Note that the ratio of the lengths of the confidence interval is

$$\frac{.0062 - .0022}{.0061 - .0031} = 1.33.$$

Based on the results in the previous exercise, what might explain why the Theil–Sen estimator provides a substantially shorter confidence interval?

MORE REGRESSION METHODS

Regression is an extremely vast topic that is difficult to cover in a single book let alone a few chapters. The goal in this chapter is to cover some modern methods that are particularly relevant in applied work, with the understanding that many issues and techniques are not discussed.

14.1 Smoothers

One way of describing a fundamental goal in regression is to say that for any given value of X, we want to estimate some (conditional) measure of location associated with Y. In Chapter 6, for example, where the standard least squares regression model was introduced [given by Equation (6.6)], the assumption is that the conditional mean of Y, given X, is $\beta_1 X + \beta_0$. More formally,

$$E(Y|X) = \beta_0 + \beta_1 X,$$

a result that is implied by assuming that the error term (ϵ) has a mean of zero. That is, the mean of Y is assumed to have a linear association with X. Often this assumption provides a reasonable summary of the data, but it is common to encounter situations where it is highly unsatisfactory. This section describes some exploratory methods, called *smoothers*, that attempt to estimate a regression line without forcing it to have a particular shape, such as a straight line. There are many smoothers, two of which are described in this section. (S-PLUS has at least six built-in smoothers.) Readers interested in more details about smoothers are referred to Hastie and Tibshirani (1990).

14.1.1 Cleveland's Smoother

Suppose we want to estimate the mean of Y given that $X = 6$ based on n pairs of observations: $(X_1, Y_1), \ldots, (X_n, Y_n)$. One strategy is to focus on the observed X values close to 6 and use the corresponding Y values to estimate the mean of Y.

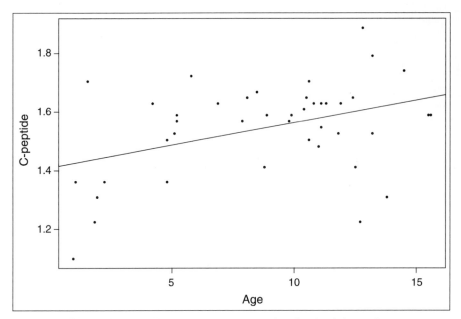

FIGURE 14.1   Least squares regression line for the diabetes data.

A specific technique for implementing this strategy was derived by Cleveland (1979) and is known as a *locally weighted running-line smoother*. The idea is that, although a regression line might not be linear over the entire range of X values, it will be approximately linear over small intervals of X. That is, for X values close to 6, say, a linear regression line might perform reasonably well and can be used to estimate Y given that $X = 6$. But for X values far from 6, some other regression estimate of Y should be used instead.

To be more concrete, look at Figure 14.1, which shows the logarithm of C-peptide concentrations in children versus their age. (The data are given in Exercise 15 of Chapter 6 and are taken from a study on diabetes in children.) The solid straight line is the least squares regression line. Cleveland's strategy begins by measuring how far away each X_i value is from the point of interest. If you want to predict or estimate the typical C-peptide concentration for a 10-year-old you measure how far away each X_i is from 10, while for an 11-year-old you measure how far away each X_i is from 11. The idea is to give less credence to those who are relatively far away from the age that is of interest. In fact, those individuals furthest away are ignored altogether. In particular, given the goal of estimating the mean of Y corresponding to some specific value for X, you measure the distance of X from each of the observed X_i values with

$$\delta_i = |X_i - X|.$$

For example, if the goal is to predict C-peptide concentrations for 10-year-old children, then $X = 10$; because $X_1 = 5.2$, $X_2 = 8.8$, and $X_3 = 10.5$, it follows that $\delta_1 = |5.2 - 10| = 4.8$, $\delta_2 = |8.8 - 10| = 1.2$, $\delta_3 = |10.5 - 10| = 0.5$, and so forth. When predicting C-peptide levels for 11-year-old children, $X = 11$, $\delta_1 = |5.2 - 11| = 5.8$, $\delta_2 = |8.8 - 11| = 2.2$, and so on.

Next, sort the δ_i values and retain the pn pairs of points that have the smallest δ_i values, where p is a number between 0 and 1. The value of p represents the proportion of points used to predict Y and is generally referred to as the *span*. For the moment, suppose $p = 1/2$. In the illustration, $n = 43$, so this means that you retain 22 of the pairs of points that have X_i values closest to $X = 10$. These 22 points are the *nearest neighbors* to $X = 10$. If you want to predict C-peptide concentrations for 11-year-old children, you retain the 22 pairs of points with X_i values closest to $X = 11$. Let δ_m be the maximum value of the δ_i values that are retained. For $X = 10$ and $p = 0.5$, $\delta_m = 2.7$. (The details are left as an exercise.) Set

$$Q_i = \frac{|X - X_i|}{\delta_m};$$

if $0 \le Q_i < 1$, set

$$w_i = \left(1 - Q_i^3\right)^3,$$

otherwise set

$$w_i = 0.$$

Finally, use *weighted least squares* to predict Y using w_i as weights (cf. Fan, 1992). That is, determine the values b_1 and b_0 that minimize

$$\sum w_i (Y_i - b_0 - b_1 X_i)^2$$

and estimate the mean of Y corresponding to X to be $\hat{Y} = b_0 + b_1 X$. Because the weights (the w_i values) change with X, generally a different regression estimate of Y is used when X is altered. Finally, let $\hat{Y}_i$ be the estimated mean of Y given that $X = X_i$ based on the method just described. Then an estimate of the regression line is obtained by the line connecting the points $(X_i, \hat{Y}_i)$ $(i = 1, \ldots, n)$ and is called a *smooth*.

The span, p, controls the raggedness of the smooth. If p is close to 1, we get a straight line stronger even when there is curvature. If p is too close to zero, an extremely ragged line is obtained instead. By choosing a value for p between .2 and .8, curvature can usually be detected and a relatively smooth regression line is obtained.

14.1.2 S-PLUS Function lowess

The built-in S-PLUS function

$$\text{lowess}(x, y, p = 2/3)$$

computes Cleveland's smoother, just described. The value for p, the span, defaults to 2/3. You can create a scatterplot of points that contains this smooth with the S-PLUS commands

$$\text{plot}(x, y)$$

$$\text{lines}(\text{lowess}(x, y))$$

If the line appears to be rather ragged, try increasing p to see what happens. If the line appears to be approximately horizontal, indicating no association, check to see what happens when p is lowered.

EXAMPLE. For the diabetes data in Figure 14.1, testing the hypothesis that the slope is zero with the conventional method in Chapter 6 returns a significance level of .008. The heteroscedastic, .95 confidence interval for the slope, based on the modified bootstrap method in Section 7.3, is $(-0.001, .029)$, and the bootstrap test of independence covered in Section 7.8 rejects at the .05 level. So although the confidence interval based on the modified bootstrap method is not quite able to reject at the .05 level (because the .95 confidence interval contains 0), there are indications that age and C-peptide concentrations are dependent, and based on Figure 14.1 it might seem that C-peptide concentrations tend to increase with age.

Now look at the solid line in Figure 14.2, which is the smooth created by lowess. Notice that it increases up to about the age of 7 and then flattens out. That is, this smooth suggests that there is a positive association up to about age seven, and then the association seems to disappear. However, some caution is warranted in this particular case. The dashed line shows the smooth when X values less than 4 are ignored; it is nearly horizontal. Because there are only six points with $X < 4$, it is difficult to know how age is related to C-peptide concentrations for very young children. The smooth suggests that generally, there is little or no association; it hints that this might not be true for children under the age of 4, but more data in this age range are needed to understand the association when $X < 4$. Generally, the ends of a smooth must be viewed with caution, because data are often sparse in these regions and there is an inherent

Continued

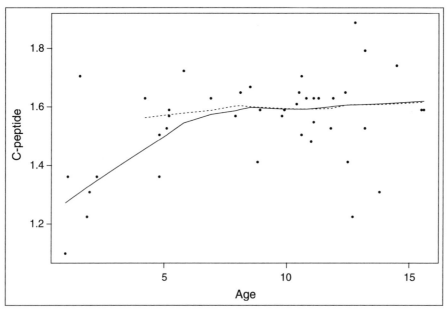

FIGURE 14.2 Two smooths for the diabetes data based on the S-PLUS function lowess. The dashed line is the smooth ignoring points with $X < 4$.

EXAMPLE. (*Continued*) bias associated with these points. For example, if $X = 3$ is the smallest observed X value, estimating $E(Y|X = 3)$ is hampered by the fact that the estimate is based on X values that are greater than or equal to 3. Ideally, a smooth would be based on X values that are both less than and greater than 3.   ■

<h2>14.2 Smooths Based on Robust Measures of Location</h2>

A natural guess at how to extend Cleveland's method to robust measures of location is to replace the weighted least squares estimator with one of the robust regression estimators covered in Chapter 13. However, this strategy is known to be highly unsatisfactory. The reason is that robust regression estimators can be insensitive to curvature, so the resulting smooth often misses curvature when in fact it exists. A better strategy is to use what is called the *running-interval smoother*. To estimate some measure of location for Y, given some value for X, a running interval smoother searches for *all* points close to the value of X that are of interest and then simply computes the measure of location based on the corresponding Y values. Note that the number of X_i values close to X will depend on the value of X. In contrast, Cleveland's method uses the k nearest points, with k fixed and chosen in advance.

To elaborate, compute MAD based on the X values and label it MAD_x. Let f be some constant that is chosen in a manner to be described and illustrated. Then the value X is said to be close to X_i if

$$|X_i - X| \leq f\left(\frac{\text{MAD}_x}{.6745}\right).$$

So for normal distributions, X is close to X_i if X is within f standard deviations of X_i. Now consider all of the Y_i values corresponding to the X_i values that are close to X. Then an estimate of the typical value of Y, given X, is the estimated measure of location based on the Y values just identified. For example, if six X_i values are identified as being close to $X = 22$ and the corresponding Y values are 2, 55, 3, 12, 19, and 21, then the estimated mean of Y, given that $X = 22$, would be the average of these six numbers: 18.7. The estimated 20% trimmed mean of Y, given that $X = 22$, would be the trimmed mean of these six values, which is 13.75.

A running-interval smoother is created as follows. For each X_i, determine which of the X_j values are close to X_i, compute a measure of location associated with the corresponding Y_j values, and label this result $\hat{Y}_i$. So we now have the following n pairs of numbers: $(X_1, \hat{Y}_1), \ldots, (X_n, \hat{Y}_n)$. The running-interval smooth is the line formed by connecting these points. The span, f, controls how ragged the line will be. As with Cleveland's method, if f is too close to 1, the smooth will be a straight line, even when there is curvature; if f is too close to zero, the result is a very ragged line.

14.2.1 S-PLUS Functions rungen and runmean

The S-PLUS function

$$\text{rungen}(x,y,\text{est}=\text{mom},\text{fr}=.8,\text{plotit}=T,\text{scat}=T,\text{pyhat}=F,\dots)$$

computes a running interval smooth assuming that there is only one predictor. (For multiple predictors, see Sections 14.2.3 and 14.2.4.) The argument est determines the measure of location that is used and defaults to MOM. The argument fr corresponds to the span (f) and defaults to .8. The function returns the $\hat{Y}_i$ values if the argument pyhat is set to T. By default, a scatterplot of the points is created with a plot of the smooth. To avoid the scatterplot, set scat=F. If there is specific interest in using a trimmed mean, the function

$$\text{runmean}(x,y,\text{fr}=.8,\text{tr}=.2,\text{pyhat}=F)$$

is a bit more convenient to use. As usual, the argument tr controls the amount of trimming and defaults to 20%.

EXAMPLE. The left panel of Figure 14.3 shows the smooth created by rungen for the reading data shown in Figures 13.2 and 13.8. The dashed line is the smooth created by lowess. In this particular case both methods give similar results, but strategically placed outliers can cause lowess to give a distinctly different impression about the association between two variables. To illustrate that curvature can be detected correctly, the right panel shows the smooth returned by runmean for 40 points generated on a computer where $Y = X^2 + \epsilon$, with both X and ϵ having standard normal distributions.   ■

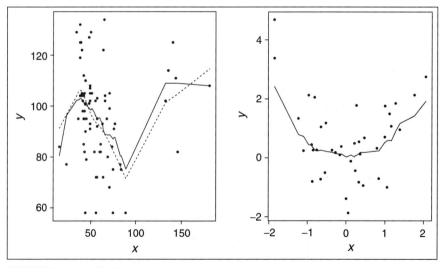

FIGURE 14.3   The left panel shows the smooths for the reading data based on lowess (the dashed line) and rungen. The right panel shows the smooths for data where $Y = X^2 + \epsilon$.

14.2.2 Prediction When X is Discrete: The S-PLUS Function rundis

In some situations the predictor X has very few values and there are multiple Y values for each X value. In this case it might be of interest to compute a measure of location for Y corresponding to each X value and to plot the results. For example, for all Y values corresponding to $X = 1$, say, compute some measure of location, do the same for $X = 2$, and so on. This is in contrast to the smooths previously described, where you search for all X values close to 2, for example, and compute a measure of location based on the corresponding Y values. For convenience, the S-PLUS function

$$\text{rundis}(x, y, \text{est=mom}, \text{plotit=T}, \text{pyhat=F}, \ldots)$$

has been supplied to perform this task.

14.2.3 Multiple Predictors

The basic idea behind the running-interval smoother can be extended to multiple predictors. First consider the case with two predictors, so we have n pairs of X values: $(X_{11}, X_{12}), \ldots, (X_{n1}, X_{n2})$. For the ith pair of predictors, (X_{i1}, X_{i2}), imagine you want to estimate the corresponding typical value of Y. For instance, if $X_{i1} = 6$ and $X_{i2} = 12$, the goal is to estimate the typical value of Y given that $(X_{i1}, X_{i2}) = (6, 12)$. Like the running-interval smoother, the strategy is to apply a measure of location to those Y values for which the corresponding (X_{j1}, X_{j2}) values are close to (X_{i1}, X_{i2}). To determine whether (X_{j1}, X_{j2}) is close to (X_{i1}, X_{i2}), we proceed in a manner that has certain similarities to the outlier detection method described in Box 13.1. For the more general case of p predictors, let $\mathbf{X}_i = (X_{i1}, \ldots, X_{ip})$ and let $\mathbf{M}$ be the MVE or MCD covariance matrix based on $\mathbf{X}_1, \ldots, \mathbf{X}_n$. Then a measure of the distance between the two points $\mathbf{X}_i$ and $\mathbf{X}_j$ is

$$D_{ij} = \sqrt{(\mathbf{X}_i - \mathbf{X}_j)\mathbf{M}^{-1}(\mathbf{X}_i - \mathbf{X}_j)'}.$$

We say that $\mathbf{X}_j$ is close to $\mathbf{X}_i$ if $D_{ij} \leq f$, where f again plays the role of the span. Generally, $f = 1$ seems to give good results; but, as usual, exceptions arise. Now, for all those $\mathbf{X}_j$ that are close to $\mathbf{X}_i$, compute a measure of location associated with the corresponding Y values and label it $\hat{Y}_i$. That is, the typical value of Y, given that $X = \mathbf{X}_i$, is estimated to be $\hat{Y}_i$.

14.2.4 S-PLUS Functions runm3d and rung3d

The S-PLUS function

$$\text{rung3d}(x, y, \text{est} = \text{mom}, \text{fr} = 1, \text{plotit} = T, \text{pyhat} = F, \ldots)$$

creates a running-interval smooth of the data based on the measure of location specified by the argument est, which defaults to MOM. Here, x is assumed to be a matrix with p columns corresponding to the p predictors. The argument fr is the span and

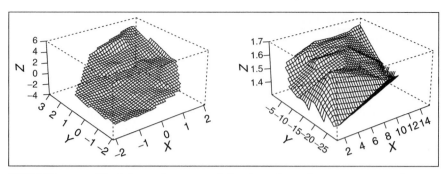

FIGURE 14.4   The left panel is a smooth for data where the true regression line is a plane. The right panel is a smooth based on data from a diabetes study.

pyhat=T will return the $\hat{Y}_i$ values. To get 20% trimmed means, use est=mean and tr=.2. The function

$$\text{runm3d}(x, y, \text{fr} = 1, \text{tr} = 0.2, \text{plotit} = T, \text{pyhat} = F)$$

is designed for trimmed means only.

EXAMPLE. The left panel of Figure 14.4 shows a smooth for 100 points generated according to $Y = X_1 + X_2 + \epsilon$, where X_1, X_2, and ϵ are independent standard normal variables. Notice that the smooth is well approximated by a plane, which is consistent with how the data were generated.   ■

EXAMPLE. The diabetes data mentioned in Section 14.1 are reconsidered, only now two predictors of C-peptide concentration are used: age and base deficit. The right panel of Figure 14.4 shows the smooth created by rung3d. In this case, it seems that a plane does not provide an adequate summary of the data.   ■

14.2.5 Seeing Curvature with More Than Two Predictors

The running-interval smoother can be used with more than two predictors, but, as is evident, visualizing curvature cannot be done in a simple manner. A variety of techniques have been proposed for dealing with this problem, and a comparison of several methods was made by Berk and Booth (1995). (They focused on predicting means, but the strategies they considered are readily extended to robust measures of location.) Generally, these methods can help, but situations can be created where any one method fails.

A simple strategy, sometimes called the *partial response plot*, is to check a smooth for each individual predictor while ignoring the others. An alternative approach is to plot the residuals versus the predicted values. Experience suggests that this strategy often can be unsatisfactory. Another strategy is based on what is called a *partial residual plot*. The idea dates back to Ezekiel (1924, p. 443) and was named a partial residual plot by Larsen and McLeary (1972). To explain it, imagine there are p predictors and that you want to check for curvature associated with predictor j. Assuming that the other predictors have a linear association with Y, fit a regression plane to the data, ignoring the jth predictor. The *partial residual plot* simply plots the resulting residuals versus X_j. A smooth applied to this plot can be used to check for curvature.

14.2.6 S-PLUS Function prplot

The S-PLUS function

$$\text{prplot}(x, y, \text{pvec}=\text{ncol}(x), \text{regfun}=\text{tsreg}, \text{fr}=.8, \text{est}=\text{mom}, \ldots)$$

creates a partial residual plot assuming that curvature is to be checked for the predictor indicated by the argument pvec. The argument x is assumed to be an n-by-p matrix. By default, it is assumed that curvature is to be checked using the data stored in the last column of the matrix x. The argument regfun indicates which regression method will be used, fr is the span used to create the smooth, and est indicates which measure of location is used with the smooth.

EXAMPLE. The model $Y = \beta_0 + \beta_1 X_1 + \beta_2 X_2 + \beta_3 X_1 X_2 + \epsilon$ is often used to investigate interactions in regression and appears to have been first suggested by Saunders (1955, 1956). One hundred vectors of observations were generated according to the model $Y = X_1 + X_2 + X_1 X_2 + \epsilon$, where X_1, X_2, and ϵ are independent standard normal random variables. The partial residual plot provides a partial check on the adequacy of this model. If the model is correct, then, in particular, a partial residual plot based on the term $X_1 X_2$ should produce a reasonably straight line. Applying the S-PLUS function prplot to the data creates the smooth shown in Figure 14.5. We see that for the bulk of the centrally located points, we do indeed get a straight line, which has a slope approximately equal to the true slope, 1.   ■

EXAMPLE. For another portion of the reading study (previously mentioned in Chapter 13), there was interest in how a measure of orthographic ability (Y) is related to a measure of sound blending (X_1) and a measure of auditory analysis (X_2). A smooth between X_1 and Y strongly indicates a linear association. Assuming that the association between X_1 and Y is linear, is it reasonable to assume that the association between Y, X_1, and X_2 is linear? That is, does the model $Y = \beta_0 + \beta_1 X_1 + \beta_2 X_2 + \epsilon$ provide an adequate summary of the data for some choice for β_0, β_1, and β_2? Figure 14.6 shows the plot created by prplot. As is evident, there seems to be a bend at approximately $X_2 = 6$, which is near the center of the X_2 values. That is, it appears that something might be wrong

Continued

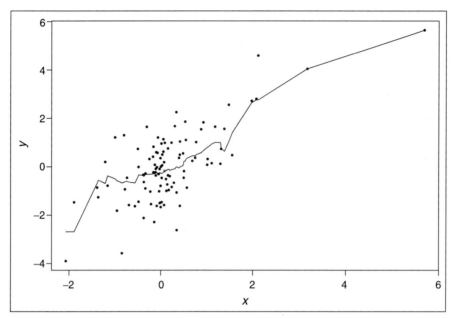

FIGURE 14.5   Output from the S-PLUS function prplot, where the goal is to check a commonly used method for modeling interactions.

EXAMPLE. (*Continued*) with this particular model. Because a smooth indicates that Y and X_1 have a linear association, it seems that there might be curvature associated with X_2. A smooth (or partial response plot) between Y and X_2 again indicates a curvilinear association. One possible way of dealing with this apparent curvature is simply to divide the data into two groups according to whether $X_2 \le 6$.   ■

14.2.7 Some Alternate Methods

The methods already covered for detecting and describing curvature when there are multiple predictors are far from exhaustive. Although complete details about other methods are not provided, it might help to mention some of the alternate strategies that have been proposed.

One approach is to assume that for some collection of functions $f_1, \ldots, f_p$,

$$Y = \beta_0 + f_1(X_1) + \cdots + f_p(X_p) + \epsilon,$$

and then to try to approximate these functions using some type of smoother. This is called a *generalized additive model*; details can be found in Hastie and Tibshirani (1990), who described an algorithm for applying it to data. (S-PLUS has built-in functions for applying this technique; see the methods listed under generalized additive models in the S-PLUS *Guide to Statistics*.)

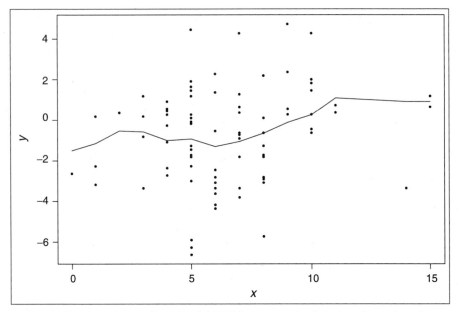

FIGURE 14.6   Output from the S-PLUS function prplot based on the reading data.

The *augmented partial residual plot* (Mallows, 1986) is like the partial residual plot, except that it includes a quadratic term for the predictor being investigated. For a generalization of this method, see Cook (1993).

Another approach is to search for nonlinear transformations of both Y and the predictors that results in an additive model. That is, search for functions $f_y, f_1, \ldots, f_p$ such that

$$f_y(Y) = f_1(X_1) + \cdots + f_p(X_p) + \epsilon$$

provides a good fit to data. An algorithm for implementing the method was derived by Breiman and Friedman (1985) and is called *alternating conditional expectations*, or *ace*. (S-PLUS has built-in functions for applying the technique.) For some refinements and extensions, see Tibshirani (1988).

14.3 Comparing the Slopes of Two Independent Groups

Consider two independent groups and imagine that for each group you have an outcome variable Y and a predictor X. For the jth group ($j = 1, 2$), assume

$$Y_j = \beta_{0j} + \beta_{1j}X_j + \epsilon_j,$$

where ϵ_j is independent of X_j and $E(\epsilon_j) = 0$. That is, for each group, the standard regression model in Section 6.2 holds. Then a common goal is to test

$$H_0 : \beta_{11} = \beta_{12}, \tag{14.1}$$

the hypothesis that the two groups have equal slopes.

A variety of methods have been proposed for testing Equation (14.1), many of which are summarized by Conerly and Mansfield (1988), but the bulk of the methods are known to be relatively unsatisfactory, including a popular technique derived by Chow (1960). Here attention is focused on comparing slopes with a percentile bootstrap method plus one of the robust estimators covered in Chapter 13. As in Chapter 13, the error terms can be heteroscedastic, and it is not assumed that the distribution for the two error terms are similar in any manner. (Methods that assume homoscedasticity or that groups have error terms with identical distributions can be highly unsatisfactory, for the basic reasons mentioned in previous chapters.) To test Equation (14.1), generate bootstrap samples from each group as described in Chapter 7, compute the slope for each group based on these bootstrap samples, and label them b_{11}^* and b_{12}^*. Next, repeat this process B times; let d_b^* be the difference between the bootstrap estimate of the slopes based on the bth bootstrap sample ($b = 1, \ldots, B$) and let $\hat{p}^*$ be the proportion of times the bootstrap estimate for the first group is less than the bootstrap estimate from the second. That is, $\hat{p}^*$ is the proportion of d_b^* values less than zero. Setting

$$\hat{p}_m^* = \min(\hat{p}^*, 1 - \hat{p}^*),$$

the estimated significance level for the hypothesis given by Equation (14.1) is $2\hat{p}_m^*$. Then reject the null hypothesis if $2\hat{p}_m^* \leq \alpha$. Putting the d_b^* values in ascending order, the $1 - \alpha$ confidence interval for the difference between the slopes, $\beta_{11} - \beta_{12}$, is $(d_{(\ell+1)}^*, d_{(u)}^*)$, where, as usual, $\ell = \alpha B/2$ and $u = B - \ell$. In essence, use the method in Section 8.8.1 (designed for measures of location) adapted to the problem at hand.

14.3.1 S-PLUS Functions reg2ci and runmean2g

The S-PLUS function

reg2ci(x1, y1, x2, y2, regfun = tsreg, nboot = 599, alpha = 0.05, plotit = T)

compares the slopes of two groups using the method just described. The data for group 1 are stored in x1 and y1, and for group 2 they are stored in x2 and y2. As usual, nboot is B, the number of bootstrap samples, regfun indicates which regression estimator is to be used and defaults to the Theil–Sen estimator, and plotit=T creates a plot of the bootstrap estimates.

To provide some visual sense of how the regression lines differ and to provide an informal check on whether both regression lines are reasonably straight, the S-PLUS function

runmean2g(x1, y1, x2, y2, fr = 1, est = mom, ...)

has been supplied. It creates a scatterplot for both groups (with a + used to indicate points that correspond to the second group) and it plots a smooth for both groups. The smooth for the first group is indicated by a solid line, and a dashed line is used for the other.

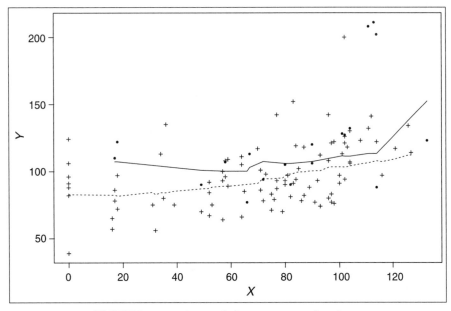

FIGURE 14.7   A smooth for two groups of students.

EXAMPLE. A controversial issue is whether teachers' expectancies influence intellectual functioning. A generic title for studies that address this issue is Pygmalion in the classroom. Rosenthal and Jacobson (1968) argue that teachers' expectancies influence intellectual functioning, and others argue that it does not. A brief summary of some of the counterarguments can be found in Snow (1995). Snow illustrates his concerns with data collected by Rosenthal, where children in grades 1 and 2 were used. Here, other issues are examined using robust regression methods. One of the analyses performed by Rosenthal involved comparing experimental children, for whom positive expectancies had been suggested to teachers, with control children, for whom no expectancies had been suggested. (The data used here are taken from Elashoff and Snow, 1970.) One measure was a reasoning IQ pretest score, and a second was a reasoning IQ posttest score. Here we consider whether the slopes of the regression lines differ for the two groups when predicting posttest scores based on pretest scores. Figure 14.7 shows the output from the S-PLUS function runmean2g. The .95 confidence interval for the difference between the slopes, returned by reg2ci and based on the Theil–Sen estimator, is $(-0.72, 0.18)$ with a significance level of .22. ■

EXAMPLE. For the example in connection with Figure 14.6, a partial residual plot suggested that there is a curvilinear relationship between a measure of

Continued

EXAMPLE. (*Continued*) orthographic ability Y and X_2, a measure of auditory analysis. In particular, there seems to be a distinct bend near $X_2 = 6$. If we split the data into two groups according to whether $X_2 \leq 6$, does this alter the association between Y and X_1, a measure of sound blending? One way of checking this possibility is to test the hypothesis that these two groups have identical slopes. The .95 confidence interval for the difference between the slopes is $(-1.5, 0.0)$ with a significance level of .09, so we fail to detect a change in the association at the .05 level.   ■

14.3.2 Comparing Correlations

The method for comparing slopes just described can be used to compare correlation coefficients simply by replacing estimates of the slopes with a measure of association that is of interest. When working with Pearson's correlation, however, a modified percentile bootstrap method is required. In particular, if there are n_1 pairs of observations for the first group and n_2 for the second, let $n = n_1 + n_2$ be the total number of observations, and set $B = 599$. Now generate bootstrap samples from both groups and for the bth bootstrap sample from each group ($b = 1, \ldots, 599$), let d_b^* be the difference between the bootstrap estimates of the correlations. Then a .95 confidence interval for $\rho_1 - \rho_2$, the difference between the population (Pearson) correlation coefficients, is

$$\left(d_{(a)}^*, d_{(c)}^* \right),$$

where for $n < 40$, $a = 7$ and $c = 593$; for $40 \leq n < 80$, $a = 8$ and $c = 592$; for $80 \leq n < 180$, $a = 11$ and $c = 588$; for $180 \leq n < 250$, $a = 14$ and $c = 585$; while for $n \geq 250$, $a = 15$ and $c = 584$. If this interval does not contain zero, reject the hypothesis of equal Pearson correlations. Adjustments (choices for a and c) for other values of α have not been determined.

14.3.3 S-PLUS Functions twocor and twopcor

The S-PLUS function

    twocor(x1, y1, x2, y2, corfun = pbcor, nboot = 599, alpha = .05, plotit = T, …)

compares the correlations corresponding to two independent groups. The argument corfun indicates which correlation is to be used and defaults to the percentage bend correlation. The argument plotit indicates whether the bootstrap estimates are to be plotted. To compare Pearson correlations, use

    twopcor(x1, y1, x2, y2).

14.4 Tests for Linearity

Smooths provide an informal check on curvature. This section describes two methods that can be used to establish curvature in a more formal manner.

14.4.1 A Method Based on a Split of the Data

The first strategy is to split the data into two groups based on the X values. If there is no curvature, then these two sets of observations should have identical slopes, which can be tested as described in Section 14.3.

EXAMPLE. Consider the diabetes data in Figure 14.2. As previously noted, a smooth suggests that there might be a positive association between age and C-peptide concentrations up to about the age of 4, or possibly 7, but then the association seems to disappear. For illustrative purposes, split the data into two groups according to whether a child is less than or greater than 7 years old. If there is no curvature, then the regression lines for these two groups should have identical slopes. Applying the S-PLUS function reg2ci to these two groups and using the default regression method (the Theil–Sen estimator), the .95 confidence interval for the difference between the slopes is (0, .125). This interval contains zero, so we fail to reject, although we come very close to detecting curvature.   ■

14.4.2 An Alternate Method

This section describes an alternate approach to detecting curvature by testing the hypothesis that there is a linear association between Y and some set of predictors. That is, the goal is to test the hypothesis that for some $\beta_0, \ldots, \beta_p$, $Y = \beta_0 + \beta_1 X_1 + \cdots + \beta_p X_p + \epsilon$. Here, it is *not* assumed that the error term is homoscedastic. The theoretical justification for the method in this section is due to Stute, Manteiga, and Quindimil (1998) and is essentially a generalization of the test of independence covered in Section 7.8. For simplicity, the method is described for the case of a single predictor only, but multiple predictors can be handled as well. Basically, proceed as was done in Section 7.8, only with $\bar{Y}$ replaced by $\hat{Y}$, where $\hat{Y}$ is the estimate of Y based on some regression estimator that assumes there is a linear association. The details are summarized in Box 14.1.

BOX 14.1 Test the Hypothesis That a Regression Line is Straight

Let $\hat{Y}$ be some regression estimate of Y. Least squares could be used, but it has been shown that this can lead to problems in terms of controlling the probability of a Type I error. So it is suggested that some robust estimator be used instead. For fixed j ($1 \leq j \leq n$), set $I_i = 1$ if $X_i \leq X_j$; otherwise $I_i = 0$,

Continued

BOX 14.1 (*Continued*) and let

$$R_j = \frac{1}{\sqrt{n}} \sum I_i(Y_i - \hat{Y}_i)$$

$$= \frac{1}{\sqrt{n}} \sum I_i r_i, \tag{14.2}$$

where $r_i = Y_i - \hat{Y}_i$ are the usual residuals. The (Kolmogorov) test statistic is the maximum absolute value of all the R_j values. That is, the test statistic is

$$D = \max |R_j|, \tag{14.3}$$

where max means that D is equal to the largest of the $|R_j|$ values. As in Chapter 7, a Cramér–von Mises test statistic can be used instead, where now

$$D = \frac{1}{n} \sum R_j^2. \tag{14.4}$$

A critical value is determined with the bootstrap method described in Section 7.8. Generate n observations from a uniform distribution and label the results $U_1, \ldots, U_n$. Next, for $i = 1, \ldots, n$, set

$$V_i = \sqrt{12}(U_i - .5),$$

$$r_i^* = r_i V_i,$$

$$Y_i^* = \hat{Y}_i + r_i^*.$$

Then based on the n pairs of points $(X_1, Y_1^*), \ldots, (X_n, Y_n^*)$, compute the test statistic and label it D^*. Repeat this process B times and label the resulting test statistics $D_1^*, \ldots, D_B^*$. Finally, put these B values in ascending order, yielding $D_{(1)}^* \leq \cdots \leq D_{(B)}^*$. The critical value is $D_{(u)}^*$, where $u = (1 - \alpha)B$ rounded to the nearest integer. That is, reject if

$$D \geq D_{(u)}^*.$$

14.4.3 S-PLUS Functions lintest and linchk

The S-PLUS function

$$\text{lintest}(x, y, \text{regfun} = \text{tsreg}, \text{nboot} = 500, \text{alpha} = .05)$$

tests the hypothesis that a regression surface is a plane using the method outlined in Box 14.1. (Execution time is fairly fast with one predictor, but it might be slow when there are multiple predictors instead.) When reading the output, the Kolmogorov test statistic is labeled dstat and its critical value is labeled critd. The Cramér–von Mises test statistic is labeled wstat. The default regression method (indicated by the argument regfun) is Theil–Sen.

For convenience, the function

linchk(x,y,sp,pv=1,regfun=tsreg,nboot=599,alpha=.05)

is supplied, which splits the data into two groups according to whether predictor pv has a value less than the value stored in the argument sp. For example,

linchk(x,y,sp=10,pv=3)

would split the data into two groups based on whether predictor 3 has a value less than 10. Then it compares the regression parameters for these two groups with the function reg2ci.

EXAMPLE. For the diabetes data shown in Figure 14.1, the Kolmogorov test statistic returned by lintest is $D = .179$; it reports a .05 critical value of .269, so you fail to reject. Note that based on results in Section 14.1, a smooth suggests that for the bulk of the data, the association is linear, with a nearly horizontal regression line. It was previously remarked, however, that this association might change for children under the age of 4. But even if a change exists, with only six children under the age of 4, detecting it is difficult at best. ■

14.5 Inferential Methods with Multiple Predictors

This section takes up the problem of making inferences about regression parameters when there are multiple predictors. We begin with the classic approach, which assumes that there is a linear relationship between the p predictors $X_1, \ldots, X_p$ and some variable Y, that Y has a normal distribution, and that there is homoscedasticity. That is,

$$Y = \beta_0 + \beta_1 X_1 + \beta_2 X_2 + \cdots + \beta_p X_p + \epsilon \qquad (14.5)$$

is assumed, where ϵ has a normal distribution with variance σ^2, and σ^2 does not depend on the values of the predictors. With one predictor ($p = 1$), Equation (14.5) reduces to the regression model given by Equation (6.6). The goal is to test

$$H_0 : \beta_1 = \cdots = \beta_p = 0, \qquad (14.6)$$

the hypothesis that all p slope parameters are zero.

Let $\hat{Y} = b_0 + b_1 X_1 + \cdots + b_p X_p$ be the least squares regression line. That is, the values $b_0, \ldots, b_p$ minimize $\sum (Y_i - \hat{Y}_i)^2$, the sum of the squared residuals. The *squared multiple correlation coefficient* is

$$R^2 = 1 - \frac{\sum \left(Y_i - \hat{Y}_i \right)^2}{\sum \left(Y_i - \bar{Y} \right)^2} \qquad (14.7)$$

and can be seen to be the squared (Pearson) correlation between Y_i and $\hat{Y}_i$. Note that when using least squares regression, R^2 is an estimate of explanatory power, as discussed in Section 13.6.8. The classic method for testing the hypothesis given by

Equation (14.6) is based on the test statistic

$$F = \left(\frac{n - p - 1}{p}\right)\left(\frac{R^2}{1 - R^2}\right). \tag{14.8}$$

Under normality and homoscedasticity, F has an F-distribution with $\nu_1 = p$ and $\nu_2 = n - p - 1$ degrees of freedom. So the null hypothesis is rejected if $F \geq f_{1-\alpha}$, the $1 - \alpha$ quantile of an F-distribution with ν_1 and ν_2 degrees of freedom.

If you have three predictors with the data stored in the S-PLUS variables x1, x2, and x3 and the outcome predictor stored in y, then the built-in S-PLUS command

$$\text{summary}(\text{lm}(y\tilde{\ }x1+x2+x3))$$

will perform the F-test just described (and it performs a Student's T-test of $H_0 : \beta_j = 0$, $j = 0, 1, \ldots, p$, for each of the $p + 1$ regression parameters).

14.5.1 A Bootstrap Method

As was the case in Chapter 6, the conventional hypothesis-testing method just described performs well in terms of controlling Type I error probabilities when Y is independent of all p predictors, which implies that there is homoscedasticity. When Y and the p predictors are dependent, the conventional F-test can be very unsatisfactory — even under normality.

A straightforward application of a particular bootstrap method can be used to test Equation (14.6), the hypothesis that the parameters in a linear regression model have a common value of zero. The method is similar in spirit to the method for comparing measures of location among dependent groups that was covered in Box 11.6 and is described in Box 14.2. Basically, generate B bootstrap estimates of the p slope parameters and then check to see how deeply the vector $\mathbf{0} = (0, \ldots, 0)$, having length p, is nested within the bootstrap values. When $p = 1$, the method in Box 14.2 is essentially a bootstrap-t method with the standard error of the regression estimators estimated by the bootstrap values. This is in contrast to the method in Section 13.5 for computing a confidence interval for the individual parameters, which uses a percentile bootstrap method instead.

BOX 14.2 Test the Hypothesis That All Slope Parameters Are Zero

First, generate a bootstrap sample as was described in Chapter 7. For the p predictors case, this means that among the n vectors of observations

$$(Y_1, X_{11}, \quad \ldots, \quad X_{1p})$$

$$\vdots$$

$$(Y_n, X_{n1}, \quad \ldots, \quad X_{np}),$$

Continued

BOX 14.2 (*Continued*)

randomly resample with replacement n vectors, yielding

$$\left(Y_1^*, X_{11}^*, \quad \ldots, \quad X_{1p}^*\right)$$

$$\vdots$$

$$\left(Y_n^*, X_{n1}^*, \quad \ldots, \quad X_{np}^*\right).$$

Let $\hat{\beta}_j^*$ be the estimate of the jth regression parameter, $j = 1, \ldots, p$. Next, repeat this process B times, yielding $\hat{\beta}_{jb}^*, j = 1, \ldots, p; b = 1, \ldots, B,$ and let

$$s_{jk} = \frac{1}{B-1} \sum_{b=1}^{B} \left(\hat{\beta}_{jb}^* - \hat{\beta}_j\right)\left(\hat{\beta}_{kb}^* - \hat{\beta}_k\right),$$

where $\hat{\beta}_j$ is the estimate of β_j based on the original data. Compute

$$d_b = \left(\hat{\beta}_b^* - \hat{\beta}\right)S^{-1}\left(\hat{\beta}_b^* - \hat{\beta}\right)',$$

where S is the matrix corresponding to s_{jk}, $\hat{\beta} = (\hat{\beta}_1, \ldots, \hat{\beta}_p),$ and $\hat{\beta}_b^* = (\hat{\beta}_{1b}^*, \ldots, \hat{\beta}_{pb}^*)$. The value of d_b measures how far away the bth bootstrap vector of estimated slope parameters is from the center of all B bootstrap values. Put the d_b values in ascending order, yielding $d_{(1)} \leq \cdots \leq d_{(B)}$. The test statistic is

$$D = (0 - \hat{\beta})S^{-1}(0 - \hat{\beta})'$$

and measures how far away the null hypothesis is from the estimated slope parameters. Reject if $D \geq d_{(u)}$, where $u = (1 - \alpha)B$, rounded to the nearest integer.

The bootstrap method in Box 14.2 has been found to perform relatively well when using robust regression estimators such as those covered in Chapter 13. Limited studies suggest that it even performs reasonably well when using least squares regression, provided $n \geq 40$. This means that when testing at the .05 level, the actual Type I error probability will be less than .05. However, a criticism is that when using least squares, the probability of a Type I error can be substantially less than .05, suggesting that power might be relatively low. For smaller sample sizes, the function lsfitci (described in Section 7.3), which uses a modified percentile bootstrap method, provides more accurate results, but it is limited to testing for a zero slope coefficient for each parameter at the $\alpha = .05$ level. That is, for each j, it tests $H_0 : \beta_j = 0$, but it does not test Equation (14.6) and it does not control the probability of at least one Type I error among all the tests that are performed. The method in Box 14.2 can be seen to differ in a crucial way from the method in Section 7.3. For the special case of a single predictor, both of these methods can be used to test $H_0 : \beta_1 = 0$. Based on the goal of controlling Type I error probabilities, little is known about their relative merits for this special case.

14.5.2 S-PLUS Function regtest

The S-PLUS function

$$\text{regtest}(x,y,\text{regfun}=\text{tsreg},\text{nboot}=600,\text{alpha}=.05,\text{grp},\text{nullvec})$$

tests the hypothesis that the regression coefficients are equal to some specified set of constants using the method in Box 14.2. By default, the hypothesis given by Equation (14.6) is tested. The argument nullvec can be used to set the hypothesized values to something other than zero. For example, if there are two predictors, the S-PLUS command

$$\text{regtest}(x,y,\text{nullvec}=c(2,4))$$

will test the hypothesis that $\beta_1 = 2$ and $\beta_2 = 4$. The argument grp can be used to indicate that a subset of the parameters is to be tested, which can include the intercept term. For example, when calling the function, setting grp=c(0,3) will test $H_0 : \beta_0 = \beta_3 = 0$, assuming the argument nullvec is not specified. The command

$$\text{regtest}(x,y,\text{grp}=c(2,4,7))$$

will test $H_0 : \beta_2 = \beta_4 = \beta_7 = 0$.

EXAMPLE. For the error term (ϵ) and each of two predictors, 30 observations were generated from a standard normal distribution and Y was determined by $Y = .33X_1 + 0X_2 + \epsilon$. Applying the standard F-test [given by Equation (14.8)] for $H_0 : \beta_1 = \beta_2 = 0$, the significance level is .23. Now suppose the same data are used, except that when $X_1 > .8$, the error term is taken to be $X_1^2 \epsilon$ rather than just ϵ. So in general the data follow the standard regression model, except when the first predictor is somewhat large, in which case there is heteroscedasticity. Now the significance level of the standard F-test is .90. If we use the S-PLUS function regtest with the default regression estimator (the Theil–Sen estimator), the significance level is .79. Situations arise where the Theil–Sen estimator substantially increases power when there is heteroscedasticity, but here it does not do much better than least squares.   ■

EXAMPLE. This example illustrates that it is possible to reject the hypothesis that the slope parameters are all equal to zero [the hypothesis given by Equation (14.6)] but fail to reject for any of the individual slope parameters. Table 14.1 shows data from a study by Hald (1952) concerning the heat evolved, in calories per gram (Y), versus the amount of each of four ingredients in the mix: tricalcium aluminate (X_1), tricalcium silicate (X_2), tetracalcium alumino ferrite (X_3), and dicalcium silicate (X_4). Consider the first and third predictors and suppose we test $H_0 : \beta_1 = \beta_3 = 0$ with the S-PLUS function regtest. Figure 14.8 shows the bootstrap estimates returned by the function regtest when using least squares.

Continued

TABLE 14.1   Hald's Cement Data

Y	X_1	X_2	X_3	X_4
78.5	7	26	6	60
74.3	1	29	15	52
104.3	11	56	8	20
87.6	11	31	8	47
95.9	7	52	6	33
109.2	11	55	9	22
102.7	3	71	17	6
72.5	1	31	22	44
93.1	2	54	18	22
115.9	21	47	4	26
83.8	1	40	23	34
113.3	11	66	9	12
109.4	10	68	8	12

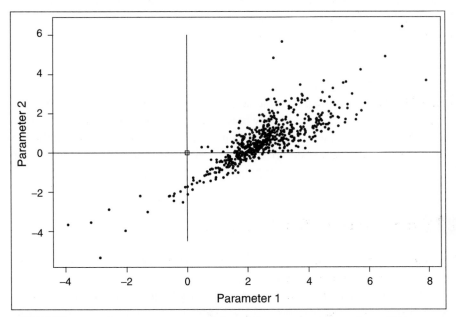

FIGURE 14.8   Bootstrap estimates of β_1 and β_3 using the cement data in Table 14.1.

EXAMPLE. (*Continued*) (That is, regfun=lsfit was used.) The significance level is .047, so in particular we would reject with $\alpha = .05$. However, if we test the individual slope parameters with the S-PLUS function lsfitci (see Section 7.3.1), the .95 confidence intervals for β_1 and β_3 are $(-0.28, 5.93)$ and $(-2.3, 3.9)$, respectively, so we fail to reject for either of the predictor variables.

Continued

EXAMPLE. (*Continued*) This phenomenon, where the omnibus test is significant but the individual tests are not, is known to occur when using the conventional F-test [given by Equation (14.8)] as well (e.g., Fairley, 1986). The reason is that when two estimators have a reasonably strong association, the resulting confidence region for the two parameters is a relatively narrow ellipse. Figure 14.8 shows a plot of the bootstrap estimates for the data under consideration and provides some indication of why this phenomenon occurs. The square, where the horizontal and vertical lines intersect, corresponds to the hypothesized values and, as is evident, is relatively far from the bulk of the bootstrap estimates. However, in order to reject $H_0 : \beta_1 = 0$ at the .05 level, which corresponds to parameter 1 in Figure 14.8, 97.5% of the bootstrap estimates would need to be either above or below the horizontal line. To reject $H_0 : \beta_3 = 0$, 97.5% of the bootstrap estimates would need to be to the right or to the left of the vertical line. Said another way, computing separate confidence intervals is essentially computing a rectangular confidence region for the two parameters under investigation. When the two estimators are approximately independent, this tends to give similar results to those obtained with the confidence region used by the S-PLUS function regtest, but otherwise it is possible for one method to reject when the other does not.   ■

EXAMPLE. Repeating the previous example using the conventional F-test, we again reject $H_0 : \beta_1 = \beta_3 = 0$, only now Student's T-test of $H_0 : \beta_1 = 0$ rejects as well. Using regtest with the Theil–Sen estimator, we reject $H_0 : \beta_1 = \beta_3 = 0$ once more and the S-PLUS function regci rejects $H_0 : \beta_1 = 0$.   ■

14.6 Identifying the Best Predictors

A problem that has received considerable attention is identifying a subset of predictors that might be used in place of the p predictors that are available. If p is large, the variance of the regression equation can be relatively large. If a subset of the p predictors can be identified that performs relatively well in some sense, not only do we get a simpler model, but we can get a regression equation with a lower variance. (For example, the variance of a sum of two variables — say, X_1 and X_2 — is $\sigma_1^2 + \sigma_2^2 + 2\rho\sigma_1\sigma_2$, where σ_1 and σ_2 are the standard deviations associated with X_1 and X_2 and ρ is Pearson's correlation. So if $\rho > 0$, the variance of the sum is larger than the variance of the individual variables.) If we have 40 predictors, surely it would be convenient if a subset of, say, five predictors could be found that could be used instead. Of particular concern in this book is subset selection when using a robust regression estimator and the number of predictors is relatively small. This is an extremely complex problem that has received relatively little attention. Based on what is known, some type of bootstrap estimate of prediction error (which is formally defined later) appears to be relatively effective, and so this approach is described here. It is stressed, however, that this area is in need of more research and perhaps some alternative strategy will be found to have practical advantages over the approach used here.

Perhaps the best-known method for selecting a subset of the predictors is *stepwise regression*, but it is known that the method can be rather unsatisfactory (e.g., Montgomery & Peck, 1992, Section 7.2.3; Derksen & Keselman, 1992), and the same is true when using a related (forward selection) method, so for brevity these techniques are not covered here. (Also see Kuo & Mallick, 1998; Huberty, 1989; Chatterjee & Hadi, 1988; cf. A.J. Miller, 1990.) Generally, methods based on R^2 (given by Equation (14.7)), F (given by Equation (14.8)) and a homoscedastic approach based on

$$C_p = \frac{1}{\hat{\sigma}^2} \sum (Y_i - \hat{Y}_i)^2 - n + 2p,$$

called Mallows' (1973) C_p criterion, cannot be recommended either (A.J. Miller, 1990).[1] Another approach is based on what is called *ridge regression*, but it suffers from problems listed by Breiman (1995). Three alternative approaches are cross-validation, bootstrap methods (such as the .632 estimator described in Box 14.3), and the so-called *nonnegative garrote technique* derived by Breiman (1995). Efron and Tibshirani (1993, Chapter 17) discuss cross-validation, but currently it seems that some type of bootstrap method is preferable, so no details are given here. (Breiman's method is appealing when the number of predictors is large. For an interesting variation of Breiman's method, see Tibshirani, 1996.) Here, henceforth, attention is restricted to methods that allow heteroscedasticity.

Imagine you observe n pairs of values $(X_1, Y_1), \ldots, (X_n, Y_n)$, you estimate the regression line to be $\hat{Y} = b_0 + b_1 X$, and now you observe a new X value, which will be labeled X_0. Based on this new X value you can, of course, estimate Y with $\hat{Y}_0 = b_0 + b_1 X_0$. That is, you do not observe the value Y_0 corresponding to X_0, but you can estimate it based on past observations. *Prediction error* refers to the discrepancy between the predicted value of Y, $\hat{Y}_0$, and the actual value of Y, Y_0, if only you could observe it. One way of measuring the typical amount of prediction error is with

$$E[(Y_0 - \hat{Y}_0)^2],$$

the expected squared difference between the observed and predicted values of Y. Of course squared error might be replaced with some other measure, but for now this issue is ignored. As is evident, the notion of prediction error is easily generalized to multiple predictors. The basic idea is that via some method we get a predicted value for Y, which we label $\hat{Y}$, and the goal is to measure the discrepancy between $\hat{Y}_0$ (the predicted value of Y based on a future collection of X values) and the actual value of Y, Y_0, if it could be observed.

A simple estimate of prediction error is the so-called *apparent error rate*, meaning you simply average the error when predicting the observed Y values with $\hat{Y}$. To elaborate,

---

1 When using C_p, $\hat{\sigma}^2$ is usually taken to be $\sum r_i^2/(n-2)$, where the residuals are based on all of the predictors under consideration.

let $Q(Y,\hat{Y})$ be some measure of the discrepancy between an observation, Y, and its predicted value, $\hat{Y}$. So squared error corresponds to

$$Q(Y,\hat{Y}) = (Y - \hat{Y})^2.$$

The goal is to estimate the typical amount of error for future observations. In symbols, the goal is to estimate

$$\eta = E[Q(Y_0, \hat{Y}_0)],$$

the expected error between a predicted value for Y, based on a future value of X, and the actual value of Y, Y_0, if it could be observed. A simple estimate of η is the *apparent error*:

$$\hat{\eta}_{ap} = \frac{1}{n} \sum Q(Y_i, \hat{Y}_i).$$

So for squared error, the apparent error is

$$\hat{\eta}_{ap} = \frac{1}{n} \sum (Y_i - \hat{Y}_i)^2,$$

the average of the squared residuals.

A practical concern is that the apparent error is biased downward because the data used to come up with a prediction rule ($\hat{Y}$) are also being used to estimate error (Efron & Tibshirani, 1993). That is, it tends to underestimate the true error rate, η. The so-called .632 bootstrap estimator, described in Box 14.3, is designed to address this problem and currently seems to be a relatively good choice for identifying the best predictors. It is stressed, however, that more research is needed when dealing with this very difficult problem, particularly when using robust methods.

BOX 14.3 How to Compute the .632 Bootstrap Estimate of η

Generate a bootstrap sample as described in Box 14.2, except rather than resample n vectors of observations, as is typically done, resample $m < n$ vectors of observations instead. (Setting $m = n$, Shao, 1995, shows that the probability of selecting the correct model may not converge to 1 as n gets large.) Here, $m = 5 \log(n)$ is used, which was derived from results reported by Shao (1995). Let $\hat{Y}_i^*$ be the estimate of Y_i based on the bootstrap sample, $i = 1, \ldots, n$. Repeat this process B times, yielding $\hat{Y}_{ib}^*$, $b = 1, \ldots, B$. Then an

Continued

BOX 14.3 (*Continued*) estimate of η is

$$\hat{\eta}_{\text{Boot}} = \frac{1}{nB} \sum_{b=1}^{B} \sum_{i=1}^{n} Q\left(Y_i, \hat{Y}_{ib}^*\right).$$

A refinement of $\hat{\eta}_{\text{Boot}}$ is to take into account whether a Y_i value is contained in the bootstrap sample used to compute $\hat{Y}_{ib}^*$. Let

$$\hat{\epsilon}_0 = \frac{1}{n} \sum_{i=1}^{n} \frac{1}{B_i} \sum_{b \in C_i} Q\left(Y_i, \hat{Y}_{ib}^*\right),$$

where C_i is the set of indices of the bth bootstrap sample not containing Y_i and B_i is the number of such bootstrap samples. Then the .632 estimate of the prediction error is

$$\hat{\eta}_{.632} = .368\hat{\eta}_{\text{ap}} + .632\hat{\epsilon}_0. \tag{14.9}$$

This estimator arises in part from a theoretical argument showing that .632 is approximately the probability that a given observation appears in a bootstrap sample of size n. [For a refinement of the .632 estimator given by Equation (14.9), see Efron & Tibshirani, 1997.]

14.6.1 S-PLUS function regpre

The S-PLUS function

$$\text{regpre}(x,y,\text{regfun} = \text{lsfit}, \text{error} = \text{sqfun}, \text{nboot} = 100,$$
$$\text{mval} = \text{round}(5 \log(\text{length}(y))),\text{model} = \text{NA})$$

estimates prediction error for a collection of models specified by the argument model, which is assumed to have list mode. For example, imagine you have three predictors and you want to consider the following models:

$$Y = \beta_0 + \beta_1 X_1 + \epsilon,$$
$$Y = \beta_0 + \beta_1 X_1 + \beta_2 X_2 + \epsilon,$$
$$Y = \beta_0 + \beta_1 X_1 + \beta_3 X_3 + \epsilon,$$
$$Y = \beta_0 + \beta_1 X_1 + \beta_2 X_2 + \beta_3 X_3 + \epsilon.$$

Then the commands

$$\text{model}< -\text{list}( )$$
$$\text{model}[[1]]< -1$$
$$\text{model}[[2]]< -c(1,2)$$

$$\text{model}[[3]] < - \text{c}(1,3)$$

$$\text{model}[[4]] < - \text{c}(1,2,3)$$

$$\text{regpre}(x,y,\text{model}=\text{model})$$

result in estimating prediction error for the four models. For example, the values in model[[3]], namely, 1 and 3, indicate that predictors 1 and 3 will be used and predictor 2 will be ignored. The argument error determines how error is measured; it defaults to squared error. Setting error=absfun will result in using absolute error.

EXAMPLE. For the Hald data in Table 14.1, if we test the hypothesis given by Equation (14.5) with the conventional F-test in Section 14.5 [given by Equation (14.8)], the significance level is less than .001, indicating that there is some association between the outcome variable and the four predictors. However, for each of the four predictors, Student's T-tests of $H_0 : \beta_j = 0$ ($j = 1, 2, 3, 4$) have significance levels .07, .5, .9, and .84, respectively. That is, we fail to reject for any specific predictor at the .05 level, yet there is evidence of some association. Now consider the eight models

$$Y = \beta_0 + \beta_1 X_1 + \epsilon,$$

$$Y = \beta_0 + \beta_2 X_2 + \epsilon,$$

$$Y = \beta_0 + \beta_3 X_3 + \epsilon,$$

$$Y = \beta_0 + \beta_4 X_4 + \epsilon,$$

$$Y = \beta_0 + \beta_1 X_1 + \beta_2 X_2 + \epsilon,$$

$$Y = \beta_0 + \beta_1 X_1 + \beta_3 X_3 + \epsilon,$$

$$Y = \beta_0 + \beta_1 X_1 + \beta_4 X_4 + \epsilon,$$

$$Y = \beta_0 + \beta_1 X_1 + \beta_2 X_2 + \beta_3 X_3 + \beta_4 X_4 + \epsilon.$$

The estimated prediction errors for these models, based on least squares regression, are 142, 94.7, 224, 94, 7.6, 219, 9.6, and 638, respectively. Notice that the full model (containing all of the predictors) has the highest prediction error, suggesting that it is the least satisfactory model considered. Model 5 has the lowest prediction error, indicating that $Y = \beta_0 + \beta_1 X_1 + \beta_2 X_2 + \epsilon$ provides the best summary of the data among the models considered.   ■

14.7 Detecting Interactions

This section illustrates the methods previously covered in this chapter in the context of detecting interactions. Consider some outcome variable, Y, and two predictors, X_1 and X_2. Roughly, the issue is whether knowing the value of X_2 modifies the association between Y and X_1. For example, for the reading study introduced in Section 13.1, there was interest in how a measure of orthographic ability (Y) is related to a measure of auditory analysis, X_1. A third variable in this study was a measure of sound blending (X_2). Does knowing the value of this third variable alter the association between Y and X_1; and if it does, how? More generally, there is interest in knowing whether a particular factor affects the magnitude of some effect size. Such factors are called *moderators* (e.g., Judd, Kenny, & McClelland, 2001).

Graphically, an interaction, in the context of regression, can be roughly described as follows. Let x_2 be any value of the second predictor variable, X_2. For example, x_2 could be the median of the X_2 values, but any other value could be used in what follows. Now consider the outcome variable, Y, and the first predictor (X_1), and imagine that we split the n pairs of points $(Y_1, X_{11}), \ldots, (Y_n, X_{n1})$ into two groups: those pairs for which the corresponding X_2 value is less than x_2 and those for which the reverse is true. No interaction means that the regression lines corresponding to these two groups are parallel. For example, if for the first group $Y = X_1^2 + \epsilon$ and for the second $Y = X_1^2 + 6 + \epsilon$, these regression lines are parallel and there is no interaction. But if for the second group $Y = X_1^2 + 8X_1 + 3 + \epsilon$, say, then the regression lines intersect (at $X_1 = -3/8$) and we say that X_2 modifies the association between Y and X_1.

A very popular method for checking and modeling interaction is with

$$Y = \beta_0 + \beta_1 X_1 + \beta_2 X_2 + \beta_3 X_1 X_2 + \epsilon. \tag{14.10}$$

That is, use the product of the two predictors to model interaction and conclude that an interaction exists if $H_0 : \beta_3 = 0$ can be rejected. (Saunders, 1955, 1956, appears to be the first to suggest this approach to detecting interactions in regression; cf. Cronbach, 1987; Baron & Kenny, 1986.) This hypothesis can be tested using methods already covered. That is, for the observations $(Y_1, X_{11}, X_{12}), \ldots, (Y_n, X_{n1}, X_{n2})$, set $X_{i3} = X_{i1} X_{i2}$ and for the model $Y = \beta_0 + \beta_1 X_1 + \beta_2 X_2 + \beta_3 X_3 + \epsilon$, test $H_0 : \beta_3 = 0$. However, it currently seems that a collection of tools is needed to address the issue of interactions in an adequate manner.

To add perspective on the product term just described, suppose we fix (or condition on) X_2. That is, we treat it as a constant. Then a little algebra shows that Equation (14.10) can be written as

$$Y = (\beta_0 + \beta_2 X_2) + (\beta_1 + \beta_3 X_2) X_1 + \epsilon.$$

So the intercept term becomes $(\beta_0 + \beta_2 X_2)$ and the slope for X_1 changes as a linear function of X_2. If $\beta_3 = 0$, then

$$Y = (\beta_0 + \beta_2 X_2) + \beta_1 X_1 + \epsilon.$$

That is, knowing X_2 alters the intercept term but not the slope. Said another way, if we split the data into two groups according to whether X_2 is less than or greater than some constant x_2, the corresponding regression lines will be parallel, consistent with the description given earlier.

There are various ways one might model interaction in a more general fashion. For example, one could replace Equation (14.10) with

$$Y = \beta_0 + \beta_1 X_1 + \beta_2 X_2 + g(X_1 X_2) + \epsilon \tag{14.11}$$

for some function g of the product. Equation (14.10) corresponds to the special case $g(X_1 X_2) = \beta_3 X_1 X_2$. A simple method for examining whether some function of the product might be useful is to create a partial residual plot as described in Section 14.2.5. This assumes, of course, that Y has a linear association with X_1 and X_2, and for the reading data, for example, we have already seen that this assumption seems dubious.

Before continuing, an exploratory tool is described that is useful when checking for a modifier variable; the method is based on a slight extension of the running-interval smoother. The basic idea is to plot the strength of the association between Y and X_1 as a function of X_2. Consider the ith observed value for X_2, X_{i2}. Briefly, identify all of the X_{j2} values that are close to X_{i2} in the manner outlined in Section 14.2. Then for the corresponding pairs of points, (X_{1j}, Y_j), compute some correlation coefficient and label it $\hat{\theta}_i$. Repeat this process for $i = 1, \ldots, n$ and then plot the pairs $(X_{i2}, \hat{\theta}_i)$. The computations are performed by the S-PLUS function runcor described in Section 14.7.1 (cf. Doksum, Blyth, Bradlow, Meng, & Zhao, 1994).

EXAMPLE. Some of the strategies that might be used to detect modifier variables are illustrated with data from a reading study previously mentioned. Again, we take as the outcome variable a measure of orhographic ability (Y), only this time the first predictor (X_1) is a measure of sound blending and the issue is whether a measure of phonological awareness (X_2) modifies the association between Y and X_1. First consider the model given by Equation (14.10), where $X_3 = X_1 X_2$. If we apply the conventional F-test to $H_0 : \beta_1 = \beta_2 = \beta_3 = 0$, the significance level is .044, but the significance levels for the three individual slope parameters are .46, .67, and .92, respectively. Estimating prediction error as described in Section 14.6, among the models

$$Y = \beta_0 + \beta_1 X_1 + \epsilon,$$
$$Y = \beta_0 + \beta_2 X_2 + \epsilon,$$
$$Y = \beta_0 + \beta_3 X_3 + \epsilon,$$
$$Y = \beta_0 + \beta_1 X_1 + \beta_2 X_2 + \epsilon,$$
$$Y = \beta_0 + \beta_1 X_1 + \beta_2 X_2 + \beta_3 X_3 + \epsilon,$$

the first has the lowest prediction error (using least squares regression and squared error). This suggests that the first predictor is important but the others are not. But before making any final decisions about the associations under study, it is important to see whether other methods paint a different picture. ■

To get some feel for the data, Figure 14.9 shows a smooth created by the S-PLUS function runm3d, described in Section 14.2.4, when using X_1 and X_2 to predict Y. Notice that the regression surface does not appear to be well approximated by a plane. That is, assuming a linear association might provide an inadequate representation of the data, contrary to what is typically assumed when investigating interactions.

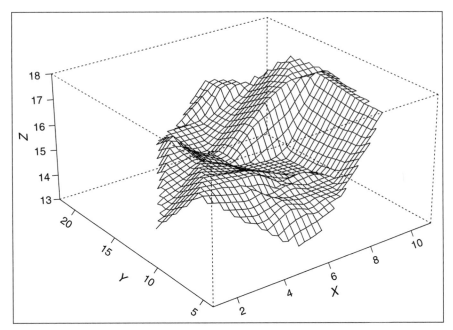

FIGURE 14.9   A smooth for two predictors used in a reading study.

Next we examine the association between X_1 and Y, ignoring X_2 for the moment. Typically it is assumed that this association is linear and so one goal is to get some sense of whether this assumption is reasonable for the problem at hand. Figure 14.10 shows a smooth of these two variables that suggests a nonlinear association. Using the S-PLUS function linchk (described in Section 14.4.3) to check for linearity, with the data split at $X_1 = 7$, the .95 confidence interval for the difference between the slopes is $(-1.5, 0)$ with a significance level of .027. The estimated slopes, using Theil–Sen, are 0 (for $X_1 < 7$) and .5. Checking for linearity with the method in Box 14.1, the test statistic reported by the S-PLUS function lintest is 7.76 and the .05 critical value is 4.27, again suggesting that the hypothesis of a linear association is false. One way to proceed from here is to incorporate a measure of association that is sensitive to monotonic relationships that are not necessarily linear. Two classic approaches are available, but the details are postponed until Chapter 15. (For an analysis of these data based on one of these approaches, see the example in Section 15.12.) It is instructive, however, to proceed under the assumption that problems with nonlinearity can ignored.

Next we plot the smooth relating the (percentage bend) correlation between Y and X_1 as a function of X_2. Figure 14.11 shows the results based on the S-PLUS function runcor (described in Section 14.7.1). The $+$ indicates the location of the median of X_2, and the quartiles are indicated by a |. Note that the smooth is fairly horizontal on the left but that there is some indication that it begins to increase around $X_2 = 12$ or perhaps $X_2 = 15$. To provide perspective, the S-PLUS function runmean2g is used to simultaneously plot a smooth between Y and X_1 based on points corresponding to $X_2 \le 12$ as well as for $X_2 > 12$. That is, runmean2g creates

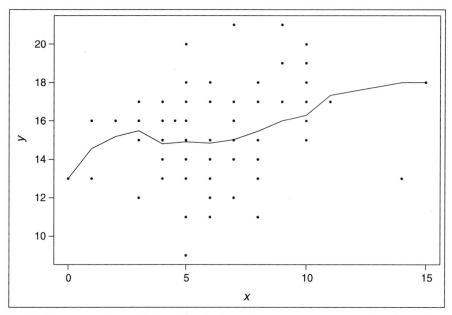

FIGURE 14.10   A smooth based on the first of the two predictors used in the reading example.

a smooth when predicting Y based on X_1 for two groups. The first group corresponds to $X_2 \leq 12$, and the second is simply $X_2 > 12$. (To facilitate this part of the analysis, the S-PLUS function regi, described in Section 14.7.1, is provided.) Figure 14.12 shows the results. Notice that for the second group (with points indicated by a +), there is only one point with $X_1 < 4$. This means that the left portion of the smooth for the second group might be relatively inaccurate, for reasons mentioned in Section 14.1. Focusing on the X values in Figure 14.12 that are greater than or equal to 4, the two smooths appear to be fairly parallel, which suggests that there is no interaction. Also, curvature between Y and X_1 now seems to be minimal. Testing the hypothesis of equal slopes with the S-PLUS function reg2ci using the Theil–Sen estimator, the significance level is .47, so we fail to reject and it might seem that it is safe to conclude that there is no interaction. The estimates of the slopes are 0 and .33.

There are, however, several concerns. The first is that for convenience, Theil–Sen was used. Checking for outliers using the MGV method and comparing the MGV regression estimate of the slope to Theil–Sen suggests that the slopes for the two groups are even more similar than indicated: The estimated slope for the second group drops slightly to .29. This might seem to support the conclusion that there is no interaction, but the MGV estimator can have a much smaller standard error, which could result in rejecting the hypothesis of equal slopes. Another concern is that the data were split according to whether $X_2 \leq 12$. Is this the optimal split for detecting an interaction? Empirical results previously described suggest splitting the data in this manner, but perhaps a slightly different split would make a practical difference. A third concern, which is always an issue when failing to reject, is whether power is sufficiently high to accept the null hypothesis of equal slopes. Addressing this problem for the

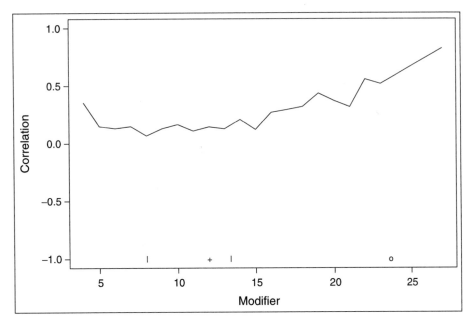

FIGURE 14.11 A smooth based on the reading data that shows the percentage bend correlation between Y and X_1 as a function of X_2.

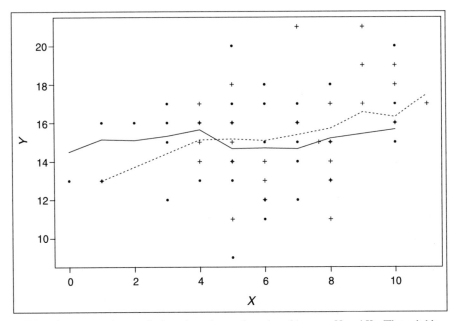

FIGURE 14.12 A smooth, based on the reading data, between Y and X_1. The solid line is based on points with $X_2 \leq 12$ and the dashed line is for $X_2 > 12$.

problem at hand is difficult at best. And finally, it is unclear whether the apparent curvilinear association between X_1 and Y can be ignored. (In Section 15.12, we will see that using a measure of association that is sensitive to this curvilinearity makes a practical difference.)

EXAMPLE. The same two predictors in the previous example are used again, but now the outcome variable of interest (Y) is a measure of word accuracy and we consider whether phonological awareness modifies the association between sound blending and Y. Here we merely split the data into two groups according to whether X_2 is less than or equal to its median. Checks on the second group indicate that a single unusual X_1 value is seriously affecting the estimated slope based on both the Theil–Sen estimator and least squares. By restricting the range in this second group so as to eliminate this seemingly aberrant point, the .95 confidence interval for the difference between the slopes, using Theil–Sen, is $(0, 4)$ with a significance level of .057 when testing the hypothesis that the slopes are equal. That is, for this particular split of the data, we are unable to reject at the .05 level the hypothesis that the two groups have unequal slopes. But we come fairly close to rejecting and concluding that phonological awareness modifies the association between sound blending and word accuracy. Testing the hypothesis that in Equation (14.10), $\beta_3 = 0$, the significance level is .14 using least squares and the conventional Student's T-test.   ■

14.7.1 S-PLUS Functions runcor, regi, and cori

The S-PLUS function

$$\text{runcor}(x, y, z, \text{fr}=1, \text{corflag}=\text{F}, \text{corfun}=\text{pbcor}, \text{plotit}=\text{T}, \text{rhat}=\text{F})$$

plots the correlation between x and y as a function of the data stored in the argument z. Setting the argument corflag to T (for true) results in using Pearson's correlation; otherwise the function uses the correlation specified by the argument corfun, which defaults to the percentage bend correlation. Setting the argument rhat to T, the function returns the values of the estimated correlations corresponding to each value stored in z. For example, if the first value in z is 6, then the first value returned in rhat is the correlation between x and y for points for which the corresponding z values are close to 6. As usual, fr is the span and plotit=F will suppress the plot.

The S-PLUS function

$$\text{regi}(x, y, z, \text{pt}=\text{median}(z), \text{fr}=.8, \text{est}=\text{mom}, \text{regfun}=\text{tsreg}, \text{testit}=\text{F}, \ldots)$$

creates two smooths. The first is based on the x and y values for which the corresponding value for z is less than the value stored in the argument pt. By default, pt is taken to be the median of the z values. The other smooth is based on the x and y values for which the corresponding value for z is greater than pt. The smooth is created with the measure of location given by the argument est, which defaults to MOM. If testit=T is used, the slopes (and intercepts) of the two regression lines, based on the function regfun (which defaults to the Theil–Sen estimator), are compared. (This is done by splitting the data for you and calling the function reg2ci.)

The S-PLUS function

$$\text{cori}(x, y, z, \text{pt}=\text{median}(z), \text{fr}=.8, \text{est}=\text{mom}, \text{corfun}=\text{pbcor}, \text{testit}=\text{F}, \ldots)$$

is like regi, except that setting testit to T causes the correlations to be compared rather than the regression slopes. By default, the percentage bend correlation is used, but the argument corfun can be used to specify some other measure of association. (For example, corfun=spear results in using Spearman's correlation, a measure of association covered in Chapter 15.)

14.8 ANCOVA

This section takes up a topic known as the *analysis of covariance*, or ANCOVA. As was the case in Chapter 8, the goal is to compare two independent groups in terms of some measure of location, but here an additional goal is to take into account the information provided by some predictor variable called a *covariate*. As a simple illustration, imagine that men and women are compared in terms of their typical score on some mathematics aptitude test and it is found that the typical male scores higher than the typical woman. However, test scores might be related to previous training in mathematics, and if we compare men and women having comparable training, now women might score higher than men. Here the covariate is previous training.

There is a standard ANCOVA method that is based on least squares regression. (See Huitema, 1980; Rutherford, 1992.) Not only does the method assume normality and homoscedasticity, but it assumes that the regression lines for the two groups being compared are parallel. As previously mentioned, there are many methods for testing the assumption that regression lines are parallel. But it is unknown how to determine whether they have enough power to detect nonparallel lines in situations where violating this assumption has practical consequences. Some least squares methods that allow nonparallel regression lines are available, a classic example being the Johnson–Neyman method (P. Johnson & Neyman, 1936). Here, however, attention is focused on modern robust methods. Unlike conventional approaches, it is not assumed that the regression line is straight. Rather, a smooth is used to approximate the regression lines and then typical values for Y, given some value for the covariate, are compared using methods covered in previous chapters.

For the jth group, let $m_j(X)$ be some population measure of location associated with Y given X. So for the first group, $m_1(6)$ might represent the population mean of Y given that $X = 6$, or it could be the population value for MOM or a 20% trimmed mean. Given X, the problem is determining how the typical value of Y in the first group compares to the typical value in the second. In the Pygmalion study introduced in Section 14.3.1, the goal might be to determine how the 20% trimmed mean of the experimental group compares to the trimmed mean of the control group, given that a student's IQ reasoning pretest score is $X = 90$. Of course, a more general goal is to determine how the trimmed means compare as X varies. A common strategy is to assume that a straight regression line can be used to predict Y from X. In the present notation, it is assumed that for the jth group,

$$m_j(X) = \beta_{0j} + \beta_{1j}X_{1j},$$

$j = 1, 2$. However, when working with robust regression methods, currently this approach to ANCOVA has been found to be relatively unsatisfactory when testing hypotheses.

A more satisfactory approach is based in part on the running-interval smooth described in Section 14.1. So in particular, it is *not* assumed that a straight line provides an adequate summary of the data; in the event it does, all indications are that the method described here continues to perform relatively well in terms of both Type I errors and power (Wilcox, 1997b). Even under normality, the conventional ANCOVA method appears to have only a minor advantage.

To elaborate on how the method is applied, first assume that an X value has been chosen with the goal of computing a confidence interval for $m_1(X) - m_2(X)$. For the jth group, let X_{ij}, $i = 1, \ldots, n_j$ be values of the predictors that are available. The value $m_j(X)$ is estimated as described in Section 14.2. That is, for fixed j, estimate $m_j(X)$ using the Y_{ij} values corresponding to the X_{ij} values that are close to X. Let $N_j(X)$ be the number of observations used to compute the estimate of $m_j(X)$. That is, $N_j(X)$ is the number of points in the jth group that are close to X, which in turn is the number of Y_{ij} values used to estimate $m_j(X)$. Provided that both $N_1(X)$ and $N_2(X)$ are not too small, a reasonably accurate confidence interval for $m_1(X) - m_2(X)$ can be computed using methods already covered. When comparing the regression lines at more than one design point, confidence intervals for $m_1(X) - m_2(X)$, having simultaneous probability coverage approximately equal to $1 - \alpha$, can be computed as described in Chapter 12.

14.8.1 S-PLUS Functions ancova, ancpb, and ancboot

The S-PLUS function

    ancova(x1,y1,x2,y2,fr1=1,fr2=1,tr=0.2,alpha=0.05,plotit=T,pts = NA)

performs ANCOVA with trimmed means as just described. The arguments x1, y1, x2, y2, tr, and alpha have their usual meaning. The arguments fr1 and fr2 are the spans used for groups 1 and 2, respectively. The argument pts can be used to specify the X values at which the two groups are to be compared. For example, pts=12 will result in comparing the trimmed mean for group 1 (based on the y1 values) to the trimmed mean of group 2 given that $X = 12$. If there is no trimming, the null hypothesis is $H_0 : E(Y_1|X = 12) = E(Y_2|X = 12)$, where Y_1 and Y_2 are the outcome variables of interest corresponding to the two groups. Using pts=c(22,36) will result in testing two hypotheses. The first is $H_0 : m_1(22) = m_2(22)$ and the second is $H_0 : m_1(36) = m_2(36)$. If no values for pts are specified, then the function picks five X values and performs the appropriate tests. The values that it picks are reported in the output illustrated later. Generally, this function controls FWE using the method in Section 12.6. If plotit=T is used, the function also creates a scatterplot and smooth for both groups, with a + and a dashed line indicating the points and the smooth, respectively, for group 2.

The function

    ancpb(x1,y1,x2,y2,est=mom,pts=NA,fr1=1,fr2=1,nboot=599,plotit=T,...)

is like the S-PLUS function ancova, except that only a percentile bootstrap method is used to test hypotheses and by default the measure of location is MOM. Now FWE is controlled as described in Section 12.7.3. In essence, the function creates groups based on the values in pts; in conjunction with the strategy behind the smooth, it creates the appropriate set of linear contrasts and then calls the function pbmcp, described in Section 12.7.4.

Finally, the function

$$\text{ancboot}(x1,y1,x2,y2,fr1=1,fr2=1,tr=0.2,nboot=599,pts=NA,plotit=T)$$

compares trimmed means using a bootstrap-t method. Now FWE is controlled as described in Section 12.7.1.

EXAMPLE. The ANCOVA methods described in this section are illustrated with the Pygmalion data described in Section 14.3.1. The goal is to compare posttest scores for the two groups, taking into account the pretest scores. If the data for the experimental group are stored in the S-PLUS matrix pyge, with the pretest scores in column 1, and the data for the control are stored in pygc, the command

$$\text{ancova}(\text{pyge}[,1],\text{pyge}[,2],\text{pygc}[,1],\text{pygc}[,2])$$

returns

```
  X  nl  n2        DIF      TEST          se      ci.low     ci.hi
 72  12  63  13.39103  1.848819    7.243016   -9.015851  35.79790
 82  16  68  14.79524  1.926801    7.678655   -8.211174  37.80165
101  14  59  22.43243  1.431114   15.674806  -26.244186  71.10905
111  12  47  23.78879  1.321946   17.995286  -35.644021  83.22161
114  12  43  21.59722  1.198906   18.014112  -37.832791  81.02724
```

The first column, headed by X, says that posttest scores are being compared given that pretest scores (X) have the values 72, 82, 101, 111, and 114. The sample sizes used to make the comparisons are given in the next two columns. For example, when $X = 72$, there are 12 observations being used from the experimental group and 63 from the control. That is, there are 12 pretest scores in the experimental group and 63 values in the control group that are close to $X = 72$. The column headed DIF contains the estimated difference between the trimmed means. For example, the estimated difference between the posttest scores, given that $X = 72$, is 13.39. The last two columns indicate the ends of the confidence intervals. These confidence intervals are designed so that FWE is approximately α. The critical value is also reported and is 3.33 for the situation here. All of the confidence intervals contain zero, and none of the test statistics exceeds the critical value, so we fail to detect any differences between posttest scores taking into account the pretest scores of these individuals. ■

Figure 14.13 shows the plot created by the S-PLUS function ancova. Note that $X = 72$ appears near the center of the plot, yet this is the smallest X value used.

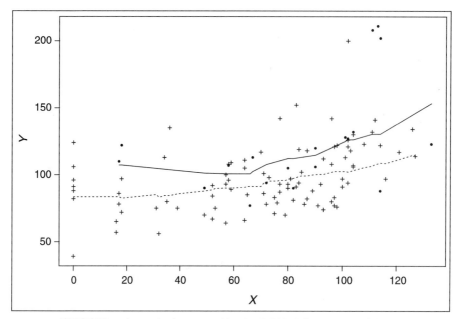

FIGURE 14.13   A plot created by the S-PLUS function ANCOVA.

The reason is that for the experimental group, there are only six cases where $X < 72$. If for example we try to compare the posttest scores given that the pretest scores are $X = 70$, there are too few individuals in the experimental group with X close to 70 to make meaningful comparisons.

If we apply the function ancpb instead, a portion of the output is

```
$mat:
        X  n1  n2
[1,]   72  12  63
[2,]   82  16  68
[3,]  101  14  59
[4,]  111  12  47
[5,]  114  12  43
```

```
      con.num     psihat  sig.level  sig.crit   ci.lower  ci.upper
[1,]        1  12.334699 0.05008347    0.0102  -4.507937  35.28625
[2,]        2   7.907925 0.10350584    0.0127  -7.200000  57.44683
[3,]        3   8.092476 0.12020033    0.0169  -5.282468  62.18519
[4,]        4   6.917874 0.18697830    0.0250  -7.025000  63.63889
[5,]        5   5.388889 0.23706177    0.0500  -5.887805  55.28488
```

Again we fail to find any differences, but note that the length of the confidence intervals are generally substantially shorter than the confidence intervals from the function ancova. For example, with $X = 114$, the length of the confidence interval here is $(55.3 + 5.9) = 61.2$. In contrast, the length of the confidence interval reported by ancova is 118.8, and the ratio of the lengths is .51. Moreover, MOM, used in conjunction with the percentile bootstrap, can provide accurate probability coverage in situations where trimmed means (with trimming less than 20%) and nonbootstrap methods are unsatisfactory.

14.9 Exercises

1. For the predictor X_2 in Table 14.1, MAD/.6745=20.756. If in Section 14.2 you take the span to be $f = 1$, what would be the estimate of Y given that $X_2 = 250$, based on a 20% trimmed mean? What would be the estimate based on MOM?

2. Exercises 1 and 2 in Chapter 13 report data on 29 lakes in Florida. Assuming that you want to predict Y (the mean annual nitrogen concentration) given TW (water retention time), plot a smooth, and comment on whether a straight regression line is advisable.

3. Repeat the previous exercise, only now use NIN (the average influent nitrogen concentration) to predict Y.

4. Again referring to Exercises 1 and 2 in Chapter 13, check for any outliers among the TW and NIN values using the MVE method, eliminate any outliers that are found, and plot a smooth for predicting Y using the S-PLUS functions rungen and runmean in Section 14.2.1. Compare the results to smooths you get when the outliers are not eliminated. Comment on why retaining the outliers among the TW and NIN values might have an adverse effect on fitting a regression plane to the data.

5. For the data in Exercises 1 and 2 of Chapter 13, check for curvature using a partial residual plot.

6. For the data in Exercises 1 and 2 of Chapter 13, eliminate any outliers among the predictor values found by the MVE method, and for the remaining data, test $H_0 : \beta_1 = \beta_2 = 0$ using the methods in Section 14.5.

7. Table 14.2 contains a portion of the data reported by Thompson and Randall-Maciver (1905) dealing with skull measurements of male Egyptians from different time periods. Here, only the data from 4000 bc and 150 ad are reported. Pool the data from both periods and create a smooth using runmean, rungen, and lowess. What do these smooths suggest about the association between X and Y? Check this possibility with the S-PLUS function lintest in Section 14.4.3.

8. For the data in Table 14.2, create a smooth using the 4000 bc data only and compare it to the smooth for the 150 ad data.

9. Compare the regression slopes in the previous exercise using the S-PLUS function reg2ci. Verify that the significance level is .04 when using the default settings.

10. Repeat the previous exercise, only now use the S-PLUS function ancova to compare the regression lines. At which X values do you get a significant difference with $\alpha = .05$?

TABLE 14.2   Skull Measurements

X (4000 BC)	138 131 132 132 143 137 130 136 134 134
	138 121 129 136 140 134 137 133 136
	131 133 135 124 134 130 135 132 129 136 138
X (150 AD)	123 131 126 134 127 138 138 126 132 135
	120 136 135 134 135 134 125 135 125
	129 136 129 126 124 127 125 128 135 129 133
Y (4000 BC)	89 92 99 96 100 89 108 93 102
	99 95 95 109 100 100 97 103 93 96
	101 102 103 93 103 104 100 93 106 114 101
Y (150 AD)	91 95 91 92 86 101 97 92 99
	92 95 101 95 93 96 95 99 96 92
	89 92 97 88 91 97 85 81 103 87 97

X is basibregmatic height of skull; Y is basialveolar length of skull.

11. Data were generated from normal distributions for X_1, X_2, and ϵ. Setting $X_3 = X_1 X_2$, consider the following models:

$$Y = \beta_0 + \beta_1 X_1 + \epsilon,$$
$$Y = \beta_0 + \beta_1 X_2 + \epsilon,$$
$$Y = \beta_0 + \beta_1 X_3 + \epsilon,$$
$$Y = \beta_0 + \beta_1 X_1 + \beta_2 X_2 + \epsilon,$$
$$Y = \beta_0 + \beta_1 X_1 + \beta_2 X_2 + \beta_3 X_3 + \epsilon.$$

The output from the S-PLUS function regpre (in Section 14.6.1) is

```
$estimates:
          apparent.error    boot.est    err.632
[1,]           1.4623288    1.639636   1.625111
[2,]           1.2262127    1.364102   1.353811
[3,]           1.0169318    1.136070   1.124203
[4,]           1.2100394    1.440440   1.422938
[5,]           0.8807802    1.121439   1.095643
```

Based on this output, which model appears to be the best summary of the data?

12. In the previous exercise, imagine that you fail to reject $H_0 : \beta_3 = 0$. Describe some reasons why it might be erroneous to conclude that there is no interaction.

13. An exploratory method for dealing with interactions is to assume that the model $Y = \beta_0 + \beta_1 X_1 + \beta_2 X_2 + \beta_3 X_1 X_2 + \epsilon$ is true and to use a partial residual plot to check the adequacy of the third term ($\beta_3 X_1 X_2$), assuming that Y has a linear association between $\beta_1 X_1$ and $\beta_2 X_2$. Describe some reasons why this approach might be unsatisfactory.

15

RANK-BASED AND NONPARAMETRIC METHODS

This chapter covers basic nonparametric and so-called rank-based methods. Generally, the techniques covered here provide a different and interesting perspective on how groups differ and how variables are related versus the methods covered in previous chapters. Often the methods in this chapter are recommended for dealing with problems that arise when sampling from nonnormal distributions, and so one goal is to try to convey their relative merits versus techniques covered in previous chapters. Many conventional rank-based methods suffer from serious practical problems when comparing groups with different distributions, but substantial progress has been made regarding how to overcome these difficulties.

15.1 Comparing Two Independent Groups

This section describes methods for comparing two independent groups. We begin with a classic technique, outline its practical problems, and then cover modern methods for dealing with these issues.

15.1.1 Wilcoxon–Mann–Whitney Test

The standard rank-based method for comparing two independent groups is called the Wilcoxon–Mann–Whitney (WMW) test. It was originally derived by Wilcoxon (1945), and later it was realized that Wilcoxon's method was the same as a procedure proposed by Mann and Whitney (1947). To describe the basic goal, imagine you randomly sample an observation from the first group and do the same for the second. Now temporarily assume that these two observations cannot have equal values. For example, if the observation from the first group is 6, it is assumed that there is zero probability that you will get 6 when sampling from the other group. Let p be the probability that a randomly sampled observation from the first group is less than a randomly sampled observation from the second. If the groups do not differ in any way, meaning that they have identical distributions, then $p = .5$. Also, p provides a perspective (a measure of effect size) on how groups differ not provided by any of the techniques covered in previous chapters. So a general goal of interest is making inferences about p based on observations we make. This includes estimating p, computing a confidence interval for p, and testing the hypothesis

$$H_0 : p = .5. \tag{15.1}$$

First consider the problem of estimating p. For illustrative purposes suppose we observe

Group 1:	30, 60, 28, 38, 42, 54
Group 2:	19, 21, 27, 73, 71, 25, 59, 61

Now focus on the first value in the first group, 30, and notice that it is less than four of the eight observations in the second group. So a reasonable estimate of p is 4/8. In a similar manner, the second observation in the first group is 60; it is less than three of the values in the second group, so a reasonable estimate of p is 3/8. These two estimates of p differ, and a natural way of combining them into a single estimate of p is to average them. More generally, if we have n_1 observations in group 1 and n_2 observations in group 2, focus on the ith observation in the first group and suppose this value is less than V_i of the observation in group 2. So based on the ith observation in group 1, an estimate of p is V_i/n_2, and we have n_1 estimates of p: $V_1/n_2, \ldots, V_{n_1}/n_2$. To combine these n_1 estimates of p into a single estimate, average them, yielding

$$\hat{p} = \frac{1}{n_1 n_2} \sum V_i. \tag{15.2}$$

As is usually done, let

$$U = n_1 n_2 \hat{p}. \tag{15.3}$$

The quantity U is called the Mann–Whitney U statistic; typically a test of $H_0 : p = .5$ is described in terms of U. If $p = .5$, it can be shown that $E(U) = n_1 n_2/2$. More generally,

$$E\left(\frac{U}{n_1 n_2}\right) = p.$$

Next, consider the problem of estimating VAR(U), the squared standard error of U. If we assume there are no tied values and that both groups have identical distributions, the classic estimate of the standard error can be derived. (By "no tied values" it is meant that each observed value occurs only once. So if we observe the value 6, for example, it never occurs again among the remaining observations.) The expression for VAR(U) is

$$\sigma_u^2 = \frac{n_1 n_2 (n_1 + n_2 + 1)}{12}.$$

This means that the null hypothesis given by Equation (15.1) can be tested with

$$Z = \frac{U - n_1 n_2/2}{\sigma_u}, \tag{15.4}$$

which has, approximately, a standard normal distribution when the assumptions are met and H_0 is true. In particular, reject if

$$|Z| \geq Z_{1-\alpha/2},$$

where $Z_{1-\alpha/2}$ is the $1 - \alpha/2$ quantile of a standard normal distribution. Hodges, Ramsey, and Wechsler (1990) suggest estimating the p-value as outlined in Box 15.1.

BOX 15.1 Computing the p-Value of the Wilcoxon–Mann–Whitney Test

Let

$$y = \frac{1}{\sigma_u}\left(U + 0.5 - \frac{n_1 n_2}{2}\right),$$

$$k = \frac{20 n_1 n_2 (n_1 + n_2 + 1)}{n_1^2 + n_2^2 + n_1 n_2 + n_1 + n_2},$$

$$S = y^2,$$

$$T_1 = S - 3,$$

$$T_2 = \frac{155 S^2 - 416 S - 195}{42},$$

$$c = 1 + \frac{T_1}{k} + \frac{T_2}{k^2}.$$

If cy is negative, the one-sided p-value is

$$P(Z \leq cy),$$

where Z is a standard normal random variable. So the p-value can be determined from Table 1 in Appendix B. If cy is positive, the one-sided p-value is

$$1 - P(Z \leq cy).$$

The two-sided p-value is

$$2[1 - P(Z \leq |cy|)].$$

EXAMPLE. Continuing the illustration using the data just following Equation (15.1), it can be seen that $\hat{p} = .479$, so $U = 23$ and

$$Z = \frac{23 - 24}{7.75} = -0.129.$$

With $\alpha = .05$, the critical value is 1.96; $|Z|$ is less than 1.96, so fail to reject. ■

Sometimes the Wilcoxon–Mann–Whitney test is described as a method for comparing medians. However, it is relatively unsatisfactory for this purpose because it is not based on a direct estimate of the population medians. For example, there are situations where power decreases as the difference between the population medians increases, and confidence intervals for the difference cannot be computed (Kendall & Stuart, 1973; Hettmansperger, 1984).

Another characterization of the Wilcoxon–Mann–Whitney method is that it tests the hypothesis that two groups have identical distributions. In symbols, if F_1 and F_2 are the distributions of the two groups being compared, the goal is to test

$$H_0 : F_1(x) = F_2(x), \tag{15.5}$$

which says that for any x, the probability that a randomly sampled observation is less than or equal to x is the same for both groups. A more accurate description of the method in Box 15.1 is that it approximates the p-value of the Wilcoxon–Mann–Whitney test when the goal is to test Equation (15.5) as opposed to testing $H_0 : p = .5$.

When tied values occur with probability zero and the goal is to test $H_0 : F_1(x) = F_2(x)$, the probability of a Type I error can be controlled exactly by computing a critical value as described, for example, in Hogg and Craig (1970, p. 373). Let

$$W = U + \frac{n_2(n_2 + 1)}{2} \tag{15.6}$$

and suppose H_0 [given by Equation (15.5)] is rejected if W is sufficiently large or small. If H_0 is rejected when

$$W \leq c_L$$

or when

$$W \geq c_U,$$

where c_L is read from Table 13 in Appendix B and

$$c_U = n_2(n_2 + n_1 + 1) - c_L,$$

then the actual probability of a Type I error will not exceed .05 under random sampling.

15.1.2 S-PLUS Function wmw

Using the data stored in the S-PLUS variables x and y, the S-PLUS function

$$\text{wmw}(x,y)$$

computes the significance level of the Wilcoxon–Mann–Whitney test as described in Box 15.1.

15.1.3 Handling Ties and Heteroscedasticity

A practical concern is that if groups differ, then under general circumstances the wrong standard error is being used by the Wilcoxon–Mann–Whitney test in

Equation (15.4), which can result in relatively poor power and an unsatisfactory confidence interval for p. Said another way, if groups have different distributions, generally σ_u^2 is the wrong standard error for U. Another problem is how to handle tied values. Currently there are two general approaches to both of these problems that appear to be relatively effective.

First consider the problem of tied values and note that if we randomly sample a single observation from both groups, there are three possible outcomes: the observation from the first group is greater than the observation from the second, the observations have identical values, and the observation from the first group is less than the observation from the second. The probabilities associated with these three mutually exclusive outcomes are labeled p_1, p_2, and p_3. In symbols, if X_{ij} represents the ith observation from the jth group, then

$$p_1 = P(X_{i1} > X_{i2}),$$
$$p_2 = P(X_{i1} = X_{i2}),$$
$$p_3 = P(X_{i1} < X_{i2}).$$

So in the notation of Section 15.1.1, $p_3 = p$. Cliff (1996) focuses on testing

$$H_0 : \delta = p_1 - p_3 = 0. \tag{15.7}$$

In the event tied values occur with probability zero, in which case $p_2 = 0$, Equation (15.7) becomes $H_0 : p_1 = p_3 = .5$, which is the same as Equation (15.1). It can be shown that another way of expressing Equation (15.7) is with

$$H_0 : p_3 + .5p_2 = .5.$$

For convenience, let $P = p_3 + .5p_2$, in which case this last equation becomes

$$H_0 : P = .5. \tag{15.8}$$

Of course, when there are no tied values, $P = p_3$. The parameter δ [in Equation (15.7)] is related to P in the following manner:

$$\delta = 1 - 2P, \tag{15.9}$$

so

$$P = \frac{1 - \delta}{2}. \tag{15.10}$$

Cliff derived a heteroscedastic confidence interval for δ, which is computed as summarized in Box 15.2. If the confidence interval for δ does not contain zero, reject $H_0 : \delta = 0$. When ties occur with probability zero, this is tantamount to rejecting $H_0 : p = .5$.

BOX 15.2 Cliff's Method for Two Independent Groups

As usual, let X_{ij} be the ith observation from the jth group, $j = 1, 2$. For the ith observation in group 1 and the bth observation in group 2, let

$$d_{ib} = \begin{cases} -1 & \text{if } X_{i1} < X_{b2}, \\ 0 & \text{if } X_{i1} = X_{b2}, \\ 1 & \text{if } X_{i1} > X_{b2}. \end{cases}$$

An estimate of $\delta = P(X_{i1} > X_{i2}) - P(X_{i1} < X_{i2})$ is

$$\hat{\delta} = \frac{1}{n_1 n_2} \sum_{i=1}^{n_1} \sum_{b=1}^{n_2} d_{ib}, \qquad (15.11)$$

the average of the d_{ib} values. Let

$$\bar{d}_{i.} = \frac{1}{n_2} \sum_b d_{ib},$$

$$\bar{d}_{.b} = \frac{1}{n_2} \sum_i d_{ib},$$

$$s_1^2 = \frac{1}{n_1 - 1} \sum_{i=1}^{n_1} (\bar{d}_{i.} - \hat{\delta})^2,$$

$$s_2^2 = \frac{1}{n_2 - 1} \sum_{b=1}^{n_2} (\bar{d}_{.b} - \hat{\delta})^2,$$

$$\tilde{\sigma}^2 = \frac{1}{n_1 n_2} \sum \sum (d_{ib} - \hat{\delta})^2.$$

Then

$$\hat{\sigma}^2 = \frac{(n_1 - 1)s_1^2 + (n_2 - 1)s_2^2 + \tilde{\sigma}^2}{n_1 n_2}$$

estimates the squared standard error of $\hat{\delta}$. Let z be the $1 - \alpha/2$ quantile of a standard normal distribution. Rather than use the more obvious confidence interval for δ, Cliff (1996, p. 140) recommends

$$\frac{\hat{\delta} - \hat{\delta}^3 \pm z\hat{\sigma}\sqrt{(1 - \hat{\delta}^2)^2 + z^2\hat{\sigma}^2}}{1 - \hat{\delta}^2 + z^2\hat{\sigma}^2}.$$

If there are no tied values, alternative heteroscedastic methods have been proposed by Mee (1990) as well as Fligner and Policello (1981). Currently it seems that for this special case, these methods offer no practical advantage over the method in Box 15.2.

15.1.4 S-PLUS function cid

The S-PLUS function

$$cid(x,y,alpha=.05),$$

written for this book, performs the calculations in Box 15.2.

15.1.5 The Brunner–Munzel Procedure

An alternative approach to both tied values and heteroscedasticity stems from Brunner and Munzel (2000). Their approach is based in part on what are called the *midranks* for handling tied values. To explain, first consider the values 45, 12, 32, 64, 13, and 25. There are no tied values and the smallest value is said to have *rank* 1, the next smallest has rank 2, and so on. A common notation for the rank corresponding to the ith observation is R_i. So in the example, the first observation is $X_1 = 45$ and its rank is $R_1 = 5$. Similarly, $X_2 = 12$ and its rank is $R_2 = 1$.

Now consider a situation where there are tied values: 45, 12, 13, 64, 13, and 25. Putting these values in ascending order yields 12, 13, 13, 25, 45, 64. So the value 12 gets a rank of 1, but there are two identical values having a rank of 2 and 3. The *midrank* is simply the average of the ranks among the tied values. Here, this means that the rank assigned to the two values equal to 13 would be $(2 + 3)/2 = 2.5$, the average of their corresponding ranks. So the ranks for all six values would be 1, 2.5, 2.5, 4, 5, 6.

Generalizing, consider

$$7, 7.5, 7.5, 8, 8, 8.5, 9, 11, 11, 11.$$

There are 10 values, so if there were no tied values, their ranks would be 1, 2, 3, 4, 5, 6, 7, 8, 9, and 10. But because there are two values equal to 7.5, their ranks are averaged, yielding a rank of 2.5 for each. There are two values equal to 8; their original ranks were 4 and 5, so their final ranks (their midranks) are both 4.5. There are three values equal to 11; their original ranks are 8, 9, and 10, the average of these ranks is 9, so their midranks are all equal to 9. So the ranks for the 10 observations are

$$1, 2.5, 2.5, 4.5, 4.5, 6, 7, 9, 9, 9.$$

Now consider testing $H_0 : P = .5$, where P is as defined in Section 15.1.3. As usual, let X_{ij} be the ith observation from the jth group ($i = 1, \ldots, n_j$; $j = 1,2$). To apply the Brunner–Munzel method, first pool all $N = n_1 + n_2$ observations and assign ranks. In the event there are tied values, ranks are averaged as just illustrated. The results for the jth group are labeled R_{ij}, $i = 1, \ldots, n_j$. That is, R_{ij} is the rank corresponding to X_{ij} among the pooled values. Let $\bar{R}_1$ be the average of the ranks corresponding to group 1 and $\bar{R}_2$ be the average for group 2. So for the jth group,

$$\bar{R}_j = \frac{1}{n_j} \sum_{i=1}^{n_j} R_{ij}.$$

Next, for the first group, rank the observations, ignoring group 2, and label the results $V_{11}, \ldots, V_{n_1 1}$. Do the same for group 2 (ignoring group 1), and label the ranks $V_{12}, \ldots, V_{n_2 2}$. The remaining calculations are shown in Box 15.3.

BOX 15.3 The Brunner–Munzel Method for Two Independent Groups

Compute

$$S_j^2 = \frac{1}{n_j - 1} \sum_{i=1}^{n_j} \left(R_{ij} - V_{ij} - \bar{R}_j + \frac{n_j + 1}{2} \right)^2,$$

$$s_j^2 = \frac{S_j^2}{(N - n_j)^2},$$

$$s_e = \sqrt{N} \sqrt{\frac{s_1^2}{n_1} + \frac{s_2^2}{n_2}},$$

$$U_1 = \left(\frac{S_1^2}{N - n_1} + \frac{S_2^2}{N - n_2} \right)^2,$$

$$U_2 = \frac{1}{n_1 - 1} \left(\frac{S_1^2}{N - n_1} \right)^2 + \frac{1}{n_2 - 1} \left(\frac{S_2^2}{N - n_2} \right)^2.$$

The test statistic is

$$W = \frac{\bar{R}_2 - \bar{R}_1}{\sqrt{N} s_e},$$

and the degrees of freedom are

$$\hat{v} = \frac{U_1}{U_2}.$$

Decision Rule
    Reject $H_0 : P = .5$ if $|W| \geq t$, where t is the $1 - \alpha/2$ quantile of a Student's T-distribution with $\hat{v}$ degrees of freedom. An estimate of P is

$$\hat{P} = \frac{1}{n_1} \left(R_2 - \frac{n_2 + 1}{2} \right) = \frac{1}{N} \left(\bar{R}_2 - \bar{R}_1 \right) + \frac{1}{2}.$$

The estimate of δ is

$$\hat{\delta} = 1 - 2\hat{P}.$$

An approximate $1 - \alpha$ confidence interval for P is

$$\hat{P} \pm t s_e.$$

There is a connection between the method just described and the Wilcoxon–Mann–Whitney test that is worth mentioning:

$$U = n_2 \bar{R}_2 - \frac{n_2(n_2 + 1)}{2}.$$

That is, if you sum the ranks of the second group (which is equal to $n_2\bar{R}_2$) and subtract $n_2(n_2 + 1)/2$, you get the Wilcoxon–Mann–Whitney U statistic given by Equation (15.3). Many books describe the Wilcoxon–Mann–Whitney method in terms of U rather than the approach used here.

Note that both the Cliff and Brunner–Munzel rank-based methods offer protection against low power due to outliers. If, for example, the largest observation among a batch of numbers is increased from 12 to 1 million, its rank does not change. But how should one choose between rank-based methods covered here versus the robust methods in Chapter 8? If our only criterion is high power, both perform well, with weak evidence that in practice, robust methods are a bit better. But the more important point is that they provide different information about how groups compare. Some authorities argue passionately that as a measure of effect size, P and δ, as defined in this section, reflect what is most important and what we want to know. Others argue that measures of location also provide useful information; they reflect what is typical and provide a sense of the magnitude of the difference between groups that is useful and not provided by rank-based methods. The only certainty is that at present, there is no agreement about which approach should be preferred or even if it makes any sense to ask the question of which is better.

Often Cliff's method gives similar results to the Brunner–Munzel technique. But when the probability of a tied value is high and there are relatively few outcomes possible (i.e., there are few possible X values that can be observed), the Brunner–Munzel procedure can have a higher Type I error probability than Cliff. Based on a very limited comparison of the two methods, the author has found that generally there seems to be little separating Cliff's approach from Brunner–Munzel. However, situations can be constructed where, with many tied values, Cliff's approach seems to be better at guaranteeing an actual Type I error probability less than the nominal α level; and when testing at the .05 level, Cliff's method seems to do an excellent job of avoiding actual Type I error probabilities less than .04. In contrast, the Brunner–Munzel method can have an actual Type I error rate close to .07 when tied values are common and sample sizes are small. This suggests the possibility that the Brunner–Munzel method might have more power and reject when Cliff's method does not, but the issue of how these two methods compare needs closer scrutiny.

15.1.6 S-PLUS function bmp

The S-PLUS function

$$\text{bmp(x,y,alpha=.05)}$$

performs the Brunner–Munzel method. It returns the p-value (or significance level) when testing $H_0 : P = .5$ plus an estimate of P labeled phat and a confidence interval for P labeled ci.p (an estimate of δ, labeled d.hat, is returned as well).

EXAMPLE. Table 8.3 reports data from a study of hangover symptoms among sons of alcoholics versus a control. Note that there are many tied values among these data. In the second group, for example, 14 of the 20 values are zero. Welch's test for means has a significance level of .14, Yuen's test has a significance level of .076, the Brunner–Munzel method has a significance level of .042, and its .95 confidence interval for P is $(.167, .494)$. Cliff's method also rejects at the .05 level, the .95 confidence interval for δ being $(0.002, 0.60)$. ■

15.1.7 The Kolmogorov–Smirnov Test

Yet another way of testing the hypothesis that two independent groups have identical distributions is with the so-called Kolmogorov–Smirnov test. This test forms the basis of the shift function in Section 8.11.2. Like the WMW test in Section 15.1.1, exact control over the probability of a Type I error can be had by assuming random sampling only. When there are no tied values, the method in Kim and Jennrich (1973) can be used to compute the exact significance level. With tied values, the exact significance level can be computed with a method derived by Schroër and Trenkler (1995). The S-PLUS function supplied to perform the Kolmogorov–Smirnov test has an option for computing the exact significance level, but the details of the method are not given here. (Details can be found in Wilcox, 1997a.)

To apply the Kolmogorov–Smirnov test, let $\hat{F}_1(x)$ be the proportion of observations in group 1 that are less than or equal to x, and let $\hat{F}_2(x)$ be the corresponding proportion for group 2. Let

$$U_i = |\hat{F}_1(X_{i1}) - \hat{F}_2(X_{i1})|,$$

$i = 1, \ldots, n_1$. In other words, for X_{i1}, the ith observation in group 1, compute the proportion of observations in group 1 that are less than or equal to X_{i1}, do the same for group 2, take the absolute value of the difference, and label the result U_i. Repeat this process for the observations in group 2 and label the results

$$V_i = |\hat{F}_1(X_{i2}) - \hat{F}_2(X_{i2})|,$$

$i = 1, \ldots, n_2$. The Kolmogorov–Smirnov test statistic is

$$KS = \max\{U_1, \ldots, U_{n_1}, V_1, \ldots, V_{n_2}\}, \tag{15.12}$$

the largest of the pooled U and V values. For large sample sizes, an approximate critical value when $\alpha = .05$ is

$$1.36\sqrt{\frac{n_1 + n_2}{n_1 n_2}}.$$

Reject when KS is greater than or equal to the critical value. When there are no tied values, the Kolmogorov–Smirnov test can have relatively high power; but with ties, its power can be relatively low.

15.1.8 S-PLUS Function ks

The S-PLUS function

$$ks(x,y,w{=}F,sig{=}T)$$

performs the Kolmogorov–Smirnov test. The argument w can be used to invoke a weighted version of the Kolmogorov–Smirnov test not covered here. (See, for example, Wilcox, 1997a.) By default, w=F, meaning that the version described here will be used. With sig=T, the exact critical value will be used. With large sample sizes, computing the exact critical value can result in high execution time. Setting sig=F avoids this problem, but now only the approximate critical value with $\alpha = .05$ can be used.

EXAMPLE. For the data in Table 8.3, the function ks returns

```
$test:
[1] 0.35

$critval:
[1] 0.4300698

$siglevel:
[1] 0.03942698
```

This says that the Kolmogorov–Smirnov test statistic is KS = 0.35, the approximate .05 critical value is 0.43, which is greater than KS, yet the exact significance level, assuming only random sampling, is .039. ■

15.2 Comparing More Than Two Groups

15.2.1 The Kruskall–Wallis Test

The best-known rank-based method for comparing multiple groups is the Kruskall–Wallis test. The goal is to test

$$H_0 : F_1(x) = F_2(x) = \cdots = F_J(x), \tag{15.13}$$

the hypothesis that J independent groups have identical distributions. The method begins by pooling all $N = \sum n_j$ observations and assigning ranks. In symbols, if X_{ij} is the ith observation in the jth group, let R_{ij} be its rank among the pooled data. When there are tied values, use midranks, as described in connection with the Brunner–Munzel method. Next, sum the ranks for each group. In symbols, compute

$$R_j = \sum_{i=1}^{n_j} R_{ij},$$

$(j = 1, \ldots, J)$. Letting

$$S^2 = \frac{1}{N-1} \left(\sum_{j=1}^{J} \sum_{i=1}^{n_j} R_{ij}^2 - \frac{N(N+1)^2}{4} \right),$$

the test statistic is

$$T = \frac{1}{S^2} \left(-\frac{N(N+1)^2}{4} + \sum \frac{R_j^2}{n_j} \right).$$

If there are no ties, S^2 simplifies to

$$S^2 = \frac{N(N+1)}{12},$$

and T becomes

$$T = -3(N+1) + \frac{12}{N(N+1)} \sum \frac{R_j^2}{n_j}.$$

The hypothesis of identical distributions is rejected if $T \geq c$, where c is some appropriate critical value. For small sample sizes, exact critical values are available from Iman, Quade, and Alexander (1975). For large sample sizes, the critical value is approximately equal to the $1 - \alpha$ quantile of a chi-squared distribution with $J - 1$ degrees of freedom.

EXAMPLE. Table 15.1 shows data for three groups and the corresponding ranks. For example, after pooling all $N = 10$ values, $X_{11} = 40$ has a rank of $R_{11} = 1$, the value 56 has a rank of 6, and so forth. The sum of the ranks corresponding to each group are $R_1 = 1 + 6 + 2 = 9, R_2 = 3 + 7 + 8 = 18$, and $R_3 = 9 + 10 + 5 + 4 = 28$. The number of groups is $J = 3$, so the degrees of freedom are $\nu = 2$; from Table 3 in Appendix B, the critical value is approximately $c = 5.99$ with $\alpha = .05$. Because there are no ties among the N observations,

$$T = -3(10+1) + \frac{12}{10 \times 11} \left(\frac{9^2}{3} + \frac{18^2}{3} + \frac{28^2}{4} \right) = 3.109.$$

Because $3.109 < 5.99$, fail to reject. That is, you are unable to detect a difference among the distributions. ■

15.2.2 The BDM Method

The Kruskall–Wallis test performs relatively well when the null hypothesis of identical distributions is true, but concerns arise when the null hypothesis is false. In particular, the Kruskall–Wallis test is homoscedastic, which might affect power.

TABLE 15.1 Hypothetical Data Illustrating the Kruskall–Wallis Test

Group 1		Group 2		Group 3	
X_{i1}	R_{i1}	X_{i2}	R_{i2}	X_{i3}	R_{i3}
40	1	45	3	61	9
56	6	58	7	65	10
42	2	60	8	55	5
				47	4

A heteroscedastic analog of the Kruskall–Wallis test that allows tied values was derived by Brunner, Dette, and Munk (1997), which will be called the BDM method. Again the goal is to test the hypothesis that all J groups have identical distributions. The basic idea is that if J independent groups have identical distributions and we assign ranks to the pooled data as was done in the Kruskall–Wallis test, then for each group the average of the ranks should be approximately equal. (This greatly oversimplifies the technical issues.) To apply it, compute the ranks of the pooled data as was done in connection with the Kruskall–Wallis test. As before, let $N = \sum n_j$ be the total number of observations. The remaining calculations are relegated to Box 15.4. The values in the vector $\mathbf{Q}$ in Box 15.4 are called the *relative effects* and reflect the average ranks among the groups, which provide some sense of how the groups compare.

BOX 15.4 BDM Heteroscedastic Rank-Based ANOVA Method

Let $\bar{R}_j$ be the average of the pooled ranks corresponding to the jth group. So if R_{ij} is the rank of X_{ij} after the data are pooled, then

$$\bar{R}_j = \frac{1}{n_j} \sum_{i=1}^{n_j} R_{ij}.$$

Let

$$\mathbf{Q} = \frac{1}{N} \left(\bar{R}_1 - \frac{1}{2}, \ldots, \bar{R}_J - \frac{1}{2} \right).$$

For the jth group, compute

$$s_j^2 = \frac{1}{N^2 (n_j - 1)} \sum_{i=1}^{n_j} (R_{ij} - \bar{R}_j)^2,$$

and let

$$\mathbf{V} = N \, \mathrm{diag} \left\{ \frac{s_1^2}{n_1}, \ldots, \frac{s_J^2}{n_J} \right\}.$$

Continued

BOX 15.4 (*Continued*)
Let **I** be a *J*-by-*J* identity matrix, let **J** be a *J*-by-*J* matrix of 1's, and set

$$M = I - \frac{1}{J}J.$$

(The diagonal entries in **M** have a common value, a property required to satisfy certain theoretical restrictions.) The test statistic is

$$F = \frac{N}{\text{tr}(M_{11}V)}QMQ', \tag{15.14}$$

where tr indicates trace and **Q**′ is the transpose of the matrix **Q**. (See Appendix C for how the trace and transpose of a matrix are defined.)

Decision Rule
    Reject if $F \geq f$, where f is the $1 - \alpha$ quantile of an *F*-distribution with

$$\nu_1 = \frac{M_{11}[\text{tr}(V)]^2}{\text{tr}(MVMV)} \qquad \nu_2 = \frac{[\text{tr}(V)]^2}{\text{tr}(V^2 \Lambda)}$$

degrees of freedom and

$$\Lambda = \text{diag}\{(n_1 - 1)^{-1}, \ldots, (n_J - 1)^{-1}\}.$$

An alternative heteroscedastic method, one that assumes there are no tied values, was derived by Rust and Fligner (1984). In the event there are no tied values, it is unknown how the Rust–Fligner and BDM methods compare. Perhaps the use of midranks in conjunction with the Rust–Fligner procedure performs reasonably well when tied values occur, but it seems that this issue has not been investigated. (The S-PLUS function rfanova(x), written for this book, performs the Rust–Fligner technique but is not described because currently it seems that the BDM method suffices.)

15.2.3   S-PLUS Function bdm

The S-PLUS function

$$bdm(x)$$

has been supplied to perform the BDM rank-based ANOVA described in Box 15.4. Here, x can have list mode or it can be a matrix with columns corresponding to groups. The function returns the value of the test statistic, the degrees of freedom, the vector of relative effects, which is labeled q.hat, and the significance level.

EXAMPLE. For the schizophrenia data in Table 9.1, the S-PLUS function bmd returns a significance level of .040. The relative effect sizes (the **Q** values) are

Continued

EXAMPLE. (*Continued*) reported as

```
$ output$q.hat:
          [,1]
[1,]    0.4725
[2,]    0.4725
[3,]    0.3550
[4,]    0.7000
```

So the conclusion is that the distributions associated with these four groups differ, and we see that the average of the ranks among the pooled data is smallest for group 3 and highest for group 4. This is consistent with the means. That is, group 3 has the smallest mean and group 4 the largest. The same is true when using a 20% trimmed mean or MOM. ■

15.3 Multiple Comparisons Among Independent Groups

One way of extending the Cliff and the Brunner–Munzel methods when comparing all pairs of J groups, $J > 2$, is to proceed in the manner used to derive Dunnett's T3 (described in Section 12.3.1). In particular, use a critical value from Table 10 in Appendix B (which reports some quantiles of the Studentized maximum modulus distribution). Here, assuming all pairwise comparisons are to be made, there are a total of $C = (J^2 - J)/2$ hypotheses to be tested. First consider an extension of the Brunner–Munzel method. When comparing group j to group k, simply perform the calculations in Section 15.1.5, ignoring the other groups. Let $\hat{v}_{jk}$ be the resulting degrees of freedom, let s_{ejk} be the corresponding value of s_e, and let $\hat{P}_{jk}$ be the estimate of P. So when there are no ties, P_{jk} is the probability that a randomly sampled observation from group j is less than a sampled observation from group k. The confidence interval for P_{jk} is

$$\hat{P}_{jk} \pm cs_{ejk},$$

where c is the critical value read from Table 10 in Appendix B, which depends on C and $\hat{v}_{jk}$.

As for extending Cliff's method, the same strategy can be used. That is, proceed as described in Box 15.2, but when computing a confidence interval for δ, replace z with a critical value read from Table 10 with degrees of freedom $v = \infty$. For example, if there are four groups and all pairwise comparisons are to be performed, the total number of hypotheses to be tested is $C = 6$, so if $\alpha = .05$, the critical value is 2.63. That is, replace z with 2.63 in Box 15.2.

15.3.1 S-PLUS Functions cidmul and bmpmul

The S-PLUS function

$$\text{bmpmul(x,alpha=.05)}$$

performs the extension of the bmp method in Box 15.3 when the goal is to perform all pairwise comparisons among the groups stored in x, where x has list mode or is a matrix with columns corresponding to groups. The S-PLUS function

$$\text{cidmul}(x, \text{alpha}=.05)$$

is like bmpmul, except that Cliff's method for making inferences about δ is used instead. Both of these functions are limited to $\alpha = .05$ and .01. If the argument alpha has any value other than .05, $\alpha = .01$ is assumed.

15.3.2 Multiple Comparisons Based on BDM

Rather than perform multiple comparisons based on estimates of P, another approach is to compare each pair of groups in terms of their distributions. That is, for the jth and kth groups, test

$$H_0 : F_j(x) = F_k(x),$$

with the goal of controlling FWE among all the hypotheses to be tested. For each pair of groups, the significance level associated with this hypothesis can be computed as described in Box 15.4. Here, FWE is controlled using either Rom's method, which is described in Section 12.8.2, or the Benjamini–Hochberg technique in Section 12.3.3.

15.3.3 S-PLUS r1mcp

The S-PLUS function

$$\text{r1mcp}(x, \text{alpha}=.05, \text{bhop}=F)$$

performs all pairwise comparisons of J independent groups, each comparison being based on the BDM method in Section 15.2.2; if bhop=F, then FWE is controlled via Rom's procedure. Here, the argument alpha corresponds to the desired value for FWE. Setting bhop=T, the Benjamini and Hochberg method (described in Section 12.3.3) is used to control FWE.

15.4 Two-Way Designs

The BDM method described in Section 15.3.2 can be extended to a two-way design. (For an extension to a mixed design where one factor is fixed and the other is random, see Brunner & Dette, 1992.) Following the notation in Chapter 10, now we have J levels associated with the first factor and K levels for the other, for a total of JK independent groups. The observations are represented by X_{ijk}, $i = 1, \ldots, n_{jk}$; $j = 1, \ldots, J$; and $k = 1, \ldots, K$. Let $F_{jk}(x)$ be the distribution associated with the jth and kth levels. So, for example, $F_{23}(6)$ is the probability that for the second level of the first factor and the third level of the second factor, a randomly sampled observation has a value less than or equal to 6.

Pool all of the observations and assign ranks. In the event of tied values, ranks are averaged as in Section 15.1.5. For convenience, let $L = JK$ represent the total number of groups and assume the first K groups correspond to level 1 of factor A, the next K correspond to level 2 of factor A, and so on. So the groups are arranged as described in Section 10.3.1. Furthermore, the sample size for the ℓth group is n_ℓ. For each of the L groups, compute the average of the corresponding ranks, label the results $\bar{R}_1, \ldots, \bar{R}_L$, and let

$$s_\ell^2 = \frac{1}{N^2(n_\ell - 1)} \sum_{i=1}^{n_\ell} (R_{i\ell} - \bar{R}_\ell)^2,$$

where $N = \sum n_\ell$ is the total sample size and $R_{i\ell}$ is the rank of the ith observation in the ℓth group. Main effects and interactions are tested as described in Box 15.5.

BOX 15.5 BDM Two-Way, Heteroscedastic Rank-Based ANOVA Method

Set

$$V = N \, \mathrm{diag} \left\{ \frac{s_1^2}{n_1}, \ldots, \frac{s_L^2}{n_L} \right\}.$$

Let I_J be a J-by-J identity matrix, let H_J be a J-by-J matrix of 1's, and let

$$P_J = I_J - \frac{1}{J} H_J, \qquad M_A = P_J \otimes \frac{1}{K} H_K,$$

$$M_B = \frac{1}{J} H_J \otimes P_K, \qquad M_{AB} = P_J \otimes P_K.$$

(The notation $\otimes$ refers to the right Kronecker product, which is described in Appendix C.) As in Box 15.4, let

$$Q = \frac{1}{N} \left(\bar{R}_1 - \frac{1}{2}, \ldots, \bar{R}_L - \frac{1}{2} \right)$$

be the relative effects. The test statistics are:

$$F_A = \frac{N}{\mathrm{tr}(M_{A11}V)} Q M_A Q', \qquad F_B = \frac{N}{\mathrm{tr}(M_{B11}V)} Q M_B Q',$$

$$F_{AB} = \frac{N}{\mathrm{tr}(M_{AB11}V)} Q M_{AB} Q'.$$

Decision rules

For factor A, reject if $F_A \geq f$, where f is the $1 - \alpha$ quantile of an F-distribution with degrees of freedom

$$\nu_1 = \frac{M_{A11}[\mathrm{tr}(V)]^2}{\mathrm{tr}(M_A V M_A V)}, \qquad \nu_2 = \frac{[\mathrm{tr}(V)]^2}{\mathrm{tr}(V^2 \Lambda)},$$

Continued

BOX 15.5 (*Continued*)
where

$$\mathbf{\Lambda} = \text{diag}\{(n_1 - 1)^{-1}, \dots, (n_L - 1)^{-1}\}.$$

Here M_{A11} is the first diagonal element of the matrix $\mathbf{M}_A$. (By design, all of the diagonal elements of $\mathbf{M}_A$ have a common value.) For factor B, reject if $F_B \geq f$, where

$$\nu_1 = \frac{M_{B11}[\text{tr}(\mathbf{V})]^2}{\text{tr}(\mathbf{M}_B\mathbf{V}\mathbf{M}_B\mathbf{V})}.$$

(The value for ν_2 remains the same.) As for the hypothesis of no interactions, reject if $F_{AB} \geq f$, where now

$$\nu_1 = \frac{M_{AB11}[\text{tr}(\mathbf{V})]^2}{\text{tr}(\mathbf{M}_{AB}\mathbf{V}\mathbf{M}_{AB}\mathbf{V})}$$

and ν_2 is the same value used to test for main effects.

Before continuing it might help to be more precise about how the null hypotheses are formulated. The basic idea stems from Akritas and Arnold (1994) and the particular case of a two-way design was addressed by Akritas, Arnold, and Brunner (1997). For any value x, let

$$\bar{F}_{j.}(x) = \frac{1}{K}\sum_{k=1}^{K} F_{jk}(x)$$

be the average of the distributions among the K levels of factor B corresponding to the jth level of factor A.

EXAMPLE. Consider a 2-by-2 design and suppose that $F_{11}(6) = .5$ and $F_{12}(6) = .3$. That is, for the first level of factor A and the first level of factor B, the probability that an observation is less than 6 is .5. For the first level of factor A and the second level of factor B, this probability is .3. So $\bar{F}_{1.}(6) = (.5 + .3)/2 = .4$. More generally, for any x,

$$\bar{F}_{1.}(x) = \frac{F_{11}(x) + F_{12}(x)}{2}.$$

■

The hypothesis of no main effects for factor A is

$$H_0 : \bar{F}_{1.}(x) = \bar{F}_{2.}(x) = \cdots = \bar{F}_{J.}(x).$$

for any x. Letting

$$\bar{F}_{.k}(x) = \frac{1}{J}\sum_{j=1}^{J} F_{jk}(x)$$

be the average of the distributions for the kth level of factor B, the hypothesis of no main effects for factor B is

$$H_0 : \bar{F}_{.1}(x) = \bar{F}_{.2}(x) = \cdots = \bar{F}_{.K}(x).$$

As for interactions, first consider a 2-by-2 design. Then no interaction is taken to mean that for any x,

$$F_{11}(x) - F_{12}(x) = F_{21}(x) - F_{22}(x),$$

which has a certain similarity to how no interaction based on means was defined in Chapter 10. Here, no interaction in a J-by-K design means that for any two rows and any two columns, there is no interaction as just described. From a technical point of view, a convenient way of stating the hypothesis of no interactions among all JK groups is with

$$H_0 : F_{jk}(x) - \bar{F}_{j.}(x) - \bar{F}_{.k}(x) + \bar{F}_{..}(x) = 0,$$

for any x, all j $(j = 1, \ldots, J)$ and all k $(k = 1, \ldots, K)$, where

$$\bar{F}_{..}(x) = \frac{1}{JK} \sum_{j=1}^{J} \sum_{k=1}^{K} F_{jk}(x).$$

15.4.1 S-PLUS Function bdm2way

The S-PLUS function

$$bdm2way(J, K, x)$$

performs the two-way ANOVA method described in Box 15.5.

15.5 Multiple Comparisons in a Two-Way Design

One approach when dealing with multiple comparisons among the levels of each factor in a two-way design is to perform each comparison of interest using the BDM method and control FWE using Rom's method or perhaps the Benjamini–Hochberg technique described in Chapter 12.

To elaborate, first consider factor A and imagine that all pairwise comparisons among the J levels are to be performed. This means that for any two levels — say, j and j' — the goal is to test

$$H_0 : \bar{F}_{j.} = \bar{F}_{j'.}$$

and simultaneously to control FWE among all pairwise comparisons. As noted in Chapter 12, if all pairwise comparisons are to be made among the J levels of factor A, there are a total of $C = (J^2 - J)/2$ hypotheses to be tested. To compare level j to level j', compute F and v_2 as indicated in Box 15.4, ignoring all other groups. (That is, perform the calculations in Box 15.4 as if you were testing the hypothesis of no main effect for factor A in a 2-by-K design, in which case $v_1 = 1$.) Then for each of the individual hypotheses, make a decision about whether to reject using the significance levels as described in Section 12.3.3. (That is, use the Rom or Benjamini–Hochberg method.) The levels of factor B can be compared in a similar manner using the methods in Section 12.8 to control FWE.

As for interactions, a basic approach is to test the hypothesis of no interaction for any two levels of factor A and factor B and to repeat this for all pairs of rows and all pairs of columns. That is, for every $j < j'$ and $k < k'$, test

$$F_{jk}(x) - F_{jk'}(x) = F_{j'k}(x) - F_{j'k'}(x).$$

Again Rom's method or the Benjamini–Hochberg technique can be used to control FWE (the familywise Type I error rate).

15.5.1 S-PLUS Function r2mcp

The S-PLUS function

$$r2mcp(J,K,x,grp=NA,alpha=.05,bhop=F)$$

performs the multiple comparisons method just described. The groups are assumed to be arranged as in Section 10.3.1; if not arranged in this manner, the argument grp can be used to address this problem as illustrated in Section 10.3.1. The default value for bhop causes critical significance levels to be computed using Rom's method; bhop=T results in using the Benjamini and Hochberg method instead.

15.5.2 The Patel–Hoel Approach to Interactions

Patel and Hoel (1973) proposed an alternative approach to interactions in a 2-by-2 design that can be extended to a multiple comparison method for a J-by-K design, even when there are tied values. To describe the basic idea, first consider a 2-by-2 design where X_{ijk} is the ith observation randomly sampled from the jth level of factor A and the kth level of factor B. Temporarily assume ties never occur and let

$$p_{11,12} = P(X_{i11} < X_{i12}).$$

In words, X_{i11} represents a randomly sampled observation from level 1 of factors A and B, X_{i12} is a randomly sampled observation from level 1 of factor A and level 2 of factor B, and $p_{11,12}$ is the probability that X_{i11} is less than X_{i12}. Note that ignoring level 2 of factor A, levels 1 and 2 of factor B can be compared by testing $H_0 : p_{11,12} = 0$, as described in Sections 15.1.3 and 15.1.5. The Patel–Hoel definition of no interaction is that $p_{11,12} = p_{21,22}$. That is, the probability that an observation is smaller under level 1 of factor B than under level 2 is the same for both levels of factor A. In the event ties can occur, then define

$$p_{11,12} = P(X_{i11} \leq X_{i12}) + \frac{1}{2}P(X_{i11} = X_{i12}),$$

$$p_{21,22} = P(X_{i21} \leq X_{i22}) + \frac{1}{2}P(X_{i21} = X_{i22}),$$

and the hypothesis of no interaction is

$$H_0 : p_{11,12} = p_{21,22}.$$

Again, temporarily ignore level 2 of factor A and note that the 2 independent groups corresponding to the 2 levels of factor B can be compared in terms of δ, as defined in Section 15.1.3. Let δ_1 represent δ when focusing on level 1 of factor A with level 2 ignored, and let $\hat{\delta}_1$ be the estimate of δ as given by Equation (15.11). An estimate of its squared standard error is computed as indicated in Box 15.2 and will be labeled $\hat{\sigma}_1^2$. Similarly, let δ_2 be δ when focusing on level 2 of factor A with level 1 ignored, and denote its estimate with $\hat{\delta}_2$, which has an estimated squared standard error $\hat{\sigma}_2^2$. It can be seen that the null hypothesis of no interaction just defined corresponds to

$$H_0 : \Delta = \frac{\delta_2 - \delta_1}{2} = 0.$$

Moreover, the results in Box 15.2 can be used to estimate $p_{11,12} - p_{21,22}$ and compute an appropriate $1 - \alpha$ confidence interval. The estimate is

$$\hat{\Delta} = \frac{\hat{\delta}_2 - \hat{\delta}_1}{2},$$

the estimated squared standard error of $\hat{\Delta}$ is

$$S^2 = \frac{1}{4}(\hat{\sigma}_1^2 + \hat{\sigma}_2^2),$$

where $\hat{\sigma}_j^2$ is the value of $\hat{\sigma}^2$ in Box 15.2 corresponding to the jth level of factor A, and a $1 - \alpha$ confidence interval for Δ is

$$\hat{\Delta} \pm z_{1-\alpha/2}S,$$

where $z_{1-\alpha/2}$ is the $1 - \alpha/2$ quantile of a standard normal distribution. The hypothesis of no interaction is rejected if this confidence interval does not contain zero.

There remains the problem of controlling FWE for the more general case of a J-by-K design. Here, an analog of Dunnett's T3 method is used, but Rom's method or the Benjamini–Hochberg approach are other possibilities. When working with levels j and j' of factor A and levels k and k' of factor B, we represent the parameter Δ by $\Delta_{jj'kk'}$, its estimate is labeled $\hat{\Delta}_{jj'kk'}$, and the estimated squared standard error is denoted by $S^2_{jj'kk'}$. For every $j < j'$ and $k < k'$, the goal is to test

$$H_0 : \Delta_{jj'kk'} = 0.$$

The total number of hypotheses to be tested is

$$C = \frac{J^2 - J}{2} \times \frac{K^2 - K}{2}.$$

The critical value, c, is read from Table 10 in Appendix B with degrees of freedom $\nu = \infty$. The confidence interval for $\Delta_{jj'kk'}$ is

$$\hat{\Delta}_{jj'kk'} \pm cS_{jj'kk'},$$

and the hypothesis of no interaction, corresponding to levels j and j' of factor A and levels k and k' of factor B, is rejected if this confidence interval does not contain zero.

Yet another approach to testing the hypothesis of no interactions can be based on the results in Section 15.1.5. The method just described seems to perform well in simulations (Wilcox, 2000b), so currently there seems to be little motivation for extending the method in Section 15.1.5 to the problem at hand.

15.5.3 S-PLUS Function rimul

The S-PLUS function

$$rimul(J,K,x,p=J*K,grp=c(1{:}p))$$

performs the test for interactions just described. (The fourth argument, p=J*K, is not important in applied work; it is used to deal with certain conventions in S-PLUS.) The groups are assumed to be arranged as in Section 10.3.1, and the argument grp is used as illustrated in Section 10.3.1.

EXAMPLE. Table 10.1 shows data on weight gains in rats under four conditions: source of protein (beef versus cereal), which is factor A, and amount of protein (factor B), which has two levels: low and high. Assume that for beef, the data for low protein is stored in x[[1]], and for high protein it is stored in x[[2]]. As for cereal, low-and high-protein data are stored in x[[3]] and x[[4]], respectively. Then the S-PLUS command rimul(2,2,x) returns:

```
$test:
           Factor A   Factor A  Factor B  Factor B  delta
[1,]           1          2         1         2     0.31

ci.lower                  ci.upper
-0.02708888             0.6470889
```

Let $p_{11,12}$ represent the probability that for beef, the weight gain for a randomly sampled rat will be smaller under a low-versus high-protein diet. Similarly, for cereal, let $p_{21,22}$ represent the probability that weight gain is smaller under a low-protein diet. The estimate of $\Delta = p_{11,12} - p_{21,22}$ is listed under delta and is 0.31, and a .95 confidence interval for Δ is $(-0.027, 0.647)$. So the hypothesis of no interaction is not rejected at the $\alpha = .05$ level. ■

15.6 Comparing Two Dependent Groups

15.6.1 The Sign Test

A simple method for comparing dependent groups is the so-called *sign test*. In essence it is based on making inferences about the probability of success associated with a binomial distribution, which is discussed in Section 4.13, but here we elaborate a bit on its properties.

As in Chapter 11, let $(X_{11}, X_{12}), \ldots, (X_{n1}, X_{n2})$ be a random sample of n pairs of observations. That is, X_{ij} is the ith observation from the jth group. Primarily for convenience, it is temporarily assumed that tied values never occur. Let p be the probability that for a randomly sampled pair of observations, the observation from group 1 is less than the observation from group 2. In symbols,

$$p = P(X_{i1} < X_{i2}).$$

Letting $D_i = X_{i1} - X_{i2}$, an estimate of p is simply the proportion of D_i values that are less than zero. More formally, let $V_i = 1$ if $D_i < 0$; otherwise $V_i = 0$. Then an estimate of p is

$$\hat{p} = \frac{1}{n} \sum V_i. \tag{15.15}$$

Because $X = \sum V_i$ has a binomial probability function, results in Section 4.13 provide a confidence interval for p. If this interval does not contain zero, reject

$$H_0 : p = .5$$

and conclude that the groups differ. If $p > .5$, group 1 is more likely to have a lower observed value than group 2; if $p < .5$, the reverse is true.

Now consider a situation where ties can occur. Given the goal of making inferences about p, one strategy is simply to ignore or discard cases where $D_i = 0$. So if among n pairs of observations, there are N D_i values not equal to zero, then an estimate of p is

$$\hat{p} = \frac{1}{N} \sum V_i, \tag{15.16}$$

where V_i is defined as before.

To add perspective, look at Figure 15.1, which shows a scatterplot of the cork boring data given in Table 11.1. Also shown is a diagonal line through the origin. Points to the right of this line correspond to D_i values that are greater than zero, and points to the left are D_i values for which the reverse is true. So graphically, the sign test is based on the proportion of points to the right of this line. Points falling exactly on this line are ignored. In this particular case, $n = 28$, two points fall exactly on the line, so $N = 26$, $\hat{p} = .31$, and the .95 confidence interval for p is $(.15, .52)$; this interval contains .5, so fail to reject.

Notice that some of the methods in Chapter 11 found a difference between these two groups, in contrast to the sign test. One reason this can happen is that the sign test does not take into account how far a point is from the line in Figure 15.1 Also, it might help to note the graphical similarity between the sign test as depicted in Figure 15.1 and the graphical description of the bootstrap in Figure 11.2. The sign test is based on whether the observations are to the left or the right of the diagonal line. In contrast, the bootstrap is based on whether bootstrap estimates of the measures of location are to the left or the right of this line.

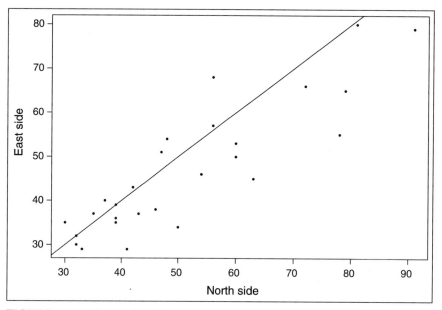

FIGURE 15.1   Scatterplot of the data in Table 11.1. In order for the sign test to reject, most of the points need to be above or below the diagonal line.

15.6.2 S-PLUS Function signt

The S-PLUS function

$$\text{signt}(x, y = NA, alpha = 0.05)$$

tests $H_0 : p = .5$ with the sign test as just described. If the argument y is not specified, it is assumed either that x is a matrix with two columns corresponding to the two dependent groups or that x has list mode. The function computes the differences $X_{i1} - X_{i2}$, eliminates all differences that are zero, leaving N values, determines the number of pairs for which $X_{i1} < X_{i2}$, $i = 1, \ldots, N$, and then calls the function binomci (see Section 4.13.1).

EXAMPLE. The output from signt based on the cork data in Figure 15.1 is

```
$phat:
[1] 0.3076923

$ci:
[1] 0.1530612 0.5179361

$n:
[1] 28

$N:
[1] 26
```

■

15.6.3 Wilcoxon Signed Rank Test

The sign test provides an interesting, useful, and reasonable perspective on how two groups differ. However, a common criticism is that its power can be low relative to other techniques that might be used. One alternative approach is the Wilcoxon signed rank test, which tests

$$H_0 : F_1(x) = F_2(x),$$

the hypothesis that two dependent groups have identical distributions. To apply it, first form difference scores as was done in conjunction with the paired T-test in Chapter 11 and discard any difference scores that are equal to zero. It is assumed that there are n difference scores not equal to zero. That is, for the ith pair of observations, compute

$$D_i = X_{i1} - X_{i2},$$

$i = 1, \ldots, n$ and each D_i value is either less than or greater than zero. Next, rank the $|D_i|$ values and let U_i denote the result for $|D_i|$. So, for example, if the D_i values are 6, -2, 12, 23, -8, then $U_1 = 2$, because after taking absolute values, 6 has a rank of 2. Similarly, $U_2 = 1$, because after taking absolute values, the second value, -2, has a rank of 1. Next set

$$R_i = U_i,$$

if $D_i > 0$; otherwise

$$R_i = -U_i.$$

Positive numbers are said to have a sign of 1 and negative numbers a sign of -1, so R_i is the value of the rank corresponding to $|D_i|$ multiplied by the sign of D_i.

If the sample size (n) is less than or equal to 40 and there are no ties among the $|D_i|$ values, the test statistic is W, the sum of the positive R_i values. For example, if $R_1 = 4, R_2 = -3, R_3 = 5, R_4 = 2$, and $R_5 = -1$, then

$$W = 4 + 5 + 2 = 11.$$

A lower critical value, c_L, is read from Table 12 in Appendix B. So for $\alpha = .05$ and $n = 5$, the critical value corresponds to $\alpha/2 = .025$ and is 0, so reject if $W \leq 0$. The upper critical value is

$$c_U = \frac{n(n+1)}{2} - c_L.$$

In the illustration, because $c_L = 0$,

$$c_U = \frac{5(6)}{2} - 0 = 15,$$

meaning that you reject if $W \geq 15$. Because $W = 11$ is between 1 and 15, fail to reject.

If there are ties among the $|D_i|$ values or the sample size exceeds 40, the test statistic is

$$W = \frac{\sum R_i}{\sqrt{\sum R_i^2}}.$$

If there are no ties, this last equation simplifies to

$$W = \frac{\sqrt{6} \sum R_i}{\sqrt{n(n+1)(2n+1)}}.$$

For a two-sided test, reject if $|W|$ equals or exceeds $z_{1-\alpha/2}$, the $1 - \alpha/2$ quantile of a standard normal distribution.

Rejecting with the signed rank test indicates that two dependent groups have different distributions. Although the signed rank test can have more power than the sign test, a criticism is that it does not provide certain details about how the groups differ. For instance, in the cork boring example, rejecting indicates that the distribution of weights differs for the north versus east side of a tree, but how might we elaborate on what this difference is? One possibility is to estimate p, the probability that the weight from the north side is less than the weight from the east side. So despite lower power, one might argue that the sign test provides a useful perspective on how groups compare.

15.6.4   S-PLUS Function wilcox.test

The built-in S-PLUS function

$$\text{wilcox.test}(x, y, \text{paired} = F, \text{exact} = T)$$

performs the Wilcoxon signed rank test just described by setting the argument paired to T, for true. (With paired=F, a one-sample version of the test is used.)

15.7   Comparing Multiple Dependent Groups

There are a variety of rank-based methods for comparing multiple dependent groups. Recent advances and techniques can be found in Agresti and Pendergast (1986); Akritas and Arnold (1994); Brunner and Denker (1994); Brunner, Munzel, and Puri (1999); and Munzel and Brunner (2000b). Here, only portions of these methods are described, plus some classic techniques that are typically covered in an introductory course.

15.7.1   Friedman's Test

The classic rank-based method for comparing multiple dependent groups is Friedman's test. The goal is to test

$$H_0 : F_1(x) = \cdots = F_J(x),$$

the hypothesis that all J dependent groups have identical distributions. The method begins by assigning ranks within rows. For example, imagine that for each individual, measures are taken at three different times, yielding

Time 1	Time 2	Time 3
9	7	12
1	10	4
8	2	1
⋮		

The ranks corresponding to the first row (the values 9, 7, and 12) are 2, 1, and 3. For the second row the ranks are 1, 3, and 2, and continuing in this manner the data become

Time 1	Time 2	Time 3
2	1	3
1	3	2
3	2	1
⋮		

Let R_{ij} be the resulting rank corresponding to X_{ij} ($i = 1, \ldots, n; j = 1, \ldots, J$). Compute

$$A = \sum_{j=1}^{J} \sum_{i=1}^{n} R_{ij}^2$$

$$R_j = \sum_{i=1}^{n} R_{ij}$$

$$B = \frac{1}{n} \sum_{j=1}^{J} R_j^2$$

$$C = \frac{1}{4} nJ(J+1)^2.$$

If there are no ties, the equation for A simplifies to

$$A = \frac{nJ(J+1)(2J+1)}{6}.$$

The test statistic is

$$F = \frac{(n-1)(B-C)}{A-B}. \tag{15.17}$$

Reject if $F \geq f_{1-\alpha}$, or if $A = B$, where $f_{1-\alpha}$ is the $1 - \alpha$ of an F-distribution with $\nu_1 = J - 1$ and $\nu_2 = (n-1)(J-1)$ degrees of freedom.

15.7.2 Agresti–Pendergast Test

A variety of improvements on Friedman's test have been proposed (e.g., Iman, 1974; Quade, 1979). One such method that currently stands out was proposed by Agresti and Pendergast (1986); it provides better control over the probability of a Type I error than does Iman's method and it can have higher power than Friedman's test. (For theoretical results on the Agresti–Pendergast test, see Kepner & Robinson, 1988.) Basically, the method tests the hypothesis of equal distributions based on the average ranks among the groups. Box 15.6 describes the computations.

BOX 15.6 The Agresti–Pendergast Method for Dependent Groups

Pool all the observations and assign ranks. Let R_{ij} be the resulting rank of the ith observation in the jth group. Compute

$$\bar{R}_j = \frac{1}{n} \sum_{i=1}^{n} R_{ij}$$

$$s_{jk} = \frac{1}{n - J + 1} \sum_{i=1}^{n} (R_{ij} - \bar{R}_j)(R_{ik} - \bar{R}_k).$$

Let the vector $\mathbf{R}'$ be defined by

$$\mathbf{R}' = (\bar{R}_1, \ldots, \bar{R}_J),$$

and let $\mathbf{C}$ be the $(J - 1)$-by-J matrix given by

$$\begin{pmatrix} 1 & -1 & 0 & \ldots & 0 & 0 \\ 0 & 1 & -1 & \ldots & 0 & 0 \\ \cdot & \cdot & \cdot & \cdot & \cdot & \cdot \\ 0 & 0 & 0 & \ldots & 1 & -1 \end{pmatrix}.$$

The test statistic is

$$F = \frac{n}{J - 1} (\mathbf{CR})'(\mathbf{CSC}')^{-1}\mathbf{CR},$$

where

$$\mathbf{S} = (s_{jk}).$$

Decision Rule
Reject if $F \geq f_{1-\alpha}$, the $1 - \alpha$ quantile of an F-distribution with $\nu_1 = J - 1$ and $\nu_2 = (J - 1)(n - 1)$ degrees of freedom.

15.7.3 S-PLUS Function apanova

The S-PLUS function

$$apanova(x)$$

performs the Agresti–Pendergast ANOVA method based on ranks.

15.8 One-Way Multivariate Methods

There are several rank-based methods for dealing with situations where there are J independent groups with K measures for each individual (e.g., Choi & Marden, 1997; Liu & Singh, 1993; Hettmansperger, Möttönen, & Oja, 1997; Munzel & Brunner, 2000a). The theoretical details of the method developed by Hettmansperger, Möttönen, and Oja assume that tied values occur with probability zero and that distributions are symmetric, so the details of their method are not given here.

15.8.1 The Munzel–Brunner Method

This section describes a one-way multivariate method derived by Munzel and Brunner (2000a). (A variation of the Munzel–Brunner method can be used in place of the Agresti–Pendergast procedure in Section 15.7.2, but the relative merits of these two techniques have not been explored.) For the jth group, there are n_j randomly sampled vectors of observations, with each vector containing K measures. Let $F_{jk}(x)$ be the distribution associated with the jth group and kth measure. So, for example, $F_{32}(6)$ is the probability that for the third group, the second variable will be less than or equal to 6 for a randomly sampled individual. For the kth measure, the goal is to test the hypothesis that all J groups have identical distributions. And the more general goal is to test the hypothesis that simultaneously, all groups have identical distributions for each of the K measures under consideration. That is, the goal is to test

$$H_0 : F_{1k}(x) = \cdots = F_{Jk}(x) \qquad \text{for all } k = 1, \ldots, K. \qquad (15.18)$$

To apply the method, begin with the first of the K measures, pool all the observations among the J groups, and assign ranks. Ties are handled in the manner described in Section 15.1.3. Repeat this process for all K measures and label the results R_{ijk}. That is, R_{ijk} is the rank of the ith observation in the jth group and for the kth measure. Let

$$\bar{R}_{jk} = \frac{1}{n_j} \sum_{i=1}^{n_j} R_{ijk},$$

be the average rank for the jth group corresponding to the kth measure. Set

$$\hat{Q}_{jk} = \frac{\bar{R}_{jk} - .5}{n},$$

where $n = \sum n_j$ is the total number of randomly sampled vectors among the J groups. The remaining calculations are summarized in Box 15.7. The $\hat{Q}$ values are called the *relative effects* and reflect the ordering of the average ranks. If, for example, $\hat{Q}_{11} < \hat{Q}_{21}$, the typical rank for variable 1 in group 1 is less than the typical rank for variable 1 in group 2. More generally, if $\hat{Q}_{jk} < \hat{Q}_{j'k}$, then based on the kth measure, the typical rank (or observed value) for group j is less than the typical rank for group j'.

BOX 15.7 The Munzel–Brunner One-Way Multivariate Method

Let

$$\hat{Q} = (\hat{Q}_{11}, \hat{Q}_{12}, \ldots, \hat{Q}_{1K}, \hat{Q}_{21}, \ldots, \hat{Q}_{JK})',$$
$$R_{ij} = (R_{ij1}, \ldots, R_{ijK})',$$
$$\bar{R}_j = (\bar{R}_{j1}, \ldots, \bar{R}_{jK})',$$
$$V_j = \frac{1}{nn_j(n_j-1)} = \sum_{i=1}^{n_j} (R_{ij} - \bar{R}_j)(R_{ij} - \bar{R}_j)',$$

where $n = \sum n_j$, and let

$$V = \text{diag}\{V_1, \ldots, V_J\}.$$

Compute the matrix M_A as described in Section 15.4. The test statistic is

$$F = \frac{n}{\text{tr}(M_A V)} \hat{Q}' M_A \hat{Q}.$$

Decision Rule
Reject if $F \geq f$, where f is the $1 - \alpha$ quantile of an F-distribution with

$$\nu_1 = \frac{(\text{tr}(M_A V))^2}{\text{tr}(M_A V M_A V)}$$

and $\nu_2 = \infty$ degrees of freedom.

15.8.2 S-PLUS Function mulrank

The S-PLUS function

$$\text{mulrank}(J, K, x)$$

performs the one-way multivariate method in Box 15.7. The data are stored in x, which can be a matrix or have list mode. If x is a matrix, the first K columns correspond to the K measures for group 1, the second K correspond to group 2, and so forth. If stored in list mode, x[[1]], ..., x[[K]] contain the data for group 1, x[[K+1]], ..., x[[2K]] contain the data for group 2, and so on.

TABLE 15.2 CGI and PGI Scores After Four Weeks of Treatment

Exercise		Clomipramine		Placebo	
CGI	PGI	CGI	PGI	CGI	PGI
4	3	1	2	5	4
1	1	1	1	5	5
2	2	2	0	5	6
2	3	2	1	5	4
2	3	2	3	2	6
1	2	2	3	4	6
3	3	3	4	1	1
2	3	1	4	4	5
5	5	1	1	2	1
2	2	2	0	4	4
5	5	2	3	5	5
2	4	1	0	4	4
2	1	1	1	5	4
2	4	1	1	5	4
6	5	2	1	3	4

EXAMPLE. Table 15.2 summarizes data (reported by Munzel & Brunner, 2000a) from a psychiatric clinical trial where three methods are compared for treating individuals with panic disorder. The three methods are exercise, clomipramine, and placebo. The two measures of effectiveness were a clinical global impression (CGI) and the patient's global impression (PGI). The test statistic is $F = 12.7$ with $v_1 = 2.83$ and a significance level less than .001. The relative effects are:

```
$q.hat:
             [,1]            [,2]
[1,]     0.5074074      0.5096296
[2,]     0.2859259      0.2837037
[3,]     0.7066667      0.7066667
```

So among the three groups, the second group, clomipramine, has the lowest relative effects. That is, the typical ranks were lowest for this group, and the placebo group had the highest ranks on average. ■

15.8.3 The Choi–Marden Multivariate Rank Test

This section describes a multivariate analog of the Kruskal–Wallis test derived by Choi and Marden (1997). There are actually many variations of the approach they considered, but here attention is restricted to the version they focused on. As with the method in Section 15.8.1, we have K measures for each individual and there are J

independent groups. For the jth group and any vector of constants $\mathbf{x} = (x_1, \ldots, x_K)$, let

$$F_j(\mathbf{x}) = P(X_{j1} \leq x_1, \ldots, X_{jK} \leq x_K).$$

So, for example, $F_1(\mathbf{x})$ is the probability that for the first group, the first of the K measures is less than or equal to x_1, the second of the K measures is less than or equal to x_2, and so forth. The null hypothesis is that for any $\mathbf{x}$,

$$H_0 : F_1(\mathbf{x}) = \cdots = F_J(\mathbf{x}), \tag{15.19}$$

which is sometimes called the *multivariate hypothesis*, to distinguish it from Equation (15.18), which is called the *marginal hypothesis*. The multivariate hypothesis is a stronger hypothesis, in the sense that if it is true, then by implication the marginal hypothesis is true as well. For example, if the marginal distributions for both groups are standard normal distributions, the marginal hypothesis is true; but if the groups have different correlations, the multivariate hypothesis is false.

The Choi–Marden method represents an extension of a technique derived by Möttönen and Oja (1995) and is based on a generalization of the notion of a rank to multivariate data, which was also used by Chaudhuri (1996, Section 4). First consider a random sample of n observations with K measures for each individual or thing and denote the ith vector of observations by

$$\mathbf{X}_i = (X_{i1}, \ldots, X_{iK}).$$

Let

$$A_{ii'} = \sqrt{\sum_{k=1}^{K} (X_{ik} - X_{i',k})^2},$$

Here, the "rank" of the ith vector is itself a vector (having length K), given by

$$\mathbf{R}_i = \frac{1}{n} \sum_{i=1}^{n} \frac{\mathbf{X}_i - \mathbf{X}_{i'}}{A_{ii'}},$$

where

$$\mathbf{X}_i - \mathbf{X}_{i'} = (X_{i1} - X_{i'1}, \ldots, X_{iK} - X_{i'K}).$$

The remaining calculations are summarized in Box 15.8. All indications are that this method provides good control over the probability of a Type I error when ties never occur. There are no known problems when there are tied values, but this issue is in need of more research.

BOX 15.8 The Choi–Marden Method

Pool the data from all J groups and compute rank vectors as described in the text. The resulting rank vectors are denoted by $\mathbf{R}_1, \ldots, \mathbf{R}_n$, where $n = \sum n_j$

Continued

BOX 15.8 (*Continued*) is the total number of vectors among the J groups. For each of the J groups, average the rank vectors and denote the average of these vectors for the jth group by $\bar{\mathbf{R}}_j$.

Next, assign ranks to the vectors in the jth group, ignoring all other groups. We let $\mathbf{V}_{ij}$ (a column vector of length K) represent the rank vector corresponding to the ith vector of the jth group ($i = 1, \dots, n_j$; $j = 1, \dots, J$) to make a clear distinction with the ranks based on the pooled data. Compute

$$S = \frac{1}{n-J} \sum_{j=1}^{J} \sum_{i=1}^{n_j} \mathbf{V}_{ij} \mathbf{V}'_{ij},$$

where $\mathbf{V}'_{ij}$ is the transpose of $\mathbf{V}_{ij}$ (so S is a K-by-K matrix). The test statistic is

$$H = \sum_{j=1}^{J} n_j \bar{\mathbf{R}}'_j S^{-1} \bar{\mathbf{R}}_j. \tag{15.20}$$

(For $K = 1$, H does not quite reduce to the Kruskall–Wallis test statistic. In fact, H avoids a certain technical problem that is not addressed by the Kruskall–Wallis method.)

Decisions Rule
Reject if $H \geq c$, where c is the $1 - \alpha$ quantile of a chi-squared distribution with degrees of freedom $K(J - 1)$.

15.8.4 S-PLUS Function cmanova

The S-PLUS function

cmanova(J,K,x)

performs the Choi–Marden method just described. The data are assumed to be stored in x as described in Section 15.8.2.

15.9 Between-by-Within Designs

There are a variety of rank-based methods one might use in a between-by-within-subjects design, or what is called a *split-plot design*. That is, as in Section 11.9, we have a two-way design where the J levels of the first factor are independent and the K levels of the other factor are possibly dependent. For comparing the independent groups, one approach is to use the methods in Section 15.8. As for the dependent groups, one possibility is to ignore the levels of factor A and use the methods in Section 15.6. For example, if the two independent groups are males and females, and measures for every individual are taken at three different times, you could simply pool the males and females and test the hypothesis that the distributions are identical at times 1, 2, and 3.

Yet another approach is to proceed along the lines in Section 11.9, only rather than compare measures of location, hypotheses are formulated in terms of distributions, as was done in Section 15.4. So as in Section 15.4, main effects for factor A are given in terms of

$$\bar{F}_{j.}(x) = \frac{1}{K} \sum_{k=1}^{K} F_{jk}(x),$$

the average of the distributions among the K levels of factor B corresponding to the jth level of factor A. The only difference between the present situation and Section 15.4 is that the average of dependent groups is being used. The hypothesis of no main effects for factor A is

$$H_0 : \bar{F}_{1.}(x) = \bar{F}_{2.}(x) = \cdots = \bar{F}_{J.}(x)$$

for any x. Letting

$$\bar{F}_{.k}(x) = \frac{1}{J} \sum_{j=1}^{J} F_{jk}(x)$$

be the average of the distributions for the kth level of factor B, the hypothesis of no main effects for factor B is

$$H_0 : \bar{F}_{.1}(x) = \bar{F}_{.2}(x) = \cdots = \bar{F}_{.K}(x).$$

As for interactions, again proceed as before. So for a 2-by-2 design, no interaction is taken to mean that for any x,

$$F_{11}(x) - F_{12}(x) = F_{21}(x) - F_{22}(x).$$

More generally, the hypothesis of no interactions among all JK groups is

$$H_0 : F_{jk}(x) - \bar{F}_{j.}(x) - \bar{F}_{.k}(x) + \bar{F}_{..}(x) = 0,$$

for any x, all j ($j = 1, \ldots, J$) and all k ($k = 1, \ldots, K$), where

$$\bar{F}_{..}(x) = \frac{1}{JK} \sum_{j=1}^{J} \sum_{k=1}^{K} F_{jk}(x).$$

A technical difficulty is taking into account the dependence among the levels of factor B, and here the methods covered in Brunner, Domhof, and Langer (2002, Chapter 8) are used. (Beasley, 2000, suggests another approach to interactions; perhaps it has a practical advantage over the approach described here, but this issue has not been investigated as yet.)

As usual, let X_{ijk} represent the ith observation for level j of factor A and level k of factor B. Here, $i = 1, \ldots, n_j$. That is, the jth level of factor A has n_j vectors of observations, each vector containing K values. So for the jth level of factor A there are a total of $n_j K$ observations; and among all the groups, the total number of observations is denoted by N. So the total number of vectors among the J groups is $n = \sum n_j$, and the total number of observations is $N = K \sum n_j = Kn$.

Pool all N observations and assign ranks. As usual, midranks are used if there are tied values. Let R_{ijk} represent the rank associated with X_{ijk}. The remaining calculations for factor A are relegated to Box 15.9; factor B and interactions are tested as described in Box 15.10.

BOX 15.9 Main Effects for Factor A in a Between-by-Within Design

Let

$$\bar{R}_{.jk} = \frac{1}{n_j} \sum_{i=1}^{n_j} R_{ijk},$$

$$\bar{R}_{.j.} = \frac{1}{K} \sum_{k=1}^{K} \bar{R}_{.jk},$$

$$\bar{R}_{ij.} = \frac{1}{K} \sum_{k=1}^{K} R_{ijk},$$

$$\hat{\sigma}_j^2 = \frac{1}{n_j - 1} \sum_{i=1}^{n_j} (\bar{R}_{ij.} - \bar{R}_{.j.})^2,$$

$$S = \sum_{j=1}^{J} \frac{\hat{\sigma}_j^2}{n_j},$$

$$U = \sum_{j=1}^{J} \left(\frac{\hat{\sigma}_j^2}{n_j} \right)^2,$$

$$D = \sum_{j=1}^{J} \frac{1}{n_j - 1} \left(\frac{\hat{\sigma}_j^2}{n_j} \right)^2.$$

Factor A

The test statistic is

$$F_A = \frac{J}{(J-1)S} \sum_{j=1}^{J} (\bar{R}_{.j.} - \bar{R}_{...})^2,$$

Continued

BOX 15.9 (*Continued*)

where $\bar{R}_{..} = \sum \bar{R}_{.j}/J$. The degrees of freedom are

$$\nu_1 = \frac{(J-1)^2}{1 + J(J-2)U/S^2},$$

$$\nu_2 = \frac{S^2}{D}.$$

Decision Rule

Reject if $F_A \geq f$, where f is the $1 - \alpha$ quantile of an F-distribution with ν_1 and ν_2 degrees of freedom.

BOX 15.10 Interactions and Main Effects for Factor B in a

Between-by-Within Design

Factor B

Following the notation in Box 15.9, let

$$\mathbf{R}_{ij} = (R_{ij1}, \dots, R_{ijK})',$$

$$\bar{\mathbf{R}}_j = \frac{1}{n_j} \sum_{i=1}^{n_j} \mathbf{R}_{ij}, \quad \bar{\mathbf{R}}_{..} = \frac{1}{J} \sum_{j=1}^{J} \bar{\mathbf{R}}_j, \qquad n = \sum n_j \quad (\text{so } N = nK),$$

$$\mathbf{V}_j = \frac{n}{N^2 n_j (n_j - 1)} \sum_{i=1}^{n_j} (\mathbf{R}_{ij} - \bar{\mathbf{R}}_j)(\mathbf{R}_{ij} - \bar{\mathbf{R}}_j)'.$$

So $\mathbf{V}_j$ is a K-by-K matrix of covariances based on the ranks. Let

$$\mathbf{S} = \frac{1}{J^2} \sum_{j=1}^{J} \mathbf{V}_j$$

and let $\mathbf{P}_K$ be defined as in Box 15.5.

The test statistic is

$$F_B = \frac{n}{N^2 \text{tr}(\mathbf{P}_K \mathbf{S})} \sum_{k=1}^{K} (\bar{R}_{..k} - \bar{R}_{...})^2.$$

The degrees of freedom are

$$\nu_1 = \frac{(\text{tr}(\mathbf{P}_K \mathbf{S}))^2}{\text{tr}(\mathbf{P}_K \mathbf{S} \mathbf{P}_K \mathbf{S})}, \qquad \nu_2 = \infty.$$

Continued

BOX 15.10 (*Continued*)

Interactions

Let V be the block diagonal matrix based on the matrices $V_j, j = 1, \ldots, J$. (See Appendix C for a definition of a block diagonal matrix.) Letting M_{AB} be defined as in Box 15.5, the test statistic is

$$F_{AB} = \frac{n}{N^2 \text{tr}(M_{AB}V)} \sum_{j=1}^{J} \sum_{k=1}^{K} (\bar{R}_{.jk} - \bar{R}_{.j.} - \bar{R}_{..k} + \bar{R}_{...})^2.$$

The degrees of freedom are

$$\nu_1 = \frac{(\text{tr}(M_{AB}V))^2}{\text{tr}(M_{AB}VM_{AB}V)}, \qquad \nu_2 = \infty.$$

Decision Rule

Reject if $F_A \geq f$ (or if $F_{AB} \geq f$), where f is the $1 - \alpha$ quantile of an F-distribution with ν_1 and ν_2 degrees of freedom.

15.9.1 S-PLUS Function bwrank

The S-PLUS function

$$\text{bwrank}(J,K,x)$$

performs a between-by-within ANOVA based on ranks using the method just described. In addition to testing hypotheses as indicated in Boxes 15.9 and 15.10, the function returns the average ranks $(\bar{R}_{.jk})$ associated with all JK groups as well as the relative effects, $(\bar{R}_{.jk} - .5)/N$.

EXAMPLE. Lumley (1996) reports data on shoulder pain after surgery; the data are from a study by Jorgensen et al. (1995). Table 15.3 shows a portion of the results where two treatment methods are used and measures of pain are taken at three different times. The output from bwrank is

```
$test.A:
[1] 12.87017

$sig.A:
[1] 0.001043705

$test.B:
[1] 0.4604075

$sig.B:
[1] 0.5759393
```

Continued

EXAMPLE. (*Continued*)

```
$test.AB:
[1] 8.621151

$sig.AB:
[1] 0.0007548441

$avg.ranks:
          [,1]        [,2]        [,3]
[1,]   58.29545    48.40909    39.45455
[2,]   66.70455    82.36364    83.04545

$rel.effects:
          [,1]        [,2]        [,3]
[1,]   0.4698817   0.3895048   0.3167036
[2,]   0.5382483   0.6655580   0.6711013
```

Continued

TABLE 15.3   Shoulder Pain Data

Active treatment			No active treatment		
Time 1	Time 2	Time 3	Time 1	Time 2	Time 3
1	1	1	5	2	3
3	2	1	1	5	3
3	2	2	4	4	4
1	1	1	4	4	4
1	1	1	2	3	4
1	2	1	3	4	3
3	2	1	3	3	4
2	2	1	1	1	1
1	1	1	1	1	1
3	1	1	1	5	5
1	1	1	1	3	2
2	1	1	2	2	3
1	2	2	2	2	1
3	1	1	1	1	1
2	1	1	1	1	1
1	1	1	5	5	5
1	1	1	3	3	3
2	1	1	5	4	4
4	4	2	1	3	3
4	4	4			
1	1	1			
1	1	1			

1 = low, 5 = high.

> **EXAMPLE.** (*Continued*)
> So treatment methods are significantly different and there is a significant inter-
> action, but no significant difference is found over time. Note that the average
> ranks and relative effects suggest that a disordinal interaction might exist. In
> particular, for group 1 (the active treatment group), time 1 has higher average
> ranks than time 2, and the reverse is true for the second group. However, the
> Wilcoxon signed rank test fails to reject at the .05 level when comparing time 1
> to time 2 for both groups. When comparing time 1 to time 3 for the first group,
> again using the Wilcoxon signed rank test, you reject at the .05 level, but a
> nonsignificant result is obtained for group 2. So again a disordinal interaction
> appears to be a possibility, but the empirical evidence is not compelling.   ■

15.9.2 Multiple Comparisons Based on Comparing Distributions

Multiple comparisons for a between-by-within design can be performed in essentially
the same way used to compare all independent groups as described in Section 15.5.
For example, when dealing with factor A, simply compare level j to level j', ignoring
the other levels. For all pairwise comparisons among the J levels of factor A, FWE
is controlled using Rom's method or the Benjamini–Hochberg technique, and the
same is done for factor B and the collection of all interactions corresponding to any
two levels of factor A and any two levels of factor B.

15.9.3 S-PLUS Function bwrmcp

The S-PLUS function

$$bwrmcp(J,K,x,grp=NA,alpha=.05,bhop=F)$$

performs all pairwise multiple comparisons using the method for a between-by-within-
subjects design described in the previous section. The value for alpha indicates the
FWE for all pairwise comparisons among the levels of factor A, as well as all pairwise
comparisons among the levels of factor B and all interactions. So if $J = 3$ and $K = 4$,
the default value for FWE when performing all three comparisons among any two
levels of factor A will be .05. There are a total of 18 interactions that would be tested,
and again the default FWE will be .05 among these 18 hypotheses.

15.9.4 Multiple Comparisons When Using a Patel–Hoel Approach to Interactions

Rather than compare distributions when dealing with a between-by-within design,
one could use a simple analog of the Patel–Hoel approach instead. First consider a
two-by-two design and focus on level 1 of factor A. Then the two levels of factor B
are dependent and can be compared with the sign test described in Section 15.6.1.
In essence, inferences are being made about p_1, the probability that for a randomly
sampled pair of observations, the observation from level 1 of factor B is less than the

corresponding observation from level 2. Of course, for level 2 of factor A, we can again compare levels 1 and 2 of factor B with the sign test. Now we let p_2 be the probability that for a randomly sampled pair of observations, the observation from level 1 of factor B is less than the corresponding observation from level 2. Then as in Section 15.5.2, no interaction can be defined as $p_1 = p_2$.

The hypothesis of no interaction,

$$H_0 : p_1 = p_2,$$

is just the hypothesis that two independent binomials have equal probabilities of success, which can be tested using one of the methods described in Section 8.13. Here, Beal's method (described in Box 8.4) is used rather than the Storer–Kim method in Box 8.3, because it currently seems that Beal's method provides more accurate control over FWE for the problem at hand, execution time can be much lower when sample sizes are large, and, unlike the Storer–Kim procedure, Beal's method provides confidence intervals. There are various ways FWE might be controlled using the methods in Chapter 12. However, it seems that they result in an actual FWE that can be substantially smaller than the nominal level, so the following modification is used.

Among a collection of techniques considered by Wilcox (2001c) for controlling FWE, the following method was found to be relatively effective. Let q be the critical value read from Table 10 in Appendix B with degrees of freedom $\nu = \infty$. Recall that in Table 10, C is the total number of hypotheses to be tested. Assuming that all pairs of rows and columns are to be considered when checking for interactions,

$$C = \frac{J^2 - J}{2} \times \frac{K^2 - K}{2}.$$

Let Z be a standard normal random variable (so $P(Z \le z)$ can be read from Table 1 in Appendix B). Then if FWE is to be α, test each of the C hypotheses at the α_a level, where

- If $(J, K) = (5, 2)$, then $\alpha_a = 2(1 - P(Z \le q))$.
- If $(J, K) = (3, 2)$, $(4, 2)$, or $(2, 3)$, then $\alpha_a = 3(1 - P(Z \le q))$.
- For all other J and K values, $\alpha_a = 4(1 - P(Z \le q))$.

These adjusted α values appear to work well when the goal is to achieve FWE less than or equal to .05. Whether this remains the case with FWE equal to .01 is unknown. For $C > 28$ and FWE equal to .05, use

$$q = 2.383904C^{1/10} - .202.$$

(Of course, for $C = 1$, no adjustment is necessary; simply use Beal's method as described in Chapter 8.)

Tied values are handled in the same manner as with the sign test in Section 15.6.1: Pairs of observations with identical values are simply discarded. So among the remaining observations, for every pair of observations, the observation from level 1 of factor B, for example, is either less than or greater than the corresponding value from level 2.

A criticism of the method in this section is that compared to the test of no interaction in Section 15.9.2, power is generally lower. However, the method in Section 15.9.2 tests hypotheses about entire distributions based on average ranks.

The method in this section adds information about how the groups differ that is not supplied merely by comparing distributions. The situation is similar to how the sign test compares to the Wilcoxon signed rank test. The sign test generally has lower power, but it provides a direct estimate of the probability that an observation in the first group is less than an observation in the second. This probability adds perspective that is not available when attention is restricted to the Wilcoxon signed rank test. The method in this section deals directly with how p_1 compares to p_2, which provides a characterization how groups differ that we do not get when comparing groups based on the average ranks.

A variation of the approach in this section is where, for level 1 of factor B, p_1 is the probability that an observation from level 1 of factor A is less than an observation from level 2. Similarly, p_2 is now defined in terms of the two levels of factor A when working with level 2 of factor B. However, the details of how to implement this approach have not been studied.

15.9.5 S-PLUS Function sisplit

The method just described for testing hypotheses of no interaction can be applied with the S-PLUS function

$$\text{sisplit}(J,K,x)$$

This function assumes $\alpha = .05$; other values are not allowed.

15.10 Rank-Based Correlations

As in previous chapters, imagine that we have n randomly sampled pairs of observations, which are labeled $(X_1, Y_1), \ldots, (X_n, Y_n)$. Two goals are determining whether these two measures are dependent and, if they are, characterizing what this dependence is like. There are two well-known rank-based measures of association aimed at accomplishing these goals: Kendalls' tau and Spearman's rho.

15.10.1 Kendall's tau

Kendall's tau is based on the following idea. Consider two pairs of observations, which are labeled in the usual way as (X_1, Y_1) and (X_2, Y_2). For convenience, assume that $X_1 < X_2$. If $Y_1 < Y_2$, then these two pairs of numbers are said to be concordant. That is, if Y increases as X increases or if Y decreases as X decreases, we have concordant pairs of observations. If two pairs of observations are not concordant, they are said to be discordant.

Roughly, among all pairs of points, Kendall's tau is just the average number that are concordant minus the average number that are discordant. If the measures X and Y are independent, then this difference should be approximately equal to zero. To describe its computation in a more formal manner, let $K_{ij} = 1$ if the ith and jth pairs of observations are concordant; otherwise $K_{ij} = -1$. Then Kendall's tau is given by

$$\hat{\tau} = \frac{2 \sum_{i<j} K_{ij}}{n(n - 1)} \qquad (15.21)$$

and has a value between -1 and 1. If $\hat{\tau}$ is positive, there is a tendency for Y to increase with X — possibly in a nonlinear fashion — and if $\hat{\tau}$ is negative, the reverse is true.

The population analog of $\hat{\tau}$ is labeled τ and can be shown to be zero when X and Y are independent. To test

$$H_0 : \tau = 0,$$

compute

$$\sigma_\tau^2 = \frac{2(2n+5)}{9n(n-1)},$$

$$Z = \frac{\hat{\tau}}{\sigma_\tau},$$

and reject if

$$|Z| \geq z_{1-\frac{\alpha}{2}},$$

where $z_{1-\alpha/2}$ is the $1 - \alpha/2$ quantile of a standard normal distribution (which can be read from Table 1).

Cliff (1996) suggests computing a confidence interval for τ by estimating the variance of $\hat{\tau}$ and applying Laplace's strategy. There are various ways this might be done, and a method that seems to be relatively effective is outlined in Box 15.11. (Also see J. D. Long & Cliff, 1997.) Another approach is to use the heteroscedastic bootstrap method in Section 13.6.4, which can be done with the S-PLUS function corb in Section 13.6.5. Direct comparisons between the heteroscedastic bootstrap method and the method in Box 15.11 have not been made. An educated guess is that the bootstrap method is better for general use, but this issue is in need of further study.

BOX 15.11 Confidence Interval for Kendall's tau

For the ith and hth pair of observations, set

$$U_{ih} = \text{sign}(X_i - X_h),$$

where $\text{sign}(X)$ is 1, 0, or -1 according to whether X is greater than, equal to, or less than zero. Let

$$W_{ih} = \text{sign}(Y_i - Y_h),$$

$$t_{ih} = U_{ih}W_{ih},$$

$$t_{i.} = \frac{1}{n-1}\sum_h t_{ih},$$

$$V_1 = \frac{1}{n-1}\sum_i (t_{i.} - \hat{\tau})^2,$$

Continued

BOX 15.11 (*Continued*)

$$V_2 = \frac{1}{[n(n-1)]-1}\left[\left(\sum\sum t_{ib}^2\right) - n(n-1)\hat{\tau}^2\right],$$

$$s_t^2 = \frac{1}{n(n-1)}(4(n-2)V_1 + 2V_2).$$

The $1 - \alpha$ confidence interval for τ is

$$\hat{\tau} \pm z_{1-\alpha/2}s_t,$$

where z is the $1 - \alpha/2$ quantile of a standard normal distribution (read from Table 1 in Appendix B).

For a general discussion about handling tied values when dealing with Kendall's tau, readers are referred to Cliff (1996), who provides an excellent summary of various approaches. Here it is merely remarked that different strategies lead to different measures of association, which include the Goodman–Kruskal γ, Somers' d, and Yule's Q.

EXAMPLE. Imagine we observe the 10 values $X = .1, .2, \ldots, 1$ and that $Y = X^2$. Then there is a perfect monotonic increasing relationship between X and Y, $\hat{\tau} = 1$, and Pearson's correlation is $r = .975$. So in this particular case there is little separating the two coefficients. However, Kendall's tau provides some protection against missing an association due to one or more outliers. For example, if the largest Y value is increased from 1 to 10, again $\hat{\tau} = 1$ but now $r = .59$ with a significance level (based on Student's T) equal to .07. Increasing the largest Y value to 50, $r = .54$.

As just indicated, Kendall's tau provides protection against outliers among the X values; it does the same among the Y values, but it does not take into account the overall structure of the data. That is, it does not address the concerns raised in Section 13.3. ■

EXAMPLE. Following the second illustration in Section 13.3, 20 points were generated according to the model $Y = X + \epsilon$, where both X and ϵ have standard normal distributions. The points are shown in Figure 13.4. Kendall's tau is estimated to be .368 and when testing $H_0 : \tau = 0$, the significance level is .023 based on the method in Box 15.11. Now we add two aberrant points at $(X, Y) = (2.1, -2.4)$; a scatterplot of the points now appears as shown in Figure 13.5. The estimate of Kendall's tau drops to .13 and the significance level increases to .398 — these two unusual points have a substantial impact on Kendall's tau. ■

15.10.2 Kendall's tau and the Theil–Sen Estimator

There is a connection between Kendall's tau and the Theil–Sen regression estimator in Section 13.2.1 that is worth mentioning. As was done in Chapter 13, let b_1 be some estimate of the slope. A general approach for determining b_1 from data is to take it to be the value that results in a zero correlation between the n pairs of points Y_i and $Y_i - b_1 X_i$. If the correlation used is Kendall's tau, b_1 is the Theil–Sen estimate in Chapter 13.

15.10.3 S-PLUS Function tau

The S-PLUS function

$$tau(x,y)$$

has been supplied for computing Kendall's tau and testing $H_0 : \tau = 0$. The function returns an estimate of τ plus the significance level. To compute a confidence interval for τ, it is suggested that the S-PLUS function corb in Section 13.6.5 be used.

15.10.4 Spearman's rho

Spearman's rho, labeled r_s, is just Pearson's correlation based on the ranks associated with X versus the ranks associated with Y. Under independence, the population analog of r_s, ρ_s, is zero. Also, like Kendall's tau, Spearman's rho is exactly equal to 1 if there is a monotonic increasing relationship between X and Y. That is, Y is a strictly increasing function of X. And $\rho_s = -1$ if the association is monotonic decreasing instead.

The usual approach to testing

$$H_0 : \rho_s = 0$$

is based on

$$T = \frac{r_s \sqrt{n-2}}{\sqrt{1-r_s^2}}.$$

When there is independence, T has, approximately, a Student's T-distribution with $\nu = n - 2$ degrees of freedom. So reject and conclude there is an association if $|T| \geq t$, where t is the $1 - \alpha/2$ quantile of a Student's T-distribution with $n - 2$ degrees of freedom.

Like Kendall's tau, Spearman's rho provides protection against outliers among the X values or among the Y values, but it does not take into account the overall structure of the data. That is, a few unusual points, properly placed, can have a substantial influence on its value.

> **EXAMPLE.** In the last example it was illustrated that Kendall's tau can be influenced by a few unusual values. We repeat this illustration with Spearman's rho. For the original 20 values, $r_s = .54$ with a significance level of .014. But when the two outliers are added at $(X, Y) = (2.1, -2.4)$, now $r_s = .16$ with a significance level of .48. ■

15.10.5 S-PLUS Function spear

The S-PLUS function

$$spear(x,y)$$

computes Spearman's rho and returns the significance level when testing $H_0 : \rho_s = 0$ as described in Section 15.10.3. A confidence interval for ρ_s can be computed with the S-PLUS function corb in Section 13.6.5.

15.11 Comparing Rank-Based Correlations

Section 14.3.2 describes a bootstrap method for comparing the correlations associated with two independent groups that allows heteroscedasticity. The computations are performed with the S-PLUS function twocor in Section 14.3.3. It is noted that this method can be used to compare rank-based correlations as well. That is, we have two independent groups with two measures for each participant. Letting ρ_{s1} and ρ_{s2} represent the population Spearman correlations, the S-PLUS function computes a confidence interval for $\rho_{s1} - \rho_{s2}$, so, in particular, the hypothesis

$$H_0 : \rho_{s1} = \rho_{s2}$$

can be tested.

EXAMPLE. In Section 14.7, one of the examples was aimed at investigating whether the association between two variables from a reading study (orthographic ability, Y, and a measure of sound blending, X_1) is modified by a third variable, phonological awareness (X_2). A smooth of the data, using both X_1 and X_2 as predictors, is shown in Figure 14.9. Attempts at establishing that X_2 modifies the association failed, but it was noted that there seems to be a nonlinear association between Y and X_1. Rank-based correlations are sensitive to monotonic associations that are not necessarily linear, and it was suggested that using a rank-based correlation with these data might make a difference in the conclusions reached. If we use the S-PLUS function runcor (in Section 14.7.1) with the argument corfun set equal to spear (so that Spearman's correlation is used), we see that the association between X_1 and Y appears to change around $X_2 = 14$. Using the S-PLUS function twocor to compare Spearman's correlation between X_1 and Y when $X_2 < 14$, versus $X_2 \geq 14$, the .95 confidence interval for the difference between these two correlations is $(-0.775, -0.012)$ with a significance level of .043. So there is empirical evidence that X_2 modifies the association between X_1 and Y. ■

15.12 Rank-Based Regression

Yet another approach to regression, beyond those covered in Chapters 6 and 13, is to minimize sum function of the ranks of the residuals. In simple regression, for example, with n pairs of values, $(X_1, Y_1), \ldots, (X_n, Y_n)$, consider any choice for the slope and intercept, say, b_1 and b_0. As in previous chapters, $\hat{Y}_i = b_1 X_i + b_0$ is the predicted value of Y based on X_i. Least squares chooses the slope and intercept so as to minimize the sum of squared residuals, $\sum r_i^2$, where as usual $r_i = Y_i - \hat{Y}_i$. A rank-based approach simply replaces the residuals with their ranks, and often some function of the ranks is used instead (e.g., Hössjer, 1994). Here attention is focused on a slight variation of this method, which is called the Wilcoxon R estimate. The method begins by choosing the slope (b_1) so as to minimize

$$\sum_{i=1}^{n} = a(R(Y_i - b_1 X_i))(Y_i - b_1 X_i), \tag{15.22}$$

where $R(Y_i - b_1 X_i)$ is the rank of $Y_i - b_1 X_i$ among $Y_1 - b_1 X_1, \ldots, Y_n - b_1 X_n$,

$$a(i) = \phi\left(\frac{i}{n+1}\right),$$

and

$$\phi(u) = \sqrt{12}\left(u - \frac{1}{2}\right).$$

Once b_1 is determined, one way of estimating the intercept is with

$$b_0 = \text{med}\{Y_i - b_1 X_i\},$$

the median of the values $Y_1 - b_1 X_1, \ldots, Y_n - b_1 X_n$. (Readers interested in a theoretical treatment of this estimator are referred to Hettmansperger & McKean, 1998.)

15.12.1 S-PLUS Function wreg

The S-PLUS function

$$\text{wreg}(x, y)$$

computes an estimate of the slope and intercept using the Wilcoxon R estimate just described. A negative feature of this function is that execution time can be somewhat high compared to some of the other estimators in Chapter 13.

EXAMPLE. For the reading data in Figure 13.18, wreg estimates the slope to be zero, which is close to the least squares estimate of -0.02 as well as the Coakley–Hettmansperger M-estimate of -0.04. It is fairly evident, however, that the six most right points are outliers. If these six points are ignored, wreg estimates the slope to be -0.53. The Theil–Sen estimate of the slope is -0.6, again ignoring the outliers; but even with the outliers, the Theil–Sen estimate is -0.28. So switching to the rank-based estimate does not eliminate the possibility that a few points can dominate the estimate of the slope and give a relatively poor fit to the bulk of the points. ■

EXAMPLE. For the star data in Figure 13.12, wreg estimates the slope to be -0.477 versus the Theil–Sen estimate of 1.73, which gives a better fit to the bulk of the observations. Again, ignoring the outliers by restricting X to be greater than 4.01, a good fit to the data is obtained with the rank-based regression method.   ■

EXAMPLE. As with many of the other estimators covered in Chapter 13, restricting the range of X values by checking for outliers among the X values and eliminating any that are found is no guarantee that a good fit to the data will be obtained with wreg. For example, in Figure 13.5, the data were generated having a slope of 1 and then two unusual points were added, but no outliers are found among the X values using the methods in Chapter 3. The function wreg estimates the slope to be .498.   ■

As stressed in Chapter 13, it currently seems that no single regression estimator is ideal under all circumstances, and in applied work it seems that multiple methods are required, at least in the preliminary stages of data analysis. The rank-based regression estimator described here does not perform well in some situations where other estimators give excellent results, but it remains among the group of estimators that seem to have practical value.

One appeal of the Wilcoxon R estimator is that it appears to perform reasonably well when there are multiple predictors and the goal is to detect curvature using a partial residual plot (as described in Section 14.2.5). McKean and Sheather (2000) provide results on this issue and compare the use of the Wilcoxon R estimator to partial residual plots based on least squares.

Yet another issue is testing hypotheses when using the Wilcoxon R estimator. One possibility is the bootstrap method in Section 14.5. It is possible to avoid the bootstrap using results in Hettmansperger and McKean (1998), but it seems the relative merits of this approach, versus the bootstrap method have not been investigated.

15.12.2 Other Rank-Based Estimators

For completeness, there are other rank-based estimators that are briefly mentioned here. Jaeckel (1972) suggests estimating the slope by minimizing

$$\sum_{i<j} |(Y_i - b_1 X_i) - (Y_j - b_1 X_j)|,$$

which turns out to be tantamount to minimizing a function of the ranks of the residuals. A generalization of this method was derived by Naranjo and Hettmansperger (1994). An appeal of their method is that it appears to control Type I error probabilities relatively well without resorting to the bootstrap. A negative feature is that when there are leverage points (outliers among the X values), power can be low versus least squares (Wilcox, 1995). For yet another rank-based approach, see Cliff (1996).

15.13 The Rank-Transform Method

It should be noted that a simple approach when dealing with rank-based methods is to replace the observations by their ranks and apply a method for means. There are situations where this strategy gives reasonable results, but there are general conditions where it performs poorly. For criticisms of the method, see Blair, Sawilowski, and Higgens (1987); Akritas (1990); G. Thompson and Ammann (1990); and G. Thompson (1991). For more details, see McKean and Vidmar (1994); Hettmansperger and McKean (1998, Section 4.7); and Brunner, Domhof, and Langer (2002, Section 5.8). Headrick and Rotou (2001) studied this approach when dealing with multiple regression and found it to be unsatisfactory.

15.14 Exercises

1. Two methods for reducing shoulder pain after laparoscopic surgery were compared by Jorgensen et al. (1995). The data were

Group 1:	1 2 1 1 1 1 1 1 1 1 2 4 1 1
Group 2:	3 3 4 3 1 2 3 1 1 5 4

   Verify that all of the methods in Section 15.1 reject at the .05 level. Although the Kolmogorov–Smirnov test rejects with $\alpha = .05$, why might you suspect that the Kolmogorov–Smirnov test will have relatively low power in this particular situation? Check your results using the S-PLUS functions provided.

2. Imagine two groups of cancer patients are compared, the first group having a rapidly progressing form of the disease and the other having a slowly progressing form instead. At issue is whether psychological factors are related to the progression of cancer. The outcome measure is one where highly negative scores indicated a tendency to present the appearance of serenity in the presence of stress. The results are

Group 1:	-25 -24 -22 -22 -21 -18 -18 -18 -18 -17 -16 -14 -14 -13 -13 -13 -13 -9 -8 -7 -5 1 3 7 7
Group 2:	-21 -18 -16 -16 -16 -14 -13 -13 -12 -11 -11 -11 -9 -9 -9 -9 -7 -6 -3 -2 3 10

   Verify that the Wilcoxon–Mann–Whitney test rejects at the .05 level but that none of the other methods in Section 15.1 reject. Check your results using the S-PLUS functions provided. What might explain this?

3. Chapter 14 mentions data from a study regarding four skull measurements from five different time periods. If we compare the five groups based on these four measures with the S-PLUS function mulrank in Section 15.8.2, the output is as follows.

```
$test.stat:
[1] 4.197179

$nul:
[1] 14.13007

$sig.level:
[1] 1.717545e-07

$N:
[1] 150

$q.hat:
              [,1]         [,2]         [,3]         [,4]
[1,]     0.3605556    0.5682222    0.6316667    0.4626667
[2,]     0.3881111    0.4988889    0.6518889    0.4422222
[3,]     0.5380000    0.5635556    0.4756667    0.4568889
[4,]     0.5898889    0.4855556    0.3880000    0.5961111
[5,]     0.6234444    0.3837778    0.3527778    0.5421111
```

Interpret the results.

4. If a significant result is obtained with the S-PLUS function cmanova in Section 15.8.4, is this evidence that we should also reject the hypothesis tested by mulrank in Section 15.8.2?

5. Imagine that you get a significant result when using the ANOVA F-test in Chapter 9. If we increase the largest observation among the J groups, it is generally the case that eventually we will no longer reject. Is the same thing true when using the rank-based methods in Section 15.2?

6. Two independent groups are given different cold medicines and the goal is to compare reaction times. Suppose that the decreases in reaction times when taking drug A versus B are as follows.

A:	1.96,	2.24,	1.71,	2.41,	1.62,	1.93	
B:	2.11,	2.43,	2.07,	2.71,	2.50,	2.84,	2.88

Compare these two groups with the Mann–Whitney–Wilcoxon test using Equation (15.4). What is your estimate of the probability that a randomly sampled participant receiving drug A will have less of a reduction in reaction time than a randomly sampled participant receiving drug B?

7. Repeat the previous exercise, only now use the Cliff as well as the Brunner–Munzel methods.

8. For two dependent groups you get

Group 1:	10	14	15	18	20	29	30	40
Group 2:	40	8	15	20	10	8	2	3

Compare the two groups with the sign test and the Wilcoxon signed rank test with $\alpha = .05$. Verify that according to the sign test, $\hat{p} = .36$, that the .95

confidence interval for p is (.13, .69), and that the Wilcoxon signed rank test has an approximate significance level of .46.

9. For two dependent groups you get

Group 1:	86	71	77	68	91	72	77	91	70	71	88	87
Group 2:	88	77	76	64	96	72	65	90	65	80	81	72

Apply the Wilcoxon signed rank test with $\alpha = .05$. Verify that $W = .7565$ and that you fail to reject.

10. A developing nation is trying to improve its ability to grow its own crops. Four methods of growing corn are being considered, and you have been asked to help determine whether it makes a difference which method is used. To find out, four adjacent plots of land are used to grow the corn, and the yield is measured at the end of the season. This process is repeated in 12 locations located throughout the country, and the results are as shown in Table 15.4. Because results for adjacent plots of land might not be independent, you decide to use Friedman's test. Verify that a significant result among the four methods is obtained with $\alpha = .05$.

11. Repeat the previous exercise using the Agresti–Pendergast procedure. (Use the S-PLUS function apanova.)

12. Use the Agresti–Pendergast procedure to perform all pairwise comparisons of the groups in the previous two exercises. Use Rom's method so that the familywise error rate does not exceed 0.05. Verify that a significant result is obtained only when comparing groups 1 and 3.

13. For the following pairs of observations, test $H_0 : \tau = 0$ with $\alpha = .05$.

Time 1:	10	16	15	20
Time 2:	25	8	18	9

Verify that $\hat{\tau} = -0.667$, $Z = -1.36$, so you fail to reject.

TABLE 15.4   Data for Exercise 10

A	B	C	D
10	7	6	4
11	6	9	5
9	5	5	10
8	2	7	9
10	5	4	8
6	6	6	7
4	11	5	12
6	9	2	8
9	4	5	9
10	3	5	4
9	4	8	3
6	2	3	6

14. Repeat the previous exercise, except use Spearman's rho instead. Verify that $r_s = -0.8$ and that again you fail to reject.

15. Section 15.8.3 makes a distinction between the multivariate versus the marginal hypothesis. Describe a situation where the marginal hypothesis is true but the multivariate hypothesis is false.

16. Apply the S-PLUS function mulrank to the data in Table 15.2, and verify the results in the example of Section 15.8.2.

17. Using the S-PLUS function cmanova, apply the Choi–Marden method to the data in Table 15.2 and verify that the significance level is less than .01.

18. Apply the S-PLUS function bwrank to the data in Table 15.2. Verify that the significance level associated with the main effect for factor A is less than .001 but that you get a highly nonsignificant result for factor B and the hypothesis of no interaction.

19. Using the S-PLUS function bwrmcp, verify that for the data in Table 15.2, all pairwise comparisons among the levels of factor A are significant when the goal is to have FWE equal to .05.

20. Verify that no interaction is found using the method in Section 15.9.4 with the data in Table 15.2.

21. In the previous three exercises, beyond random sampling, what is being assumed about the variables when making inferences about the levels of factor B and interactions?

22. Verify that for the data in Table 15.3, the S-PLUS function bwrmcp returns

```
$Factor.A:
        Level   Level   test.stat    sig.level   sig.crit
[1,]        1       2   12.87017   0.001043705       0.05

$Factor.B:
        Level   Level   test.stat    sig.level   sig.crit
[1,]        1       2   0.3048713   0.5808447     0.0250
[2,]        1       3   0.2224258   0.6371979     0.0500
[3,]        2       3   0.5858703   0.4440207     0.0169

$Factor.AB:
        Lev.A     Lev.A    Lev.B    Lev.B     test.stat
[1,]        1         2        1        2      6.520209
[2,]        1         2        1        3     11.170399
[3,]        1         2        2        3      3.728463

    sig.level    sig.crit
0.0106656917      0.0250
0.0008311581      0.0169
0.0534928817      0.0500
```

23. Based on the results of the previous exercise, can you conclude that based on the average ranks among the groups, there is a disordinal interaction?

24. Look at the data in Table 15.3 and comment on why the method in Section 15.9.4 would be expected to have relatively low power.

25. Verify that for the data in Table 15.3, time 1 and time 2 among the active treatment group have a Kendall's tau equal to .31 with a significance level of .04.

26. For the data in Table 15.3, compute a .95 confidence interval for Spearman's rho using the data for time 1 and time 2 among the active treatment group. Use the S-PLUS function corb in Section 13.6.5. Verify that the .95 confidence interval is (0.036, 0.84), so again you reject $H_0 : \rho_s = 0$ at the .05 level.

27. For the data used in the last exercise, plot the points and comment on the association and the robustness of Spearman's rho and Kendall's tau.

28. Verify that for the data in Table 15.3, the Choi–Marden method has a significance level less than .001.

29. Verify that the hypothesis given by Equation (15.7) is the same as the hypothesis given by Equation (15.8).

APPENDIX A

Solutions to Selected Exercises

Chapter 2

3. $\mu = 3$, $\sigma^2 = 1.6$. 4. Smaller, $\sigma^2 = 1.3$. 5. Larger. 7. (a) .3, (b) .03/.3, (c) .09/.3, (d) .108/.18. 8. Yes. For example, probability of a high income given that they are under 30 is .03/.3 = .1, which is equal to the probability of a high income. 9. (a) 1253/3398, (b) 757/1828, (c) 757/1253, (d) no, (e) 1831/3398. 10. Median = 2.5. The .1 quantile — say, y — is given by $(y - 1) * (1/3) = .1$, so the .1 quantile is 1.3. The .9 quantile is 3.7. 11. (a) $4 \times (1/5)$, (b) $1.5 \times (1/5)$, (c) $2 \times (1/5)$, (d) $2.2 \times (1/5)$, (e) 0. 12. Median = $-.5$. The .25 quantile, y, is given by $(y - (-3)) \times 1/5 = .25$, so $y = -1.75$. 13. (a) $(c - (-1)) \times (1/2) = .9$, so $c = .8$, (b) $(c - (-1)) \times (1/2) = .95$, $c = .9$, (c) $(c - (-1)) \times (1/2) = .01$, $c = -.98$. 14. (a) .9, (b) $c = .95$, (c) $c = .99$. 15. (a) 0, (b) .25, (c) .5. 16. (a) 0, (b) 1/6, (c) 2/3, (d) 1/6. 17. $y \times (1/60) = .8$, $y = 48$. 18. (a) 0.0668, (b) 0.0062, (c) 0.0062, (d) .683. 19. (a) 0.691, (b) 0.894, (c) .77. 20. (a) .31, (b) .885, (c) 0.018, (d).221. 21. (a) -2.33, (b) 1.93, (c) -0.174, (d) .3. 22. (a) 1.43, (b) -0.01, (c) 1.7, (d) 1.28. 23. (a) .133, (b) .71, (c) .133, (d) .733. 24. (a) .588, (b) .63, (c) .71, (d) .95. 26. $c = 1.96$. 27. 1.28. 28. .16. 29. 84.45. 30. $1 - .91$. 31. .87. 32. .001. 33. .68. 34. .95. 35. .115. 36. .043. 37. Yes. 39. No, for small departures from normality this probability can be close to 1. 40. No, for reasons similar to those in the previous exercise. 41. Yes. 42. Yes. 43. $\mu = 2.3$, $\sigma = .9$, and $P(\mu - \sigma \leq X \leq \mu + \sigma) = .7$. 44. (a) $.75^5$, (b) $.25^5$, (c) $1 - .25^5 - 5(.75)(.25)^4$. 45. (a) .586, (b) .732, (c) $1 - .425$, (d) $1 - .274$. 46. .4(25), .4(.6)(25), .4, .4(.6)/25.

Chapter 3

3. Two. 7. $n = 88$. 10. 9. 11. Yes. 12. One. 13. 20%. 17. 98, 350, 370, and 475. 18. $\sum (x - \bar{X})^2 (f_x/n)$. 25. The lower and upper quartiles are approximately 900 and 1300, respectively. So the standard boxplot rule would declare a value an outlier if it is less than $900 - 1.5(1300 - 900)$ or greater than $1300 + 1.5(1300 - 900)$. 26. .1.

Chapter 4

2. 1.28, 1.75, 2.32. 6. 3. 45±1.96. 4. 45±2.58. 5. Yes, upper end of confidence interval is 1158. 6. (a) $65 \pm 1.96(22)/\sqrt{12}$, (b) $185 \pm 1.96(10)/\sqrt{22}$, (c) $19 \pm 1.96(30)/\sqrt{50}$. 8. 9 and 8/10. 9. 2.7 and 1.01/12. 10. 2.7. 11. 1.01. 12. 94.3/8 and $\sqrt{94.3/8}$. 13. 32. No. 14. 93,663.52/12. 15. They inflate the standard error. 16. 94.3/8, $\sqrt{94.3/8}$. 17. No. 18. 10.9/25; small departures from normality can inflate the standard error. 20. (a) .0228, (b) .159. 21. (a) .16, (b) .023, (c) $.977 - .028$. 22. .023. 24. $.933 - .067$. 25. (a) .055, (b) .788, (c) .992, (d) $.788 - .055$. 26. (a) .047, (b) .952, (c) $1 - .047$, (d) $.952 - .047$. 27. Sample from a heavy-tailed distribution. 28. Sampling from a light-tailed distribution, the distribution of the sample mean will be well approximated by the central limit theorem. 29. (a) $26 \pm 2.26(9)/\sqrt{10}$, (b) $132 \pm 2.09(20)/\sqrt{18}$, (c) $52 \pm 2.06(12)/\sqrt{25}$. 31. (161.4, 734.7). 32. (10.7, 22.4). 35. (a) $52 \pm 2.13\sqrt{12}/(.6\sqrt{24})$, (b) $10 \pm 2.07\sqrt{30}/(.6\sqrt{36})$. 37. (160.4, 404.99). 38. Outliers. 39. Outliers. 41. No.

Chapter 5

1. $Z = -1.265$, fail to reject. 2. Fail to reject. 3. (74.9, 81.1). 4. .103. 5. .206. 6. $Z = -14$, reject. 7. Reject. 8. (118.6, 121.4). 9. Yes, because $\bar{X}$ is consistent with H_0. 10. $Z = 10$, reject. 11. $Z = 2.12$, reject. 19. Increase α. 20. (a) $T = 1$, fail to reject. (b) $T = .5$, fail to reject. (c) $T = 2.5$, reject. 22. (a) $T = .8$, fail to reject. (b) $T = .4$, fail to reject. (c) $T = 2$, fail to reject. 24. $T = .39$, fail to reject. 25. $T = -2.61$, reject. 26. (a) $T_t = .596$, fail to reject. (b) $T_t = .298$, fail to reject. (c) $T_t = .894$, fail to reject. 28. $T_t = -3.1$, reject. 29. $T_t = .129$, fail to reject.

Chapter 6

4. .87. 6. $b_1 = -0.0355$, $b_0 = 39.93$. 7. $b_1 = .0039$, $b_0 = .485$. 10. One concern is that $X = 600$ lies outside the range of X values used to compute the least squares regression line. 9. $b_1 = -0.0754$, $b_0 = -1.253$. 11. $r = -.366$. Not significantly different from zero, so can't be reasonably certain about whether ρ is positive or negative. 14. Health improves as vitamin intake increases, but health deteriorates with too much vitamin A. That is, there is a nonlinear association. 17. Extrapolation can be misleading.

Chapter 7

1. (8.1, 14.6), the middle 80%. 4. The one based on trimcibt, which uses the bootstrap-t method. 7. The one based on trimpb, which uses the percentile bootstrap method. 10. In a bootstrap sample, outliers are more likely to inflate the 20% trimmed mean if no Winsorizing is done. 12. The amount of Winsorizing equals the amount of trimming, which might mean inaccurate probability coverage. 13. (7, 14.5). Even though the one-step M-estimator removes outliers, situations arise where using a median might yield a substantially shorter confidence interval. 17. There might be a nonlinear association. 18. Restricting the range of the X values can reveal an association that is not otherwise detected. 19. Now you reject, before you did not

reject, so again restricting the range of X can reveal an association that is otherwise missed. **20.** The conventional method appears to have accurate probability coverage in this particular situation.

Chapter 8

3. $T = (45 - 36)/\sqrt{11.25(.083333)} = 9.3$, $\nu = 48$, reject. **4.** $W = (45 - 36)/\sqrt{8/20 + 24/10} = 10.49$, reject. **5.** Welch's test might have more power. **6.** $T = 3.79$, $\nu = 38$, reject. **7.** $W = 3.8$, $\nu = 38$, reject. **8.** With equal sample sizes and equal sample variances, T and W give exactly the same result. This suggests that if the sample variances are approximately equal, it makes little difference which method is used. But if the sample variances differ enough, Welch's method is generally more accurate. **9.** $h_1 = 16$, $h_2 = 10$, $d_1 = 2.4$, $d_2 = 6$, $\nu = 16.1$, so $t = 2.12$. $T_y = 2.07$, fail to reject. **10.** .99 confidence interval is $(-2.5, 14.5)$. **11.** $\nu = 29$, $t = 2.045$, CI is $(1.38, 8.62)$, reject. **12.** CI is $(1.39, 8.6)$, reject. **13.** $W = 1.2$, $\hat{\nu} = 11.1$, $t = 2.2$, fail to reject. **14.** $T_y = 1.28$, $\hat{\nu} = 9.6$, fail to reject. **15.** No, power might be low. **16.** .95 CI is $(-7.4, .2)$, fail to reject. **17.** Fail to reject. **20.** The data indicate that the distributions differ, so the confidence interval based on Student's T might be inaccurate. **24.** The distributions differ, so some would argue that by implication the means in particular differ. **26.** Power. **28.** One concern is that the second group appears to be more skewed than the first, suggesting that probability coverage, when using means, might be poor. Another concern is that the first group has an outlier. **31.** If the tails of the distributions are sufficiently heavy, medians have smaller standard errors and hence can have more power. **33.** An improper estimate of the standard error is being used if extreme observations are discarded and methods for means are applied to the data that remain. See Section 4.9.1.

Chapter 9

17. Set allp=F, but this might lower power. A better approach would be to use the method in Section 12.7.3. **19.** $(MSBG - MSWG)(J - 1)/n$. **22.** Using allp=F might lower power, but this is not an issue here because you reject.

Chapter 10

2. There are main effects for both factors A and B but no interaction. **3.** There is a disordinal interaction. **4.** Row 1: 10, 20, 30; Row 2: 20, 30, 40; Row 3: 30, 40, 50. **5.**

Source	SS	DF	MS	F
A	800	2	400	3.00
B	600	3	200	1.5
INTER	1200	6	200	1.5
WITHIN	4800	36	133.3	

This is a 3-by-4 design. The total number of observations is $N = 48$. None of the hypotheses is rejected.

7. The means suggest that there is a disordinal interaction by rows. That is, method 1 is best for females, but method 2 is best for males. Although no interaction was detected, accepting the null hypothesis is not warranted, because power might be too low to detect a true interaction.
9. Disordinal interactions means that interpreting main effects may not be straightforward. 10. No, more needs to be done to establish that a disordinal interaction exists.

Chapter 11

4. One possible reason is that these measures of location differ. That is, they are testing different hypotheses; depending on how the groups differ, one approach could have more power than the other.
5. Currently it seems that trimmed means based on difference scores usually provide more power.

Chapter 12

1. Does not control the familywise error rate (the probability of at least one Type I error).
2. MSWG $= 11.6$. (a) $T = |15 - 10|/\sqrt{11.6(1/20 + 1/20)} = 4.64$, $\nu = 100 - 5 = 95$, reject. (b) $T = |15 - 10|/\sqrt{11.6(1/20 + 1/20)/2} = 6.565$, $q = 3.9$, reject. (c) $W = (15 - 10)/\sqrt{4/20 + 9/20} = 6.2$, $\hat{\nu} = 33$, $c = 2.99$, reject. (d) $(15 - 10)/\sqrt{.5(4/20 + 9/20)} = 8.77$, $q = 4.1$, reject. (e) $f = 2.47$, $S = \sqrt{4(2.47)(11.6)(1/20 + 1/20)} = 3.39$, reject. (f) $f = 2.66$, $A = 4(1 + 3/33)2.66 = 11.61$, reject.
3. MSWG $= 8$. (a) $T = |20 - 12|/\sqrt{8(1/10 + 1/10)} = 6.325$, $\nu = 50 - 5 = 45$, reject. (b) $T = |20 - 12|/\sqrt{8(1/10 + 1/10)/2} = 8.94$, $q = 4.01$, reject. (c) $W = (20 - 12)/\sqrt{5/10 + 6/10} = 7.63$, $\hat{\nu} = 37.7$, $c = 2.96$, reject. (d) $(20 - 12)/\sqrt{.5(5/10 + 6/10)} = 10.79$, $q = 4.06$, reject. (e) $f = 2.58$, $S = \sqrt{4(2.58)(8)(1/10 + 1/10)} = 4.06$, reject. (f) $f = 2.62$, $A = 11.3$, reject.
4. Reject if the significance level is less than or equal to $.05/6 = .0083$. So the fourth test is significant.
5. Fourth test, having significance level .001. 6. None is rejected. 7. All are rejected. 8. MSWG goes up and eventually you will no longer reject when comparing groups one and two. 9. The same problem occurs as in Exercise 8. 16. Significant results are obtained when comparing groups 1 to 3, 1 to 4, 1 to 5, as well as 2 to 5. 17. One possibility is that the boxplot rule is missing outliers. Using the rule based on MAD and M in Section 3.4.2, group 3 is found to have four outliers, but none of the boxplot rules in Chapter 3 finds an outlier in this group. Comparing the standard errors of the means versus the standard errors when using MOM, sometimes MOM has a smaller standard error, but sometimes the reverse is true. Skewness can affect power when comparing means, even when there are no outliers, and in general the differences among the MOM population values might be larger than those among the population means.

Chapter 13

6. Both outliers among the Y values and heteroscedasticity can affect the confidence intervals. The only known method that effectively determines whether robust methods make a difference is simply to try the methods and compare the results. The conventional .95 confidence interval for the slope, using the method in Section 6.3.1, is $(-4.23, 0.285)$ with a significance level of .083. Using the heteroscedastic bootstrap method in Chapter 7 it is $(-4.62, 0.41)$. Using Theil–Sen it is $(-4.26, 0.26)$, and for MGV it is $(-4.55, 0.37)$. So all four methods fail to reject. 8. When computing a confidence interval, the length of the confidence interval can change substantially, depending on which estimator is used. That is, power can depend to a large extent on which regression method is employed. 10. Eliminating too many points can mask the true association among the bulk of the observations.

Chapter 14

1. All but three of the X_3 values are close to 50. The three that are not are observations 1, 2, and 7. Excluding the corresponding Y values, the 20% trimmed mean of the remaining Y values is 99.9 and the value of MOM is 98.5. 5. It appears that the association might be nonlinear. However, lintest fails to reject the hypothesis of a linear association. 8. $X = 131$ and 133.

Chapter 15

6. $\hat{p} = .9$. Significance level is .015. 7. Using Cliff's method, the .95 confidence interval for δ is $(-0.966, -0.21)$, so reject with $\alpha = .05$. The Brunner–Munzel method returns a significance level of .00066. 15. For the first group, suppose there are two variables, each having standard normal distributions with a correlation of zero. Imagine for the second group that, again, the two variables each have standard normal distributions, only now $\rho = .5$. Then the marginal hypothesis is true but the multivariate hypothesis is false. 21. A basic assumption is that it is meaningful to compare the variables. For example, if for every individual we measure height and weight, we could analyze the data with the methods in Section 15.9, but the analysis is meaningless when dealing with factor B or interactions. We could, however, compare groups, based on these measures, using the methods in Section 15.8. 23. No. The average ranks need to be compared for the appropriate groups. 24. There are many tied values. 27. The significant results associated with Spearman's rho and Kendall's tau are due to the two points where both time 1 and time 2 scores are equal to 4. These two points appear in the upper right corner of a scatterplot. Ignoring these two points, both Kendall's tau and Spearman's rho are no longer significant.

APPENDIX B

Tables

TABLE 1 Standard Normal Distribution

z	$P(Z \leq z)$	z	$P(Z \leq z)$	z	$P(Z \leq z)$	z	$P(Z \leq z)$
-3.00	0.0013	-2.99	0.0014	-2.98	0.0014	-2.97	0.0015
-2.96	0.0015	-2.95	0.0016	-2.94	0.0016	-2.93	0.0017
-2.92	0.0018	-2.91	0.0018	-2.90	0.0019	-2.89	0.0019
-2.88	0.0020	-2.87	0.0021	-2.86	0.0021	-2.85	0.0022
-2.84	0.0023	-2.83	0.0023	-2.82	0.0024	-2.81	0.0025
-2.80	0.0026	-2.79	0.0026	-2.78	0.0027	-2.77	0.0028
-2.76	0.0029	-2.75	0.0030	-2.74	0.0031	-2.73	0.0032
-2.72	0.0033	-2.71	0.0034	-2.70	0.0035	-2.69	0.0036
-2.68	0.0037	-2.67	0.0038	-2.66	0.0039	-2.65	0.0040
-2.64	0.0041	-2.63	0.0043	-2.62	0.0044	-2.61	0.0045
-2.60	0.0047	-2.59	0.0048	-2.58	0.0049	-2.57	0.0051
-2.56	0.0052	-2.55	0.0054	-2.54	0.0055	-2.53	0.0057
-2.52	0.0059	-2.51	0.0060	-2.50	0.0062	-2.49	0.0064
-2.48	0.0066	-2.47	0.0068	-2.46	0.0069	-2.45	0.0071
-2.44	0.0073	-2.43	0.0075	-2.42	0.0078	-2.41	0.0080
-2.40	0.0082	-2.39	0.0084	-2.38	0.0087	-2.37	0.0089
-2.36	0.0091	-2.35	0.0094	-2.34	0.0096	-2.33	0.0099
-2.32	0.0102	-2.31	0.0104	-2.30	0.0107	-2.29	0.0110
-2.28	0.0113	-2.27	0.0116	-2.26	0.0119	-2.25	0.0122
-2.24	0.0125	-2.23	0.0129	-2.22	0.0132	-2.21	0.0136
-2.20	0.0139	-2.19	0.0143	-2.18	0.0146	-2.17	0.0150
-2.16	0.0154	-2.15	0.0158	-2.14	0.0162	-2.13	0.0166
-2.12	0.0170	-2.11	0.0174	-2.10	0.0179	-2.09	0.0183
-2.08	0.0188	-2.07	0.0192	-2.06	0.0197	-2.05	0.0202
-2.04	0.0207	-2.03	0.0212	-2.02	0.0217	-2.01	0.0222
-2.00	0.0228	-1.99	0.0233	-1.98	0.0239	-1.97	0.0244
-1.96	0.0250	-1.95	0.0256	-1.94	0.0262	-1.93	0.0268
-1.92	0.0274	-1.91	0.0281	-1.90	0.0287	-1.89	0.0294
-1.88	0.0301	-1.87	0.0307	-1.86	0.0314	-1.85	0.0322
-1.84	0.0329	-1.83	0.0336	-1.82	0.0344	-1.81	0.0351
-1.80	0.0359	-1.79	0.0367	-1.78	0.0375	-1.77	0.0384
-1.76	0.0392	-1.75	0.0401	-1.74	0.0409	-1.73	0.0418
-1.72	0.0427	-1.71	0.0436	-1.70	0.0446	-1.69	0.0455
-1.68	0.0465	-1.67	0.0475	-1.66	0.0485	-1.65	0.0495
-1.64	0.0505	-1.63	0.0516	-1.62	0.0526	-1.61	0.0537
-1.60	0.0548	-1.59	0.0559	-1.58	0.0571	-1.57	0.0582
-1.56	0.0594	-1.55	0.0606	-1.54	0.0618	-1.53	0.0630
-1.52	0.0643	-1.51	0.0655	-1.50	0.0668	-1.49	0.0681
-1.48	0.0694	-1.47	0.0708	-1.46	0.0721	-1.45	0.0735
-1.44	0.0749	-1.43	0.0764	-1.42	0.0778	-1.41	0.0793
-1.40	0.0808	-1.39	0.0823	-1.38	0.0838	-1.37	0.0853

continued

TABLE 1   *continued*

z	$P(Z \le z)$	z	$P(Z \le z)$	z	$P(Z \le z)$	z	$P(Z \le z)$
−1.36	0.0869	−1.35	0.0885	−1.34	0.0901	−1.33	0.0918
−1.32	0.0934	−1.31	0.0951	−1.30	0.0968	−1.29	0.0985
−1.28	0.1003	−1.27	0.1020	−1.26	0.1038	−1.25	0.1056
−1.24	0.1075	−1.23	0.1093	−1.22	0.1112	−1.21	0.1131
−1.20	0.1151	−1.19	0.1170	−1.18	0.1190	−1.17	0.1210
−1.16	0.1230	−1.15	0.1251	−1.14	0.1271	−1.13	0.1292
−1.12	0.1314	−1.11	0.1335	−1.10	0.1357	−1.09	0.1379
−1.08	0.1401	−1.07	0.1423	−1.06	0.1446	−1.05	0.1469
−1.04	0.1492	−1.03	0.1515	−1.02	0.1539	−1.01	0.1562
−1.00	0.1587	−0.99	0.1611	−0.98	0.1635	−0.97	0.1662
−0.96	0.1685	−0.95	0.1711	−0.94	0.1736	−0.93	0.1762
−0.92	0.1788	−0.91	0.1814	−0.90	0.1841	−0.89	0.1867
−0.88	0.1894	−0.87	0.1922	−0.86	0.1949	−0.85	0.1977
−0.84	0.2005	−0.83	0.2033	−0.82	0.2061	−0.81	0.2090
−0.80	0.2119	−0.79	0.2148	−0.78	0.2177	−0.77	0.2207
−0.76	0.2236	−0.75	0.2266	−0.74	0.2297	−0.73	0.2327
−0.72	0.2358	−0.71	0.2389	−0.70	0.2420	−0.69	0.2451
−0.68	0.2483	−0.67	0.2514	−0.66	0.2546	−0.65	0.2578
−0.64	0.2611	−0.63	0.2643	−0.62	0.2676	−0.61	0.2709
−0.60	0.2743	−0.59	0.2776	−0.58	0.2810	−0.57	0.2843
−0.56	0.2877	−0.55	0.2912	−0.54	0.2946	−0.53	0.2981
−0.52	0.3015	−0.51	0.3050	−0.50	0.3085	−0.49	0.3121
−0.48	0.3156	−0.47	0.3192	−0.46	0.3228	−0.45	0.3264
−0.44	0.3300	−0.43	0.3336	−0.42	0.3372	−0.41	0.3409
−0.40	0.3446	−0.39	0.3483	−0.38	0.3520	−0.37	0.3557
−0.36	0.3594	−0.35	0.3632	−0.34	0.3669	−0.33	0.3707
−0.32	0.3745	−0.31	0.3783	−0.30	0.3821	−0.29	0.3859
−0.28	0.3897	−0.27	0.3936	−0.26	0.3974	−0.25	0.4013
−0.24	0.4052	−0.23	0.4090	−0.22	0.4129	−0.21	0.4168
−0.20	0.4207	−0.19	0.4247	−0.18	0.4286	−0.17	0.4325
−0.16	0.4364	−0.15	0.4404	−0.14	0.4443	−0.13	0.4483
−0.12	0.4522	−0.11	0.4562	−0.10	0.4602	−0.09	0.4641
−0.08	0.4681	−0.07	0.4721	−0.06	0.4761	−0.05	0.4801
−0.04	0.4840	−0.03	0.4880	−0.02	0.4920	−0.01	0.4960
0.01	0.5040	0.02	0.5080	0.03	0.5120	0.04	0.5160
0.05	0.5199	0.06	0.5239	0.07	0.5279	0.08	0.5319
0.09	0.5359	0.10	0.5398	0.11	0.5438	0.12	0.5478
0.13	0.5517	0.14	0.5557	0.15	0.5596	0.16	0.5636
0.17	0.5675	0.18	0.5714	0.19	0.5753	0.20	0.5793
0.21	0.5832	0.22	0.5871	0.23	0.5910	0.24	0.5948
0.25	0.5987	0.26	0.6026	0.27	0.6064	0.28	0.6103

continued

TABLE 1   *continued*

z	$P(Z \leq z)$	z	$P(Z \leq z)$	z	$P(Z \leq z)$	z	$P(Z \leq z)$
0.29	0.6141	0.30	0.6179	0.31	0.6217	0.32	0.6255
0.33	0.6293	0.34	0.6331	0.35	0.6368	0.36	0.6406
0.37	0.6443	0.38	0.6480	0.39	0.6517	0.40	0.6554
0.41	0.6591	0.42	0.6628	0.43	0.6664	0.44	0.6700
0.45	0.6736	0.46	0.6772	0.47	0.6808	0.48	0.6844
0.49	0.6879	0.50	0.6915	0.51	0.6950	0.52	0.6985
0.53	0.7019	0.54	0.7054	0.55	0.7088	0.56	0.7123
0.57	0.7157	0.58	0.7190	0.59	0.7224	0.60	0.7257
0.61	0.7291	0.62	0.7324	0.63	0.7357	0.64	0.7389
0.65	0.7422	0.66	0.7454	0.67	0.7486	0.68	0.7517
0.69	0.7549	0.70	0.7580	0.71	0.7611	0.72	0.7642
0.73	0.7673	0.74	0.7703	0.75	0.7734	0.76	0.7764
0.77	0.7793	0.78	0.7823	0.79	0.7852	0.80	0.7881
0.81	0.7910	0.82	0.7939	0.83	0.7967	0.84	0.7995
0.85	0.8023	0.86	0.8051	0.87	0.8078	0.88	0.8106
0.89	0.8133	0.90	0.8159	0.91	0.8186	0.92	0.8212
0.93	0.8238	0.94	0.8264	0.95	0.8289	0.96	0.8315
0.97	0.8340	0.98	0.8365	0.99	0.8389	1.00	0.8413
1.01	0.8438	1.02	0.8461	1.03	0.8485	1.04	0.8508
1.05	0.8531	1.06	0.8554	1.07	0.8577	1.08	0.8599
1.09	0.8621	1.10	0.8643	1.11	0.8665	1.12	0.8686
1.13	0.8708	1.14	0.8729	1.15	0.8749	1.16	0.8770
1.17	0.8790	1.18	0.8810	1.19	0.8830	1.20	0.8849
1.21	0.8869	1.22	0.8888	1.23	0.8907	1.24	0.8925
1.25	0.8944	1.26	0.8962	1.27	0.8980	1.28	0.8997
1.29	0.9015	1.30	0.9032	1.31	0.9049	1.32	0.9066
1.33	0.9082	1.34	0.9099	1.35	0.9115	1.36	0.9131
1.37	0.9147	1.38	0.9162	1.39	0.9177	1.40	0.9192
1.41	0.9207	1.42	0.9222	1.43	0.9236	1.44	0.9251
1.45	0.9265	1.46	0.9279	1.47	0.9292	1.48	0.9306
1.49	0.9319	1.50	0.9332	1.51	0.9345	1.52	0.9357
1.53	0.9370	1.54	0.9382	1.55	0.9394	1.56	0.9406
1.57	0.9418	1.58	0.9429	1.59	0.9441	1.60	0.9452
1.61	0.9463	1.62	0.9474	1.63	0.9484	1.64	0.9495
1.65	0.9505	1.66	0.9515	1.67	0.9525	1.68	0.9535
1.69	0.9545	1.70	0.9554	1.71	0.9564	1.72	0.9573
1.73	0.9582	1.74	0.9591	1.75	0.9599	1.76	0.9608
1.77	0.9616	1.78	0.9625	1.79	0.9633	1.80	0.9641
1.81	0.9649	1.82	0.9656	1.83	0.9664	1.84	0.9671
1.85	0.9678	1.86	0.9686	1.87	0.9693	1.88	0.9699
1.89	0.9706	1.90	0.9713	1.91	0.9719	1.92	0.9726

continued

TABLE 1   *continued*

z	$P(Z \leq z)$	z	$P(Z \leq z)$	z	$P(Z \leq z)$	z	$P(Z \leq z)$
1.93	0.9732	1.94	0.9738	1.95	0.9744	1.96	0.9750
1.97	0.9756	1.98	0.9761	1.99	0.9767	2.00	0.9772
2.01	0.9778	2.02	0.9783	2.03	0.9788	2.04	0.9793
2.05	0.9798	2.06	0.9803	2.07	0.9808	2.08	0.9812
2.09	0.9817	2.10	0.9821	2.11	0.9826	2.12	0.9830
2.13	0.9834	2.14	0.9838	2.15	0.9842	2.16	0.9846
2.17	0.9850	2.18	0.9854	2.19	0.9857	2.20	0.9861
2.21	0.9864	2.22	0.9868	2.23	0.9871	2.24	0.9875
2.25	0.9878	2.26	0.9881	2.27	0.9884	2.28	0.9887
2.29	0.9890	2.30	0.9893	2.31	0.9896	2.32	0.9898
2.33	0.9901	2.34	0.9904	2.35	0.9906	2.36	0.9909
2.37	0.9911	2.38	0.9913	2.39	0.9916	2.40	0.9918
2.41	0.9920	2.42	0.9922	2.43	0.9925	2.44	0.9927
2.45	0.9929	2.46	0.9931	2.47	0.9932	2.48	0.9934
2.49	0.9936	2.50	0.9938	2.51	0.9940	2.52	0.9941
2.53	0.9943	2.54	0.9945	2.55	0.9946	2.56	0.9948
2.57	0.9949	2.58	0.9951	2.59	0.9952	2.60	0.9953
2.61	0.9955	2.62	0.9956	2.63	0.9957	2.64	0.9959
2.65	0.9960	2.66	0.9961	2.67	0.9962	2.68	0.9963
2.69	0.9964	2.70	0.9965	2.71	0.9966	2.72	0.9967
2.73	0.9968	2.74	0.9969	2.75	0.9970	2.76	0.9971
2.77	0.9972	2.78	0.9973	2.79	0.9974	2.80	0.9974
2.81	0.9975	2.82	0.9976	2.83	0.9977	2.84	0.9977
2.85	0.9978	2.86	0.9979	2.87	0.9979	2.88	0.9980
2.89	0.9981	2.90	0.9981	2.91	0.9982	2.92	0.9982
2.93	0.9983	2.94	0.9984	2.95	0.9984	2.96	0.9985
2.97	0.9985	2.98	0.9986	2.99	0.9986	3.00	0.9987

Note: This table was computed with IMSL subroutine ANORIN.

TABLE 2   Binomial Probability Function (Values of Entries are $P(X \leq k)$)

$n = 5$

k	.05	.1	.2	.3	.4	p .5	.6	.7	.8	.9	.95
0	0.774	0.590	0.328	0.168	0.078	0.031	0.010	0.002	0.000	0.000	0.000
1	0.977	0.919	0.737	0.528	0.337	0.188	0.087	0.031	0.007	0.000	0.000
2	0.999	0.991	0.942	0.837	0.683	0.500	0.317	0.163	0.058	0.009	0.001
3	1.000	1.000	0.993	0.969	0.913	0.813	0.663	0.472	0.263	0.081	0.023
4	1.000	1.000	1.000	0.998	0.990	0.969	0.922	0.832	0.672	0.410	0.226

continued

TABLE 2 *continued*

$n = 6$

k	.05	.1	.2	.3	.4	p .5	.6	.7	.8	.9	.95
0	0.735	0.531	0.262	0.118	0.047	0.016	0.004	0.001	0.000	0.000	0.000
1	0.967	0.886	0.655	0.420	0.233	0.109	0.041	0.011	0.002	0.000	0.000
2	0.998	0.984	0.901	0.744	0.544	0.344	0.179	0.070	0.017	0.001	0.000
3	1.000	0.999	0.983	0.930	0.821	0.656	0.456	0.256	0.099	0.016	0.002
4	1.000	1.000	0.998	0.989	0.959	0.891	0.767	0.580	0.345	0.114	0.033
5	1.000	1.000	1.000	0.999	0.996	0.984	0.953	0.882	0.738	0.469	0.265

$n = 7$

k	.05	.1	.2	.3	.4	p .5	.6	.7	.8	.9	.95
0	0.698	0.478	0.210	0.082	0.028	0.008	0.002	0.000	0.000	0.000	0.000
1	0.956	0.850	0.577	0.329	0.159	0.062	0.019	0.004	0.000	0.000	0.000
2	0.996	0.974	0.852	0.647	0.420	0.227	0.096	0.029	0.005	0.000	0.000
3	1.000	0.997	0.967	0.874	0.710	0.500	0.290	0.126	0.033	0.003	0.000
4	1.000	1.000	0.995	0.971	0.904	0.773	0.580	0.353	0.148	0.026	0.004
5	1.000	1.000	1.000	0.996	0.981	0.938	0.841	0.671	0.423	0.150	0.044
6	1.000	1.000	1.000	1.000	0.998	0.992	0.972	0.918	0.790	0.522	0.302

$n = 8$

k	.05	.1	.2	.3	.4	p .5	.6	.7	.8	.9	.95
0	0.663	0.430	0.168	0.058	0.017	0.004	0.001	0.000	0.000	0.000	0.000
1	0.943	0.813	0.503	0.255	0.106	0.035	0.009	0.001	0.000	0.000	0.000
2	0.994	0.962	0.797	0.552	0.315	0.145	0.050	0.011	0.001	0.000	0.000
3	1.000	0.995	0.944	0.806	0.594	0.363	0.174	0.058	0.010	0.000	0.000
4	1.000	1.000	0.990	0.942	0.826	0.637	0.406	0.194	0.056	0.005	0.000
5	1.000	1.000	0.999	0.989	0.950	0.855	0.685	0.448	0.203	0.038	0.006
6	1.000	1.000	1.000	0.999	0.991	0.965	0.894	0.745	0.497	0.187	0.057
7	1.000	1.000	1.000	1.000	0.999	0.996	0.983	0.942	0.832	0.570	0.337

$n = 9$

k	.05	.1	.2	.3	.4	p .5	.6	.7	.8	.9	.95
0	0.630	0.387	0.134	0.040	0.010	0.002	0.000	0.000	0.000	0.000	0.000
1	0.929	0.775	0.436	0.196	0.071	0.020	0.004	0.000	0.000	0.000	0.000
2	0.992	0.947	0.738	0.463	0.232	0.090	0.025	0.004	0.000	0.000	0.000
3	0.999	0.992	0.914	0.730	0.483	0.254	0.099	0.025	0.003	0.000	0.000
4	1.000	0.999	0.980	0.901	0.733	0.500	0.267	0.099	0.020	0.001	0.000
5	1.000	1.000	0.997	0.975	0.901	0.746	0.517	0.270	0.086	0.008	0.001
6	1.000	1.000	1.000	0.996	0.975	0.910	0.768	0.537	0.262	0.053	0.008
7	1.000	1.000	1.000	1.000	0.996	0.980	0.929	0.804	0.564	0.225	0.071
8	1.000	1.000	1.000	1.000	1.000	0.998	0.990	0.960	0.866	0.613	0.370

continued

TABLE 2 *continued*

$n = 10$

k	.05	.1	.2	.3	.4	*p* .5	.6	.7	.8	.9	.95
0	0.599	0.349	0.107	0.028	0.006	0.001	0.000	0.000	0.000	0.000	0.000
1	0.914	0.736	0.376	0.149	0.046	0.011	0.002	0.000	0.000	0.000	0.000
2	0.988	0.930	0.678	0.383	0.167	0.055	0.012	0.002	0.000	0.000	0.000
3	0.999	0.987	0.879	0.650	0.382	0.172	0.055	0.011	0.001	0.000	0.000
4	1.000	0.998	0.967	0.850	0.633	0.377	0.166	0.047	0.006	0.000	0.000
5	1.000	1.000	0.994	0.953	0.834	0.623	0.367	0.150	0.033	0.002	0.000
6	1.000	1.000	0.999	0.989	0.945	0.828	0.618	0.350	0.121	0.013	0.001
7	1.000	1.000	1.000	0.998	0.988	0.945	0.833	0.617	0.322	0.070	0.012
8	1.000	1.000	1.000	1.000	0.998	0.989	0.954	0.851	0.624	0.264	0.086
9	1.000	1.000	1.000	1.000	1.000	0.999	0.994	0.972	0.893	0.651	0.401

$n = 15$

k	.05	.1	.2	.3	.4	*p* .5	.6	.7	.8	.9	.95
0	0.463	0.206	0.035	0.005	0.000	0.000	0.000	0.000	0.000	0.000	0.000
1	0.829	0.549	0.167	0.035	0.005	0.000	0.000	0.000	0.000	0.000	0.000
2	0.964	0.816	0.398	0.127	0.027	0.004	0.000	0.000	0.000	0.000	0.000
3	0.995	0.944	0.648	0.297	0.091	0.018	0.002	0.000	0.000	0.000	0.000
4	0.999	0.987	0.836	0.515	0.217	0.059	0.009	0.001	0.000	0.000	0.000
5	1.000	0.998	0.939	0.722	0.403	0.151	0.034	0.004	0.000	0.000	0.000
6	1.000	1.000	0.982	0.869	0.610	0.304	0.095	0.015	0.001	0.000	0.000
7	1.000	1.000	0.996	0.950	0.787	0.500	0.213	0.050	0.004	0.000	0.000
8	1.000	1.000	0.999	0.985	0.905	0.696	0.390	0.131	0.018	0.000	0.000
9	1.000	1.000	1.000	0.996	0.966	0.849	0.597	0.278	0.061	0.002	0.000
10	1.000	1.000	1.000	0.999	0.991	0.941	0.783	0.485	0.164	0.013	0.001
11	1.000	1.000	1.000	1.000	0.998	0.982	0.909	0.703	0.352	0.056	0.005
12	1.000	1.000	1.000	1.000	1.000	0.996	0.973	0.873	0.602	0.184	0.036
13	1.000	1.000	1.000	1.000	1.000	1.000	0.995	0.965	0.833	0.451	0.171
14	1.000	1.000	1.000	1.000	1.000	1.000	1.000	0.995	0.965	0.794	0.537

$n = 20$

k	.05	.1	.2	.3	.4	*p* .5	.6	.7	.8	.9	.95
0	0.358	0.122	0.012	0.001	0.000	0.000	0.000	0.000	0.000	0.000	0.000
1	0.736	0.392	0.069	0.008	0.001	0.000	0.000	0.000	0.000	0.000	0.000
2	0.925	0.677	0.206	0.035	0.004	0.000	0.000	0.000	0.000	0.000	0.000
3	0.984	0.867	0.411	0.107	0.016	0.001	0.000	0.000	0.000	0.000	0.000
4	0.997	0.957	0.630	0.238	0.051	0.006	0.000	0.000	0.000	0.000	0.000
5	1.000	0.989	0.804	0.416	0.126	0.021	0.002	0.000	0.000	0.000	0.000
6	1.000	0.998	0.913	0.608	0.250	0.058	0.006	0.000	0.000	0.000	0.000
7	1.000	1.000	0.968	0.772	0.416	0.132	0.021	0.001	0.000	0.000	0.000
8	1.000	1.000	0.990	0.887	0.596	0.252	0.057	0.005	0.000	0.000	0.000

continued

TABLE 2 *continued*

$n = 20$

						p					
k	.05	.1	.2	.3	.4	.5	.6	.7	.8	.9	.95
9	1.000	1.000	0.997	0.952	0.755	0.412	0.128	0.017	0.001	0.000	0.000
10	1.000	1.000	0.999	0.983	0.872	0.588	0.245	0.048	0.003	0.000	0.000
11	1.000	1.000	1.000	0.995	0.943	0.748	0.404	0.113	0.010	0.000	0.000
12	1.000	1.000	1.000	0.999	0.979	0.868	0.584	0.228	0.032	0.000	0.000
13	1.000	1.000	1.000	1.000	0.994	0.942	0.750	0.392	0.087	0.002	0.000
14	1.000	1.000	1.000	1.000	0.998	0.979	0.874	0.584	0.196	0.011	0.000
15	1.000	1.000	1.000	1.000	1.000	0.994	0.949	0.762	0.370	0.043	0.003
16	1.000	1.000	1.000	1.000	1.000	0.999	0.984	0.893	0.589	0.133	0.016
17	1.000	1.000	1.000	1.000	1.000	1.000	0.996	0.965	0.794	0.323	0.075
18	1.000	1.000	1.000	1.000	1.000	1.000	0.999	0.992	0.931	0.608	0.264
19	1.000	1.000	1.000	1.000	1.000	1.000	1.000	0.999	0.988	0.878	0.642

$n = 25$

						p					
k	.05	.1	.2	.3	.4	.5	.6	.7	.8	.9	.95
0	0.277	0.072	0.004	0.000	0.000	0.000	0.000	0.000	0.000	0.000	0.000
1	0.642	0.271	0.027	0.002	0.000	0.000	0.000	0.000	0.000	0.000	0.000
2	0.873	0.537	0.098	0.009	0.000	0.000	0.000	0.000	0.000	0.000	0.000
3	0.966	0.764	0.234	0.033	0.002	0.000	0.000	0.000	0.000	0.000	0.000
4	0.993	0.902	0.421	0.090	0.009	0.000	0.000	0.000	0.000	0.000	0.000
5	0.999	0.967	0.617	0.193	0.029	0.002	0.000	0.000	0.000	0.000	0.000
6	1.000	0.991	0.780	0.341	0.074	0.007	0.000	0.000	0.000	0.000	0.000
7	1.000	0.998	0.891	0.512	0.154	0.022	0.001	0.000	0.000	0.000	0.000
8	1.000	1.000	0.953	0.677	0.274	0.054	0.004	0.000	0.000	0.000	0.000
9	1.000	1.000	0.983	0.811	0.425	0.115	0.013	0.000	0.000	0.000	0.000
10	1.000	1.000	0.994	0.902	0.586	0.212	0.034	0.002	0.000	0.000	0.000
11	1.000	1.000	0.998	0.956	0.732	0.345	0.078	0.006	0.000	0.000	0.000
12	1.000	1.000	1.000	0.983	0.846	0.500	0.154	0.017	0.000	0.000	0.000
13	1.000	1.000	1.000	0.994	0.922	0.655	0.268	0.044	0.002	0.000	0.000
14	1.000	1.000	1.000	0.998	0.966	0.788	0.414	0.098	0.006	0.000	0.000
15	1.000	1.000	1.000	1.000	0.987	0.885	0.575	0.189	0.017	0.000	0.000
16	1.000	1.000	1.000	1.000	0.996	0.946	0.726	0.323	0.047	0.000	0.000
17	1.000	1.000	1.000	1.000	0.999	0.978	0.846	0.488	0.109	0.002	0.000
18	1.000	1.000	1.000	1.000	1.000	0.993	0.926	0.659	0.220	0.009	0.000
19	1.000	1.000	1.000	1.000	1.000	0.998	0.971	0.807	0.383	0.033	0.001
20	1.000	1.000	1.000	1.000	1.000	1.000	0.991	0.910	0.579	0.098	0.007
21	1.000	1.000	1.000	1.000	1.000	1.000	0.998	0.967	0.766	0.236	0.034
22	1.000	1.000	1.000	1.000	1.000	1.000	1.000	0.991	0.902	0.463	0.127
23	1.000	1.000	1.000	1.000	1.000	1.000	1.000	0.998	0.973	0.729	0.358
24	1.000	1.000	1.000	1.000	1.000	1.000	1.000	1.000	0.996	0.928	0.723

TABLE 3 Percentage Points of the Chi-Squared Distribution

ν	$\chi^2_{.005}$	$\chi^2_{.01}$	$\chi^2_{.025}$	$\chi^2_{.05}$	$\chi^2_{.10}$
1	0.0000393	0.0001571	0.0009821	0.0039321	0.0157908
2	0.0100251	0.0201007	0.0506357	0.1025866	0.2107213
3	0.0717217	0.1148317	0.2157952	0.3518462	0.5843744
4	0.2069889	0.2971095	0.4844186	0.7107224	1.0636234
5	0.4117419	0.5542979	0.8312111	1.1454763	1.6103077
6	0.6757274	0.8720903	1.2373447	1.6353836	2.2041321
7	0.9892554	1.2390423	1.6898699	2.1673594	2.8331099
8	1.3444128	1.6464968	2.1797333	2.7326374	3.4895401
9	1.7349329	2.0879011	2.7003908	3.3251143	4.1681604
10	2.1558590	2.5582132	3.2469759	3.9403019	4.8651857
11	2.6032248	3.0534868	3.8157606	4.5748196	5.5777788
12	3.0738316	3.5705872	4.4037895	5.2260313	6.3037949
13	3.5650368	4.1069279	5.0087538	5.8918715	7.0415068
14	4.0746784	4.6604300	5.6287327	6.5706167	7.7895403
15	4.6009169	5.2293501	6.2621403	7.2609539	8.5467529
16	5.1422071	5.8122101	6.9076681	7.9616566	9.3122330
17	5.6972256	6.4077673	7.5641880	8.6717682	10.0851974
18	6.2648115	7.0149183	8.2307510	9.3904572	10.8649368
19	6.8439512	7.6327391	8.9065247	10.1170273	11.6509628
20	7.4338474	8.2603989	9.5907822	10.8508148	12.4426041
21	8.0336685	8.8972015	10.2829285	11.5913391	13.2396393
22	8.6427155	9.5425110	10.9823456	12.3380432	14.0414886
23	9.2604370	10.1957169	11.6885223	13.0905151	14.8479385
24	9.8862610	10.8563690	12.4011765	13.8484344	15.6587067
25	10.5196533	11.5239716	13.1197433	14.6114349	16.4734497
26	11.1602631	12.1981506	13.8439331	15.3792038	17.2919159
27	11.8076019	12.8785095	14.5734024	16.1513977	18.1138763
28	12.4613495	13.5647125	15.3078613	16.9278717	18.9392395
29	13.1211624	14.2564697	16.0470886	17.7083893	19.7678223
30	13.7867584	14.9534760	16.7907562	18.4926147	20.5992126
40	20.7065582	22.1642761	24.4330750	26.5083008	29.0503540
50	27.9775238	29.7001038	32.3561096	34.7638702	37.6881561
60	35.5294037	37.4848328	40.4817810	43.1865082	46.4583282
70	43.2462311	45.4230499	48.7503967	51.7388763	55.3331146
80	51.1447754	53.5226593	57.1465912	60.3912201	64.2818604
90	59.1706543	61.7376862	65.6405029	69.1258850	73.2949219
100	67.3031921	70.0493622	74.2162018	77.9293976	82.3618469

continued

TABLE 3   *continued*

ν	$\chi^2_{.900}$	$\chi^2_{.95}$	$\chi^2_{.975}$	$\chi^2_{.99}$	$\chi^2_{.995}$
1	2.7056	3.8415	5.0240	6.6353	7.8818
2	4.6052	5.9916	7.3779	9.2117	10.5987
3	6.2514	7.8148	9.3486	11.3465	12.8409
4	7.7795	9.4879	11.1435	13.2786	14.8643
5	9.2365	11.0707	12.8328	15.0870	16.7534
6	10.6448	12.5919	14.4499	16.8127	18.5490
7	12.0171	14.0676	16.0136	18.4765	20.2803
8	13.3617	15.5075	17.5355	20.0924	21.9579
9	14.6838	16.9191	19.0232	21.6686	23.5938
10	15.9874	18.3075	20.4837	23.2101	25.1898
11	17.2750	19.6754	21.9211	24.7265	26.7568
12	18.5494	21.0263	23.3370	26.2170	28.2995
13	19.8122	22.3627	24.7371	27.6882	29.8194
14	21.0646	23.6862	26.1189	29.1412	31.3193
15	22.3077	24.9970	27.4883	30.5779	32.8013
16	23.5421	26.2961	28.8453	31.9999	34.2672
17	24.7696	27.5871	30.1909	33.4087	35.7184
18	25.9903	28.8692	31.5264	34.8054	37.1564
19	27.2035	30.1434	32.8523	36.1909	38.5823
20	28.4120	31.4104	34.1696	37.5662	39.9968
21	29.6150	32.6705	35.4787	38.9323	41.4012
22	30.8133	33.9244	36.7806	40.2893	42.7958
23	32.0069	35.1725	38.0757	41.6384	44.1812
24	33.1962	36.4151	39.3639	42.9799	45.5587
25	34.3815	37.6525	40.6463	44.3142	46.9280
26	35.5631	38.8852	41.9229	45.6418	48.2899
27	36.7412	40.1134	43.1943	46.9629	49.6449
28	37.9159	41.3371	44.4608	48.2784	50.9933
29	39.0874	42.5571	45.7223	49.5879	52.3357
30	40.2561	43.7730	46.9792	50.8922	53.6721
40	51.8050	55.7586	59.3417	63.6909	66.7660
50	63.1670	67.5047	71.4201	76.1538	79.4899
60	74.3970	79.0820	83.2977	88.3794	91.9516
70	85.5211	90.5283	95.0263	100.4409	104.2434
80	96.5723	101.8770	106.6315	112.3434	116.3484
90	107.5600	113.1425	118.1392	124.1304	128.3245
100	118.4932	124.3395	129.5638	135.8203	140.1940

Note: This table was computed with IMSL subroutine CHIIN.

TABLE 4   Percentage Points of Student's T-Distribution

ν	$t_{.9}$	$t_{.95}$	$t_{.975}$	$t_{.99}$	$t_{.995}$	$t_{.999}$
1	3.078	6.314	12.706	31.821	63.6567	318.313
2	1.886	2.920	4.303	6.965	9.925	22.327
3	1.638	2.353	3.183	4.541	5.841	10.215
4	1.533	2.132	2.776	3.747	4.604	7.173
5	1.476	2.015	2.571	3.365	4.032	5.893
6	1.440	1.943	2.447	3.143	3.707	5.208
7	1.415	1.895	2.365	2.998	3.499	4.785
8	1.397	1.856	2.306	2.897	3.355	4.501
9	1.383	1.833	2.262	2.821	3.245	4.297
10	1.372	1.812	2.228	2.764	3.169	4.144
12	1.356	1.782	2.179	2.681	3.055	3.930
15	1.341	1.753	2.131	2.603	2.947	3.733
20	1.325	1.725	2.086	2.528	2.845	3.552
24	1.318	1.711	2.064	2.492	2.797	3.467
30	1.310	1.697	2.042	2.457	2.750	3.385
40	1.303	1.684	2.021	2.423	2.704	3.307
60	1.296	1.671	2.000	2.390	2.660	3.232
120	1.289	1.658	1.980	2.358	2.617	3.160
∞	1.282	1.645	1.960	2.326	2.576	3.090

Entries were computed with IMSL subroutine TIN.

TABLE 5   Percentage Points of the F-Distribution, $\alpha = .10$

ν_2	ν_1								
	1	2	3	4	5	6	7	8	9
1	39.86	49.50	53.59	55.83	57.24	58.20	58.91	59.44	59.86
2	8.53	9.00	9.16	9.24	9.29	9.33	9.35	9.37	9.38
3	5.54	5.46	5.39	5.34	5.31	5.28	5.27	5.25	5.24
4	4.54	4.32	4.19	4.11	4.05	4.01	3.98	3.95	3.94
5	4.06	3.78	3.62	3.52	3.45	3.40	3.37	3.34	3.32
6	3.78	3.46	3.29	3.18	3.11	3.05	3.01	2.98	2.96
7	3.59	3.26	3.07	2.96	2.88	2.83	2.79	2.75	2.72
8	3.46	3.11	2.92	2.81	2.73	2.67	2.62	2.59	2.56
9	3.36	3.01	2.81	2.69	2.61	2.55	2.51	2.47	2.44
10	3.29	2.92	2.73	2.61	2.52	2.46	2.41	2.38	2.35

continued

TABLE 5 *continued*

					ν_1				
ν_2	1	2	3	4	5	6	7	8	9
11	3.23	2.86	2.66	2.54	2.45	2.39	2.34	2.30	2.27
12	3.18	2.81	2.61	2.48	2.39	2.33	2.28	2.24	2.21
13	3.14	2.76	2.56	2.43	2.35	2.28	2.23	2.20	2.16
14	3.10	2.73	2.52	2.39	2.31	2.24	2.19	2.15	2.12
15	3.07	2.70	2.49	2.36	2.27	2.21	2.16	2.12	2.09
16	3.05	2.67	2.46	2.33	2.24	2.18	2.13	2.09	2.06
17	3.03	2.64	2.44	2.31	2.22	2.15	2.10	2.06	2.03
18	3.01	2.62	2.42	2.29	2.20	2.13	2.08	2.04	2.00
19	2.99	2.61	2.40	2.27	2.18	2.11	2.06	2.02	1.98
20	2.97	2.59	2.38	2.25	2.16	2.09	2.04	2.00	1.96
21	2.96	2.57	2.36	2.23	2.14	2.08	2.02	1.98	1.95
22	2.95	2.56	2.35	2.22	2.13	2.06	2.01	1.97	1.93
23	2.94	2.55	2.34	2.21	2.11	2.05	1.99	1.95	1.92
24	2.93	2.54	2.33	2.19	2.10	2.04	1.98	1.94	1.91
25	2.92	2.53	2.32	2.18	2.09	2.02	1.97	1.93	1.89
26	2.91	2.52	2.31	2.17	2.08	2.01	1.96	1.92	1.88
27	2.90	2.51	2.30	2.17	2.07	2.00	1.95	1.91	1.87
28	2.89	2.50	2.29	2.16	2.06	2.00	1.94	1.90	1.87
29	2.89	2.50	2.28	2.15	2.06	1.99	1.93	1.89	1.86
30	2.88	2.49	2.28	2.14	2.05	1.98	1.93	1.88	1.85
40	2.84	2.44	2.23	2.09	2.00	1.93	1.87	1.83	1.79
60	2.79	2.39	2.18	2.04	1.95	1.87	1.82	1.77	1.74
120	2.75	2.35	2.13	1.99	1.90	1.82	1.77	1.72	1.68
∞	2.71	2.30	2.08	1.94	1.85	1.77	1.72	.167	1.63

					ν_1				
ν_2	10	12	15	20	24	30	40	60	120	∞
1	60.19	60.70	61.22	61.74	62.00	62.26	62.53	62.79	63.06	63.33
2	9.39	9.41	9.42	9.44	9.45	9.46	9.47	9.47	9.48	9.49
3	5.23	5.22	5.20	5.19	5.18	5.17	5.16	5.15	5.14	5.13
4	3.92	3.90	3.87	3.84	3.83	3.82	3.80	3.79	3.78	3.76
5	3.30	3.27	3.24	3.21	3.19	3.17	3.16	3.14	3.12	3.10
6	2.94	2.90	2.87	2.84	2.82	2.80	2.78	2.76	2.74	2.72
7	2.70	2.67	2.63	2.59	2.58	2.56	2.54	2.51	2.49	2.47
8	2.54	2.50	2.46	2.42	2.40	2.38	2.36	2.34	2.32	2.29
9	2.42	2.38	2.34	2.30	2.28	2.25	2.23	2.21	2.18	2.16
10	2.32	2.28	2.24	2.20	2.18	2.16	2.13	2.11	2.08	2.06

continued

TABLE 5 *continued*

ν_2					ν_1					
	10	12	15	20	24	30	40	60	120	∞
11	2.25	2.21	2.17	2.12	2.10	2.08	2.05	2.03	2.00	1.97
12	2.19	2.15	2.10	2.06	2.04	2.01	1.99	1.96	1.93	1.90
13	2.14	2.10	2.05	2.01	1.98	1.96	1.93	1.90	1.88	1.85
14	2.10	2.05	2.01	1.96	1.94	1.91	1.89	1.86	1.83	1.80
15	2.06	2.02	1.97	1.92	1.90	1.87	1.85	1.82	1.79	1.76
16	2.03	1.99	1.94	1.89	1.87	1.84	1.81	1.78	1.75	1.72
17	2.00	1.96	1.91	1.86	1.84	1.81	1.78	1.75	1.72	1.69
18	1.98	1.93	1.89	1.84	1.81	1.78	1.75	1.72	1.69	1.66
19	1.96	1.91	1.86	1.81	1.79	1.76	1.73	1.70	1.67	1.63
20	1.94	1.89	1.84	1.79	1.77	1.74	1.71	1.68	1.64	1.61
21	1.92	1.87	1.83	1.78	1.75	1.72	1.69	1.66	1.62	1.59
22	1.90	1.86	1.81	1.76	1.73	1.70	1.67	1.64	1.60	1.57
23	1.89	1.84	1.80	1.74	1.72	1.69	1.66	1.62	1.59	1.55
24	1.88	1.83	1.78	1.73	1.70	1.67	1.64	1.61	1.57	1.53
25	1.87	1.82	1.77	1.72	1.69	1.66	1.63	1.59	1.56	1.52
26	1.86	1.81	1.76	1.71	1.68	1.65	1.61	1.58	1.54	1.50
27	1.85	1.80	1.75	1.70	1.67	1.64	1.60	1.57	1.53	1.49
28	1.84	1.79	1.74	1.69	1.66	1.63	1.59	1.56	1.52	1.48
29	1.83	1.78	1.73	1.68	1.65	1.62	1.58	1.55	1.51	1.47
30	1.82	1.77	1.72	1.67	1.64	1.61	1.57	1.54	1.50	1.46
40	1.76	1.71	1.66	1.61	1.57	1.54	1.51	1.47	1.42	1.38
60	1.71	1.66	1.60	1.54	1.51	1.48	1.44	1.40	1.35	1.29
120	1.65	1.60	1.55	1.48	1.45	1.41	1.37	1.32	1.26	1.19
∞	1.60	1.55	1.49	1.42	1.38	1.34	1.30	1.24	1.17	1.00

Note: Entries in this table were computed with IMSL subroutine FIN.

TABLE 6 Percentage Points of the F-Distribution, $\alpha = .05$

ν_2					ν_1				
	1	2	3	4	5	6	7	8	9
1	161.45	199.50	215.71	224.58	230.16	233.99	236.77	238.88	240.54
2	18.51	19.00	19.16	19.25	19.30	19.33	19.35	19.37	19.38
3	10.13	9.55	9.28	9.12	9.01	8.94	8.89	8.85	8.81
4	7.71	6.94	6.59	6.39	6.26	6.16	6.09	6.04	6.00
5	6.61	5.79	5.41	5.19	5.05	4.95	4.88	4.82	4.77
6	5.99	5.14	4.76	4.53	4.39	4.28	4.21	4.15	4.10
7	5.59	4.74	4.35	4.12	3.97	3.87	3.79	3.73	3.68
8	5.32	4.46	4.07	3.84	3.69	3.58	3.50	3.44	3.39
9	5.12	4.26	3.86	3.63	3.48	3.37	3.29	3.23	3.18
10	4.96	4.10	3.71	3.48	3.33	3.22	3.14	3.07	3.02

continued

TABLE 6 *continued*

ν_2	ν_1 1	2	3	4	5	6	7	8	9
11	4.84	3.98	3.59	3.36	3.20	3.09	3.01	2.95	2.90
12	4.75	3.89	3.49	3.26	3.11	3.00	2.91	2.85	2.80
13	4.67	3.81	3.41	3.18	3.03	2.92	2.83	2.77	2.71
14	4.60	3.74	3.34	3.11	2.96	2.85	2.76	2.70	2.65
15	4.54	3.68	3.29	3.06	2.90	2.79	2.71	2.64	2.59
16	4.49	3.63	3.24	3.01	2.85	2.74	2.66	2.59	2.54
17	4.45	3.59	3.20	2.96	2.81	2.70	2.61	2.55	2.49
18	4.41	3.55	3.16	2.93	2.77	2.66	2.58	2.51	2.46
19	4.38	3.52	3.13	2.90	2.74	2.63	2.54	2.48	2.42
20	4.35	3.49	3.10	2.87	2.71	2.60	2.51	2.45	2.39
21	4.32	3.47	3.07	2.84	2.68	2.57	2.49	2.42	2.37
22	4.30	3.44	3.05	2.82	2.66	2.55	2.46	2.40	2.34
23	4.28	3.42	3.03	2.80	2.64	2.53	2.44	2.37	2.32
24	4.26	3.40	3.01	2.78	2.62	2.51	2.42	2.36	2.30
25	4.24	3.39	2.99	2.76	2.60	2.49	2.40	2.34	2.28
26	4.23	3.37	2.98	2.74	2.59	2.47	2.39	2.32	2.27
27	4.21	3.35	2.96	2.73	2.57	2.46	2.37	2.31	2.25
28	4.20	3.34	2.95	2.71	2.56	2.45	2.36	2.29	2.24
29	4.18	3.33	2.93	2.70	2.55	2.43	2.35	2.28	2.22
30	4.17	3.32	2.92	2.69	2.53	2.42	2.33	2.27	2.21
40	4.08	3.23	2.84	2.61	2.45	2.34	2.25	2.18	2.12
60	4.00	3.15	2.76	2.53	2.37	2.25	2.17	2.10	2.04
120	3.92	3.07	2.68	2.45	2.29	2.17	2.09	2.02	1.96
∞	3.84	3.00	2.60	2.37	2.21	2.10	2.01	1.94	1.88

ν_2	ν_1 10	12	15	20	24	30	40	60	120	∞
1	241.88	243.91	245.96	248.00	249.04	250.08	251.14	252.19	253.24	254.3
2	19.40	19.41	19.43	19.45	19.45	19.46	19.47	19.48	19.49	19.50
3	8.79	8.74	8.70	8.66	8.64	8.62	8.59	8.57	8.55	8.53
4	5.97	5.91	5.86	5.80	5.77	5.74	5.72	5.69	5.66	5.63
5	4.73	4.68	4.62	4.56	4.53	4.50	4.46	4.43	4.40	4.36
6	4.06	4.00	3.94	3.87	3.84	3.81	3.77	3.74	3.70	3.67
7	3.64	3.57	3.51	3.44	3.41	3.38	3.34	3.30	3.27	3.23
8	3.35	3.28	3.22	3.15	3.12	3.08	3.04	3.00	2.97	2.93
9	3.14	3.07	3.01	2.94	2.90	2.86	2.83	2.79	2.75	2.71
10	2.98	2.91	2.85	2.77	2.74	2.70	2.66	2.62	2.58	2.54

continued

TABLE 6 *continued*

ν_2	ν_1 10	12	15	20	24	30	40	60	120	∞
11	2.85	2.79	2.72	2.65	2.61	2.57	2.53	2.49	2.45	2.40
12	2.75	2.69	2.62	2.54	2.51	2.47	2.43	2.38	2.34	2.30
13	2.67	2.60	2.53	2.46	2.42	2.38	2.34	2.30	2.25	2.21
14	2.60	2.53	2.46	2.39	2.35	2.31	2.27	2.22	2.18	2.13
15	2.54	2.48	2.40	2.33	2.29	2.25	2.20	2.16	2.11	2.07
16	2.49	2.42	2.35	2.28	2.24	2.19	2.15	2.11	2.06	2.01
17	2.45	2.38	2.31	2.23	2.19	2.15	2.10	2.06	2.01	1.96
18	2.41	2.34	2.27	2.19	2.15	2.11	2.06	2.02	1.97	1.92
19	2.38	2.31	2.23	2.16	2.11	2.07	2.03	1.98	1.93	1.88
20	2.35	2.28	2.20	2.12	2.08	2.04	1.99	1.95	1.90	1.84
21	2.32	2.25	2.18	2.10	2.05	2.01	1.96	1.92	1.87	1.81
22	2.30	2.23	2.15	2.07	2.03	1.98	1.94	1.89	1.84	1.78
23	2.27	2.20	2.13	2.05	2.00	1.96	1.91	1.86	1.81	1.76
24	2.25	2.18	2.11	2.03	1.98	1.94	1.89	1.84	1.79	1.73
25	2.24	2.16	2.09	2.01	1.96	1.92	1.87	1.82	1.77	1.71
26	2.22	2.15	2.07	1.99	1.95	1.90	1.85	1.80	1.75	1.69
27	2.20	2.13	2.06	1.97	1.93	1.88	1.84	1.79	1.73	1.67
28	2.19	2.12	2.04	1.96	1.91	1.87	1.82	1.77	1.71	1.65
29	2.18	2.10	2.03	1.94	1.90	1.85	1.81	1.75	1.70	1.64
30	2.16	2.09	2.01	1.93	1.89	1.84	1.79	1.74	1.68	1.62
40	2.08	2.00	1.92	1.84	1.79	1.74	1.69	1.64	1.58	1.51
60	1.99	1.92	1.84	1.75	1.70	1.65	1.59	1.53	1.47	1.39
120	1.91	1.83	1.75	1.66	1.61	1.55	1.50	1.43	1.35	1.25
∞	1.83	1.75	1.67	1.57	1.52	1.46	1.39	1.32	1.22	1.00

Note: Entries in this table were computed with IMSL subroutine FIN.

TABLE 7 Percentage Points of the F-Distribution, $\alpha = .025$

ν_2	ν_1 1	2	3	4	5	6	7	8	9
1	647.79	799.50	864.16	899.59	921.85	937.11	948.22	956.66	963.28
2	38.51	39.00	39.17	39.25	39.30	39.33	39.36	39.37	39.39
3	17.44	16.04	15.44	15.10	14.88	14.74	14.63	14.54	14.47
4	12.22	10.65	9.98	9.61	9.36	9.20	9.07	8.98	8.90
5	10.01	8.43	7.76	7.39	7.15	6.98	6.85	6.76	6.68

continued

TABLE 7 *continued*

ν_2	1	2	3	4	5	6	7	8	9
6	8.81	7.26	6.60	6.23	5.99	5.82	5.70	5.60	5.52
7	8.07	6.54	5.89	5.52	5.29	5.12	5.00	4.90	4.82
8	7.57	6.06	5.42	5.05	4.82	4.65	4.53	4.43	4.36
9	7.21	5.71	5.08	4.72	4.48	4.32	4.20	4.10	4.03
10	6.94	5.46	4.83	4.47	4.24	4.07	3.95	3.85	3.78
11	6.72	5.26	4.63	4.28	4.04	3.88	3.76	3.66	3.59
12	6.55	5.10	4.47	4.12	3.89	3.73	3.61	3.51	3.44
13	6.41	4.97	4.35	4.00	3.77	3.60	3.48	3.39	3.31
14	6.30	4.86	4.24	3.89	3.66	3.50	3.38	3.29	3.21
15	6.20	4.77	4.15	3.80	3.58	3.41	3.29	3.20	3.12
16	6.12	4.69	4.08	3.73	3.50	3.34	3.22	3.12	3.05
17	6.04	4.62	4.01	3.66	3.44	3.28	3.16	3.06	2.98
18	5.98	4.56	3.95	3.61	3.38	3.22	3.10	3.01	2.93
19	5.92	4.51	3.90	3.56	3.33	3.17	3.05	2.96	2.88
20	5.87	4.46	3.86	3.51	3.29	3.13	3.01	2.91	2.84
21	5.83	4.42	3.82	3.48	3.25	3.09	2.97	2.87	2.80
22	5.79	4.38	3.78	3.44	3.22	3.05	2.93	2.84	2.76
23	5.75	4.35	3.75	3.41	3.18	3.02	2.90	2.81	2.73
24	5.72	4.32	3.72	3.38	3.15	2.99	2.87	2.78	2.70
25	5.69	4.29	3.69	3.35	3.13	2.97	2.85	2.75	2.68
26	5.66	4.27	3.67	3.33	3.10	2.94	2.82	2.73	2.65
27	5.63	4.24	3.65	3.31	3.08	2.92	2.80	2.71	2.63
28	5.61	4.22	3.63	3.29	3.06	2.90	2.78	2.69	2.61
29	5.59	4.20	3.61	3.27	3.04	2.88	2.76	2.67	2.59
30	5.57	4.18	3.59	3.25	3.03	2.87	2.75	2.65	2.57
40	5.42	4.05	3.46	3.13	2.90	2.74	2.62	2.53	2.45
60	5.29	3.93	3.34	3.01	2.79	2.63	2.51	2.41	2.33
120	5.15	3.80	3.23	2.89	2.67	2.52	2.39	2.30	2.22
∞	5.02	3.69	3.12	2.79	2.57	2.41	2.29	2.19	2.11

ν_2	10	12	15	20	24	30	40	60	120	∞
1	968.62	976.71	984.89	993.04	997.20	1001	1006	1010	1014	1018
2	39.40	39.41	39.43	39.45	39.46	39.46	39.47	39.48	39.49	39.50
3	14.42	14.33	14.26	14.17	14.13	14.08	14.04	13.99	13.95	13.90
4	8.85	8.75	8.66	8.56	8.51	8.46	8.41	8.36	8.31	8.26
5	6.62	6.53	6.43	6.33	6.28	6.23	6.17	6.12	6.07	6.02
6	5.46	5.37	5.27	5.17	5.12	5.06	5.01	4.96	4.90	4.85
7	4.76	4.67	4.57	4.47	4.41	4.36	4.31	4.25	4.20	4.14
8	4.30	4.20	4.10	4.00	3.95	3.89	3.84	3.78	3.73	3.67
9	3.96	3.87	3.77	3.67	3.61	3.56	3.51	3.45	3.39	3.33
10	3.72	3.62	3.52	3.42	3.37	3.31	3.26	3.20	3.14	3.08

continued

TABLE 7 *continued*

ν_2					ν_1					
	10	12	15	20	24	30	40	60	120	∞
11	3.53	3.43	3.33	3.23	3.17	3.12	3.06	3.00	2.94	2.88
12	3.37	3.28	3.18	3.07	3.02	2.96	2.91	2.85	2.79	2.72
13	3.25	3.15	3.05	2.95	2.89	2.84	2.78	2.72	2.66	2.60
14	3.15	3.05	2.95	2.84	2.79	2.73	2.67	2.61	2.55	2.49
15	3.06	2.96	2.86	2.76	2.70	2.64	2.59	2.52	2.46	2.40
16	2.99	2.89	2.79	2.68	2.63	2.57	2.51	2.45	2.38	2.32
17	2.92	2.82	2.72	2.62	2.56	2.50	2.44	2.38	2.32	2.25
18	2.87	2.77	2.67	2.56	2.50	2.44	2.38	2.32	2.26	2.19
19	2.82	2.72	2.62	2.51	2.45	2.39	2.33	2.27	2.20	2.13
20	2.77	2.68	2.57	2.46	2.41	2.35	2.29	2.22	2.16	2.09
21	2.73	2.64	2.53	2.42	2.37	2.31	2.25	2.18	2.11	2.04
22	2.70	2.60	2.50	2.39	2.33	2.27	2.21	2.14	2.08	2.00
23	2.67	2.57	2.47	2.36	2.30	2.24	2.18	2.11	2.04	1.97
24	2.64	2.54	2.44	2.33	2.27	2.21	2.15	2.08	2.01	1.94
25	2.61	2.51	2.41	2.30	2.24	2.18	2.12	2.05	1.98	1.91
26	2.59	2.49	2.39	2.28	2.22	2.16	2.09	2.03	1.95	1.88
27	2.57	2.47	2.36	2.25	2.19	2.13	2.07	2.00	1.93	1.85
28	2.55	2.45	2.34	2.23	2.17	2.11	2.05	1.98	1.91	1.83
29	2.53	2.43	2.32	2.21	2.15	2.09	2.03	1.96	1.89	1.81
30	2.51	2.41	2.31	2.20	2.14	2.07	2.01	1.94	1.87	1.79
40	2.39	2.29	2.18	2.07	2.01	1.94	1.88	1.80	1.72	1.64
60	2.27	2.17	2.06	1.94	1.88	1.82	1.74	1.67	1.58	1.48
120	2.16	2.05	1.95	1.82	1.76	1.69	1.61	1.53	1.43	1.31
∞	2.05	1.94	1.83	1.71	1.64	1.57	1.48	1.39	1.27	1.00

Note: Entries in this table were computed with IMSL subroutine FIN.

TABLE 8   Percentage Points of the F-Distribution, $\alpha = .01$

ν_2					ν_1				
	1	2	3	4	5	6	7	8	9
1	4052	4999	5403	5625	5764	5859	5928	5982	6022
2	98.50	99.00	99.17	99.25	99.30	99.33	99.36	99.37	99.39
3	34.12	30.82	29.46	28.71	28.24	27.91	27.67	27.50	27.34
4	21.20	18.00	16.69	15.98	15.52	15.21	14.98	14.80	14.66
5	16.26	13.27	12.06	11.39	10.97	10.67	10.46	10.29	10.16
6	13.75	10.92	9.78	9.15	8.75	8.47	8.26	8.10	7.98
7	12.25	9.55	8.45	7.85	7.46	7.19	6.99	6.84	6.72
8	11.26	8.65	7.59	7.01	6.63	6.37	6.18	6.03	5.91
9	10.56	8.02	6.99	6.42	6.06	5.80	5.61	5.47	5.35
10	10.04	7.56	6.55	5.99	5.64	5.39	5.20	5.06	4.94

continued

TABLE 8 *continued*

v_2	1	2	3	4	5	6	7	8	9
					v_1				
11	9.65	7.21	6.22	5.67	5.32	5.07	4.89	4.74	4.63
12	9.33	6.93	5.95	5.41	5.06	4.82	4.64	4.50	4.39
13	9.07	6.70	5.74	5.21	4.86	4.62	4.44	4.30	4.19
14	8.86	6.51	5.56	5.04	4.69	4.46	4.28	4.14	4.03
15	8.68	6.36	5.42	4.89	4.56	4.32	4.14	4.00	3.89
16	8.53	6.23	5.29	4.77	4.44	4.20	4.03	3.89	3.78
17	8.40	6.11	5.18	4.67	4.34	4.10	3.93	3.79	3.68
18	8.29	6.01	5.09	4.58	4.25	4.01	3.84	3.71	3.60
19	8.18	5.93	5.01	4.50	4.17	3.94	3.77	3.63	3.52
20	8.10	5.85	4.94	4.43	4.10	3.87	3.70	3.56	3.46
21	8.02	5.78	4.87	4.37	4.04	3.81	3.64	3.51	3.40
22	7.95	5.72	4.82	4.31	3.99	3.76	3.59	3.45	3.35
23	7.88	5.66	4.76	4.26	3.94	3.71	3.54	3.41	3.30
24	7.82	5.61	4.72	4.22	3.90	3.67	3.50	3.36	3.26
25	7.77	5.57	4.68	4.18	3.85	3.63	3.46	3.32	3.22
26	7.72	5.53	4.64	4.14	3.82	3.59	3.42	3.29	3.18
27	7.68	5.49	4.60	4.11	3.78	3.56	3.39	3.26	3.15
28	7.64	5.45	4.57	4.07	3.75	3.53	3.36	3.23	3.12
29	7.60	5.42	4.54	4.04	3.73	3.50	3.33	3.20	3.09
30	7.56	5.39	4.51	4.02	3.70	3.47	3.30	3.17	3.07
40	7.31	5.18	4.31	3.83	3.51	3.29	3.12	2.99	2.89
60	7.08	4.98	4.13	3.65	3.34	3.12	2.95	2.82	2.72
120	6.85	4.79	3.95	3.48	3.17	2.96	2.79	2.66	2.56
∞	6.63	4.61	3.78	3.32	3.02	2.80	2.64	2.51	2.41

v_2	10	12	15	20	24	30	40	60	120	∞
					v_1					
1	6056	6106	6157	6209	6235	6261	6287	6313	6339	6366
2	99.40	99.42	99.43	99.45	99.46	99.46	99.47	99.48	99.49	99.50
3	27.22	27.03	26.85	26.67	26.60	26.50	26.41	26.32	26.22	26.13
4	14.55	14.37	14.19	14.02	13.94	13.84	13.75	13.65	13.56	13.46
5	10.05	9.89	9.72	9.55	9.46	9.38	9.30	9.20	9.11	9.02
6	7.87	7.72	7.56	7.40	7.31	7.23	7.15	7.06	6.97	6.88
7	6.62	6.47	6.31	6.16	6.07	5.99	5.91	5.82	5.74	5.65
8	5.81	5.67	5.52	5.36	5.28	5.20	5.12	5.03	4.95	4.86
9	5.26	5.11	4.96	4.81	4.73	4.65	4.57	4.48	4.40	4.31
10	4.85	4.71	4.56	4.41	4.33	4.25	4.17	4.08	4.00	3.91

continued

TABLE 8 *continued*

ν_2	ν_1 10	12	15	20	24	30	40	60	120	∞
11	4.54	4.40	4.25	4.10	4.02	3.94	3.86	3.78	3.69	3.60
12	4.30	4.16	4.01	3.86	3.78	3.70	3.62	3.54	3.45	3.36
13	4.10	3.96	3.82	3.66	3.59	3.51	3.43	3.34	3.25	3.17
14	3.94	3.80	3.66	3.51	3.43	3.35	3.27	3.18	3.09	3.00
15	3.80	3.67	3.52	3.37	3.29	3.21	3.13	3.05	2.96	2.87
16	3.69	3.55	3.41	3.26	3.18	3.10	3.02	2.93	2.84	2.75
17	3.59	3.46	3.31	3.16	3.08	3.00	2.92	2.83	2.75	2.65
18	3.51	3.37	3.23	3.08	3.00	2.92	2.84	2.75	2.66	2.57
19	3.43	3.30	3.15	3.00	2.92	2.84	2.76	2.67	2.58	2.49
20	3.37	3.23	3.09	2.94	2.86	2.78	2.69	2.61	2.52	2.42
21	3.31	3.17	3.03	2.88	2.80	2.72	2.64	2.55	2.46	2.36
22	3.26	3.12	2.98	2.83	2.75	2.67	2.58	2.50	2.40	2.31
23	3.21	3.07	2.93	2.78	2.70	2.62	2.54	2.45	2.35	2.26
24	3.17	3.03	2.89	2.74	2.66	2.58	2.49	2.40	2.31	2.21
25	3.13	2.99	2.85	2.70	2.62	2.54	2.45	2.36	2.27	2.17
26	3.09	2.96	2.81	2.66	2.58	2.50	2.42	2.33	2.23	2.13
27	3.06	2.93	2.78	2.63	2.55	2.47	2.38	2.29	2.20	2.10
28	3.03	2.90	2.75	2.60	2.52	2.44	2.35	2.26	2.17	2.06
29	3.00	2.87	2.73	2.57	2.49	2.41	2.33	2.23	2.14	2.03
30	2.98	2.84	2.70	2.55	2.47	2.39	2.30	2.21	2.11	2.01
40	2.80	2.66	2.52	2.37	2.29	2.20	2.11	2.02	1.92	1.80
60	2.63	2.50	2.35	2.20	2.12	2.03	1.94	1.84	1.73	1.60
120	2.47	2.34	2.19	2.03	1.95	1.86	1.76	1.66	1.53	1.38
∞	2.32	2.18	2.04	1.88	1.79	1.70	1.59	1.47	1.32	1.00

Note: Entries in this table were computed with IMSL subroutine FIN.

TABLE 9 Studentized Range Statistic, q, for $\alpha = .05$

ν	J (number of groups) 2	3	4	5	6	7	8	9	10	11
3	4.50	5.91	6.82	7.50	8.04	8.48	8.85	9.18	9.46	9.72
4	3.93	5.04	5.76	6.29	6.71	7.05	7.35	7.60	7.83	8.03
5	3.64	4.60	5.22	5.68	6.04	6.33	6.59	6.81	6.99	7.17
6	3.47	4.34	4.89	5.31	5.63	5.89	6.13	6.32	6.49	6.65
7	3.35	4.17	4.69	5.07	5.36	5.61	5.82	5.99	6.16	6.30
8	3.27	4.05	4.53	4.89	5.17	5.39	5.59	5.77	5.92	6.06
9	3.19	3.95	4.42	4.76	5.03	5.25	5.44	5.59	5.74	5.87
10	3.16	3.88	4.33	4.66	4.92	5.13	5.31	5.47	5.59	5.73

continued

TABLE 9   *continued*

ν	\multicolumn{10}{c}{J (number of groups)}
	2	3	4	5	6	7	8	9	10	11
11	3.12	3.82	4.26	4.58	4.83	5.03	5.21	5.36	5.49	5.61
12	3.09	3.78	4.19	4.51	4.76	4.95	5.12	5.27	5.39	5.52
13	3.06	3.73	4.15	4.45	4.69	4.88	5.05	5.19	5.32	5.43
14	3.03	3.70	4.11	4.41	4.64	4.83	4.99	5.13	5.25	5.36
15	3.01	3.67	4.08	4.37	4.59	4.78	4.94	5.08	5.20	5.31
16	3.00	3.65	4.05	4.33	4.56	4.74	4.90	5.03	5.15	5.26
17	2.98	3.63	4.02	4.30	4.52	4.70	4.86	4.99	5.11	5.21
18	2.97	3.61	4.00	4.28	4.49	4.67	4.83	4.96	5.07	5.17
19	2.96	3.59	3.98	4.25	4.47	4.65	4.79	4.93	5.04	5.14
20	2.95	3.58	3.96	4.23	4.45	4.62	4.77	4.90	5.01	5.11
24	2.92	3.53	3.90	4.17	4.37	4.54	4.68	4.81	4.92	5.01
30	2.89	3.49	3.85	4.10	4.30	4.46	4.60	4.72	4.82	4.92
40	2.86	3.44	3.79	4.04	4.23	4.39	4.52	4.63	4.73	4.82
60	2.83	3.40	3.74	3.98	4.16	4.31	4.44	4.55	4.65	4.73
120	2.80	3.36	3.68	3.92	4.10	4.24	4.36	4.47	4.56	4.64
∞	2.77	3.31	3.63	3.86	4.03	4.17	4.29	4.39	4.47	4.55
\multicolumn{11}{c}{$\alpha = .01$}
2	14.0	19.0	22.3	24.7	26.6	28.2	29.5	30.7	31.7	32.6
3	8.26	10.6	12.2	13.3	14.2	15.0	15.6	16.2	16.7	17.8
4	6.51	8.12	9.17	9.96	10.6	11.1	11.5	11.9	12.3	12.6
5	5.71	6.98	7.81	8.43	8.92	9.33	9.67	9.98	10.24	10.48
6	5.25	6.34	7.04	7.56	7.98	8.32	8.62	8.87	9.09	9.30
7	4.95	5.92	6.55	7.01	7.38	7.68	7.94	8.17	8.37	8.55
8	4.75	5.64	6.21	6.63	6.96	7.24	7.48	7.69	7.87	8.03
9	4.59	5.43	5.96	6.35	6.66	6.92	7.14	7.33	7.49	7.65
10	4.49	5.28	5.77	6.14	6.43	6.67	6.88	7.06	7.22	7.36
11	4.39	5.15	5.63	5.98	6.25	6.48	6.68	6.85	6.99	7.13
12	4.32	5.05	5.51	5.84	6.11	6.33	6.51	6.67	6.82	6.94
13	4.26	4.97	5.41	5.73	5.99	6.19	6.38	6.53	6.67	6.79
14	4.21	4.89	5.32	5.63	5.88	6.08	6.26	6.41	6.54	6.66
15	4.17	4.84	5.25	5.56	5.80	5.99	6.16	6.31	6.44	6.55
16	4.13	4.79	5.19	5.49	5.72	5.92	6.08	6.22	6.35	6.46
17	4.10	4.74	5.14	5.43	5.66	5.85	6.01	6.15	6.27	6.38
18	4.07	4.70	5.09	5.38	5.60	5.79	5.94	6.08	6.20	6.31
19	4.05	4.67	5.05	5.33	5.55	5.73	5.89	6.02	6.14	6.25
20	4.02	4.64	5.02	5.29	5.51	5.69	5.84	5.97	6.09	6.19
24	3.96	4.55	4.91	5.17	5.37	5.54	5.69	5.81	5.92	6.02
30	3.89	4.45	4.80	5.05	5.24	5.40	5.54	5.65	5.76	5.85
40	3.82	4.37	4.69	4.93	5.10	5.26	5.39	5.49	5.60	5.69
60	3.76	4.28	4.59	4.82	4.99	5.13	5.25	5.36	5.45	5.53
120	3.70	4.20	4.50	4.71	4.87	5.01	5.12	5.21	5.30	5.37
∞	3.64	4.12	4.40	4.60	4.76	4.88	4.99	5.08	5.16	5.23

Note: The values in this table were computed with the IBM SSP subroutines DQH32 and DQG32.

TABLE 10   Studentized Maximum Modulus Distribution

ν	α	C (Number of Tests Performed)								
		2	3	4	5	6	7	8	9	10
2	.05	5.57	6.34	6.89	7.31	7.65	7.93	8.17	8.83	8.57
	.01	12.73	14.44	15.65	16.59	17.35	17.99	18.53	19.01	19.43
3	.05	3.96	4.43	4.76	5.02	5.23	5.41	5.56	5.69	5.81
	.01	7.13	7.91	8.48	8.92	9.28	9.58	9.84	10.06	10.27
4	.05	3.38	3.74	4.01	4.20	4.37	4.50	4.62	4.72	4.82
	.01	5.46	5.99	6.36	6.66	6.89	7.09	7.27	7.43	7.57
5	.05	3.09	3.39	3.62	3.79	3.93	4.04	4.14	4.23	4.31
	.01	4.70	5.11	5.39	5.63	5.81	5.97	6.11	6.23	6.33
6	.05	2.92	3.19	3.39	3.54	3.66	3.77	3.86	3.94	4.01
	.01	4.27	4.61	4.85	5.05	5.20	5.33	5.45	5.55	5.64
7	.05	2.80	3.06	3.24	3.38	3.49	3.59	3.67	3.74	3.80
	.01	3.99	4.29	4.51	4.68	4.81	4.93	5.03	5.12	5.19
8	.05	2.72	2.96	3.13	3.26	3.36	3.45	3.53	3.60	3.66
	.01	3.81	4.08	4.27	4.42	4.55	4.65	4.74	4.82	4.89
9	.05	2.66	2.89	3.05	3.17	3.27	3.36	3.43	3.49	3.55
	.01	3.67	3.92	4.10	4.24	4.35	4.45	4.53	4.61	4.67
10	.05	2.61	2.83	2.98	3.10	3.19	3.28	3.35	3.41	3.47
	.01	3.57	3.80	3.97	4.09	4.20	4.29	4.37	4.44	4.50
11	.05	2.57	2.78	2.93	3.05	3.14	3.22	3.29	3.35	3.40
	.01	3.48	3.71	3.87	3.99	4.09	4.17	4.25	4.31	4.37
12	.05	2.54	2.75	2.89	3.01	3.09	3.17	3.24	3.29	3.35
	.01	3.42	3.63	3.78	3.89	3.99	4.08	4.15	4.21	4.26
14	.05	2.49	2.69	2.83	2.94	3.02	3.09	3.16	3.21	3.26
	.01	3.32	3.52	3.66	3.77	3.85	3.93	3.99	4.05	4.10
16	.05	2.46	2.65	2.78	2.89	2.97	3.04	3.09	3.15	3.19
	.01	3.25	3.43	3.57	3.67	3.75	3.82	3.88	3.94	3.99
18	.05	2.43	2.62	2.75	2.85	2.93	2.99	3.05	3.11	3.15
	.01	3.19	3.37	3.49	3.59	3.68	3.74	3.80	3.85	3.89
20	.05	2.41	2.59	2.72	2.82	2.89	2.96	3.02	3.07	3.11
	.01	3.15	3.32	3.45	3.54	3.62	3.68	3.74	3.79	3.83
24	.05	2.38	2.56	2.68	2.77	2.85	2.91	2.97	3.02	3.06
	.01	3.09	3.25	3.37	3.46	3.53	3.59	3.64	3.69	3.73
30	.05	2.35	2.52	2.64	2.73	2.80	2.87	2.92	2.96	3.01
	.01	3.03	3.18	3.29	3.38	3.45	3.50	3.55	3.59	3.64
40	.05	2.32	2.49	2.60	2.69	2.76	2.82	2.87	2.91	2.95
	.01	2.97	3.12	3.22	3.30	3.37	3.42	3.47	3.51	3.55
60	.05	2.29	2.45	2.56	2.65	2.72	2.77	2.82	2.86	2.90
	.01	2.91	3.06	3.15	3.23	3.29	3.34	3.38	3.42	3.46
∞	.05	2.24	2.39	2.49	2.57	2.63	2.68	2.73	2.77	2.79
	.01	2.81	2.93	3.02	3.09	3.14	3.19	3.23	3.26	3.29

continued

TABLE 10   *continued*

ν	α	C 11	12	13	14	15	16	17	18	19
2	.05	8.74	8.89	9.03	9.16	9.28	9.39	9.49	9.59	9.68
	.01	19.81	20.15	20.46	20.75	20.99	20.99	20.99	20.99	20.99
3	.05	5.92	6.01	6.10	6.18	6.26	6.33	6.39	6.45	6.51
	.01	10.45	10.61	10.76	10.90	11.03	11.15	11.26	11.37	11.47
4	.05	4.89	4.97	5.04	5.11	5.17	5.22	5.27	5.32	5.37
	.01	7.69	7.80	7.91	8.01	8.09	8.17	8.25	8.32	8.39
5	.05	4.38	4.45	4.51	4.56	4.61	4.66	4.70	4.74	4.78
	.01	6.43	6.52	6.59	6.67	6.74	6.81	6.87	6.93	6.98
6	.05	4.07	4.13	4.18	4.23	4.28	4.32	4.36	4.39	4.43
	.01	5.72	5.79	5.86	5.93	5.99	6.04	6.09	6.14	6.18
7	.05	3.86	3.92	3.96	4.01	4.05	4.09	4.13	4.16	4.19
	.01	5.27	5.33	5.39	5.45	5.50	5.55	5.59	5.64	5.68
8	.05	3.71	3.76	3.81	3.85	3.89	3.93	3.96	3.99	4.02
	.01	4.96	5.02	5.07	5.12	5.17	5.21	5.25	5.29	5.33
9	.05	3.60	3.65	3.69	3.73	3.77	3.80	3.84	3.87	3.89
	.01	4.73	4.79	4.84	4.88	4.92	4.96	5.01	5.04	5.07
10	.05	3.52	3.56	3.60	3.64	3.68	3.71	3.74	3.77	3.79
	.01	4.56	4.61	4.66	4.69	4.74	4.78	4.81	4.84	4.88
11	.05	3.45	3.49	3.53	3.57	3.60	3.63	3.66	3.69	3.72
	.01	4.42	4.47	4.51	4.55	4.59	4.63	4.66	4.69	4.72
12	.05	3.39	3.43	3.47	3.51	3.54	3.57	3.60	3.63	3.65
	.01	4.31	4.36	4.40	4.44	4.48	4.51	4.54	4.57	4.59
14	.05	3.30	3.34	3.38	3.41	3.45	3.48	3.50	3.53	3.55
	.01	4.15	4.19	4.23	4.26	4.29	4.33	4.36	4.39	4.41
16	.05	3.24	3.28	3.31	3.35	3.38	3.40	3.43	3.46	3.48
	.01	4.03	4.07	4.11	4.14	4.17	4.19	4.23	4.25	4.28
18	.05	3.19	3.23	3.26	3.29	3.32	3.35	3.38	3.40	3.42
	.01	3.94	3.98	4.01	4.04	4.07	4.10	4.13	4.15	4.18
20	.05	3.15	3.19	3.22	3.25	3.28	3.31	3.33	3.36	3.38
	.01	3.87	3.91	3.94	3.97	3.99	4.03	4.05	4.07	4.09
24	.05	3.09	3.13	3.16	3.19	3.22	3.25	3.27	3.29	3.31
	.01	3.77	3.80	3.83	3.86	3.89	3.91	3.94	3.96	3.98
30	.05	3.04	3.07	3.11	3.13	3.16	3.18	3.21	3.23	3.25
	.01	3.67	3.70	3.73	3.76	3.78	3.81	3.83	3.85	3.87
40	.05	2.99	3.02	3.05	3.08	3.09	3.12	3.14	3.17	3.18
	.01	3.58	3.61	3.64	3.66	3.68	3.71	3.73	3.75	3.76
60	.05	2.93	2.96	2.99	3.02	3.04	3.06	3.08	3.10	3.12
	.01	3.49	3.51	3.54	3.56	3.59	3.61	3.63	3.64	3.66
∞	.05	2.83	2.86	2.88	2.91	2.93	2.95	2.97	2.98	3.01
	.01	3.32	3.34	3.36	3.38	3.40	3.42	3.44	3.45	3.47

continued

TABLE 10 *continued*

		C								
ν	α	20	21	22	23	24	25	26	27	28
2	.05	9.77	9.85	9.92	10.00	10.07	10.13	10.20	10.26	10.32
	.01	22.11	22.29	22.46	22.63	22.78	22.93	23.08	23.21	23.35
3	.05	6.57	6.62	6.67	6.71	6.76	6.80	6.84	6.88	6.92
	.01	11.56	11.65	11.74	11.82	11.89	11.97	12.07	12.11	12.17
4	.05	5.41	5.45	5.49	5.52	5.56	5.59	5.63	5.66	5.69
	.01	8.45	8.51	8.57	8.63	8.68	8.73	8.78	8.83	8.87
5	.05	4.82	4.85	4.89	4.92	4.95	4.98	5.00	5.03	5.06
	.01	7.03	7.08	7.13	7.17	7.21	7.25	7.29	7.33	7.36
6	.05	4.46	4.49	4.52	4.55	4.58	4.60	4.63	4.65	4.68
	.01	6.23	6.27	6.31	6.34	6.38	6.41	6.45	6.48	6.51
7	.05	4.22	4.25	4.28	4.31	4.33	4.35	4.38	4.39	4.42
	.01	5.72	5.75	5.79	5.82	5.85	5.88	5.91	5.94	5.96
8	.05	4.05	4.08	4.10	4.13	4.15	4.18	4.19	4.22	4.24
	.01	5.36	5.39	5.43	5.45	5.48	5.51	5.54	5.56	5.59
9	.05	3.92	3.95	3.97	3.99	4.02	4.04	4.06	4.08	4.09
	.01	5.10	5.13	5.16	5.19	5.21	5.24	5.26	5.29	5.31
10	.05	3.82	3.85	3.87	3.89	3.91	3.94	3.95	3.97	3.99
	.01	4.91	4.93	4.96	4.99	5.01	5.03	5.06	5.08	5.09
11	.05	3.74	3.77	3.79	3.81	3.83	3.85	3.87	3.89	3.91
	.01	4.75	4.78	4.80	4.83	4.85	4.87	4.89	4.91	4.93
12	.05	3.68	3.70	3.72	3.74	3.76	3.78	3.80	3.82	3.83
	.01	4.62	4.65	4.67	4.69	4.72	4.74	4.76	4.78	4.79
14	.05	3.58	3.59	3.62	3.64	3.66	3.68	3.69	3.71	3.73
	.01	4.44	4.46	4.48	4.50	4.52	4.54	4.56	4.58	4.59
16	.05	3.50	3.52	3.54	3.56	3.58	3.59	3.61	3.63	3.64
	.01	4.29	4.32	4.34	4.36	4.38	4.39	4.42	4.43	4.45
18	.05	3.44	3.46	3.48	3.50	3.52	3.54	3.55	3.57	3.58
	.01	4.19	4.22	4.24	4.26	4.28	4.29	4.31	4.33	4.34
20	.05	3.39	3.42	3.44	3.46	3.47	3.49	3.50	3.52	3.53
	.01	4.12	4.14	4.16	4.17	4.19	4.21	4.22	4.24	4.25
24	.05	3.33	3.35	3.37	3.39	3.40	3.42	3.43	3.45	3.46
	.01	4.00	4.02	4.04	4.05	4.07	4.09	4.10	4.12	4.13
30	.05	3.27	3.29	3.30	3.32	3.33	3.35	3.36	3.37	3.39
	.01	3.89	3.91	3.92	3.94	3.95	3.97	3.98	4.00	4.01
40	.05	3.20	3.22	3.24	3.25	3.27	3.28	3.29	3.31	3.32
	.01	3.78	3.80	3.81	3.83	3.84	3.85	3.87	3.88	3.89
60	.05	3.14	3.16	3.17	3.19	3.20	3.21	3.23	3.24	3.25
	.01	3.68	3.69	3.71	3.72	3.73	3.75	3.76	3.77	3.78
∞	.05	3.02	3.03	3.04	3.06	3.07	3.08	3.09	3.11	3.12
	.01	3.48	3.49	3.50	3.52	3.53	3.54	3.55	3.56	3.57

Note: This table was computed using the FORTRAN program described in Wilcox (1986b).

TABLE 11   Percentage Points, h, of the Range of J Independent t Variates

α	$\nu = 5$	$\nu = 6$	$\nu = 7$	$\nu = 8$	$\nu = 9$	$\nu = 14$	$\nu = 19$	$\nu = 24$	$\nu = 29$	$\nu = 39$	$\nu = 59$
					$J = 2$ groups						
.05	3.63	3.45	3.33	3.24	3.18	3.01	2.94	2.91	2.89	2.85	2.82
.01	5.37	4.96	4.73	4.51	4.38	4.11	3.98	3.86	3.83	3.78	3.73
					$J = 3$ groups						
.05	4.49	4.23	4.07	3.95	3.87	3.65	3.55	3.50	3.46	3.42	3.39
.01	6.32	5.84	5.48	5.23	5.07	4.69	5.54	4.43	4.36	4.29	4.23
					$J = 4$ groups						
.05	5.05	4.74	4.54	4.40	4.30	4.03	3.92	3.85	3.81	3.76	3.72
.01	7.06	6.40	6.01	5.73	5.56	5.05	4.89	4.74	4.71	4.61	4.54
					$J = 5$ groups						
.05	5.47	5.12	4.89	4.73	4.61	4.31	4.18	4.11	4.06	4.01	3.95
.01	7.58	6.76	6.35	6.05	5.87	5.33	5.12	5.01	4.93	4.82	4.74
					$J = 6$ groups						
.05	5.82	5.42	5.17	4.99	4.86	4.52	4.38	4.30	4.25	4.19	4.14
.01	8.00	7.14	6.70	6.39	6.09	5.53	5.32	5.20	5.12	4.99	4.91
					$J = 7$ groups						
.05	6.12	5.68	5.40	5.21	5.07	4.70	4.55	4.46	4.41	4.34	4.28
.01	8.27	7.50	6.92	6.60	6.30	5.72	5.46	5.33	5.25	5.16	5.05
					$J = 8$ groups						
.05	6.37	5.90	5.60	5.40	5.25	4.86	4.69	4.60	4.54	4.47	4.41
.01	8.52	7.73	7.14	6.81	6.49	5.89	5.62	5.45	5.36	5.28	5.16
					$J = 9$ groups						
.05	6.60	6.09	5.78	5.56	5.40	4.99	4.81	4.72	4.66	4.58	4.51
.01	8.92	7.96	7.35	6.95	6.68	6.01	5.74	5.56	5.47	5.37	5.28
					$J = 10$ groups						
.05	6.81	6.28	5.94	5.71	5.54	5.10	4.92	4.82	4.76	4.68	4.61
.01	9.13	8.14	7.51	7.11	6.83	6.10	5.82	5.68	5.59	5.46	5.37

Source: Reprinted, with permission, from R. Wilcox, "A table of percentage points of the range of independent t variables," *Technometrics* 25: 201–204, 1983.

TABLE 12   Lower Critical Values for the One-Sided Wilcoxon Signed Rank Test

n	$\alpha = .005$	$\alpha = .01$	$\alpha = .025$	$\alpha = .05$
4	0	0	0	0
5	0	0	0	1
6	0	0	1	3
7	0	1	3	4
8	1	2	4	6
9	2	4	6	9
10	4	6	9	11
11	6	8	11	14
12	8	10	14	18
13	10	13	18	22

continued

TABLE 12 *continued*

n	$\alpha = .005$	$\alpha = .01$	$\alpha = .025$	$\alpha = .05$
14	13	16	22	26
15	16	20	26	31
16	20	24	30	36
17	24	28	35	42
18	28	33	41	48
19	33	38	47	54
20	38	44	53	61
21	44	50	59	68
22	49	56	67	76
23	55	63	74	84
24	62	70	82	92
25	69	77	90	101
26	76	85	111	125
27	84	94	108	120
28	92	102	117	131
29	101	111	127	141
30	110	121	138	152
31	119	131	148	164
32	129	141	160	176
33	139	152	171	188
34	149	163	183	201
35	160	175	196	214
36	172	187	209	228
37	184	199	222	242
38	196	212	236	257
39	208	225	250	272
40	221	239	265	287

Entries were computed as described in Hogg and Craig, 1970, p. 361.

TABLE 13 Lower Critical Values, c_L, for the One-Sided Wilcoxon–Mann–Whitney Test

n_2	$\alpha = .025$							
	$n_1 = 3$	$n_1 = 4$	$n_1 = 5$	$n_1 = 6$	$n_1 = 7$	$n_1 = 8$	$n_1 = 9$	$n_1 = 10$
3	6	6	7	8	8	9	9	10
4	10	11	12	13	14	15	15	16
5	16	17	18	19	21	22	23	24
6	23	24	25	27	28	30	32	33
7	30	32	34	35	37	39	41	43
8	39	41	43	45	47	50	52	54
9	48	50	53	56	58	61	63	66
10	59	61	64	67	70	73	76	79

continued

TABLE 13   *continued*

n_2				$\alpha = .005$				
	$n_1 = 3$	$n_1 = 4$	$n_1 = 5$	$n_1 = 6$	$n_1 = 7$	$n_1 = 8$	$n_1 = 9$	$n_1 = 10$
3	10	10	10	11	11	12	12	13
4	15	15	16	17	17	18	19	20
5	21	22	23	24	25	26	27	28
6	28	29	30	32	33	35	36	38
7	36	38	39	41	43	44	46	48
8	46	47	49	51	53	55	57	59
9	56	58	60	62	65	67	69	72
10	67	69	72	74	77	80	83	85

Entries were determined with the algorithm in Hogg and Craig, 1970, p. 373.

APPENDIX C

Basic Matrix Algebra

A matrix is a two dimensional array of numbers or variables having r rows and c columns.

EXAMPLE.

$$\begin{pmatrix} 32 & 19 & 67 \\ 11 & 21 & 99 \\ 25 & 56 & 10 \\ 76 & 39 & 43 \end{pmatrix}$$

is a matrix having four rows and three columns. ■

The matrix is said to be square if $r = c$ (the number of rows equals the number of columns). A matrix with $r = 1$ ($c = 1$) is called a row (column) vector.

A common notation for a matrix is $X = (x_{ij})$, meaning that X is a matrix where x_{ij} is the value in the ith row and jth column. For the matrix just shown, the value in the first row and first column is $x_{11} = 32$ and the value in the third row and second column is $x_{32} = 56$.

EXAMPLE. A commonly encountered square matrix is the correlation matrix. That is, for every individual, we have p measures with r_{ij} being Pearson's correlation between the ith and jth measures. Then the correlation matrix is $R = (r_{ij})$. If $p = 3$, $r_{12} = .2$, $r_{13} = .4$, and $r_{23} = .3$, then

$$R = \begin{pmatrix} 1 & .2 & .4 \\ .2 & 1 & .3 \\ .4 & .3 & 1 \end{pmatrix}.$$

(The correlation of a variable with itself is 1.) ■

The transpose of a matrix is just the matrix obtained when the rth row becomes the rth column. More formally, the transpose of the matrix $X = (x_{ij})$ is

$$X' = (x_{ji}),$$

which has c rows and r columns.

EXAMPLE. The transpose of the matrix

$$X = \begin{pmatrix} 23 & 91 \\ 51 & 29 \\ 63 & 76 \\ 11 & 49 \end{pmatrix}$$

is

$$X' = \begin{pmatrix} 23 & 51 & 63 & 11 \\ 91 & 29 & 76 & 49 \end{pmatrix}.$$

■

The matrix X is said to be symmetric if $X = X'$. That is, $x_{ij} = x_{ji}$.

The diagonal of an r-by-r (square) matrix refers to x_{ii}, $i = 1, \ldots, r$. A diagonal matrix is an r-by-r matrix where the off-diagonal elements (the x_{ij}, $i \neq j$) are zero. An important special case is the identity matrix, which has 1's along the diagonal and zeros elsewhere. For example,

$$\begin{pmatrix} 1 & 0 & 0 \\ 0 & 1 & 0 \\ 0 & 0 & 1 \end{pmatrix}$$

is the identity matrix when $r = c = 3$. A common notation for the identity matrix is I.

Two $r \times c$ matrices, X and Y, are said to be equal if for every i and j, $x_{ij} = y_{ij}$. That is, every element in X is equal to the corresponding element in Y.

The sum of two matrices having the same number of rows and columns is

$$z_{ij} = x_{ij} + y_{ij}.$$

EXAMPLE.

$$\begin{pmatrix} 1 & 3 \\ 4 & -1 \\ 9 & 2 \end{pmatrix} + \begin{pmatrix} 8 & 2 \\ 4 & 9 \\ 1 & 6 \end{pmatrix} = \begin{pmatrix} 9 & 5 \\ 8 & 8 \\ 10 & 8 \end{pmatrix}.$$

■

Multiplication of a matrix by a scalar — say, a — is

$$a\mathbf{X} = (ax_{ij}).$$

That is, every element of the matrix $\mathbf{X}$ is multiplied by a.

EXAMPLE.

$$2\begin{pmatrix} 8 & 2 \\ 4 & 9 \\ 1 & 6 \end{pmatrix} = \begin{pmatrix} 16 & 4 \\ 8 & 18 \\ 2 & 12 \end{pmatrix}.$$

■

For an n-by-p matrix (meaning we have p measures for each of n individuals), the sample mean is

$$\bar{\mathbf{X}} = (\bar{X}_1, \ldots, \bar{X}_p),$$

the vector of the sample means corresponding to the p measures. That is,

$$\bar{X}_j = \sum_{i=1}^{n} X_{ij}, \quad j = 1, \ldots, p.$$

If $\mathbf{X}$ is an r-by-c matrix and $\mathbf{Y}$ is a c-by-t matrix (the number of columns for $\mathbf{X}$ is the same as the number of rows for $\mathbf{Y}$), the product of $\mathbf{X}$ and $\mathbf{Y}$ is the r-by-t matrix $\mathbf{Z} = \mathbf{XY}$, where

$$z_{ij} = \sum_{k=1}^{c} x_{ik} y_{kj}.$$

EXAMPLE.

$$\begin{pmatrix} 8 & 2 \\ 4 & 9 \\ 1 & 6 \end{pmatrix} \begin{pmatrix} 5 & 3 \\ 2 & 1 \end{pmatrix} = \begin{pmatrix} 44 & 26 \\ 38 & 21 \\ 17 & 9 \end{pmatrix}$$

■

EXAMPLE. Consider a random sample of n observations, $X_1, \ldots, X_n$, and let $\mathbf{J}$ be a row matrix of 1's. That is, $\mathbf{J} = (1, 1, \ldots, 1)$. Letting $\mathbf{X}$ be a column matrix containing $X_1, \ldots, X_n$, then

$$\sum X_i = \mathbf{JX}.$$

Continued

EXAMPLE. (*Continued*)
The sample mean is

$$\bar{X} = \frac{1}{n}JX.$$

The sum of the squared observations is

$$\sum X_i^2 = X'X.$$

■

Let X be an n-by-p matrix of p measures taken on n individuals. Then X_i is the ith row (vector) in the matrix X and $(X_i - \bar{X})'$ is a p-by-1 matrix consisting of the ith row of X minus the sample mean. Moreover, $(X_i - \bar{X})'(X_i - \bar{X})$ is a p-by-p matrix. The (sample) covariance matrix is

$$S = \frac{1}{n-1} \sum_{i=1}^{n} (X_i - \bar{X})'(X_i - \bar{X}).$$

That is, $S = (s_{jk})$, where s_{jk} is the covariance between the jth and kth measures. When $j = k$, s_{jk} is the sample variance corresponding to the jth variable under study.

For any square matrix X, the matrix X^{-1} is said to be the inverse of X if

$$XX^{-1} = I,$$

the identity matrix. If an inverse exists, X is said to be *nonsingular*; otherwise it is *singular*. The inverse of a nonsingular matrix can be computed with the S-PLUS built-in function

solve(m),

where m is any S-PLUS variable having matrix mode, with the number of rows equal to the number of columns.

EXAMPLE. Consider the matrix

$$\begin{pmatrix} 5 & 3 \\ 2 & 1 \end{pmatrix}.$$

Storing it in the S-PLUS variable m, the command solve(m) returns

$$\begin{pmatrix} -1 & 3 \\ 2 & -5 \end{pmatrix}.$$

It is left as an exercise to verify that multiplying these two matrices together yields I.

■

EXAMPLE. It can be shown that the matrix

$$\begin{pmatrix} 2 & 5 \\ 2 & 5 \end{pmatrix}$$

does not have an inverse. The S-PLUS function solve, applied to this matrix, reports that the matrix appears to be singular.   ■

Consider any r-by-c matrix X, and let k indicate any square submatrix. That is, consider the matrix consisting of any k rows and any k columns taken from X. The *rank* of X is equal to the largest k for which a k-by-k submatrix is nonsingular.

The *trace* of a square matrix is just the sum of the diagonal elements and is often denoted by tr. For example, if

$$A = \begin{pmatrix} 5 & 3 \\ 2 & 1 \end{pmatrix},$$

then

$$\mathrm{tr}(A) = 5 + 1 = 6.$$

The notation

$$\mathrm{diag}\{x_1, \ldots, x_n\}$$

refers to a *diagonal* matrix with the values $x_1, \ldots, x_n$ along the diagonal. For example,

$$\mathrm{diag}\{4,5,2\} = \begin{pmatrix} 4 & 0 & 0 \\ 0 & 5 & 0 \\ 0 & 0 & 2 \end{pmatrix}.$$

A *block diagonal* matrix refers to a matrix where the diagonal elements are themselves matrices.

EXAMPLE. If

$$V_1 = \begin{pmatrix} 9 & 2 \\ 4 & 15 \end{pmatrix} \quad \text{and} \quad V_2 = \begin{pmatrix} 11 & 32 \\ 14 & 29 \end{pmatrix},$$

then

$$\mathrm{diag}(V_1, V_2) = \begin{pmatrix} 9 & 2 & 0 & 0 \\ 4 & 15 & 0 & 0 \\ 0 & 0 & 11 & 32 \\ 0 & 0 & 14 & 29 \end{pmatrix}.$$

■

Let A be an $m_1 \times n_1$ matrix, and let B be an $m_2 \times n_2$ matrix. The (right) Kronecker product of A and B is the $m_1 m_2 \times n_1 n_2$ matrix

$$A \otimes B = \begin{pmatrix} a_{11}B & a_{12}B & \cdots & a_{1n_1}B \\ a_{21}B & a_{22}B & \cdots & a_{2n_1}B \\ \vdots & \vdots & \vdots & \vdots \\ a_{m_1 1}B & a_{m_1 2}B & \cdots & a_{m_1 n_1}B \end{pmatrix}$$

A matrix X^- is said to be a *generalized inverse* of the matrix X if

1. XX^- is symmetric.
2. X^-X is symmetric.
3. $XX^-X = X$.
4. $X^-XX^- = X^-$.

A method for computing a generalized inverse can be found in Graybill (1983).

REFERENCES

Agresti, A. (1990). *Categorical Data Analysis*. New York: Wiley.

Agresti, A. (1996). *An Introduction to Categorical Data Analysis*. New York: Wiley.

Agresti, A., & Pendergast, J. (1986). Comparing mean ranks for repeated measures data. *Communications in Statistics—Theory and Methods, 15,* 1417–1433.

Akritas, M. G. (1990). The rank transform method in some two-factor designs. *Journal of the American Statistical Association, 85,* 73–78.

Akritas, M. G., & Arnold, S. F. (1994). Fully nonparametric hypotheses for factorial designs I: Multivariate repeated measures designs. *Journal of the American Statistical Association, 89,* 336–343.

Akritas, M. G., Arnold, S. F., & Brunner, E. (1997). Nonparametric hypotheses and rank statistics for unbalanced factorial designs. *Journal of the American Statistical Association, 92,* 258–265.

Alexander, R. A., & Govern, D. M. (1994). A new and simpler approximation for ANOVA under variance heterogeneity. *Journal of Educational Statistics, 19,* 91–101.

Algina, J., Oshima, T. C., & Lin, W.-Y. (1994). Type I error rates for Welch's test and James's second-order test under nonnormality and inequality of variance when there are two groups. *Journal of Educational and Behavioral Statistics, 19,* 275–291.

Andersen, E. B. (1997). *Introduction to the Statistical Analysis of Categorical Data*. New York: Springer.

Andrews, D. F., Bickel, P. J., Hampel, F. R., Huber, P. J., Rogers, W. H., & Tukey, J. W. (1972). *Robust Estimates of Location: Survey and Advances*. Princeton University Press, Princeton, NJ.

Asiribo, O., & Gurland, J. (1989). Some simple approximate solutions to the Behrens–Fisher problem. *Communications in Statistics—Theory and Methods, 18,* 1201–1216.

Bahadur, R., & Savage, L. (1956). The nonexistence of certain statistical procedures in nonparametric problems. *Annals of Statistics, 25,* 1115–1122.

Barnett, V., & Lewis, T. (1994). *Outliers in Statistical Data*. New York: Wiley.

Baron, R. M., & Kenny, D. A. (1986). The moderator–mediator variable distinction in social psychological research: Conceptual, strategic, and statistical considerations. *Journal of Personality and Social Psychology 51,* 1173–1182.

Barrett, J. P. (1974). The coefficient of determination—Some limitations. *Annals of Statistics, 28,* 19–20.

Basu, S., & DasGupta, A. (1995). Robustness of standard confidence intervals for location parameters under departure from normality. *Annals of Statistics, 23,* 1433–1442.

Beal, S. L. (1987). Asymptotic confidence intervals for the difference between two binomial parameters for use with small samples. *Biometrics, 43,* 941–950.

Beasley, T. M. (2000). Nonparametric tests for analyzing interactions among intrablock ranks in multiple group repeated measures designs. *Journal of Educational and Behavioral Statistics, 25,* 20–59.

Belsley, D. A., Kuh, E., & Welsch, R. E. (1980). *Regression Diagnostics: Identifying Influential Data and Sources of Collinearity*. New York: Wiley.

Benjamini, Y., & Hochberg, Y. (1995). Controlling the false discovery rate: A practical and powerful approach to multiple testing. *Journal of the Royal Statistical Society, B, 57,* 289–300.

Benjamini, Y., & Hochberg, Y. (2000). On the adaptive control of the false discovery rate in multiple testing with independent statistics. *Journal of Educational and Behavioral Statistics, 25,* 60–83.

Berger, R. L. (1993). More powerful tests from confidence interval p values. *American Statistician, 50,* 314–318.

Berk, K. N., & Booth, D. E. (1995). Seeing a curve in multiple regression. *Technometrics, 37,* 385–398.

Bernhardson, C. (1975). Type I error rates when multiple comparison procedures follow a significant *F* test of ANOVA. *Biometrics, 31,* 719–724.

Bianco, A., Boente, G., & di Rienzo, J. (2000). Some results for robust GM-estimators in heteroscedastic regression models. *Journal of Statistical Planning and Inference, 89,* 215–242.

Bickel, P. J., & Lehmann, E. L. (1975). Descriptive statistics for nonparametric models II. Location. *Annals of Statistics, 3,* 1045–1069.

Bishop, T., & Dudewicz, E. J. (1978). Exact analysis of variance with unequal variances: Test procedures and tables. *Technometrics, 20,* 419–430.

Blair, R. C., Sawilowski, S. S., & Higgens, J. J. (1987). Limitations of the rank transform statistic in tests for interactions. *Communications in Statistics—Simulation and Computation, 16,* 1133–1145.

Blyth, C. R. (1986). Approximate binomial confidence limits. *Journal of the American Statistical Association, 81,* 843–855.

Boik, R. J. (1981). A priori tests in repeated measures designs: Effects of nonsphericity. *Psychometrika, 46,* 241–255.

Boik, R. J. (1987). The Fisher–Pitman permutation test: A nonrobust alternative to the normal theory *F* test when variances are heterogeneous. *British Journal of Mathematical and Statistical Psychology, 40,* 26–42.

Boos, D. B., & Hughes-Oliver, J. M. (2000). How large does *n* have to be for *Z* and *t* intervals? *American Statistician, 54,* 121–128.

Boos, D. B., & Zhang, J. (2000). Monte Carlo evaluation of resampling-based hypothesis tests. *Journal of the American Statistical Association, 95,* 486–492.

Booth, J. G., & Sarkar, S. (1998). Monte Carlo approximation of bootstrap variances. *American Statistician, 52,* 354–357.

Box, G. E. P. (1953). Non-normality and tests on variances. *Biometrika, 40,* 318–335.

Bradley, J. V. (1978) Robustness? *British Journal of Mathematical and Statistical Psychology, 31,* 144–152.

Breiman, L. (1995). Better subset regression using the nonnegative garrote. *Technometrics, 37,* 373–384.

Breiman, L., & Friedman, J. H. (1985). Estimating optimal transformations for multiple regression and correlation (with discussion). *Journal of the American Statistical Association, 80,* 580–619.

Brown, C., & Mosteller, F. (1991). Components of variance. In D. Hoaglin, F. Mosteller, & J. Tukey (Eds.), *Fundamentals of Exploratory Analysis of Variance,* pp. 193–251. New York: Wiley.

Brown, M. B., & Forsythe, A. (1974a). The small-sample behavior of some statistics which test the equality of several means. *Technometrics, 16,* 129–132.

Brown, M. B., & Forsythe, A. (1974b). Robust tests for the equality of variances. *Journal of the American Statistical Association, 69,* 364–367.

Brunner, E., & Denker, M. (1994). Rank statistics under dependent observations and applications to factorial designs. *Journal of Statistical Planning and Inference, 42,* 353–378.

Brunner, E., & Dette, H. (1992). Rank procedures for the two-factor mixed model. *Journal of the American Statistical Association, 87,* 884–888.

Brunner, E., Dette, H., & Munk, A. (1997). Box-type approximations in nonparametric factorial designs. *Journal of the American Statistical Association, 92,* 1494–1502.

Brunner, E., Domhof, S., & Langer, F. (2002). *Nonparametric Analysis of Longitudinal Data in Factorial Experiments.* New York: Wiley.

Brunner, E., & Munzel, U. (2000). The nonparametric Behrens–Fisher problem: Asymptotic theory and small-sample approximation. *Biometrical Journal 42,* 17–25.

Brunner, E., Munzel, U., & Puri, M. L. (1999). Rank-score tests in factorial designs with repeated measures. *Journal of Multivariate Analysis, 70,* 286–317.

Carling, K. (2000). Resistant outlier rules and the non-Gaussian case. *Computational Statistics & Data Analysis, 33,* 249–258.

Carroll, R. J., & Ruppert, D. (1982). Robust estimation in heteroscedastic linear models. *Annals of Statistics, 10,* 429–441.

Carroll, R. J., & Ruppert, D. (1988). *Transformation and Weighting in Regression.* New York: Chapman and Hall.

Chatterjee, S., & Hadi, A. S. (1988). *Sensitivity Analysis in Linear Regression Analysis.* New York: Wiley.

Chaudhuri, P. (1996). On a geometric notion of quantiles for multivariate data. *Journal of the American Statistical Association, 91,* 862–872.

Chen, L. (1995). Testing the mean of skewed distributions. *Journal of the American Statistical Association, 90,* 767–772.

Chen, S., & Chen, H. J. (1998). Single-stage analysis of variance under heteroscedasticity. *Communications in Statistics—Simulation and Computation 27,* 641–666.

Chernick, M. R. (1999). *Bootstrap Methods: A Practitioner's Guide.* New York: Wiley.

Choi, K., & Marden, J. (1997). An approach to multivariate rank tests in multivariate analysis of variance. *Journal of the American Statistical Association, 92,* 1581–1590.

Chow, G. C. (1960). Tests of equality between sets of coefficients in two linear regressions. *Econometrika, 28,* 591–606.

Clark, S. (1999). *Towards the Edge of the Universe: A Review of Modern Cosmology.* New York: Springer.

Clemons, T. E., & Bradley Jr., E. L. (2000). A nonparametric measure of the overlapping coefficient. *Computational Statistics & Data Analysis, 34,* 51–61.

Cleveland, W. S. (1979). Robust locally weighted regression and smoothing scatterplots. *Journal of the American Statistical Association, 74,* 829–836.

Cleveland, W. S. (1985). *The Elements of Graphing Data.* Summit, NJ: Hobart Press.

Cliff, N. (1996). *Ordinal Methods for Behavioral Data Analysis.* Mahwah, NJ: Erlbaum.

Coakley, C. W., & Hettmansperger, T. P. (1993). A bounded influence, high breakdown, efficient regression estimator. *Journal of the American Statistical Association, 88,* 872–880.

Cochran, W. G., & Cox, G. M. (1950). *Experimental Design.* New York: Wiley.

Coe, P. R., & Tamhane, A. C. (1993). Small-sample confidence intervals for the difference, ratio, and odds ratio of two success probabilities. *Communications in Statistics—Simulation and Computation, 22,* 925–938.

Cohen, J. (1977). *Statistical Power Analysis for the Behavioral Sciences.* New York: Academic Press.

Cohen, J. (1994). The earth is round ($p < .05$). *American Psychologist, 49,* 997–1003.

Cohen, M., Dalal, S. R., & Tukey, J. W. (1993). Robust, smoothly heterogeneous variance regression. *Applied Statistics, 42,* 339–353.

Coleman, J. S. (1964). *Introduction to Mathematical Sociology.* New York: Free Press.

Comrey, A. L. (1985). A method for removing outliers to improve factor analytic results. *Multivariate Behavioral Research, 20,* 273–281.

Conerly, M. D., & Mansfield, E. R. (1988). An approximate test for comparing heteroscedastic regression models. *Journal of the American Statistical Association, 83,* 811–817.

Conover, W., Johnson, M., & Johnson, M. (1981). A comparative study of tests for homogeneity of variances, with applications to the outer continental shelf bidding data. *Technometrics, 23,* 351–361.

Cook, R. D. (1993). Exploring partial residuals plots. *Technometrics, 35,* 351–362.

Cook, R. D., & Weisberg, S. (1992). *Residuals and Influence in Regression.* New York: Chapman and Hall.

Cressie, N. A. C., & Whitford, H. J. (1986). How to use the two sample *t*-test. *Biometrical Journal, 28,* 131–148.

Cronbach, L. J. (1987). Statistical tests for moderator variables: Flaws in analyses recently proposed. *Psychological Bulletin, 102,* 414–417.

Cronbach, L. J., Gleser, G. C., Nanda, H., & Rajaratnam, N. (1972). *The Dependability of Behavioral Measurements.* New York: Wiley.

Crowder, M. J., & Hand, D. J. (1990). *Analysis of Repeated Measures.* London: Chapman.

Croux, C., Rousseeuw, P. J., & Hössjer, O. (1994). Generalized S-estimators. *Journal of the American Statistical Association, 89,* 1271–1281.

Dana, E. (1990). Salience of the self and salience of standards: Attempts to match self to standard. Unpublished Ph.D. dissertation, University of Southern California.

Dantzig, G. (1940). On the non-existance of tests of "Student's" hypothesis having power functions independent of σ. *Annals of Mathematical Statistics, 11,* 186.

Davies, L., & Gather, U. (1993). The identification of multiple outliers (with discussion). *Journal of the American Statistical Association, 88,* 782–792.

Davison, A. C., & Hinkley, D. V. (1997). *Bootstrap Methods and Their Application.* Cambridge: Cambridge University Press.

Dawson, M. E., Schell, A. M., Hazlett, E. A., Nuechterlein, K. H., & Filion, D. L. (2000). On the clinical and cognitive meaning of impaired sensorimotor gating in schizophrenia. *Psychiatry Research, 96,* 187–197.

Deming, W. E. (1943). *Statistical Adjustment of Data.* New York: Wiley.

Derksen, S., & Keselman, H. J. (1992). Backward, forward and stepwise automated subset selection algorithms: Frequency of obtaining authentic and noise variables. *British Journal of Mathematical and Statistical Psychology, 45,* 265–282.

Dielman, T., Lowry, C., & Pfaffenberger, R. (1994). A comparison of quantile estimators. *Communications in Statistics—Simulation and Computation, 23,* 355–371.

Dietz, E. J. (1989). Teaching regression in a nonparametric statistics course. *American Statistician, 43*, 35–40.

Diggle, P. J., Liang, K.-Y., & Zeger, S. L. (1994). *Analysis of Longitudinal Data*. Oxford: Oxford University Press.

Doksum, K. A. (1974). Empirical probability plots and statistical inference for nonlinear models in the two-sample case. *Annals of Statistics, 2*, 267–277.

Doksum, K. A. (1977). Some graphical methods in statistics. A review and some extensions. *Statistica Neerlandica, 31*, 53–68.

Doksum, K. A., Blyth, S., Bradlow, E., Meng, X., & Zhao, H. (1994). Correlation curves as local measures of variance explained by regression. *Journal of the American Statistical Association, 89*, 571–582.

Doksum, K. A., & Samarov, A. (1995). Nonparametric estimation of global functionals and a measure of the explanatory power of covariates in regression. *Annals of Statistics, 23*, 1443–1473.

Doksum, K. A., & Sievers, G. L. (1976). Plotting with confidence: Graphical comparisons of two populations. *Biometrika, 63*, 421–434.

Doksum, K. A., & Wong, C.-W. (1983). Statistical tests based on transformed data. *Journal of the American Statistical Association, 78*, 411–417.

Donner, A., & Wells, G. (1986). A comparison of confidence interval methods for the intraclass correlation coefficient. *Biometrics, 42*, 401–412.

Donoho, D. L., & Gasko, M. (1992). Breakdown properties of location estimates based on halfspace depth and projected outlyingness. *Annals of Statistics, 20*, 1803–1827.

Drouet, D., & Kotz, S. (2001). *Correlation and Dependence*. London: Imperial College Press.

Duncan, G. T., & Layard, M. W. (1973). A Monte-Carlo study of asymptotically robust tests for correlation. *Biometrika, 60*, 551–558.

Dunnett, C. W. (1980a). Pairwise multiple comparisons in the unequal variance case. *Journal of the American Statistical Association, 75*, 796–800.

Dunnett, C. W. (1980b). Pairwise multiple comparisons in the homogeneous variance, unequal sample size case. *Journal of the American Statistical Association, 75*, 796–800.

Dunnett, C. W., & Tamhane, A. C. (1992). A step-up multiple test procedure. *Journal of the American Statistical Association, 87*, 162–170.

Efron, B., & Tibshirani, R. J. (1993). *An Introduction to the Bootstrap*. New York: Chapman and Hall.

Efron, B., & Tibshirani, R. J. (1997). Improvements on cross-validation: The .632+ bootstrap method. *Journal of the American Statistical Association, 92*, 548–560.

Elashoff, J. D., & Snow, R. E. (1970). A case study in statistical inference: Reconsideration of the Rosenthal–Jacobson data on teacher expectancy. Technical report no. 15, School of Education, Stanford University, Stanford, CA.

Emerson, J. D., & Hoaglin, D. C. (1983). Stem-and-leaf displays. In D. C. Hoaglin, F. Mosteller, & J. W. Tukey (Eds.), *Understanding Robust and Exploratory Data Analysis*, pp. 7–32. New York: Wiley.

Ezekiel, M. (1924). A method for handling curvilinear correlation for any number of variables. *Journal of the American Statistical Association, 19*, 431–453.

Fairly, D. (1986). Cherry trees with cones? *American Statistician, 40*, 138–139.

Fan, J. (1992). Design-adaptive nonparametric regression. *Journal of the American Statistical Association, 87*, 998–1004.

Fears, T. R., Benichou, J., & Gail, M. H. (1996). A reminder of the fallibility of the Wald Statistic. *Journal of the American Statistical Association, 94*, 226–227.

Ferretti, N., Kelmansky, D., Yohai, V. J., & Zamar, R. (1999). A class of locally and globally robust regression estimates. *Journal of the American Statistical Association, 94*, 174–188.

Finner, H., & Roters, M. (2002). Multiple hypotheses testing and expected number of type I errors. *Annals of Statistics, 30*, 220–238.

Fisher, R. A. (1935). The fiducial argument in statistical inference. *Annals of Eugenics, 6*, 391–398.

Fisher, R. A. (1941). The asymptotic approach to Behren's integral, with further tables for the d test of significance. *Annals of Eugenics, 11*, 141–172.

Fleiss, J. L. (1981). *Statistical Methods for Rates and Proportions*, 2nd ed. New York: Wiley.

Fligner, M. A., & Policello II, G. E. (1981). Robust rank procedures for the Behrens–Fisher problem. *Journal of the American Statistical Association, 76*, 162–168.

Freedman, D., & Diaconis, P. (1981). On the histogram as density estimator: L_2 theory. *Z. Wahrsche. verw. Ge., 57*, 453–476.

Freedman, D., & Diaconis, P. (1982). On inconsistent M-estimators. *Annals of Statistics, 10*, 454–461.

Frigge, M., Hoaglin, D. C., & Iglewicz, B. (1989). Some implementations of the Boxplot. *American Statistician*, *43*, 50–54.

Fung, W.-K. (1993). Unmasking outliers and leverage points: A confirmation. *Journal of the American Statistical Association*, *88*, 515–519.

Games, P. A., & Howell, J. (1976). Pairwise multiple comparison procedures with unequal *n*'s and/or variances: A Monte Carlo study. *Journal of Educational Statistics*, *1*, 113–125.

Giltinan, D. M., Carroll, R. J., & Ruppert, D. (1986). Some new estimation methods for weighted regression when there are possible outliers. *Technometrics*, *28*, 219–230.

Goldberg, K. M., & Iglewicz, B. (1992). Bivariate extensions of the boxplot. *Technometrics*, *34*, 307–320.

Graybill, F. A. (1983). *Matrices with Applications in Statistics*. Belmont, CA: Wadsworth.

Hald, A. (1952). *Statistical Theory with Engineering Applications*. New York: Wiley.

Hald, A. (1998). *A History of Mathematical Statistics from 1750 to 1930*. New York: Wiley.

Hall, P. (1988a). On symmetric bootstrap confidence intervals. *Journal of the Royal Statistical Society, Series B*, *50*, 35–45.

Hall, P. (1988b). Theoretical comparison of bootstrap confidence intervals. *Annals of Statistics*, *16*, 927–953.

Hall, P., & Hall, D. (1995). *The Bootstrap and Edgeworth Expansion*. New York: Springer Verlag.

Hall, P., & Sheather, S. J. (1988). On the distribution of a Studentized quantile. *Journal of the Royal Statistical Society, Series B*, *50*, 380–391.

Hampel, F. R., Ronchetti, E. M., Rousseeuw, P. J., & Stahel, W. A. (1986). *Robust Statistics: The Approach Based on Influence Functions*. New York: Wiley.

Hand, D. J., & Crowder, M. J. (1996). *Practical Longitudinal Data Analysis*. London: Chapman and Hall.

Harrell, F. E., & Davis, C. E. (1982). A new distribution-free quantile estimator. *Biometrika*, *69*, 635–640.

Hastie, T. J., & Tibshirani, R. J. (1990). *Generalized Additive Models*. New York: Chapman and Hall.

Hawkins, D. M., & Olive, D. (1999). Applications and algorithms for least trimmed sum of absolute deviations regression. *Computational Statistics & Data Analysis*, *28*, 119–134.

Hayter, A. (1984). A proof of the conjecture that the Tukey–Kramer multiple comparison procedure is conservative. *Annals of Statistics*, *12*, 61–75.

Hayter, A. (1986). The maximum familywise error rate of Fisher's least significant difference test. *Journal of the American Statistical Association*, *81*, 1000–1004.

He, X., & Portnoy, S. (1992). Reweighted LS estimators converge at the same rate as the initial estimator. *Annals of Statistics*, *20*, 2161–2167.

He, X., Simpson, D. G., & Portnoy, S. L. (1990). Breakdown robustness of tests. *Journal of the American Statistical Association*, *85*, 446–452.

Headrick, T. C., & Rotou, O. (2001). An investigation of the rank transformation in multiple regression. *Computational Statistics & Data Analysis*, *38*, 203–215.

Hettmansperger, T. P. (1984). *Statistical Inference Based on Ranks*. New York: Wiley.

Hettmansperger, T. P., & McKean, J. W. (1977). A robust alternative based on ranks to least squares in analyzing linear models. *Technometrics*, *19*, 275–284.

Hettmansperger, T. P., & McKean, J. W. (1998). *Robust Nonparametric Statistical Methods*. London: Arnold.

Hettmansperger, T. P., Möttönen, J., & Oja, H. (1997). Affine-invariant one-sample signed-rank tests. *Journal of the American Statistical Association*, *92*, 1591–1600.

Hettmansperger, T. P., & Sheather, S. J. (1986). Confidence intervals based on interpolated order statistics. *Statistics and Probability Letters 4*, 75–79.

Hewett, J. E., & Spurrier, J. D. (1983). A survey of two-stage tests of hypotheses: Theory and application. *Communications in Statistics—Theory and Methods*, *12*, 2307–2425.

Hoaglin, D. C., & Iglewicz, B. (1987). Fine-tuning some resistant rules for outlier labeling. *Journal of the American Statistical Association*, *82*, 1147–1149.

Hoaglin, D. C., Iglewicz, B., & Tukey, J. W. (1986). Performance of some resistant rules for outlier labeling. *Journal of the American Statistical Association*, *81*, 991–999.

Hochberg, Y. (1975). Simultaneous inference under Behrens–Fisher conditions: A two sample approach. *Communications in Statistics*, *4*, 1109–1119.

Hochberg, Y. (1988). A sharper Bonferroni procedure for multiple tests of significance. *Biometrika*, *75*, 800–802.

Hodges, J. L., Ramsey, P. H., & Wechsler, S. (1990). Improved significance probabilities of the Wilcoxon test. *Journal of Educational Statistics*, *15*, 249–265.

Hoenig, J. M., & Heisey, D. M. (2001). The abuse of power: The pervasive fallacy of power calculations for data analysis. *American Statistician*, *55*, 19–24.

Hogg, R. V., & Craig, A. T. (1970). *Introduction to Mathematical Statistics*. New York: Macmillan.

Hollander, M., & Sethuraman, J. (1978). Testing for agreement between two groups of judges. *Biometrika, 65*, 403–411.

Hommel, G. (1988). A stagewise rejective multiple test procedure based on a modified Bonferroni test. *Biometrika, 75*, 383–386.

Hosmane, B. S. (1986). Improved likelihood ratio tests and Pearson chi-square tests for independence in two-dimensional tables. *Communications Statistics—Theory and Methods, 15*, 1875–1888.

Hössjer, O. (1994). Rank-based estimates in the linear model with high breakdown point. *Journal of the American Statistical Association, 89*, 149–158.

Huber, P. J. (1964). Robust estimation of location parameters. *Annals of Mathematical Statistics, 35*, 73–101.

Huber, P. J. (1981). *Robust Statistics*. New York: Wiley.

Huber, P. J. (1993). Projection pursuit and robustness. In S. Morgenthaler, E. Ronchetti, & W. Stahel (Eds.), *New Directions in Statistical Data Analysis and Robustness*. Boston: Birkhäuser Verlag.

Huberty, C. J. (1989). Problems with stepwise methods—Better alternatives. *Advances in Social Science Methodology, 1*, 43–70.

Huitema, B. E. (1980). *The Analysis of Covariance and Alternatives*. New York: Wiley.

Hussain, S. S., & Sprent, P. (1983). Nonparametric regression. *Journal of the Royal Statistical Society, 146*, 182–191.

Huynh, H., & Feldt, L. S. (1976). Estimation of the Box correction for degrees of freedom from sample data in randomized block and split-plot designs. *Journal of Educational Statistics, 1*, 69–82.

Hyndman, R. B., & Fan, Y. (1996). Sample quantiles in social packages. *American Statistician, 50*, 361–365.

Iman, R. L. (1974). A power study of a rank transform for the two-way classification model when interactions may be present. *Canadian Journal of Statistics, 2*, 227–239.

Iman, R. L., Quade, D., & Alexander, D. A. (1975). Exact probability levels for the Kruskal–Wallis test. *Selected Tables in Mathematical Statistics, 3*, 329–384.

Jaeckel, L. A. (1972). Estimating regression coefficients by minimizing the dispersion of residuals. *Annals of Mathematical Statistics, 43*, 1449–1458.

James, G. S. (1951). The comparison of several groups of observations when the ratios of the population variances are unknown. *Biometrika, 38*, 324–329.

Jeyaratnam, S., & Othman, A. R. (1985). Test of hypothesis in one-way random effects model with unequal error variances. *Journal of Statistical Computation and Simulation, 21*, 51–57.

Johansen, S. (1980). The Welch–James approximation to the distribution of the residual sum of squares in a weighted linear regression. *Biometrika, 67*, 85–93.

Johnson, N. J. (1978). Modifed *t*-tests and confidence intervals for asymmetrical populations. *Journal of the American Statistical Association, 73*, 536–576.

Johnson, P., & Neyman, J. (1936). Tests of certain linear hypotheses and their application to some educational problems. *Statistical Research Memoirs, 1*, 57–93.

Jones, R. H. (1993). *Longitudinal Data with Serial Correlation: A State-Space Approach*. London: Chapman & Hall.

Jorgensen, J., Gilles, R. B., Hunt, D. R., Caplehorn, J. R. M., & Lumley, T. (1995). A simple and effective way to reduce postoperative pain after laparoscopic cholecystectomy. *Australia and New Zealand Journal of Surgery, 65*, 466–469.

Judd, C. M., Kenny, D. A., & McClelland, G. H. (2001). Estimating and testing mediation and moderation in within-subjects designs. *Psychological Methods, 6*, 115–134.

Kaiser, L., & Bowden, D. (1983). Simultaneous confidence intervals for all linear contrasts of means with heterogeneous variances. *Communications in Statistics—Theory and Methods, 12*, 73–88.

Kallenberg, W. C. M., & Ledwina, T. (1999). Data-driven rank tests for independence. *Journal of the American Statistical Association, 94*, 285–310.

Kendall, M. G., & Stuart, A. (1973). *The Advanced Theory of Statistics*, Vol. 2. New York: Hafner.

Kepner, J. L., & Robinson, D. H. (1988). Nonparametric methods for detecting treatment effects in repeated-measures designs. *Journal of the American Statistical Association, 83*, 456–461.

Keselman, H. J., Algina, J., Boik, R. J., & Wilcox, R. R. (1999). New approaches to the analysis of repeated measurements. In B. Thompson (Ed.), *Advances in Social Science Methodology, 5*, 251–268. Greenwich, CT: JAI Press.

Keselman, H. J., Algina, J., Wilcox, R. R., & Kowalchuk, R. K. (2000). Testing repeated measures hypotheses when covariance matrices are heterogeneous: Revisiting the robustness of the Welch–James test again. *Educational and Psychological Measurement, 60*, 925–938.

Keselman, H. J., & Wilcox, R. R. (1999). The "improved" Brown and Forsythe test for mean equality: Some things can't be fixed. *Communications in Statistics—Simulation and Computation, 28,* 687–698.

Keselman, H. J., Wilcox, R. R., Taylor, J., & Kowalchuk, R. K. (2000). Tests for mean equality that do not require homogeneity of variances: Do they really work? *Communications in Statistics—Simulation and Computation, 29,* 875–895.

Keselman, J. C., Cribbie, R., & Holland, B. (1999). The pairwise multiple comparison multiplicity problem: An alternative approach to familywise and comparisonwise Type I error control. *Psychological Methods, 4,* 58–69.

Keselman, J. C., Rogan, J. C., Mendoza, J. L., & Breen, L. J. (1980). Testing the validity conditions of repeated measures *F*-tests. *Psychological Bulletin, 87,* 479–481.

Kim, P. J., & Jennrich, R. I. (1973). Tables of the exact sampling distribution of the two-sample Kolmogorov–Smirnov criterion, D_{mn}, $m \leq n$. In H. L. Harter, & D. B. Owen (Eds.), *Selected Tables in Mathematical Statistics,* Vol. I. Providence, RI: American Mathematical Society.

Kirk, R. E. (1995). *Experimental Design.* Monterey, CA: Brooks/Cole.

Kramer, C. (1956). Extension of multiple range test to group means with unequal number of replications. *Biometrics, 12,* 307–310.

Krause, A., & Olson, M. (2000). *The Basics of S and S-PLUS.* New York: Springer.

Krutchkoff, R. G. (1988). One-way fixed-effects analysis of variance when the error variances may be unequal. *Journal of Statistical Computation and Simulation, 30,* 259–271.

Kuo, L., & Mallick, B. (1998). Variable selection for regression models. *Sankhya, Series B, 60,* 65–81.

Larsen, W. A., & McCleary, S. J. (1972). The use of partial residual plots in regression analysis. *Technometrics, 14,* 781–790.

Lax, D. A. (1985). Robust estimators of scale: Finite sample performance in long-tailed symmetric distributions. *Journal of the American Statistical Association, 80,* 736–741.

Le, C. T. (1994). Some tests of linear trend of variances. *Communications in Statistics—Theory and Methods, 23,* 2269–2282.

Li, G. (1985). Robust regression. In D. Hoaglin, F. Mosteller, & J. Tukey (Eds.), *Exploring Data Tables, Trends, and Shapes,* pp. 281–343. New York: Wiley.

Liu, R. G., Parelius, J. M., & Singh, K. (1999). Multivariate analysis by data depth. *Annals of Statistics, 27,* 783–840.

Liu, R. G., & Singh, K. (1993). A quality index based on data depth and multivariate rank tests. *Journal of the American Statistical Association, 88,* 257–262.

Liu, R. G., & Singh, K. (1997). Notions of limiting *P* values based on data depth and bootstrap. *Journal of the American Statistical Association, 92,* 266–277.

Lloyd, C. J. (1999). *Statistical Analysis of Categorical Data.* New York: Wiley.

Loh, W.-Y. (1987). Does the correlation coefficient really measure the degree of clustering around a line? *Journal of Educational Statistics, 12,* 235–239.

Long, J. D., & Cliff, N. (1997). Confidence intervals for Kendall's tau. *British Journal of Mathematical and Statistical Psychology, 50,* 31–42.

Long, J. S., & Ervin, L. H. (2000). Using heteroscedasticity consistent standard errors in the linear regression model. *American Statistician, 54,* 217–224.

Lord, F. M., & Novick, M. R. (1968). *Statistical Theories of Mental Test Scores.* Reading, MA: Addison-Wesley.

Ludbrook, J., & Dudley, H. (1998). Why permutation tests are superior to *t*- and *F*-tests in biomedical research. *American Statistician, 52,* 127–132.

Lumley, T. (1996). Generalized estimating equations for ordinal data: A note on working correlation structures. *Biometrics, 52,* 354–361.

Lunneborg, C. E. (2000). *Data Analysis by Resampling: Concepts and Applications.* Pacific Grove, CA: Duxbury.

Lyon, J. D., & Tsai, C.-L. (1996). A comparison of tests for homogeneity. *Statistician, 45,* 337–350.

Mallows, C. L. (1973). Some comments on C_p. *Technometrics, 15,* 661–675.

Mallows, C. L. (1986). Augmented partial residuals. *Technometrics, 28,* 313–319.

Mann, H. B., & Whitney, D. R. (1947). On a test of whether one of two random variables is stochastically larger than the other. *Annals of Mathematical Statistics, 18,* 50–60.

Mardia, K. V., Kent, J. T., & Bibby, J. M. (1979). *Multivariate Analysis.* San Diego, CA: Academic Press.

Markowski, C. A., & Markowski, E. P. (1990). Conditions for the effectiveness of a preliminary test of variance. *American Statistician, 44,* 322–326.

Maronna, R., & Morgenthaler, S. (1986). Robust regression through robust covariances. *Communications in Statistics—Theory and Methods, 15,* 1347–1365.

Maronna, R., Yohai, V. J., & Zamar, R. (1993). Bias-robust regression estimation: A partial survey. In S. Morgenthaler, E. Ronchetti, & W. A. Stahel (Eds.), *New Directions in Statistical Data Analysis and Robustness.* Boston: Fuller Verlag.

Matuszewski, A., & Sotres, D. (1986). A simple test for the Behrens–Fisher problem. *Computational Statistics and Data Analysis, 3,* 241–249.

McKean, J. W., & Schrader, R. M. (1984). A comparison of methods for studentizing the sample median. *Communications in Statistics—Simulation and Computation, 13,* 751–773.

McKean, J. W., & Sheather, S. J. (2000). Partial residual plots based on robust fits. *Technometrics, 42,* 249–261.

McKean, J. W., & Vidmar, T. J. (1994). A comparison of two rank-based methods for the analysis of linear models. *American Statistician, 48,* 220–229.

Mee, R. W. (1990). Confidence intervals for probabilities and tolerance regions based on a generalization of the Mann–Whitney statistic. *Journal of the American Statistical Association, 85,* 793–800.

Mehrotra, D. V. (1997). Improving the Brown–Forsythe solution to the generalized Behrens–Fisher problem. *Communications in Statistics—Simulation and Computation, 26,* 1139–1145.

Miller, A. J. (1990). *Subset Selection in Regression.* London: Chapman and Hall.

Miller, R. G. (1974). The jackknife—A review. *Biometrika, 61.*

Miller, R. G. (1976). Least squares regression with censored data. *Biometrika, 63,* 449–464.

Montgomery, D. C., & Peck, E. A. (1992). *Introduction to Linear Regression Analysis.* New York: Wiley.

Mooney, C. Z., & Duval, R. D. (1993). *Bootstrapping: A Nonparametric Approach to Statistical Inference.* Newbury Park, CA: Sage.

Morgenthaler, S., & Tukey, J. W. (1991). *Configural Polysampling.* New York: Wiley.

Moser, B. K., Stevens, G. R., & Watts, C. L. (1989). The two-sample *t*-test versus Satterthwaite's approximate *F*-test. *Communications in Statistics—Theory and Methods, 18,* 3963–3975.

Möttönen, J., & Oja, H. (1995). Multivariate spatial sign and rank methods. *Nonparametric Statistics, 5,* 201–213.

Muirhead, R. J. (1982). *Aspects of Multivariate Statistical Theory.* New York: Wiley.

Müller, H.-G. (1988). *Nonparametric Regression Analysis.* New York: Springer-Verlag.

Munzel, U. (1999). Linear rank score statistics when ties are present. *Statistics and Probability Letters, 41,* 389–395.

Munzel, U., & Brunner, E. (2000a). Nonparametric test in the unbalanced multivariate one-way design. *Biometrical Journal, 42,* 837–854.

Munzel, U., & Brunner, E. (2000b). Nonparametric methods in multivariate factorial designs. *Journal of Statistical Planning and Inference, 88,* 117–132.

Naranjo, J. D., & Hettmansperger, T. P. (1994). Bounded influence rank regression. *Journal of the Royal Statistical Society, B, 56,* 209–220.

Olejnik, S., Li, J., Supattathum, S., & Huberty, C. J. (1997). Multiple testing and statistical power with modified Bonferroni procedures. *Journal of Educational and Behavioral Statistics, 22,* 389–406.

Pagurova, V. I. (1968). On a comparison of means of two normal samples. *Theory of Probability and Its Applications, 13,* 527–534.

Parrish, R. S. (1990). Comparison of quantile estimators in normal sampling. *Biometrics, 46,* 247–257.

Patel, K. M., & Hoel, D. G. (1973). A nonparametric test for interactions in factorial experiments. *Journal of the American Statistical Association, 68,* 615–620.

Pedersen, W. C., Miller, L. C., Putcha-Bhagavatula, A. D., & Yang, Y. (2002). Evolved sex differences in sexual strategies: The long and the short of it. *Psychological Science, 13,* 157–161.

Peña, D., & Prieto, F. J. (2001). Multivariate outlier detection and robust covariance matrix estimation. *Technometrics, 43,* 286–299.

Piepho, H.-P. (1997). Tests for equality of dispersion in bivariate samples—Review and empirical comparison. *Journal of Statistical Computation and Simulation, 56,* 353–372.

Poon, W. Y., Lew, S. F., & Poon, Y. S. (2000). A local-influence approach to identifying multiple multivariate outliers. *British Journal of Mathematical and Statistical Psychology, 53,* 255–273.

Powers, D. A., & Xie, Y. (1999). *Statistical Methods for Categorical Data Analysis.* San Diego, CA: Academic Press.

Pratt, J. W. (1968). A normal approximation for binomial, *F*, beta, and other common, related tail probabilities, I. *Journal of the American Statistical Association, 63,* 1457–1483.

Price, R. M., & Bonett, D. G. (2001). Estimating the variance of the median. *Journal of Statistical Computation and Simulation, 68,* 295–305.

Quade, D. (1979). Using weighted rankings in the analysis of complete blocks with additive block effects. *Journal of the American Statistical Association, 74,* 680–683.

Raine, A., Buchsbaum, M., & LaCasse, L. (1997). Brain abnormalities in murderers indicated by positron emission tomography. *Biological Psychiatry, 42,* 495–508.

Ramsey, P. H. (1980). Exact Type I error rates for robustness of Student's *t*-test with unequal variances. *Journal of Educational Statistics, 5,* 337–349.

Rao, C. R. (1948). Tests of significance in multivariate analysis. *Biometrika, 35,* 58–79.

Rao, P. S., Kaplan, J., & Cochran, W. G. (1981). Estimators for the one-way random effects model with unequal error variances. *Journal of the American Statistical Association, 76,* 89–97.

Rasmussen, J. L. (1989). Data transformation, Type I error rate and power. *British Journal of Mathematical and Statistical Psychology, 42,* 203–211.

Rocke, D. M. (1996). Robustness properties of *S*-estimators of multivariate location and shape in high dimensions. *Annals of Statistics, 24,* 1327–1345.

Rocke, D. M., & Woodruff, D. L. (1996). Identification of outliers in multivariate data. *Journal of the American Statistical Association, 91,* 1047–1061.

Rogan, J. C., Keselman, H. J., & Mendoza, J. L. (1979). Analysis of repeated measurements. *British Journal of Mathematical and Statistical Psychology, 32,* 269–286.

Rom, D. M. (1990). A sequentially rejective test procedure based on a modified Bonferroni inequality. *Biometrika, 77,* 663–666.

Rosenthal, R., & Jacobson, L. (1968). *Pygmalion in the Classroom: Teacher Expectations and Pupil's Intellectual Development.* New York: Holt, Rinehart and Winston.

Rousseeuw, P. J., & Hubert, M. (1999). Regression depth. *Journal of the American Statistical Association, 94,* 388–402.

Rousseeuw, P. J., & Leroy, A. M. (1987). *Robust Regression & Outlier Detection.* New York: Wiley.

Rousseeuw, P. J., & Ruts, I. (1996). AS 307: Bivariate location depth. *Applied Statistics, 45,* 516–526.

Rousseeuw, P. J., Ruts, I., & Tukey, J. W. (1999). The bagplot: A bivariate boxplot. *American Statistician, 53,* 382–387.

Rousseeuw, P. J., & Struyf, A. (1998). Computing location depth and regression depth in higher dimensions. *Statistical Computations, 8,* 193–203.

Rousseeuw, P. J., & van Driesen, K. (1999). A fast algorithm for the minimum covariance determinant estimator. *Technometrics, 41,* 212–223.

Rousseeuw, P. J., & van Zomeren, B. C. (1990). Unmasking multivariate outliers and leverage points (with discussion). *Journal of the American Statistical Association, 85,* 633–639.

Rust, S. W., & Fligner, M. A. (1984). A modification of the Kruskal–Wallis statistic for the generalized Behrens–Fisher problem. *Communications in Statistics—Theory and Methods, 13,* 2013–2027.

Rutherford, A. (1992). Alternatives to traditional analysis of covariance. *British Journal of Mathematical and Statistical Psychology, 45,* 197–223.

Salk, L. (1973). The role of the heartbeat in the relations between mother and infant. *Scientific American, 235,* 26–29.

Sackrowitz, H., & Samuel-Cahn, E. (1999). *P* values as random variables—Expected *P* values. *American Statistician, 53,* 326–331.

Sarkar, S. K. (2002). Some results on false discovery rate in stepwise multiple testing procedures. *Annals of Statistics, 30,* 239–257.

Saunders, D. R. (1955). The "moderator variable" as a useful tool in prediction. In Proceedings of the 1954 Invitational Conference on Testing Problems (pp. 54–58). Princeton, NJ: Educational Testing Service.

Saunders, D. R. (1956). Moderator variables in prediction. *Educational and Psychological Measurement, 16,* 209–222.

Scariano, S. M., & Davenport, J. M. (1986). A four-moment approach and other practical solutions to the Behrens–Fisher problem. *Communications in Statistics—Theory and Methods, 15,* 1467–1501.

Scheffé, H. (1959). *The Analysis of Variance.* New York: Wiley.

Schenker, N., & Gentleman, J. F. (2001). On judging the significance of differences by examining the overlap between confidence intervals. *American Statistician, 55,* 182–186.

Schrader, R. M., & Hettmansperger, T. P. (1980). Robust analysis of variance. *Biometrika, 67,* 93–101.

Schroër, G., & Trenkler, D. (1995). Exact and randomization distributions of Kolmogorov–Smirnov tests two or three samples. *Computational Statistics and Data Analysis, 20,* 185–202.

Scott, D. W. (1979). On optimal and data-based histograms. *Biometrika, 66,* 605–610.

Scott, W. A. (1955). Reliability of content analysis: The case of nominal scale coding. *Public Opinion Quarterly, 19,* 321–325.

Searle, S. R. (1971). *Linear Models*. New York: Wiley.

Sen, P. K. (1968). Estimate of the regression coefficient based on Kendall's tau. *Journal of the American Statistical Association, 63*, 1379–1389.

Serfling, R. J. (1980). *Approximation Theorems of Mathematical Statistics*. New York: Wiley.

Shaffer, J. P. (1974). Bidirectional unbiased procedures. *Journal of the American Statistical Association, 69*, 437–439.

Shao, J. (1995). Bootstrap model selection. *Journal of the American Statistical Association, 91*, 655–665.

Shao, J., & Tu, D. (1995). *The Jackknife and the Bootstrap*. New York: Springer-Verlag.

Sheather, S. J., & McKean, J. W. (1987). A comparison of testing and confidence intervals for the median. *Statistical Probability Letters, 6*, 31–36.

Silverman, B. W. (1986). *Density Estimation for Statistics and Data Analysis*. New York: Chapman and Hall.

Singh, K. (1998). Breakdown theory for bootstrap quantiles. *Annals of Statistics, 26*, 1719–1732.

Small, C. G. (1990). A survey of multidimensional medians. *International Statistical Review, 58*, 263–277.

Smith, C. A. B. (1956). Estimating genetic correlations. *Annals of Human Genetics, 44*, 265–284.

Snedecor, G. W., & Cochran, W. (1967). *Statistical Methods*, 6th ed. Ames, IA: University Press.

Snow, R. E. (1995). Pygmalion and Intelligence? *Current Directions in Psychological Science, 4*, 169–172.

Sockett, E. B., Daneman, D., Clarson, C., & Ehrich, R. M. (1987). Factors affecting and patterns of residual insulin secretion during the first year of type I (insulin-dependent) diabetes mellitus in children. *Diabetes, 30*, 453–459.

Staudte, R. G., & Sheather, S. J. (1990). *Robust Estimation and Testing*. New York: Wiley.

Stein, C. (1945). A two-sample test for a linear hypothesis whose power is independent of the variance. *Annals of Statistics, 16*, 243–258.

Storer, B. E., & Kim, C. (1990). Exact properties of some exact test statistics for comparing two binomial proportions. *Journal of the American Statistical Association, 85*, 146–155.

Stromberg, A. J. (1993). Computation of high-breakdown nonlinear regression parameters. *Journal of the American Statistical Association, 88*, 237–244.

Stute, W., Manteiga, W. G., & Quindimil, M. P. (1998). Bootstrap approximations in model checks for regression. *Journal of the American Statistical Association, 93*, 141–149.

Sutton, C. D. (1993). Computer-intensive methods for tests about the mean of an asymmetrical distribution. *Journal of the American Statistical Association, 88*, 802–810.

Tamhane, A. C. (1977). Multiple comparisons in model I one-way ANOVA with unequal variances. *Communications in Statistics—Theory and Methods, 6*, 15–32.

Theil, H. (1950). A rank-invariant method of linear and polynomial regression analysis. *Indagationes Mathematicae, 12*, 85–91.

Thompson, A., & Randall-Maciver, R. (1905). *Ancient Races of the Thebaid*. Oxford: Oxford University Press.

Thompson, G. L. (1991). A unified approach to rank tests for multivariate and repeated measures designs. *Journal of the American Statistical Association, 86*, 410–419.

Thompson, G. L., & Ammann, L. P. (1990). Efficiencies of interblock rank statistics for repeated measures designs. *Journal of the American Statistical Association, 85*, 519–528.

Tibshirani, R. (1988). Estimating transformations for regression via additivity and variance stabilization. *Journal of the American Statistical Association, 83*, 394–405.

Tibshirani, R. (1996). Regression shrinkage and selection via the lasso. *Journal of the Royal Statistical Society, B, 58*, 267–288.

Tryon, W. W. (2001). Evaluating statistical difference, equivalence, and indeterminacy using inferential confidence intervals: An integrated alternative method of conducting null hypothesis statistical tests. *Psychological Methods, 6*, 371–386.

Tukey, J. W. (1960). A survey of sampling from contaminated normal distributions. In I. Olkin et al. (Eds.), *Contributions to Probability and Statistics*. Stanford, CA: Stanford University Press.

Tukey, J. W. (1977). *Exploratory Data Analysis*. Reading, MA: Addison-Wesley.

Tukey, J. W., & McLaughlin, D. H. (1963). Less vulnerable confidence and significance procedures for location based on a single sample: Trimming/Winsorization 1. *Sankhya A, 25*, 331–352.

Vonesh, E. (1983). Efficiency of repeated measures designs versus completely randomized designs based on multiple comparisons. *Communications in Statistics—Theory and Methods, 12*, 289–302.

Wald, A. (1955). Testing the difference between the means of two normal populations with unknown standard deviations. In T. W. Anderson et al. (Eds.), *Selected Papers in Statistics and Probability by Abraham Wald*. New York: McGraw-Hill.

Wechsler, D. (1958). *The Measurement and Appraisal of Adult Intelligence*. Baltimore: Williams and Wilkins.

Weerahandi, S. (1995). ANOVA under unequal error variances. *Biometrics, 51*, 589–599.

Welch, B. L. (1938). The significance of the difference between two means when the population variances are unequal. *Biometrika, 29*, 350–362.

Welch, B. L. (1951). On the comparison of several mean values: An alternative approach. *Biometrika, 38*, 330–336.

Westfall, P. (1988). Robustness and power of tests for a null variance ratio. *Biometrika, 75*, 207–214.

Westfall, P. H., & Young, S. S. (1993). *Resampling-Based Multiple Testing*. New York: Wiley.

Wilcox, R. R. (1983). A table of percentage points of the range of independent t variables. *Technometrics, 25*, 201–204.

Wilcox, R. R. (1992). An improved method for comparing variances when distributions have nonidentical shapes. *Computational Statistics & Data Analysis, 13*, 163–172.

Wilcox, R. R. (1993a). Some results on the Tukey–McLaughlin and Yuen methods for trimmed means when distributions are skewed. *Biometrical Journal, 36*, 259–273.

Wilcox, R. R. (1993b). Comparing one-step M-estimators of location when there are more than two groups. *Psychometrika, 58*, 71–78.

Wilcox, R. R. (1994a). A one-way random-effects model for trimmed means. *Psychometrika, 59*, 289–306.

Wilcox, R. R. (1994b). The percentage bend correlation coefficient. *Psychometrika, 59*, 601–616.

Wilcox, R. R. (1995). Some small-sample results on a bounded influence rank regression method. *Communications in Statistics—Theory and Methods, 24*, 881–888.

Wilcox, R. R. (1996a). Estimation in the simple linear regression model when there is heteroscedasticity of unknown form. *Communications in Statistics—Theory and Methods, 25*, 1305–1324.

Wilcox, R. R. (1996b). Confidence intervals for the slope of a regression line when the error term has nonconstant variance. *Computational Statistics & Data Analysis, 22*, 89–98.

Wilcox, R. R. (1996c). *Statistics for the Social Sciences*. San Diego, CA: Academic Press.

Wilcox, R. R. (1997a). *Introduction to Robust Estimation and Hypothesis Testing*. San Diego, CA: Academic Press.

Wilcox, R. R. (1997b). ANOVA based on comparing a robust measure of location at empirically determined design points. *British Journal of Mathematical and Statistical Psychology, 50*, 93–103.

Wilcox, R. R. (1998). Simulation results on extensions of the Theil–Sen regression estimator. *Communications in Statistics—Simulation and Computation, 27*, 1117–1126.

Wilcox, R. R. (2000a). Some exploratory methods for studying curvature in robust regression. *Biometrical Journal, 42*, 335–347.

Wilcox, R. R. (2000b). Rank-based tests for interactions in a two-way design when there are ties. *British Journal of Mathematical and Statistical Psychology, 53*, 145–153.

Wilcox, R. R. (2001a). *Fundamentals of Modern Statistical Methods: Substantially Increasing Power and Accuracy*. New York: Springer.

Wilcox, R. R. (2001b). Robust regression estimators that reduce contamination bias and have high efficiency when there is heteroscedasticity. Unpublished technical report, Dept. of Psychology, University of Southern California.

Wilcox, R. R. (2001c). Rank-based multiple comparisons for interactions in a split-plot design. Unpublished technical report, Dept. of Psychology, University of Southern California.

Wilcox, R. R. (2001d). Pairwise comparisons of trimmed means for two or more groups. *Psychometrika, 66*, 343–256.

Wilcox, R. R. (2002). Inferences based on a skipped correlation coefficient. Unpublished technical report, Dept. of Psychology, University of Southern California.

Wilcox, R. R. (2002). Comparing the variances of independent groups. *British Journal of Mathematical and Statistical Psychology, 55*, 169–176.

Wilcox, R. R., & Charlin, V. (1986). Comparing medians: A Monte Carlo study. *Journal of Educational Statistics, 11*, 263–274.

Wilcox, R. R., Charlin, V., & Thompson, K. L. (1986). New Monte Carlo results on the robustness of the ANOVA F, W, and F^* statistics. *Communications in Statistics—Simulation and Computation, 15*, 933–944.

Wilcox, R. R., & Keselman, H. J. (2002). Power analyses when comparing trimmed means. *Journal of Modern Statistical Methods, 1*, 24–31.

Wilcox, R. R., Keselman, H. J., Muska, J., & Cribbie, R. (2000). Repeated measures ANOVA: Some new results on comparing trimmed means and means. *British Journal of Mathematical and Statistical Psychology, 53*, 69–82.

Wilcox, R. R., & Muska, J. (1999). Measuring effect size: A nonparametric analogue of ω^2. *British Journal of Mathematical and Statistical Psychology, 52*, 93–110.

Wilcox, R. R., & Muska, J. (2001). Inferences about correlations when there is heteroscedasticity. *British Journal of Mathematical and Statistical Psychology, 54,* 39–47.

Wilcox, R. R., & Muska, J. (2002). Comparing correlation coefficients. *Communications in Statistics—Simulation and Computation, 31,* 49–59.

Wilcoxon, F. (1945). Individual comparisons by ranking methods. *Biometrics, 1,* 80–83.

Williams, V. S. L., Jones, L. V., & Tukey, J. W. (1999). Controlling error in multiple comparisons, with examples from state-to-state differences in educational achievement. *Journal of Educational and Behavioral Statistics, 24,* 42–69.

Wisnowski, J. W., Montgomery, D. C., & Simpson, J. R. (2001). A comparative analysis of multiple outlier detection procedures in the linear regression model. *Computational Statistics & Data Analysis, 36,* 351–382.

Woodruff, D. L., & Rocke, D. M. (1994). Computable robust estimation of multivariate location and shape in high dimension using compound estimators. *Journal of the American Statistical Association, 89,* 888–896.

Wu, C. F. J. (1986). Jackknife, bootstrap, and other resampling methods in regression analysis. *The Annals of Statistics, 14,* 1261–1295.

Yu, M. C., & Dunn, O. J. (1986). Robust test for the equality of two correlations: A Monte Carlo study. *Educational and Psychological Measurement, 42,* 987–1004.

Yuen, K. K. (1974). The two-sample trimmed *t* for unequal population variances. *Biometrika, 61,* 165–170.

INDEX